AF738472

Homotopical Methods in Geometry and Physics

CONTEMPORARY MATHEMATICS

841

Homotopical Methods in Geometry and Physics

International research conference
Homotopical Methods in Geometry and Physics
March 21–25, 2022
Evanston, IL

Gus Schrader
Boris Tsygan
Editors

AMS AMERICAN MATHEMATICAL SOCIETY

EDITORIAL COMMITTEE

Michael Loss, Chair

John Etnyre Angela Gibney Alex Kontorovich

2020 *Mathematics Subject Classification*. Primary 18M60, 14D23, 81T45, 14N35, 16E40.

Library of Congress Cataloging-in-Publication Data

Names: Homotopical methods in geometry and physics (Conference) (2022 : Evanston, Ill.) author | Schrader, Gus Knight, 1988– editor | Tsygan, Boris editor | Getzler, Ezra, 1962– honoree

Title: Homotopical methods in geometry and physics / Gus Schrader, Boris Tsygan.

Description: Providence, Rhode Island : American Mathematical Society, [2026] | Series: Contemporary mathematics, 0271-4132 ; volume 841 | The conference was held at Northwestern University on March 21–25, 2022.

Identifiers: LCCN 2026003133 | ISBN 9781470481940 paperback | ISBN 9781470485719 ebook

Subjects: LCSH: Homotopy theory | Geometry, Algebraic | Mathematical physics | AMS: Algebraic geometry – Families, fibrations – Stacks and moduli problems | Quantum theory – Quantum field theory; related classical field theories – Topological field theories | Algebraic geometry – Projective and enumerative geometry – Gromov-Witten invariants, quantum cohomology, Gopakumar-Vafa invariants, Donaldson-Thomas invariants | Associative rings and algebras – Homological methods – (Co)homology of rings and algebras (e.g. Hochschild, cyclic, dihedral, etc.)

Classification: LCC QA612.7 .H633 2026

LC record available at https://lccn.loc.gov/2026003133

Contemporary Mathematics ISSN: 0271-4132 (print); ISSN: 1098-3627 (online)

DOI: https://doi.org/10.1090/conm/841

Color graphic policy. Any graphics created in color will be rendered in grayscale for the printed version unless color printing is authorized by the Publisher. In general, color graphics will appear in color in the online version.

Copying and reprinting. Individual readers of this publication, and nonprofit libraries acting for them, are permitted to make fair use of the material, such as to copy select pages for use in teaching or research. Permission is granted to quote brief passages from this publication in reviews, provided the customary acknowledgment of the source is given.

Republication, systematic copying, or multiple reproduction of any material in this publication is permitted only under license from the American Mathematical Society. Requests for permission to reuse portions of AMS publication content are handled by the Copyright Clearance Center. For more information, please visit `www.ams.org/publications/pubpermissions`.

Send requests for translation rights and licensed reprints to `reprint-permission@ams.org`.

© 2026 by the American Mathematical Society. All rights reserved.
The American Mathematical Society retains all rights
except those granted to the United States Government.
Printed in the United States of America.

♾ The paper used in this book is acid-free and falls within the guidelines
established to ensure permanence and durability.
Visit the AMS home page at `https://www.ams.org/`

10 9 8 7 6 5 4 3 2 1 31 30 29 28 27 26

Contents

Foreword

This is a volume of proceedings of the conference entitled "Homotopical Methods in Geometry and Physics," in honor of Ezra Getzler's sixtieth birthday. The conference was held at Northwestern University on March 21–25, 2022.

The topic of much of the conference, as well as of much of Ezra's work, is the algebra, geometry, and topology of spaces of data parametrizing mathematical objects of certain type, up to symmetries. Those spaces are usually called moduli spaces; when the object in question is a collection of points in a given space, they are also called configuration spaces. Moduli spaces and configuration spaces play a central and fundamental role in much of mathematics and mathematical physics. In an important particular case, they often parametrize operations in algebraic structures.

More precisely, it had been understood since the 1960s that the axioms for classical algebraic structures, such as associative, commutative, or Lie algebras, as well as associative monoids, categories, etc., are too restrictive for applications and in many cases should be replaced by their homotopy versions. Configuration spaces arise very naturally in this context. For example, an associative monoid carries one natural ternary operation $(ab)c = a(bc)$. A homotopy associative monoid carries operations parametrized by a point t of the unit interval, or by a triple of points $(0, t, 1)$. The two operations $(ab)c$ and $a(bc)$ correspond to $t = 0$ and $t = 1$. Equivalently, one can say that operations are parametrized by configurations of three points $t_0 \leq t_1 \leq t_2$ on the real line, not all of them equal, up to translation and positive dilation. One can, of course, identify this space with a closed interval. Operations of valency more than 3 are parametrized by higher dimensional polyhedra called *associahedra*.

Operations of this sort are organized into *operads* (if they take several arguments and produce one output) or *PROPS* (if they take several arguments and produce several outputs). The way in which different points merge into one reflect the ways in which the operations are composed. It was understood in the early 1990s, in particular by Getzler and Jones [*Operads, homotopy algebra, and iterated integrals for double loop spaces*, Preprint, `arXiv:hep-th/9403055`, (1994)], that operads of compactified configuration spaces in d-dimensional real vector space, not just on the line, are central to studying algebraic structures arising in mathematical physics and noncommutative geometry. Those operads are called Fulton–MacPherson operads. Sometimes it is easier to understand the composition of operations by representing them by confidurations of little discs, not points. Fulton–MacPherson operads, along with little d-discs operads and various other related constructions (see below), are versions of the $\mathbb{E}_d$ operad.

With this in mind, let us briefly describe the contents of the volume. We will order our comments thematically rather than alphabetically. The article of Beuckelmann and Moerdijk provides a catalog of various versions of the $\mathbb{E}_d$ operad. (As we saw, those different versions come naturally, and it is important to understand all of them). The article of Bottman and Mavrides concerns the geometry of double associahedra. These are polyhedra that describe operations in homotopy two-monoids (i.e., monoids in monoids), objects very close to $\mathbb{E}_2$-algebras (via Dunn's additivity theorem $\mathbb{E}_1 \otimes \mathbb{E}_1 \xrightarrow{\sim} \mathbb{E}_2$).

The contribution of Tsygan deals with the topic of algebraic operations in noncommutative geometry, where a manifold is replaced by an associative algebra. Multivector fields and differential forms get replaced by the complexes of Hochschild (co)chains of the algebra. In classical geometry there is Cartan calculus between multivectors and forms that is subject to simple algebraic identities called Cartan formulas. In noncommutative geometry these formulas hold up to homotopy. This can be deduced from the action of a certain extension of the $\mathbb{E}_2$ operad. This procedure is highly inexplicit and not canonical. However, a large part of the structure can be carried out in a canonical way. As a consequence, one constructs a flat superconnection on the periodic cyclic complex of a family of algebras. This generalizes Getzler's original work [*Cartan homotopy formulas and the Gauss–Manin connection in cyclic homology*, Israel Math. Conf. Proc., **7** (1993), 65–78 MR1261901].

The work of Dancso, Hogan, and Robertson studies the relation between the Grothendieck–Teichmüller group and the Kashiwara–Vergne group. The former acts by symmetries on the $\mathbb{E}_2$ operad (more precisely, on one of its realizations, constructed via parenthesized braids); the latter acts on solutions of the Kashiwara–Vergne problem which is a fundamental problem in the theory of Lie algebras.

The next two contributions deal with geometry and topology of moduli spaces that are close relatives of configuration spaces. The article of Davison studies the moduli space of coherent sheaves of finite length on $\mathbb{C}^2$. The equivariant homology of this moduli space carries an associative Hall-type product, which is a special case of the 'cohomological Hall algebra' associated by Kontsevich and Soibelman to a quiver with potential. Davison gives a full description of this algebra in terms of an affinizized, deformed version of a certain Lie algebra attached to the quiver known as the BPS Lie algebra.

The work of Kapranov and Schechtman concerns space of zero-cycles on $\mathbb{C}$ with valued in a commutative monoid L, and the category of perverse sheaves on this space. They obtain a concrete description of this sheaf category as the representation category of a certain PROB associated to L, which captures the identities satisfied by the natural maps among tensor powers $A^{\otimes n}$ of an L-graded braided bialgebra A.

The article of Biswas and Looijenga is devoted to moduli spaces of objects other than configurations, namely of representations of a discrete group Γ in a complex Lie group G. The authors construct a holomorphic n-form on the moduli stack with values in the cohomology $H^n(\Gamma, \mathbb{C})$ for a G-invariant polynomial of degree n on the Lie algebra of G. When $n = 2$ and Γ is the fundamental group of an oriented surface, one recovers Goldman's classical construction of the symplectic structure on the moduli space of local systems on the surface.

The work of Wolfson deals with another aspect of geometry and topology of moduli spaces. Namely, as we mentioned above, moduli spaces parametrize objects up to symmetries. The quotients that arise are often quite complicated and should be treated as higher stacks. Behrend and Getzler [*Geometric higher groupoids and categories*, Progr. Math., **310** (2017), 1–45 Birkhäuser/Springer Cham MR3821920] showed that higher stacks can be studied in the context of descent categories, or categories with covers. Wolfson develops the context of a principle bundle over such an object.

The contribution of Aleshkin and Liu deals with another application of moduli spaces. Namely, invariants of symplectic manifolds are obtained by constructing moduli spaces of holomorphic maps to that manifold from a Riemann surface, and then counting points of this moduli space when it happens to be of dimension zero. There are profound correspondences between various cases of these invariants. Among them is the open-closed correspondence between invariants obtained by counting holomorphic maps of discs to a threefold and holomorphic maps of spheres into a fourfold. The authors find a new open-closed correspondence. In it, the open part is as above but for more general threefolds; the closed part is obtained by counting stable holomorphic curves in the critical locus of an algebraic function w on the stack $V//G$ obtained by GIT reduction of a complex affine space by an algebraic group. The two invariants satisfy the same differential equation of Picard–Fuchs type. This equation is the same as for a Riemann–Roch type invariant of matrix factorizations corresponding to the function w. This provides an intriguing link to periodic cyclic homology of the category of matrix factorizations, and the Getzler–Gauss–Manin connection on it (see above). This is one of several interesting interconnections between the topics in this volume.

We would like to thank to all who attended and contributed to the lively scientific atmosphere of the conference, making it a warm and fitting celebration of Ezra Getzler's many contributions to mathematics and to our community.

Gus Schrader
Boris Tsygan

Bibliography

[1] Kai Behrend and Ezra Getzler, *Geometric higher groupoids and categories*, Geometry, analysis and probability, Progr. Math., vol. 310, Birkhäuser/Springer, Cham, 2017, pp. 1–45, DOI 10.1007/978-3-319-49638-2_1. MR3821920

[2] E. Getzler and J. D. S. Jones, *Operads, homotopy algebra and iterated integrals for double loop spaces*, arXiv:hep-th/9403055

[3] Ezra Getzler, *Cartan homotopy formulas and the Gauss-Manin connection in cyclic homology*, Quantum deformations of algebras and their representations (Ramat-Gan, 1991/1992; Rehovot, 1991/1992), Israel Math. Conf. Proc., vol. 7, Bar-Ilan Univ., Ramat Gan, 1993, pp. 65–78. MR1261901

Contemporary Mathematics
Volume **841**, 2026
https://doi.org/10.1090/conm/841/16883

Open/closed correspondence and extended LG/CY correspondence for quintic threefolds

Konstantin Aleshkin and Chiu-Chu Melissa Liu

Dedicated to Professor Ezra Getzler on the occasion of his sixtieth birthday

ABSTRACT. We show that Walcher's disk potential for the quintic threefold can be represented as a central charge of a specific Gauged Linear Sigma Model which we call the extended quintic GLSM. This representation provides an open/closed correspondence for the quintic threefold since the central charge is a generating function of closed genus-zero GLSM invariants. We also explain how open Landau-Ginzburg/Calabi-Yau correspondence and open mirror symmetry for the quintic are compatible with wall-crossing and mirror symmetry of the extended GLSM, respectively.

1. Introduction

Open/closed correspondence proposed by Mayr [**Ma**] relates genus-zero A-model topological open strings on a toric Calabi-Yau threefold X to genus-zero A-model topological closed strings on a toric Calabi-Yau fourfold $\widetilde{X}$. More precisely, the correspondence relates (i) genus-zero open Gromov-Witten (GW) invariants counting holomorphic disks in a symplectic toric Calabi-Yau threefold X with boundaries in an Aganagic-Vafa Lagrangian A-brane L, and (ii) genus-zero (closed) GW invariants counting holomorphic spheres in a symplectic toric Calabi-Yau fourfold $\widetilde{X}$ determined by the pair (X, L). The correspondence has been proved in full generality and also generalized to toric Calabi-Yau 3-orbifolds by S. Yu and the second named author [**LY22, LY**]. Open string mirror symmetry relates disk invariants of X to relative periods of a holomorphic 3-form on the Calabi-Yau threefold $X^\vee$ mirror to X; closed string mirror symmetry relates genus-zero GW invariants of $\widetilde{X}$ to period integrals of a holomorphic 4-form on the Calabi-Yau fourfold $\widetilde{X}^\vee$ mirror $\widetilde{X}$. There is a B-model open/closed correspondence which is compatible with the A-model open/closed correspondence and mirror symmetry [**LY**]. The extended Picard-Fuchs equations satisfied by the disk potential (together with components of the I-function of X) can be identified with the Picard-Fuchs equations satisfied by components of the I-function of the toric Calabi-Yau fourfold $\widetilde{X}$ [**Ma, LM, FL13, FLT22**].

2020 *Mathematics Subject Classification.* Primary 14N35, 53D45, 14J33.
Key words and phrases. Gromov-Witten invariants, Calabi-Yau threefolds, FJRW invariants, Landau-Ginzberg models, matrix factorizations, gauge linear sigma models, mirror symmetry.
The second author was supported in part by NSF Grant DMS-1564497.

©2026 American Mathematical Society

Toric Calabi-Yau manifolds/orbifolds are non-compact. Open GW invariants in (i) and closed GW invariants in (ii) are defined and computed by torus localization. In this paper we extend open/closed correspondence to compact Calabi-Yau threefolds. Solomon [**So**] defined open GW invariants counting holomorphic disks in a compact symplectic manifold (X,ω) with boundaries in a Lagrangian submanifold $L \subset X$ under the assumption that $L = X^\phi$ is the fixed locus of an anti-symplectic involution $\phi : X \to X$, is relatively spin, and $\dim_{\mathbb{R}} L = 2$ or 3. Solomon's disk invariants coincide with Welschinger's invariants counting real pseudo-holomorphic curves in the strongly positive case [**We05, We**]. When $X = X_5$ is the quintic Calabi-Yau threefold and $X^\phi = \mathbb{R}X_5$ is the real quintic, a mirror formula for disk invariants of the pair $(X_5, \mathbb{R}X_5)$ was conjectured by Walcher [**Wa07**] and proved in [**PSW08**], see Section 2.3.1.

It is well-known that components of the I-function of the quintic threefold satisfy the degree 4 Picard-Fuchs differential equation (2.8). Walcher's B-model disk potential $\mathcal{T}^{\mathrm{CY}}$ satisfies a weaker degree 5 differential equation (2.15) together with components of the I-function of X_5. The extended Picard-Fuchs equation (2.15) is not the Picard-Fuchs equation of any Calabi-Yau variety. In this paper we notice that Equation (2.15) is the equation satisfied by components of the I-function of a particular Gauged Linear Sigma Model (GLSM) which we call the extended quintic GLSM, see Section 5. See [**Wi93**] for the introduction of GLSM in physics, [**FJR18, CFGKS, FK**] for a mathematical theory via algebraic geometry, and [**TX18,TX21,TX16,TX20,TX17,TX**] for a mathematical theory via symplectic geometry.

GLSM is a GW type enumerative theory that generalizes the classical theory to the case where the target is a (not necessarily smooth) critical locus of a regular function on a GIT quotient DM stack of a vector space. In particular case where the critical locus is a smooth complete intersection DM stack, GLSM invariants are GW invariants of the critical locus up to a sign [**CFGKS**].

We show that the disk potential $\mathcal{T}^{\mathrm{CY}}$ is equal to the extended GLSM central charge [**AL23**] associated to a certain matrix factorization up to a constant factor (Proposition 5.1). Geometry of the extended GLSM depends on the choice of the superpotential. We show that there is a natural one-parametric family of superpotentials deforming the classical quintic superpotential. For the generic value of the parameter the critical locus is a smooth DM stack which is a product $X_5 \times B\mu_2$ of the quintic threefold X_5 and the classifying space $B\mu_2$ of $\mu_2 = \{\pm 1\}$. In this case, the GLSM central charges [**AL23**] reproduce the components of the I-functions of X_5. However, for the degenerate choice of the parameter the critical locus becomes larger and singular, and in this case there is an additional central charge corresponding to $\mathcal{T}^{\mathrm{CY}}$.

GLSM setting is very convenient for studying the wall-crossing problems in the target geometry such as Landau-Ginzburg/Calabi-Yau (LG/CY) correspondence. For the quintic threefold it has been worked out by Chiodo-Ruan [**CR10**]. Walcher [**Wa07**] performed analytic continuation of the disk potential $\mathcal{T}^{\mathrm{CY}}$ in the CY phase and obtained a disk potential $\mathcal{T}^{\mathrm{LG}}$ in the LG phase. This leads to a mirror conjecture on genus-zero open FJRW invariants, see Section 2.3.3. Genus-zero open FJRW invariants have been constructed for the affine LG models (x^r, μ_r) [**BCT22, BCT24**], $(x^r + y^s, \mu_r \times \mu_s)$ [**GKT**], and (after the first version of this paper appeared in arXiv) $(x_1^r + \cdots + x_m^r, \mu_r)$ [**TZ1, TZ2**]; the case $(x_1^5 + \cdots + x_5^5, \mu_5)$ corresponds to the quintic threefold.

In Section 5.3, we show that the extended (open and closed) LG/CY correspondence corresponds to the usual wall-crossing for closed invariants of the extended quintic GLSM [**AL23**]. In this case, the GIT stability condition ζ is a real number. The positive phase $\zeta > 0$ corresponds to the CY phase, while the negative phase $\zeta < 0$ corresponds to the LG phase; they are separated by the wall $\zeta = 0$. In Section 6.3 and Section 6.4, we provide the mirror integral representations for the disk potentials $\mathcal{T}^{\mathrm{CY}}$ and $\mathcal{T}^{\mathrm{LG}}$, respectively, from the extended GLSM viewpoint.

Buryak-Clader-Tessler [**BCT24**] establish an open-closed correspondence between open r-spin theory and the closed extended r-spin theory constructed in [**BCT19**]. In work in progress [**Ni**], S. Nill provides a different extension of genus-zero FJRW theory of $(x_1^5 + \cdots + x_5^5, \mu_5)$ by integration of the fifth power of Witten's top Chern class of the closed extended 5-spin theory.

Outline. In Section 2, we review the LG/CY correspondence and the disk potential for the quintic threefold. Section 3 is an introduction to GLSM I-functions and central charges following [**AL23**]. In particular, we review the Higgs-Coulomb correspondence and wall crossing for central charges. The wall crossing is demonstrated on the quintic example in Section 4. In Section 5 we introduce the extended GLSM, compute I-functions and central charges and show the open/closed correspondence for the quintic threefold. Finally, in Section 6 we explain the B-model side of the story. First, we review the B-model representation of the disk potential following [**Wa07**]. Then we provide the B-model of the extended GLSM and show how the quintic disk potential appears as a usual period of the mirror extended GLSM.

2. LG/CY correspondence and mirror symmetry

2.1. LG/CY correspondence. An affine Landau-Ginzburg (LG) model is a pair (W, G), where the superpotential $W : \mathbb{C}^N \to \mathbb{C}$ is a non-degenerate, quasi-homogeneous polynomial, and G is an admissible subgroup of the finite abelian group $G_W := \{\gamma \in (\mathbb{C}^*)^N : W(\gamma x) = W(x)\}$. Fan-Jarvis-Ruan-Witten (FJRW) invariants of (W, G) are virtual counts of sections of orbifold line bundles over orbicurves which are solutions of the (perturbed) Witten equations defined by the superpotential W. FJRW theory [**FJR07**] can be viewed as a mathematical theory of A-model topological strings on affine LG models.

Gromov-Witten (GW) invariants are virtual counts of parametrized holomorphic curves in Kähler manifolds, or more generally parametrized holomorphic orbicurves in Kähler orbifolds. GW theory can be viewed as a mathematical theory of A-model topological strings on Kähler manifolds/orbifolds.

Given an affine LG model (W, G), the quasi-homogeneity of W implies that the equation $W = 0$ defines a hypersurface X_W in a weighted projective space, and the non-degeneracy of W implies that X_W has at most quotient singularities. Suppose that X_W is Calabi-Yau in the sense that $c_1(X_W) = 0 \in H^2(X_W; \mathbb{Q})$. Then $\mathcal{X}_W := [X_W/\widetilde{G}]$ is a compact Calabi-Yau orbifold, where $\widetilde{G}$ is the quotient of G by a cyclic subgroup $\langle J_W \rangle$ which acts trivially on X_W. The Landau-Ginzburg/Calabi-Yau (LG/CY) correspondence relates FJRW theory of the affine LG model (W, G) and (orbifold) GW theory of the Calabi-Yau orbifold $\mathcal{X}_W$.

2.2. LG/CY correspondence for quintic threefolds. Chiodo-Ruan [**CR10**] proved genus-zero LG/CY correspondence when $W = W_5 := x_1^5 + \cdots + x_5^5$ is the Fermat quintic polynomial on $\mathbb{C}^5$ and $G = \mu_5$ is the group of 5-th roots of unity, acting diagonally on $\mathbb{C}^5$. In this case $\widetilde{G}$ is trivial, and

$$\mathcal{X}_W = X_W = X_5 := \{W_5(x) = 0\} \subset \mathbb{P}^4$$

is the Fermat quintic threefold, which is a non-singular projective Calabi-Yau threefold. Indeed, all smooth quintic threefolds (i.e. degree five hypersurfaces in $\mathbb{P}^4$) have the same GW theory, so Chiodo-Ruan proved genus-zero LG/CY correspondence for all (smooth) quintic Calabi-Yau threefolds.

2.2.1. *GW invariants.* Given any positive integer d, let $\overline{\mathcal{M}}_{0,0}(X_5, d)$ and $\overline{\mathcal{M}}_{0,0}(\mathbb{P}^4, d)$ be moduli spaces of genus-0, 0-pointed, degree d stable maps to X_5 and to $\mathbb{P}^4$, respectively. The genus-zero, degree d GW invariant of X_5 is

$$N_d^{\mathrm{CY}} := \int_{[\overline{\mathcal{M}}_{0,0}(X_5,d)]^{\mathrm{vir}}} 1 = \int_{[\overline{\mathcal{M}}_{0,0}(\mathbb{P}^4,d)]} e(E_d) \in \mathbb{Q}$$

where E_d is an orbifold vector bundle over the compact complex orbifold $\overline{\mathcal{M}}_{0,0}(\mathbb{P}^4, d)$ whose fiber over the moduli point $[f : C \to \mathbb{P}^4]$ is $H^0(C, f^*\mathcal{O}_{\mathbb{P}^4}(1))$, and $e(E_d) = c_{5d+1}(E_d)$ is the Euler class of E_d; we have $\operatorname{rank} E_d = \dim \overline{\mathcal{M}}_{0,0}(\mathbb{P}^4, d) = 5d + 1$. The genus-zero GW potential of X_5 is

$$F_0^{\mathrm{CY}}(Q) := \frac{5}{6}(\log Q)^3 + \sum_{d=1}^{\infty} N_d^{\mathrm{CY}} Q^d.$$

2.2.2. *The CY mirror theorem.* The (small) J-function of GW theory of X_5 is

$$J^{\mathrm{CY}}(Q,z) = z e^{(\log Q)H/z}\left(1 + \sum_{d=1}^{\infty} Q^d \sum_{k=0}^{3}\left(\int_{[\overline{\mathcal{M}}_{0,1}(X_5,d)]^{\mathrm{vir}}} \frac{\mathrm{ev}^* H^{3-k}}{z(z-\psi)}\right)\frac{H^k}{5}\right) \tag{2.1}$$

where $H \in H^2(X_5)$ is the restriction of the hyperplane class in $H^2(\mathbb{P}^4)$ and $\psi = c_1(\mathbb{L})$ is the first Chern class of a line bundle $\mathbb{L}$ on $\overline{\mathcal{M}}_{0,1}(X_5, d)$ whose fiber over the moduli point $[f : (C, x) \to X_5]$ is the cotangent line T_x^*C. Let $D = Q\frac{\partial}{\partial Q}$. Then $J^{\mathrm{CY}}(Q, z)$ can be expressed in terms of F_0^{CY}:

$$\begin{aligned} J^{\mathrm{CY}}(Q,z) =& z + (\log Q)H + \frac{1}{5}DF_0^{\mathrm{CY}}(Q)z^{-1}H^2 \\ &+ \Big(\frac{1}{5}(\log Q)DF_0^{\mathrm{CY}}(Q) - \frac{2}{5}F_0^{\mathrm{CY}}(Q)\Big)z^{-2}H^3. \end{aligned} \tag{2.2}$$

The (small) I-function is given explicitly by

$$I^{\mathrm{CY}}(q,z) = z e^{(\log q)H/z}\Big(1 + \sum_{d=1}^{\infty} q^d \frac{\prod_{m=1}^{5d}(5H + mz)}{\prod_{m=1}^{d}(H+mz)^5}\Big) = z\sum_{k=0}^{3} I_k^{\mathrm{CY}}(q)\frac{H^k}{z^k} \tag{2.3}$$

where $H^4 = 0$. In particular,

$$I_0^{\mathrm{CY}}(q) = 1 + \sum_{d=1}^{\infty}\frac{(5d)!}{(d!)^5}q^d, \quad \frac{I_1^{\mathrm{CY}}(q)}{I_0^{\mathrm{CY}}(q)} = \log q + f^{\mathrm{CY}}(q),$$

where

$$f^{\mathrm{CY}}(q) = \frac{1}{I_0^{\mathrm{CY}}(q)}\sum_{d=1}^{\infty} q^d \frac{(5d)!}{(d!)^5}\Big(\sum_{m=1}^{5d}\frac{5}{m} - 5\sum_{m=1}^{d}\frac{1}{m}\Big) \in q\mathbb{Q}[\![q]\!].$$

I^{CY} and J^{CY} are functions of one complex variable q and Q respectively. They take values in the 4-dimensional complex vector space

$$\mathcal{H}_{\mathrm{CY}} = \mathbb{C}z \oplus \mathbb{C}H \oplus \mathbb{C}z^{-1}H^2 \oplus \mathbb{C}z^{-2}H^3$$

equipped with the linear symplectic structure

$$(z^{1-k}H^k, z^{1-\ell}H^\ell) = 5(-1)^k \delta_{k+\ell,3}.$$

The following mirror theorem was first proved by Givental [**Gi96**] and Lian-Liu-Yau [**LLY97**] and implies the mirror conjecture on $D^3F_0^{\mathrm{CY}}(Q)$ predicted by Candelas, de la Ossa, Green, Parkes [**CdGP**].

THEOREM 2.1 (mirror theorem for the CY threefold X_5).

$$J^{\mathrm{CY}}(Q,z) = \frac{I^{\mathrm{CY}}(q,z)}{I_0^{\mathrm{CY}}(q)} \quad \textit{under the mirror map} \quad \log Q = \frac{I_1^{\mathrm{CY}}(q)}{I_0^{\mathrm{CY}}(q)}. \tag{2.4}$$

Ciocan-Fontanine and Kim [**CK14**] introduce a family of J-functions J_ϵ^{CY} where $\epsilon \in \mathbb{Q}_{>0}$, by $\mathbb{C}^*$-localization on ϵ-stable quasimap graph spaces. The I-function and J-function correspond to the limits $\epsilon \to 0+$ and $\epsilon \to \infty$, respectively:

$$I^{\mathrm{CY}} = J^{\mathrm{CY},0+}, \quad J^{\mathrm{CY}} = J^{\mathrm{CY},\infty}.$$

Theorem 2.1 is a consequence of their quasimap wall-crossing formula [**CK14**, **CK20**]. This is also known as ϵ-wall-crossing.

2.2.3. *FJRW invariants.* A genus-g, n-pointed stable 5-spin curve is a triple $((C,z_1,\dots,z_n),\mathcal{L},\rho)$, where $(C,z_1,\dots,z_n)$ is a genus-g, n-pointed twisted stable curve,[1] $\mathcal{L}$ is a line bundle on C which defines a representable morphism from C to $B\mathbb{C}^*$, and $\rho: \mathcal{L}^{\otimes 5} \to \omega_C^{\log} = \omega_C(z_1+\cdots+z_n)$ is an isomorphism. Given $m_1, \dots, m_n \in \{1,2,3,4\}$, let $\overline{\mathcal{M}}^{1/5}_{g,(m_1,\dots,m_n)}$ denote the moduli of genus-g, n-pointed stable 5-spin curves $((C,z_1,\dots,z_n),\mathcal{L},\rho)$ such that $z_i = B\mu_5$ and $L_{z_i} = (T_{z_i}C)^{\otimes m_i}$ for $i=1,\dots,n$. Then $\overline{\mathcal{M}}^{1/5}_{g,(m_1,\dots,m_n)}$ is nonempty if and only if

$$2g-2+\sum_{i=1}^n (1-m_i) \in 5\mathbb{Z}$$

in this case, it is a proper smooth Deligne-Mumford stack of dimension $3g-3+n$. In particular,

$$M_n := \overline{\mathcal{M}}_{0,(\underbrace{2,\dots,2}_{n})}$$

is nonempty if and only if $(n-3)/5 \in \mathbb{Z}_{\geq 0}$, and in this case there is a rank $(n-3)/5$ vector bundle E_n over M_n whose fiber over the moduli point $((C,z_1,\dots,z_n),\mathcal{L},\rho)$ is $H^1(C,\mathcal{L})$. The genus-0, n-pointed primary FJRW invariants of the affine LG model (W_5,μ_5) are

$$\theta_n := \begin{cases} \displaystyle\int_{[M_n]} e(E_n)^5, & \text{if } \dfrac{n-3}{5} \in \mathbb{Z}_{\geq 0}; \\ 0, & \text{otherwise.} \end{cases}$$

Let

$$F_0^{\mathrm{LG}}(\tau) := \sum_{n=0}^\infty \theta_n \frac{\tau^n}{n!} = \sum_{\ell=0}^\infty \theta_{3+5\ell} \frac{\tau^{3+5\ell}}{(3+5\ell)!}$$

[1] In particular, C is a Deligne-Mumford curve with at most nodal singularities and is a scheme away from the marked points z_i and nodes.

2.2.4. *The LG mirror theorem.* The state space $H_{\rm LG}$ of FJRW theory of (W_5,μ_5) is isomorphic to $H^*(X_5;\mathbb{C})$ as a graded vector space over $\mathbb{C}$. In particular, for $k\in\{0,1,2,3\}$,

$$H^{2k}_{\rm LG}\cong H^{2k}(X_5;\mathbb{C})=\mathbb{C}H^k\cong\mathbb{C}.$$

The (small) J-function of FJRW theory of (W_5,μ_5) is

$$J^{\rm LG}(\tau,z)=z\Big(\phi_0+\sum_{n=1}^\infty\frac{\tau^n}{n!}\sum_{k=0}^3\langle\frac{\phi_{3-k}}{z-\psi},\phi_0,\underbrace{\phi_1,\dots,\phi_1}_{n\text{ times}}\rangle_{0,n+2}\phi_k\Big) \tag{2.5}$$

where ϕ_k is a specific $\mathbb{C}$-basis of $H^{2k}_{\rm LG}\cong\mathbb{C}$. The function $J^{\rm LG}(\tau,z)$ can be expressed in terms of the generating function $F_0^{\rm LG}(\tau)$:

$$J^{\rm LG}(\tau,z)=z\phi_0+\tau\phi_1+\frac{\partial F_0^{\rm LG}}{\partial\tau}(\tau)z^{-1}\phi_2+\Big(\tau\frac{\partial F_0^{\rm LG}}{\partial\tau}-2F_0^{\rm LG}(\tau)\Big)z^{-2}\phi_3. \tag{2.6}$$

The (small) I-function $I^{\rm LG}(t,z)$ is given explicitly by

$$I^{\rm LG}(t,z)=z\sum_{d=0}^\infty\frac{t^{d+1}}{d!}\frac{\Gamma(\frac{d+1}{5})^5}{\Gamma(\{\frac{d+1}{5}\})^5}\frac{\phi_{5\{\frac{d}{5}\}}}{z^{5\{\frac{d}{5}\}}}=z\sum_{k=0}^3 I_k^{\rm LG}(t)\frac{\phi_k}{z^k}. \tag{2.7}$$

$$I_0^{\rm LG}(t)=t+\sum_{\ell=1}^\infty\frac{t^{5m+1}}{(5m)!}\frac{\Gamma(\frac15+m)^5}{\Gamma(\frac15)^5},$$

$$\frac{I_1^{\rm LG}(t)}{I_0^{\rm LG}(t)}=\frac{t^2}{I_0^{\rm LG}(t)}\Big(1+\sum_{m=1}^\infty\frac{t^{5m}}{(5m+1)!}\frac{\Gamma(\frac25+m)^5}{\Gamma(\frac25)^5}\Big)=t(1+f^{\rm LG}(t))$$

where $f^{\rm LG}(t)\in t^5\mathbb{Q}[\![t^5]\!]$.

$I^{\rm LG}(t,z)$ and $J^{\rm LG}(\tau,z)$ are functions of one (complex) variable, taking values in the 4-dimensional complex vector space

$$\mathcal{H}_{\rm LG}=\mathbb{C}z\phi_0\oplus\mathbb{C}\phi_1\oplus\mathbb{C}z^{-1}\phi_2\oplus\mathbb{C}z^{-2}\phi_3$$

equipped with the linear symplectic structure

$$(z^{1-k}\phi_k,z^{1-\ell}\phi_\ell)=(-1)^k\delta_{k+\ell,3}.$$

The results in [**CR10**] imply the following mirror theorem, which provides a mirror formula of $F_0^{\rm LG}(\tau)$.

THEOREM 2.2 (mirror theorem for the LG model (W_5,μ_5)).

$$J^{\rm LG}(\tau,z)=\frac{I^{\rm LG}(t,z)}{I_0^{\rm LG}(t)}\quad\textit{under the mirror map}\quad\tau=\frac{I_1^{\rm LG}(t)}{I_0^{\rm LG}(t)}.$$

Ross-Ruan [**RR17**] introduce a family of J-functions $J^{{\rm LG},\epsilon}$ where $\epsilon\in\mathbb{Q}_{>0}$, which interpolate the I-function and the J-function:

$$I^{\rm LG}=J^{{\rm LG},0+},\quad J^{\rm LG}=J^{{\rm LG},\infty}.$$

Theorem 2.2.4 follows from their wall-crossing formula relating $J^{{\rm LG},\epsilon_1}$ and $J^{{\rm LG},\epsilon_2}$, so it can be viewed as a version of ϵ-wall-crossing.

2.2.5. *Picard-Fuchs equation, analytic continuation, and symplectic transform.* Let $\theta = q\frac{d}{dq}$. Then $\{I_k^{\mathrm{CY}}(q) : k = 0,1,2,3\}$ is a basis of the space of solutions to the Picard-Fuchs equation

$$\theta^4\omega = 5q(5\theta+1)(5\theta+2)(5\theta+3)(5\theta+4)\omega. \tag{2.8}$$

Let $\theta_- = t\frac{d}{dt}$. Then $\{I_k^{\mathrm{LG}}(t) : k = 0,1,2,3\}$ is a basis of the space of solutions to

$$(\theta_-)^4\omega = 5^5t^{-5}(\theta_- - 1)(\theta_- - 2)(\theta_- - 3)(\theta_- - 3)\omega. \tag{2.9}$$

Note that (2.8) and (2.9) are related by a simple change of variable $q = t^{-5}$, $\theta_- = -5\theta$.

The Picard-Fuchs equation (2.8) is satisfied by periods $\int_\gamma \Omega$ where Ω is a holomorphic 3-form on the mirror quintic $\mathcal{Q}$ and $\gamma \in H_3(\mathcal{Q};\mathbb{Z}) \cong \mathbb{Z}^4$ **[CdGP]**. See Section 6 for more details.

The I-functions $I^{\mathrm{CY}}(q,z)$ and $I^{\mathrm{LG}}(t,z)$ are related by change of variable $q = t^{-5}$, analytic continuation, and an isomorphism of complex symplectic vector spaces:

$$\phi : \mathcal{H}_{\mathrm{CY}} \xrightarrow{\cong} \mathcal{H}_{\mathrm{LG}}.$$

The LG/CY correspondence of quintic threefolds is a consequence of *global mirror symmetry* over the one-dimensional complex moduli $\mathbb{C}_t$ of the mirror quintic $\mathcal{Q}$, or equivalently the complexified Kähler moduli of the quintic threefold X_5.

2.3. Open mirror symmetry. Open GW invariants are virtual counts of parametrized holomorphic curves in Kähler manifolds with Lagrangian boundary conditions. Open GW theory can be viewed as a mathematical theory of A-model topological open strings on Kähler manifolds.

Open FJRW invariants are virtual counts of sections of orbifold line bundles over orbifold Riemann surfaces with boundaries, where the sections satisfy certain boundary conditions. Open FJRW theory can be viewed as a mathematical theory of A-model topological open strings on affine LG models.

2.3.1. *The CY open mirror theorem.* The Fermat quintic $X_5 = \{W_5(x) = 0\} \subset \mathbb{P}^4$ is defined over $\mathbb{R}$. Let $\mathbb{RP}^4 \subset \mathbb{P}^4$ be the real projective space, and let $\mathbb{R}X_5 := \mathbb{RP}^4 \cap X_5$ be the real quintic, which is diffeomorphic to $\mathbb{RP}^3$. The real quintic $\mathbb{R}X_5$ is the fixed locus of an anti-holomophic and anti-symplectic involution on the Kähler manifold X_5, so it is a totally real Lagrangian submanifold of X_5. The A-model disk potential $F_{0,1}^{\mathrm{CY}}$ of the pair $(X_5, \mathbb{R}X_5)$ is

$$F_{0,1}^{\mathrm{CY}}(Q) = \sum_{\substack{d\in\mathbb{Z}_{>0}\\ d \text{ odd}}} N_d^{\mathrm{disk}} Q^{d/2}$$

where $N_d^{\mathrm{disk}} \in \mathbb{Q}$ is the open GW invariant which counts degree d holomorphic disks in X_5 bounded by $\mathbb{R}X_5$. Near the maximal unipotent monodromy (MUM) point $q = 0$ in the complex moduli of $\mathcal{Q}$ (which corresponds to the CY/geometric phase of the Kähler moduli of X_5 under mirror symmetry), the B-model disk potential is given explicitly by

$$\mathcal{T}^{\mathrm{CY}}(q) = 2\sum_{\substack{d\in\mathbb{Z}_{>0}\\ d \text{ odd}}} \frac{(5d)!!}{(d!!)^5} q^{d/2} \tag{2.10}$$

Note that $\mathcal{T}^{\mathrm{CY}}(q) = 30\sqrt{q}(1 + h(q))$ where $h(q) \in q\mathbb{Q}[\![q]\!]$.

The following mirror formula for the A-model disk potential $F_{0,1}^{\mathrm{CY}}(Q)$ was conjectured by Walcher [**Wa07**] and proved by Pandharipande-Solomon-Walcher [**PSW08**].

THEOREM 2.3 (Open mirror theorem for $(X_5, \mathbb{R}X_5)$).

$$F_{0,1}^{\mathrm{CY}}(Q) = \frac{\mathcal{T}^{\mathrm{CY}}(q)}{I_0^{\mathrm{CY}}(q)} \quad \text{under the mirror map} \quad \log Q = \frac{I_1^{\mathrm{CY}}(q)}{I_0^{\mathrm{CY}}(q)}.$$

In the following table, we compare notations in this paper with those in [**Wa07**] and [**PSW08**].

This paper	q	$\mathcal{T}^{\mathrm{CY}}(q)$	Q	$F_{0,1}^{\mathrm{CY}}(Q)$
[**Wa07**]	z	$30\tau(z)$	$q = e^{2\pi i t}$	$\hat{\tau}(q)$
[**PSW08**]	e^t	$J(t)$	e^T	$\mathcal{F}^{disk}(T)$

2.3.2. *The inhomogeneous and extended Picard-Fuchs equations.* Let

$$\mathcal{L} := \theta^4 - 5q(5\theta+1)(5\theta+2)(5\theta+3)(5\theta+4). \tag{2.11}$$

Then the Picard-Fuch equation (2.8) can be rewritten as

$$\mathcal{L}\omega = 0 \tag{2.12}$$

The B-model disk potential $\mathcal{T}^{\mathrm{CY}}(q)$ satisfies the inhomogeneous Picard-Fuchs equation

$$\mathcal{L}\mathcal{T} = \frac{15}{8}\sqrt{q}. \tag{2.13}$$

Therefore,

$$(2\theta - 1)\mathcal{L}\mathcal{T} = 0 \tag{2.14}$$

which is equivalent to the extended Picard-Fuchs equation

$$(2\theta-1)\mathcal{L}\mathcal{T} = 5q(2\theta+1)(5\theta+1)(5\theta+2)(5\theta+3)(5\theta+4)\mathcal{T}. \tag{2.15}$$

The space of solutions to (2.15) is spanned by $\{I_k^{\mathrm{CY}} : k = 0,1,2,3\}$ and $\mathcal{T}^{\mathrm{CY}}$. Combining Theorem 2.1 (closed mirror theorem) and Theorem 2.3 (open mirror theorem), we obtain a version of *extended* mirror symmetry [**Wa08**] including both closed string and open string sectors for quintic CY threefolds.

2.3.3. *The LG open mirror conjecture.* By [**Wa07**, Section 3], the analytic continuation of $\mathcal{T}^{\mathrm{CY}}(q)$ to the orbifold point $q = \infty$ in the complex moduli of $\mathcal{Q}$ (which corresponds to the LG phase of the Kähler moduli of X_5 under mirror symmetry) is

$$\mathcal{T}^{\mathrm{LG}} + \mathcal{T}_c \tag{2.16}$$

where

$$\begin{aligned}\mathcal{T}^{\mathrm{LG}} &= -\frac{2}{3}\sum_{m=0}^{\infty}\frac{\Gamma(-3/2-5m)}{\Gamma(-3/2)}\frac{\Gamma(1/2)^5}{\Gamma(1/2-m)^5}q^{-(m+1/2)}.\\ &= -2\sum_{\substack{d>0\\ d\ \mathrm{odd}}}\frac{(d!!)^5}{d^5(5d-2)!!}q^{-d/2}\\ &= -2\sum_{\ell=0}^{\infty}\frac{((2\ell+1)!!)^5}{(2\ell+1)^5(10\ell+3)!!}t^{\frac{5}{2}+5\ell},\end{aligned}$$

and $\mathcal{T}_c$ is a linear combination of $\{I_k^{\mathrm{LG}}(t) : k = 0,1,2,3\}$

$$\mathcal{T}_c = -\sqrt{-1}\frac{\pi^3}{10}\sum_{m=1}^{4}\frac{e^{4\pi\sqrt{-1}m/5}}{\Gamma(1-m/5)^5\cos(\pi m/5)}I_{m-1}^{\mathrm{LG}}(t). \tag{2.17}$$

In particular, $\mathcal{T}^{\mathrm{LG}} = -2/3t^{5/2}+O(t^5)$. This leads to the following mirror conjecture (which is implicit in [**Wa09**]). Let $F_{0,1}^{\mathrm{LG}}(\tau)$ be the (yet-to-be-defined) generating function of genus-zero open FJRW invariants of (W_5, μ_5).

CONJECTURE 2.4 (open mirror conjecture for (W_5, μ_5)).

$$F_{0,1}^{\mathrm{LG}}(\tau) = \frac{\mathcal{T}^{\mathrm{LG}}(t)}{I_0^{\mathrm{LG}}(t)} \quad \text{under the mirror map} \quad \tau = \frac{I_1^{\mathrm{LG}}(t)}{I_0^{\mathrm{LG}}(t)}.$$

3. Gauged linear sigma models

3.1. Input data. The input data of a GLSM is a 5-tuple $(V, G, \mathbb{C}_R^*, W, \zeta)$, where:

(1) (linear space) V is a finite dimensional complex vector space.
(2) (gauge group) $G \subset \mathrm{GL}(V)$ is a complex reductive Lie group.
(3) (R symmetries) $\mathbb{C}_R^* \subset \mathrm{GL}(V)$ is a 1-parametric subgroup that commutes with G, and $G \cap \mathbb{C}_R^*$ is finite. Therefore $G \cap \mathbb{C}_R^* \simeq \mu_r$ for some $r \in \mathbb{Z}_{>0}$, where μ_r is the group of r-th roots of unity.
(4) (superpotential) $W : V \to \mathbb{C}$ is a polynomial function that is G-invariant and has weight r with respect to the grading defined by the action of $\mathbb{C}_R^*$ on V: $W(t \cdot x) = t^r W(x)$ for $t \in \mathbb{C}_R^*$ and $x \in V$.
(5) (stability condition) $\zeta \in \widehat{G} := \mathrm{Hom}(G, \mathbb{C}^*)$ is a stability condition for the geometric invariant theory (GIT) quotient stack $[V /\!\!/_\zeta G] = [V^{ss}(\zeta)/G]$.

The GIT quotient stack $\mathcal{X}_\zeta := [V /\!\!/_\zeta G]$ is a smooth Artin stack with trivial generic stabilizer (since G acts faithfully on V). We assume that $V^{ss}(\zeta) = V^s(\zeta)$, so that $\mathcal{X}_\zeta$ is an orbifold, i.e., a smooth Deligne-Mumford (DM) stack with trivial generic stabilizer. The GIT quotient variety $X_\zeta := V /\!\!/_\zeta G = V^{ss}(\zeta)/G$ is a good moduli of the stack $\mathcal{X}_\zeta$. It is equipped with a polarization L_ζ determined by the G-character ζ, and is projective over its affinization $V/_{\mathrm{aff}}G := \mathrm{Spec}\left(H^0(V, \mathcal{O}_V)^G\right)$. The superpotential $W \in H^0(V, \mathcal{O}_V)^G$ defines a regular function $w : V/_{\mathrm{aff}}G \to \mathbb{C}$. Let $w_\zeta : \mathcal{X}_\zeta \to \mathbb{C}$ denote the pullback of w under $\mathcal{X}_\zeta \to X_\zeta \to V/_{\mathrm{aff}}G$. Let

$$\mathcal{Z}_\zeta^w := Z(dw_\zeta)$$

be the critical locus of w_ζ.

The mathematical theory of the GLSM developed by Fan-Jarvis-Ruan [**FJR18**] can be viewed as a mathematical theory of A-model topological strings on the LG model $(\mathcal{X}_\zeta, w_\zeta)$. FJRW theory [**FJR07**] corresponds to the special case where G is a finite abelian group, $\mathcal{X}_\zeta = [V/G]$ and $X_\zeta = V/_{\mathrm{aff}}G$. Fan-Jarvis-Ruan [**FJR18**] define GLSM invariants in the narrow sector via Kiem-Li cosection localized virtual cycle [**KL13**], generalizing construction in [**CLL**] in the affine case. See [**KL20, CKL**] for construction of FJRW invariants in both broad and narrow sectors via cosection localization. Favero-Kim [**FK**] define general GLSM invariants via virtual factorization, generalizing construction of Polishchuk-Vaintrob [**PV16**] in the affine case and construction of Ciocan-Fontanine, Favero, Guéré, Kim, Shoemaker [**CFGKS**] in the convex hybrid case.

Tian-Xu provides a mathematical theory of the GLSM in the geometric phase via symplectic geometry. In [**TX17**] and [**TX**], V is a non-compact Kähler manifold and $W : V \to \mathbb{C}$ is a G-invariant holomorphic function.

3.2. LG quasimaps. Let $\Gamma \subset \mathrm{GL}(V)$ be the subgroup generated by G and $\mathbb{C}^*_R$, and let

$$\mathbb{C}^*_\omega := \Gamma/G = \mathbb{C}^*_R/(G \cap \mathbb{C}^*_R) = \mathbb{C}^*_R/\mu_r \simeq \mathbb{C}^*.$$

We have a short exact sequence of groups

$$1 \to G \longrightarrow \Gamma \xrightarrow{\chi} \mathbb{C}^*_\omega \to 1.$$

GLSM invariants are virtual counts of LG quasimaps. Roughly speaking, an LG quasimap to the given GLSM is a rational map from a twisted curve $(C, z_1, \dots, z_n)$ to $\mathcal{X}_\zeta = [V^{ss}_G(\zeta)/\Gamma]$ which extends to a morphism $f : C \to [V/\Gamma]$ such that the diagram

$$\begin{array}{ccc} & & [V/\Gamma] \\ & \overset{f}{\nearrow} & \downarrow \\ C & \xrightarrow{P} & B\Gamma = [\bullet/\Gamma] \\ & \underset{\omega_C^{\log}}{\searrow} & \downarrow B\chi \\ & & B\mathbb{C}^*_\omega \end{array} \tag{3.1}$$

commutes, and satisfies some stability conditions depending on $\epsilon \in \mathbb{Q}_{>0}$ and a so-called *good lift* $\tilde{\zeta}$ which is a Γ-character $\tilde{\zeta} : \Gamma \to \mathbb{C}^*$ such that $\tilde{\zeta}|_G = \zeta$ and $V^{ss}_\Gamma(\tilde{\zeta}) = V_G(\zeta)$.

For a fixed domain C, the upper half of (3.1) is equivalent to a principal Γ-bundle $P \to C$ together with a Γ-equivariant map $\tilde{f} : P \to V$, or equivalently, a principal Γ-bundle $P \to C$ together with a section u of the vector bundle $P \times_\rho V$ associated to the (faithful) representation $\rho : \Gamma \to \mathrm{GL}(V)$. The lower half of (3.1) says that the line bundle $P \times_\chi \mathbb{C}$ associated to the Γ-character $\chi : \Gamma \to \mathbb{C}^*_\omega$ is isomorphic to the log canonical line bundle $\omega_C^{\log} = \omega_C(z_1 + \cdots + z_n)$ of the marked curve $(C, z_1, \dots, z_n)$. The degree of $f : C \to [V/G]$ is an element in

$$H_2([V/\Gamma]; \mathbb{Q}) \simeq H_2(B\Gamma; \mathbb{Q}) \simeq H_2(BG; \mathbb{Q}) \oplus H_2(B\mathbb{C}^*_\omega; \mathbb{Q}).$$

where the first component is $d \in H_2(BG; \mathbb{Q}) \simeq \mathrm{Hom}(\widehat{G}, \mathbb{Q})$ and the second component is $\deg(\omega_C^{\log}) = 2g - 2 + n \in H_2(\mathbb{C}^*_\omega; \mathbb{Q}) = \mathbb{Q}$. We call $d \in \mathrm{Hom}(\widehat{G}, \mathbb{Q})$ the degree of the quasimap.

More precisely, a genus-g, n-pointed, degree d LG quasimap to the given GLSM is a 4-tuple $((C, z_1, \dots, z_n), P, \kappa, u)$, where $(C, z_1, \dots, z_n)$ is a genus-g, n-pointed twisted curve, P is principal Γ-bundle on C which defines a representable morphism $C \to B\Gamma$, $\kappa : P \times_\chi \mathbb{C} \xrightarrow{\simeq} \omega_C^{\log}$ is an isomorphism of line bundles on C, and u is a section of the vector bundle $P \times_\rho V$ such that the base locus

$$B := u^{-1}(P \times_\rho V_G^{us}(\zeta)),$$

where $V_G^{us}(\zeta) = V - V_G^{ss}(\zeta)$ is the unstable locus, is a finite set away from nodes and marked points $z_1, \ldots, z_n$.

Let $\tilde{\zeta}$ be a good lift, and let $\epsilon \in \mathbb{Q}_{>0}$. An LG quasimap is ϵ-stable with respect to $\tilde{\zeta}$ if the $\mathbb{Q}$-line bundle $\omega_C^{\log} \otimes (P \times_{\tilde{\zeta}} \mathbb{C})^\epsilon$ is ample and $\epsilon\ell(b) \leq 1$ for any point b in the base locus, where $\ell(b)$ is the length defined by $\tilde{\zeta}$.

The GLSM is called abelian if the gauge group G is abelian. In this case, we may choose coordinates $x_1, \ldots, x_N$ on $V \simeq \mathbb{C}^N$ such that both G and $\mathbb{C}_R^*$ act diagonally on V, by characters $D_i \in \widehat{G}$ and $c_i \in \widehat{\mathbb{C}}_R^* = \mathbb{Z}$, respectively. Then Γ acts on V diagonally by characters $\tilde{D}_i \in \widehat{\Gamma}$ with $\tilde{D}_i|_G = D_i$ and $\tilde{D}_i|_{\mathbb{C}_R^*} = c_i$. The G-characters D_i are called gauged charges, and $R_i := 2c_i/r \in \mathbb{Q}$ are called R-charges in the physics literature. The GIT quotient stack $\mathcal{X}_\zeta$ is a smooth toric DM stack [**BCS**] and its coarse moduli space X_ζ is a semi-projective simplicial toric variety. In this case $u = (u_1, \ldots, u_N)$ where u_i is a section of the line bundle $\mathcal{L}_i := P \times_{\tilde{D}_i} \mathbb{C}$ associated to the Γ-character $\tilde{D}_i$, i.e., $u_i \in H^0(C, \mathcal{L}_i)$. We have

$$\deg(\mathcal{L}_i) = \langle D_i, d\rangle + \frac{R_i}{2}(2g - 2 + n) \in \mathbb{Q}.$$

where $\langle D_i, d\rangle \in \mathbb{Q}$ is the pairing between $D_i \in \widehat{G}$ and $d \in \mathrm{Hom}(\widehat{G}, \mathbb{Q}) = H_2(BG; \mathbb{Q})$, and $\frac{R_i}{2}(2g - 2 + n)$ is the pairing between $\frac{R_i}{2} = \frac{c_i}{r} \in H^2(B\mathbb{C}_\omega^*; \mathbb{Q}) \simeq \mathbb{Q}$ and $2g - 2 + n \in H_2(B\mathbb{C}_\omega^*; \mathbb{Q}) \simeq \mathbb{Q}$.

3.3. LG loop spaces. Orbifold quasimap theory [**CCK**] can be viewed as a mathematical theory of the GLSM without a superpotential. The I-function in orbifold quasimap theory can be obtained by $\mathbb{C}^*$ localization on stacky loop spaces. We now assume the gauge group G is abelian, and define LG loop spaces which are analogues of the stacky loop spaces, following [**AL23**].

Given any positive integer a, let $\mathbb{P}[a, 1] = [(\mathbb{C}^2 \setminus \{0\})/\mathbb{C}^*]$, where $\mathbb{C}^*$ acts on $\mathbb{C}^2$ by weights $(a, 1)$. Let $\infty := [1, 0] \in \mathbb{P}[a, 1]$ be the image of $(1, 0)$ under the projection $\mathbb{C}^2 \setminus \{0\} \to \mathbb{P}[a, 1]$. Then $\mathbb{P}[a, 1] = \mathbb{C} \cup \{\infty\}$. The stabilizer of $(1, 0)$ is $\mu_a = \{t \in \mathbb{C}^* \mid t^a = 1\} \subset \mathbb{C}^*$, so ∞ is isomorphic to the classifying space $B\mu_a$ of μ_a. For any $m \in \mathbb{Z}$ let $\mathcal{O}_{\mathbb{P}[a,1]}(m)$ denote the line bundle over $\mathbb{P}[a, 1]$ with total space $[((\mathbb{C}^2 - 0) \times \mathbb{C})/\mathbb{C}^*$ where $\mathbb{C}^*$ acts by weights $(a, 1, m)$. Define

$$H^p\Big(\mathbb{P}^1, \mathcal{O}\big(\frac{m}{a}\big)\Big) := H^p\big(\mathbb{P}[a, 1], \mathcal{O}_{\mathbb{P}[a,1]}(m)\big).$$

Note that this is well-defined because the right hand side depends only on m/a.

Given $d \in \mathrm{Hom}(\widehat{G}, \mathbb{Q})$, define

$$V_d := \bigoplus_{i=1}^N H^0(\mathbb{P}^1, \mathcal{O}(\langle D_i, d\rangle - \frac{R_i}{2})), \quad W_d := \bigoplus_{i=1}^N H^1(\mathbb{P}^1, \mathcal{O}(\langle D_i, d\rangle - \frac{R_i}{2})),$$

where the R-charges R_i are determined by the input data of the abelian GLSM (see the last paragraph of Section 3.2). Let G act on the summand $H^p(\mathbb{P}^1, \mathcal{O}(\langle D_i, d\rangle -$

$R_i/2))$ by the G-character D_i. The degree d LG loop space L_d and the obstruction bundle $E_d \to L_d$ are

$$L_d := [V_d /\!\!/_\zeta G] = [(V_d)^{ss}_G(\zeta)/G], \quad E_d := [((V_d)^{ss}_G(\zeta) \times W_d)/G],$$

We define the set of effective classes for the stability condition $\zeta \in \widehat{G}$ to be

$$\mathrm{Eff}_\zeta = \{d \in \mathrm{Hom}(\widehat{G}, \mathbb{Q}) : [V_d /\!\!/_\zeta G] \text{ is nonempty}\}.$$

L_d is a smooth toric DM stack (whose generic stabilizer might be non-trivial), and E_d is a toric vector bundle over L_d. The action of $\mathbb{C}^*_{\mathsf{q}} \simeq \mathbb{C}^*$ on $\mathbb{P}^1$ induces a $\mathbb{C}^*_{\mathsf{q}}$-action on L_d, and E_d is a $\mathbb{C}^*_{\mathsf{q}}$-equivariant vector bundles over L_d. The virtual tangent bundle of L_d is

$$T^{\mathrm{vir}}_{L_d} = T_{L_d} - E_d.$$

There is a map $\mathrm{ev}_\infty : L_d \to I[V/G]$ given by evaluation at ∞. Let $L^\circ_d \subset L_d$ be the open substack which is the preimage of $I\mathcal{X}_\zeta = I[V^{ss}_G(\zeta)/G] \subset I[V/G]$. Then $\mathbb{C}^*_{\mathsf{q}}$ acts on L°_d. We have a disjoint union of connected components:

$$I\mathcal{X}_\zeta = \bigsqcup_{v \in \mathrm{Box}_\zeta} \mathcal{X}_{\zeta,v}$$

where Box_ζ is a finite set defined in [**BCS**]. Each element $v \in \mathrm{Box}_\zeta$ corresponds to a unique element $g(v) \in G$ such that $V^{ss}_G(\zeta)^{g(v)} := \{x \in V^{ss}_G(\zeta) : g(v) \cdot x = x\}$ is non-empty, and

$$\mathcal{X}_{\zeta,v} = [V^{ss}_G(\zeta)^{g(v)}/G].$$

Let $\mathrm{inv} : \mathrm{Box}_\zeta \to \mathrm{Box}_\zeta$ be the involution characterized by $g(\mathrm{inv}(v)) = g(v)^{-1}$. There is an involution $\mathrm{inv}_R : I\mathcal{X}_\zeta \to I\mathcal{X}_\zeta$ such that

$$\mathrm{inv}_R(\mathcal{X}_{\zeta,v}) = \mathcal{X}_{\zeta,\mathrm{inv}(v)}, \quad w_{\zeta,v} \circ \mathrm{inv}_R = -w_{\zeta,\mathrm{inv}(v)}.$$

We have

$$\mathrm{ev}_\infty : L^\circ_d \to \mathcal{X}_{\zeta,\mathrm{inv}(v(d))}$$

where $v(d) \in \mathrm{Box}_\zeta$ is determined by $d \in \mathrm{Eff}_\zeta$. Let $\mathcal{F}_d := (L^\circ_d)^{\mathbb{C}^*_{\mathsf{q}}}$ which is a connected component of $L_d^{\mathbb{C}^*_{\mathsf{q}}}$. Since $\mathbb{C}^*_{\mathsf{q}}$ acts trivially on $\mathcal{F}_d$, it acts linearly on the fibers of any $\mathbb{C}^*_{\mathsf{q}}$-equivariant vector bundle on $\mathcal{F}_d$. If V is a $\mathbb{C}^*_{\mathsf{q}}$-equivariant vector bundle over $\mathcal{F}_d$ then

$$V = \bigoplus_{k \in \mathbb{Q}} V_k = V^f \oplus V^m,$$

where V_k is the subbundle on which $\mathbb{C}^*_{\mathsf{q}}$ acts by weight k, and $V^m = \bigoplus_{k \neq 0} V_k$ (resp. $V^f = V_0$) is the moving (resp. fixed) part of V under the $\mathbb{C}^*_{\mathsf{q}}$-action. Let

$$T^1_d := T_{L_d}\big|_{\mathcal{F}_d}, \quad T^2_d := E_d\big|_{\mathcal{F}_d}.$$

Then $T^{1,f}_d = T_{\mathcal{F}_d}$ is the tangent bundle of $\mathcal{F}_d$, and $T^{1,m}_d = N_{\mathcal{F}_d/L_d}$ is the normal bundle of $\mathcal{F}_d$ in L_d. We also have $T^{2,f}_d = 0$ and $T^{2,m}_d = E_d\big|_{\mathcal{F}_d}$. The virtual tangent bundle of $\mathcal{F}_d$ is

$$T^{\mathrm{vir}}_{\mathcal{F}_d} = T^{1,f}_d - T^{2,f}_d = T_{\mathcal{F}_d} \in K_{\mathbb{C}^*_{\mathsf{q}}}(\mathcal{F}_d),$$

and the virtual normal bundle of $\mathcal{F}_d$ is

$$N^{\mathrm{vir}}_d = T^{1,m}_d - T^{2,f}_d = N_{\mathcal{F}_d/L_d} - E_d\big|_{\mathcal{F}_d} \in K_{\mathbb{C}^*_{\mathsf{q}}}(\mathcal{F}_d).$$

3.4. Virtual factorizations and virtual fundamental classes. Given any effective class $d \in \mathrm{Eff}_\varsigma$, $\mathrm{ev}_\infty : L^\circ_d \to \mathcal{X}_{\varsigma,\mathrm{inv}(v(d))}$ restricts to a closed toric embedding $\iota_{d\to\mathrm{inv}(v(d))} : \mathcal{F}_d = (L^\circ_d)^{\mathbb{C}^*_q} \to \mathcal{X}_{\varsigma,\mathrm{inv}(v(d))}$. In the remainder of this subsection, we fix an effective class $d \in \mathrm{Eff}_\varsigma$ and let $v = \mathrm{inv}(v(d))$.

Let $g(v) \in G$ be the element corresponding to $v \in \mathrm{Box}_\varsigma$. There exists a Γ-invariant subspace $H \subset V^{g(v)}$ such that

$$\mathcal{F}_d = [H /\!\!/_\varsigma G] \subset \mathcal{X}_{\varsigma,v} = [V^{g(v)} /\!\!/_\varsigma G].$$

More explicitly,

$$\begin{aligned} H &= \operatorname{Spec}\mathbb{C}[y_1,\dots,y_{n_+}] \simeq \mathbb{C}^{n_+} \subset V^{g(v)} \\ &= \operatorname{Spec}\mathbb{C}[y_1,\dots,y_{n_+},p_1,\dots,p_{n_-}] \simeq \mathbb{C}^{n_+ + n_-}, \end{aligned}$$

where $\{y_1,\dots,y_{n_+},p_1,\dots,p_{n_-}\} \subset \{x_1,\dots,x_N\}$. $\mathcal{F}_d$ is the zero locus of the section

$$\beta_d = (p_1,\dots,p_{n_-}) \in H^0(\mathcal{X}_{\varsigma,v}, B_d) \tag{3.2}$$

where B_d is a toric vector bundle over $\mathcal{X}_{\varsigma,v}$ with fiber $\operatorname{Spec}\mathbb{C}[p_1,\dots,p_{n_-}] \simeq \mathbb{C}^{n_-}$. One can show that $\iota^*_{d\to v} w_{\varsigma,v} = 0$, which implies $W|_H = 0$. Therefore, $W_v := W|_{V^{g(v)}} \in \mathbb{C}[y_1,\dots,y_{n_+},p_1,\dots,p_{n_-}]$ is in the ideal generated by $p_1,\dots,p_{n_-}$:

$$W_v(y,p) = \sum_{i=1}^{n_-} W_i(y,p)p_i.$$

$$\alpha^W_d := (-W_1,\dots,-W_{n_-}) \in H^0(\mathcal{X}_{\varsigma,v}, B^\vee_d) \tag{3.3}$$

is a section of $B^\vee_d$ (the dual of B_d) which can also be viewed as a cosection of B_d, i.e., $\alpha^W_d : B_d \to \mathcal{O}_{\mathcal{X}_{\varsigma,v}}$. We have

$$\langle \alpha^W_d, \beta_d \rangle = -w_{\varsigma,v} \in H^0\left(\mathcal{X}_{\varsigma,v}, \mathcal{O}_{\mathcal{X}_{\varsigma,v}}\right)$$

and

$$\mathcal{Z}^w_d := Z(dw_{\varsigma,v}) \cap \mathcal{F}_d = Z(\alpha^W_d) \cap Z(\beta_d).$$

In this case, the Favero-Kim virtual factorization is the following Koszul matrix factorization

$$\{\alpha^W_d, \beta_d\} = \Big[\textstyle\bigoplus_i \Lambda^{2i} B^\vee_d \underset{\partial}{\overset{\partial = i_{\beta_d} + \alpha^W_d \wedge}{\rightleftarrows}} \bigoplus_i \Lambda^{2i+1} B^\vee_d \Big] \tag{3.4}$$

where i_{β_d} is the interior product with β_d, and $\alpha^W_d \wedge$ is the wedge product with α_d.

- The cosection α^W_d depends on the superpotential $W_v : V^{g(v)} \to \mathbb{C}$ but the section β_d does not. The Koszul matrix factorization $\{\alpha^W_d, \beta_d\}$ is regular in the sense that β_d is a regular section of B_d.
- The cosection α^W_d is not unique, but different choices define quasi-isomorphic Koszul matrix factorizations. Moreover, $\iota^* B_d = N_{\mathcal{F}_d/\mathcal{X}_{\varsigma,v}}$ is the normal bundle of the closed regular embedding $\iota := \iota_{d\to v} : \mathcal{F}_d \to \mathcal{X}_{\varsigma,v}$, and

  $$\iota^*\alpha^W_d = \Big(-\frac{\partial W}{\partial p_1}(y,0), \cdots, -\frac{\partial W}{\partial p_{n_-}}(y,0) \Big) \in H^0(\mathcal{F}_d, N^\vee_{\mathcal{F}_d/\mathcal{X}_{\varsigma,v}})$$

  is independent of choice of α^W_d.

Let

$$n_d := n_- = \operatorname{rank}_{\mathbb{C}} B_d = \dim_{\mathbb{C}} \mathcal{X}_{\varsigma,v} - \dim_{\mathbb{C}} \mathcal{F}_d$$

be the codimension of $\mathcal{F}_d$ in $\mathcal{X}_{\varsigma,v}$. The Favero-Kim virtual fundamental class is

$$F_d^w := \operatorname{tdch}_{\mathcal{Z}_d^w}^{\mathcal{X}_{\varsigma,v}} \{\alpha_d^W, \beta_d\} \in \mathbb{H}_{\mathcal{Z}_d^w}^{2n_d}\Big(\mathcal{X}_{\varsigma,v}, (\Omega^\bullet_{\mathcal{X}_{\varsigma,v}}, -dw_{\varsigma,v}\wedge)\Big)$$

which is a class in the hypercohomology group of the twisted Hodge complex $(\Omega^\bullet_{\mathcal{X}_{\varsigma,v}}, -dw_{\varsigma,v}\wedge)$. The precise definition of tdch (which is quite involved) can be found in [**FK**]. In particular, when $w_{\varsigma,v} = 0$, $\{\alpha_d^W, \beta_d\} = \{0, \beta_d\}$ is the Koszul complex of the closed regular embedding $\iota : \mathcal{F}_d \to \mathcal{X}_{\varsigma,v}$, $\mathcal{Z}_d^w = \mathcal{F}_d$, and

$$F_d^w = \operatorname{tdch}_{\mathcal{F}_d}^{\mathcal{X}_{\varsigma,v}} \{0, \beta_d\} \in \mathbb{H}_{\mathcal{F}_d}^{2n_d}\Big(\mathcal{X}_{\varsigma,v}, (\Omega^\bullet_{\mathcal{X}_{\varsigma,v}}, 0)\Big).$$

We also define

$$F_d := \iota_* 1 = \tau \in H_{\mathcal{F}_d}^{2n_d}(\mathcal{X}_{\varsigma,v}) := H_{\mathcal{F}_d^{an}}^{2n_d}(\mathcal{X}_{\varsigma,v}^{an}, \underline{\mathbb{C}})$$

where $\underline{\mathbb{C}}$ is the constant sheaf on $\mathcal{X}_{\varsigma,v}^{an}$ associated to $\mathbb{C}$, and τ is the Thom class of the closed regular embedding $\iota : \mathcal{F}_d \to \mathcal{X}_{\varsigma,v}$. Recall that the Gysin map

$$\iota_* : H^k(\mathcal{F}_d) \to H_{\mathcal{F}_d}^{k+2n_d}(\mathcal{X}_{\varsigma,v})$$

is an isomorphism of $\mathbb{C}$-linear spaces. For any $a \in H^k(\mathcal{F})$,

$$\iota^* \iota_* a = e(N_{\mathcal{F}_d/\mathcal{X}_{\varsigma,v}}) \cup a$$

where $e(N_{\mathcal{F}_d/\mathcal{X}_{\varsigma,v}}) = c_{2n_d}(N_{\mathcal{F}_d/\mathcal{X}_{\varsigma,v}})$ is the Euler class (and the top Chern class) of the normal bundle $N_{\mathcal{F}_d/\mathcal{X}_{\varsigma,v}}$.

3.5. Projection formulas. The class F_d^w satisfies the following property. Let

$$a \in \mathbb{H}_{\mathcal{Y}}^k\Big(\mathcal{X}_{\varsigma,v}, (\Omega^\bullet_{\mathcal{X}_{\varsigma,v}}, dw_{\varsigma,v}\wedge)\Big)$$

where $\mathcal{Y}$ is a closed substack of $\mathcal{X}_{\varsigma,v}$. Then

$$\iota^* a \in \mathbb{H}^k_{\mathcal{Z}_d^w \cap \mathcal{Y}}\Big(\mathcal{F}_d, (\Omega^\bullet_{\mathcal{F}_d}, 0)\Big), \quad F_d^w \wedge a \in \mathbb{H}^{2n_d+k}_{\mathcal{Z}_d^w \cap \mathcal{Y}}\Big(\mathcal{X}_{\varsigma,v}, (\Omega^\bullet_{\mathcal{X}_{\varsigma,v}}, 0)\Big).$$

By the proof of [**FK**, Proposition 4.10], if $\mathcal{Z}_d^w \cap \mathcal{Y}$ is proper over $\mathbb{C}$ then

$$\int_{\mathcal{X}_{\varsigma,v}} F_d^w \wedge a = \int_{\mathcal{F}_d} \iota^* a. \tag{3.5}$$

which is zero unless $k = 2\dim_{\mathbb{C}} \mathcal{F}_d \Leftrightarrow 2n_d + k = 2\dim_{\mathbb{C}} \mathcal{X}_{\varsigma,v}$.

Similarly, let

$$a \in H_{\mathcal{Y}}^k(\mathcal{X}_{\varsigma,v})$$

where $\mathcal{Y}$ is a closed substack of $\mathcal{X}_{\varsigma,v}$. Then

$$\iota^* a \in H^k_{\mathcal{Z}_d^w \cap \mathcal{Y}}(\mathcal{F}_d), \quad F_d^w \wedge a \in H^{2n_d+k}_{\mathcal{Z}_d^w \cap \mathcal{Y}}(\mathcal{X}_{\varsigma,v})$$

If $\mathcal{Z}_d^w \cap \mathcal{Y}$ is proper over $\mathbb{C}$ then

$$\int_{\mathcal{X}_{\varsigma,v}} F_d \wedge a = \int_{\mathcal{F}_d} \iota^* a. \tag{3.6}$$

The left hand side of (3.5) (resp. (3.6)) depends only on the class of $F_d^w \wedge a$ (resp. $F_d \wedge a$) in $H_c^{2n_d+k}(\mathcal{X}_{\varsigma,v})$. The right hand side of (3.5) (resp. (3.6)) depends only on the class of $\iota^* a$ in $H_c^k(\mathcal{F}_d)$.

3.6. I-functions. We first introduce some notation. Let $\kappa := \dim_{\mathbb{C}} G$. Let $\{\xi_1^*, \dots, \xi_\kappa^*\}$ be a $\mathbb{Z}$-basis of $\widehat{G} \simeq \mathbb{Z}^\kappa$. Introduce Novikov variables $y = (y_1, \dots, y_\kappa)$, and define

$$y^d := \prod_{a=1}^{\kappa} y_a^{\langle \xi_a^*, d\rangle}, \quad d \in \mathrm{Eff}_\zeta .$$

Let $P_a \in \mathrm{Pic}(\mathcal{X}_\zeta)$ be the line bundle associated to the G-character ξ_a^*, and let $p_a := -c_1(P_a) \in H^2(\mathcal{X}_\zeta; \mathbb{Z})$. Let $\iota_v : \mathcal{X}_{\zeta,v} \to \mathcal{X}_\zeta$ be the inclusion. We define the GLSM I-function of the abelian GLSM $(V, G, \mathbb{C}_R^*, W, \zeta)$ to be

$$I_w = z \sum_{v \in \mathrm{Box}_\zeta} I_{w,v}, \quad I_{w,v} = e^{\sum_{a=1}^{\kappa} (\log y_a) \iota_v^* p_a / z} \sum_{\substack{d \in \mathrm{Eff}_\zeta \\ v(d)=v}} y^d \frac{(\mathrm{inv}_R)_* F_d^w}{e_{\mathbb{C}_q^*}(\widetilde{N}_d^{\mathrm{vir}})}$$

where $\widetilde{N}_d^{\mathrm{vir}} \in K_{\mathbb{C}_q^*}(\mathcal{X}_{\zeta,v})$ is a natural extension of $N_d^{\mathrm{vir}} \in K_{\mathbb{C}_q^*}(\mathcal{F}_d)$, $e_{\mathbb{C}_q^*}(\widetilde{N}_d^{\mathrm{vir}})$ is the $\mathbb{C}_q^*$-equivariant Euler class of $\widetilde{N}_d^{\mathrm{vir}}$, and

$$(\mathrm{inv}_R)_* F_d^w \in \mathbb{H}^{2n_d}_{\mathrm{inv}_R(\mathcal{Z}_d^w)}\Big(\mathcal{X}_{\zeta,v(d)}, (\Omega^\bullet_{\mathcal{X}_{\zeta,v(d)}}, w_{\zeta,v(d)} \wedge)\Big).$$

In particular, when $W = 0$, we obtain the following GLSM I-function of $(V, G, \mathbb{C}_R^*, 0, \zeta)$:

$$I_0 = z \sum_{v \in \mathrm{Box}_\zeta} I_{0,v}, \quad I_{0,v} = e^{\sum_{a=1}^{\kappa} (\log y_a) \iota_v^* p_a / z} \sum_{\substack{d \in \mathrm{Eff}_\zeta \\ v(d)=v}} y^d \frac{(\mathrm{inv}_R)_* F_d^0}{e_{\mathbb{C}_q^*}(\widetilde{N}_d^{\mathrm{vir}})}$$

where

$$(\mathrm{inv}_R)_* F_d^0 \in \mathbb{H}^{2n_d}_{\mathrm{inv}_R(\mathcal{F}_d)}\Big(\mathcal{X}_{\zeta,v}, (\Omega^\bullet_{\mathcal{X}_{\zeta,v}}, 0)\Big).$$

When $W = 0$, we also define the following non-equivariant I-function of $(V, G, \mathbb{C}_R^*, 0, \zeta)$:

$$I = z \sum_{v \in \mathrm{Box}_\zeta} I_v, \quad I_v = e^{\sum_a \log y_a \iota_v^* p_a / z} \sum_{\substack{d \in \mathrm{Eff}_\zeta \\ v(d)=v}} y^d \frac{(\mathrm{inv}_R)_* F_d}{e_{\mathbb{C}_q^*}(\widetilde{N}_d^{\mathrm{vir}})}$$

where

$$F_d = (\iota_{d \to \mathrm{inv}(v(d))})_* 1 \in H^{2n_d}_{\mathcal{F}_d}(\mathcal{X}_{\zeta, \mathrm{inv}(v(d))}), \quad (\mathrm{inv}_R)_* F_d \in H^{2n_d}_{\mathrm{inv}_R(\mathcal{F}_d)}(\mathcal{X}_{\zeta, v(d)}).$$

3.7. Matrix factorizations and quasimap central charges. Following [**PV11**], a *matrix factorization* for $(\mathcal{X}_\zeta, w_\zeta)$ is a pair

$$(E, \delta_E) = \Big(E_0 \underset{\delta_1}{\overset{\delta_0}{\rightleftarrows}} E_1 \Big)$$

where E_0 and E_1 are vector bundles (locally free sheaves of $\mathcal{O}_{\mathcal{X}_\zeta}$-modules) of finite rank on $\mathcal{X}_\zeta$, and

$$\delta_1 \circ \delta_0 = w_\zeta \cdot \mathrm{id}_{E_0}, \quad \delta_0 \circ \delta_1 = w_\zeta \cdot \mathrm{id}_{E_1} .$$

Let $\mathrm{MF}(\mathcal{X}_\zeta, w_\zeta)$ be the dg-category of matrix factorizations of w_ζ for $\mathcal{X}_\zeta$. There is a well-defined forgetful map

$$\mathfrak{forget} : K(\mathrm{MF}(\mathcal{X}_\zeta, w_\zeta)) \longrightarrow K(\mathcal{X}_\zeta) = K(\mathrm{Perf}(\mathcal{X}_\zeta)) \tag{3.7}$$

sending a matrix factorization $(E = E_0 \oplus E_1, \delta_E)$ to $[E_0] - [E_1]$.

In [**AL23**] we consider various versions of category of B-branes. We define the quasimap central charge of a B-brane in terms of suitable versions of I-function,

Iritani's Gamma class, and Chern character of the K-theory class of the B-brane. In this paper we will consider the following two versions.

DEFINITION 3.1 (GLSM quasimap central charges of matrix factorizations). Suppose that $\mathcal{Z}_d^w = \mathrm{Crit}(w_{\zeta,v(d)}) \cap \mathcal{F}_d$ is proper for all $d \in \mathrm{Eff}_\zeta$. Given a matrix factorization $\mathfrak{B} \in \mathrm{MF}(\mathcal{X}_\zeta, w_\zeta)$, let $[\mathfrak{B}] \in K(\mathrm{MF}(\mathcal{X}_\zeta, w_\zeta))$ be its K-theory class. The GLSM quasimap central charge of $\mathfrak{B}$ is defined to be

$$Z_w(\mathfrak{B}) = Z_w([\mathfrak{B}]) := (I_w(y,-1), \hat{\Gamma}\,\mathrm{ch}_w([\mathfrak{B}])) \tag{3.8}$$

which is a $\mathbb{C}$-valued function in y, where $\hat{\Gamma}$ is a version of Iritani's Gamma class defined in [**AL23**] and

$$\mathrm{ch}_w([\mathfrak{B}]) \in \bigoplus_{v \in \mathrm{Box}_\zeta} \mathbb{H}^{\mathrm{even}}\left(\mathcal{X}_{\zeta,v}, (\Omega^*_{\mathcal{X}_{\zeta,v}}, dw_{\zeta,v}\wedge)\right).$$

is the Chern character defined in [**CKS**].

REMARK 3.2. We use the sign convention of [**FK**] in the definition of the Chern character which is the opposite of the sign convention in [**CKS**].

DEFINITION 3.3 (non-equivariant quasimap central charges). Suppose that $\mathcal{F}_d$ is proper for all $d \in \mathrm{Eff}_\zeta$. The non-equivariant quasimap central charge of $F \in K(\mathcal{X}_\zeta)$ is defined to be

$$Z(F) := (I(y,-1), \hat{\Gamma}\,\mathrm{ch}(F)) \tag{3.9}$$

where

$$\mathrm{ch}(F) \in \bigoplus_{v \in \mathrm{Box}_\zeta} H^*(\mathcal{X}_{\zeta,v})$$

is the combinatorial Chern character defined in [**BH**, Section 5].

REMARK 3.4. When there is a good lift $\tilde{\zeta} \in \widehat{\Gamma} \otimes \mathbb{Q}$ of $\zeta \in \widehat{G} \otimes \mathbb{Q}$, one may define J^ϵ-functions J^ϵ_w and J^ϵ using ϵ-stable LG quasimap graph spaces, and define ϵ-dependent quasimap central charges Z^ϵ_w and Z^ϵ using J^ϵ_w and J^ϵ (instead of the I-functions I_w and I), respectively. Under the semi-positive assumption, the GLSM J-function $J_w = J^\infty_w$ (resp. the non-equivariant J-function $J = J^\infty$) and the GLSM I-function $I_w = J^{0+}$ (resp. the non-equivariant I-function $I = I^{0+}$) should be related by a mirror map (ϵ-wall-crossing). The central charges of a compact semi-positive toric orbifold [**Ir09, Fa20**] can be expressed in terms of the J-function of the compact orbifold and is the analogue of Z^∞.

PROPOSITION 3.5 (comparison of quasimap central charges). *Suppose that $\mathcal{F}_d$ is proper for any $d \in \mathrm{Eff}_\zeta$. For any $\mathfrak{B} \in \mathrm{MF}(\mathcal{X}_\zeta, w_\zeta)$,*

$$Z_w(\mathfrak{B}) = Z\left(\mathfrak{forget}[\mathfrak{B}]\right)$$

where $[\mathfrak{B}] \in K(\mathrm{MF}(\mathcal{X}_\zeta, w_\zeta))$ is the K-theory class of $\mathfrak{B}$, and $\mathfrak{forget}$ is the natural forgetful map (3.7).

3.8. Higgs-Coulomb correspondence. We say a GLSM $(V, G, \mathbb{C}^*_{\mathbb{R}}, W, \zeta)$ is Calabi-Yau if $G \subset \mathrm{SL}(V)$, or equivalently, the GIT quotient $X_\zeta = V /\!\!/_\zeta G$ is a Calabi-Yau variety. When the GLSM is abelian, the Calabi-Yau condition is equivalent to

$$\sum_{i=1}^N D_i = 0 \tag{3.10}$$

where $N = \dim V$.

In this case we will often call our quasimap central charges simply central charges since they correspond to (under the mirror map) central charges of D-branes represented by the matrix factorizations in the physics literature [**Ho00**].

One of the main results of [**AL23**] is the so-called Higgs-Coulomb correspondence, which means that central charges of an abelian, Calabi-Yau GLSM have explicit Mellin-Barnes type integral representations.

Let $\mathfrak{B} \in \mathrm{MF}([V/G], w) = \mathrm{MF}_G(V, W)$ be a G-equivariant matrix factorization of (V, W). There is a natural forgetful map

$$\text{(3.11)} \qquad \mathfrak{forget} : K(\mathrm{MF}([V/G], w)) \to K([V/G]) = K_G(V)$$

sending a G-equivariant matrix factorization $(E = E_0 \oplus E_1, \delta_E)$ to $[E_0] - [E_1]$.

Let $\mathfrak{g}$ denote the Lie algebra of the gauge group G, and let $\bullet$ denote a point. We have

$$K_G(V) \simeq K_G(\bullet) = \mathbb{Z}[P_1^{\pm 1}, \dots, P_\kappa^{\pm 1}], \quad G = \operatorname{Spec} \mathbb{C}[P_1^{\pm 1}, \dots, P_\kappa^{\pm 1}] \simeq (\mathbb{C}^*)^\kappa;$$
$$H_G^*(V; \mathbb{C}) \simeq H_G^*(\bullet; \mathbb{C}) = \mathbb{C}[\sigma_1, \dots, \sigma_\kappa], \quad \mathfrak{g} = \operatorname{Spec} \mathbb{C}[\sigma_1, \dots, \sigma_\kappa] \simeq \mathbb{C}^\kappa.$$

Given

$$\mathfrak{t} = \sum_{a=1}^{\kappa} t_a \xi_a^* \in \widehat{G},$$

where $t_a \in \mathbb{Z}$, let

$$\mathcal{L}_{\mathfrak{t}} = \prod_{a=1}^{\kappa} P_a^{t_a} \in \operatorname{Pic}([V/G] = \operatorname{Pic}_G(V)$$

be the associated G-equivariant line bundle on V.

Consider the $(2\pi\sqrt{-1})$-modified equivariant Chern character:

$$\text{(3.12)} \qquad \begin{aligned} &\mathrm{Ch} \;:\; K([V/G]) \to H^*_{an}([V/G]), \quad \mathrm{Ch} = (2\pi\sqrt{-1})^{\deg/2} \circ \mathrm{ch} \\ &\mathrm{Ch}(\mathcal{L}_{\mathfrak{t}}) = e^{2\pi\sqrt{-1}\langle \mathfrak{t}, \sigma\rangle} = e^{2\pi\sqrt{-1}\sum_{a=1}^{\kappa} t_a \sigma_a} \end{aligned}$$

where $H^*_{an}([V/G]) \simeq H^*_{G,an}(\bullet)$ is the analytic completion of the equivariant cohomology $H^*_G(\bullet)$ of a point $\bullet$, and $\sigma = \sum_{a=1}^{\kappa} \sigma_a \xi_a \in \mathfrak{g}$.

Assuming the Calabi-Yau condition (3.10), we define the integral density of the hemisphere partition function to be

$$\text{(3.13)} \quad \Gamma(\sigma) \cdot \mathrm{Ch}(\mathfrak{forget}([\mathfrak{B}])) = \prod_{i=1}^{N} \Gamma(\langle D_i, \sigma\rangle + \frac{R_i}{2}) \cdot \mathrm{Ch}(\mathfrak{forget}([\mathfrak{B}])) \in H^*_{an}([V/G]).$$

This expression can be treated as either an equivariant cohomology class or just a complex analytic function on κ variables $\sigma_1, \dots, \sigma_\kappa$. The (non-equivariant limit of the) hemisphere partition function of the matrix factorization $\mathfrak{B}$ is defined to be [**AL23, HR13**] an analytic function of complex variables $\theta_a = \zeta_a + 2\pi\sqrt{-1}B_a$, $1 \le a \le \kappa$:

$$\text{(3.14)} \qquad Z_{D^2}(\mathfrak{B}) := \frac{1}{2\pi\sqrt{-1}} \oint_{\delta + \sqrt{-1}\mathfrak{g}_{\mathbb{R}}} d\sigma\, \Gamma(\sigma) \cdot e^{\langle \theta, \sigma\rangle}\, \mathrm{Ch}\left(\mathfrak{forget}([\mathfrak{B}])\right),$$

where $\mathfrak{g}_{\mathbb{R}} = \bigoplus_{a=1}^{\kappa} \mathbb{R}\xi_a$, and $\delta \in \mathfrak{g}_{\mathbb{R}}$ satisfies the condition

$$\text{(3.15)} \qquad \langle D_i, \delta\rangle + \frac{R_i}{2} > 0, \quad i = 1, \dots, N.$$

LEMMA 3.6. *If there exists $\delta \in \mathfrak{g}_{\mathbb{R}}$ which satisfies* (3.15) *then $\mathcal{F}_d$ is proper for any $d \in \mathrm{Eff}_\zeta$ for any stability condition $\zeta \in \widehat{G}$ such that $V^s_G(\zeta) = V^{ss}_G(\zeta)$.*

Since $K_G(V)$ is generated by line bundles and $Z_w(\mathfrak{B})$ depends only on $\mathfrak{forget}([\mathfrak{B}])$ we can reduce the computation of $Z_w(\mathfrak{B})$ to computations of $Z(\mathcal{L}_{\mathbf{t}})$.

THEOREM 3.7 (Higgs-Coulomb correspondence [**AL23**]). *The formal power series $Z_w(\mathfrak{B}|_{\mathcal{X}_\zeta})$ has a non-zero convergence radius in the variables $\{y_a\}_{a=1}^{\kappa}$ for any $\mathfrak{B} \in \mathrm{MF}([V/G], w)$. Moreover, for any $\mathbf{t} \in \widehat{G}$ there exist a non-empty open region $U_{\mathbf{t}} \subset \mathfrak{g}^*_{\mathbb{R}}$ such that the function*

$$Z_{D^2}(\mathcal{L}_{\mathbf{t}}) := \frac{1}{2\pi\sqrt{-1}} \oint_{\delta+\sqrt{-1}\mathfrak{g}_{\mathbb{R}}} d\sigma\, \Gamma(\sigma) \cdot e^{\langle \theta, \sigma \rangle}\, \mathrm{Ch}(\mathcal{L}_{\mathbf{t}})$$

is well-defined for $\mathrm{im}(\theta) \in U_{\mathbf{t}}$.

$$Z_{D^2}(\mathcal{L}_{\mathbf{t}}) = Z(\mathcal{L}_{\mathbf{t}}|_{\mathcal{X}_\zeta})|_{\log y_a = -\theta_a}$$

for θ in a non-empty open region in $\mathfrak{g}^*$.

Higgs-Coulomb correspondence also allows to look at the Landau-Ginzburg correspondence from a different perspective. Let $C_+, C_- \subset \mathfrak{g}^*_{\mathbb{R}}$ be two chambers of the secondary fan on $\mathcal{X}_\zeta$ adjacent along codimension 1 wall perpendicular to $h \in \mathfrak{g}_{\mathbb{R}}$.

Let $\mathfrak{B} \in \mathrm{MF}([V/G], W)$. We say that $\mathfrak{B}$ satisfies the *grade restriction rule* with respect to the wall if there exist $B \in \mathfrak{g}^*_{\mathbb{R}}$ such that

$$|\langle B + \mathbf{t}, h \rangle| < \frac{1}{4} \sum_{i=1}^{N} |\langle D_i, h \rangle|$$

for all characters $\mathbf{t} \in \widehat{G}$ entering $\mathfrak{B}$ nontrivially (underlying G-representation $\mathfrak{B}$ contains a representation defined by $\mathbf{t}$).

THEOREM 3.8 (Wall-Crossing [**AL23**]). *Let $C_\pm, h$ as above and $\zeta_\pm \in \mathrm{int}(C_\pm)$. If $\mathfrak{B}$ satisfies grade restriction rule for some B, then the functions $Z_w(\mathfrak{B}|_{\mathcal{X}_{\zeta_+}})|_{\log y_a = \theta_a}$ and $Z_w(\mathfrak{B}|_{\mathcal{X}_{\zeta_-}})|_{\log y_a = \theta_a}$ are related by analytic continuation in $\{\theta_a\}$.*

4. LG/CY correspondence and wall-crossing in GLSMs

The LG/CY correspondence is a special case of wall-crossing in gauged linear sigma models (GLSMs).

4.1. GLSM of the quintic 3-fold. Let $V = \mathrm{Spec}\, \mathbb{C}[x_1, \ldots, x_5, p] \simeq \mathbb{C}^6$. The weights of the gauge group $G = \mathbb{C}^*$ and $\mathbb{C}^*_R$ are listed in the following table.

	x_1	x_2	x_3	x_4	x_5	p
G	1	1	1	1	1	-5
$\mathbb{C}^*_R$	0	0	0	0	0	1

In particular, $G \cap \mathbb{C}^*_R$ is trivial, and $\mathbb{C}_\omega = \mathbb{C}^*_R$.

Let $x = (x_1, \ldots, x_5)$. The superpotential is

$$W(x, p) = pW_5(x) = \sum_{i=1}^{5} px_i^5.$$

The stability condition $\zeta \in \mathbb{Z} - \{0\} \subset \mathbb{Z} = \widehat{G}$ is a non-zero integer.

- In the CY phase $\zeta > 0$,
$$\mathcal{X}_\zeta = \left((\mathbb{C}^5 - \{0\}) \times \mathbb{C}\right)/G = K_{\mathbb{P}^4}.$$
The critical locus of w_ζ is the Fermat quintic Calabi-Yau 3-fold:
$$\mathrm{Crit}(w_\zeta) = \{p = W_5(x) = 0\} = X_5 \subset \mathbb{P}^4 = \{p = 0\} \subset K_{\mathbb{P}^4}.$$
The GLSM invariants are, up to sign, GW invariants of X_5 [**CL**].
- In the LG phase $\zeta < 0$,
$$\mathcal{X}_\zeta = [(\mathbb{C}^5 \times (\mathbb{C} - \{0\}))/\mathbb{C}^*] = [\mathbb{C}^5/\mu_5].$$
The critical locus of w_ζ is a fat stacky point at the origin, and the reduced critical locus is
$$\mathrm{Crit}(w_\zeta)_{\mathrm{red}} = [0/\mu_5] = B\mu_5.$$
The GLSM invariants are FJRW invariants of (W_5, μ_5).

4.2. ***I*-functions in the CY phase $\zeta > 0$.**

$$I\mathcal{X}_\zeta = \mathcal{X}_\zeta = K_{\mathbb{P}^4}, \quad \mathrm{Eff}_\zeta = \{d : d \in \mathbb{Z}_{\geq 0}\}.$$
$$V_d = H^0(\mathbb{P}^1, \mathcal{O}_{\mathbb{P}^1}(d))^{\oplus 5}, \quad W_d = H^1(\mathbb{P}^1, \mathcal{O}_{\mathbb{P}^1}(-5d-1)).$$

The degree d LG loop space is
$$L_d = (V_d - 0)/\mathbb{C}^* \simeq \mathbb{P}^{5d+4}.$$
The obstruction bundle is
$$E_d = ((V_d - 0) \times W_d)/\mathbb{C}^* = \mathcal{O}_{\mathbb{P}^{5d+4}}(-5)^{\oplus 5d}.$$
For all $d \in \mathrm{Eff}_\zeta = \mathbb{Z}_{\geq 0}$,
$$\mathcal{F}_d = \mathbb{P}^4.$$

We now compute the GLSM I-function for the superpotential $W = pW_5(x)$ and for the zero superpotential $W = 0$.

When $W = pW_5(x)$, the GLSM I-function is
$$I_w(q, z) = ze^{(\log q)H/z} \sum_{d=0}^{\infty} q^d \frac{\prod_{m=1}^{5d}(-5H - mz)}{\prod_{m=1}^{d}(H + mz)^5} \mathrm{tdch}_{X_5}^{K_{\mathbb{P}^4}}\{W_5, p\}.$$
When $W = 0$, the GLSM I-function is
$$I_{w=0}(q, z) = ze^{(\log q)H/z} \sum_{d=0}^{\infty} q^d \frac{\prod_{m=1}^{5d}(-5H - mz)}{\prod_{m=1}^{d}(H + mz)^5} \mathrm{tdch}_{\mathbb{P}^4}^{K_{\mathbb{P}^4}}\{0, p\}.$$

4.3. ***I*-functions in the LG phase $\zeta < 0$.**

$$I\mathcal{X}_\zeta = \bigsqcup_{v=0}^{4} \mathcal{X}_{\zeta,v}, \quad \mathcal{X}_{\zeta,v} = \begin{cases} [\mathbb{C}^5/\mu_5], & v = 0, \\ [0/\mu_5] = B\mu_5, & v \in \{1,2,3,4\} \end{cases}.$$
$$\mathrm{Eff}_\zeta = \{d : -5d - 1 \in \mathbb{Z}_{\geq} 0\} = \left\{-\frac{k+1}{5} : k \in \mathbb{Z}_{\geq 0}\right\}.$$
$$V_{-\frac{k+1}{5}} = H^0(\mathbb{P}^1, \mathcal{O}_{\mathbb{P}^1}(k)), \quad W_{-\frac{k+1}{5}} = H^1\left(\mathbb{P}^1, \mathcal{O}_{\mathbb{P}^1}(-\frac{k+1}{5})\right)^{\oplus 5}$$

The LG loop space is

$$L_{-\frac{k+1}{5}} = (V_{-\frac{k+1}{5}} - 0)/\mathbb{C}^* = \mathbb{P}[5^{k+1}] := \mathbb{P}[\underbrace{5,\ldots,5}_{k+1}].$$

The obstruction bundle is

$$E_{-\frac{k+1}{5}} = \left((V_{-\frac{k+1}{5}} - 0) \times W_{-\frac{k+1}{5}}\right)/\mathbb{C}^* = \mathcal{O}_{\mathbb{P}[5^{k+1}]}(-1)^{\oplus 5(\lceil\frac{k+1}{5}\rceil - 1)}.$$

For all $k \in \mathbb{Z}_{\geq 0}$,

$$\mathcal{F}_{-\frac{k+1}{5}} = B\mu_5.$$

For both $W = pW_5$ and $W = 0$, the GLSM I-function

$$\begin{aligned} I_w(t,z) &= z \sum_{\substack{k\in\mathbb{Z}_{\geq 0}\\ k\notin 4+5\mathbb{Z}}} q^{-(k+1)/5} \frac{\Gamma(1-\{\frac{k+1}{5}\})^5}{\Gamma(1-\frac{k+1}{5}))^5} \frac{1}{k!} z^{5\lfloor\frac{k+1}{5}\rfloor - k} \mathbf{1}_{\{\frac{k+1}{5}\}} \\ &= z \sum_{\substack{k\in\mathbb{Z}_{\geq 0}\\ k\notin 4+5\mathbb{Z}}} q^{-(k+1)/5} \frac{\Gamma(1-\{\frac{k+1}{5}\}))^5}{\Gamma(1-\frac{k+1}{5}))^5} \frac{1}{k!} \frac{\mathbf{1}_{\{\frac{k+1}{5}\}}}{z^{5\{\frac{k+1}{5}\}-1}} \\ &= z \sum_{\substack{k\in\mathbb{Z}_{\geq 0}\\ k\notin 4+5\mathbb{Z}}} q^{-(k+1)/5} \frac{\Gamma(1-\{\frac{k+1}{5}\}))^5}{\Gamma(1-\frac{k+1}{5}))^5} \frac{1}{k!} \frac{\phi_{5\{\frac{k}{5}\}}}{z^{5\{\frac{k}{5}\}}} \end{aligned}$$

where $\phi_k = \mathbf{1}_{\frac{k+1}{5}}$, $k = 0, 1, 2, 3$.

4.4. Central charges and wall-crossing. The central charges for the quintic GLSM can be computed using the Higgs-Coulomb correspondence. Let $B \in \mathbb{R}$ and $\mathfrak{B} \in \mathrm{MF}([V/G], w)$ such that for any character $\mathbf{t} \in \widehat{G} \simeq \mathbb{Z}$ entering the underlying representation of $\mathfrak{B}$ the grade restriction rule

$$|B + \mathbf{t}| < 5/2$$

is satisfied. Then the GLSM central charges $Z_w(\mathfrak{B}|_{K_{\mathbb{P}^4}})$ and $Z_w(\mathfrak{B}|_{[\mathbb{C}^5/\mu_5]})$ are computed by the hemisphere partition function:

$$Z_{D^2}(\mathfrak{B}) = -\frac{1}{2\pi\sqrt{-1}} \int_{\sqrt{-1}\mathbb{R}+\delta} d\sigma\, \Gamma(\sigma)^5 \Gamma(1-5\sigma) q^{-\sigma}\, \mathrm{Ch}(\mathfrak{forget}[\mathfrak{B}]), \tag{4.1}$$

where $\delta \in (0, 1/5)$ and $q = \exp(\zeta + 2\pi\sqrt{-1}B)$.

REMARK 4.1. The hemisphere partition function provides the analytic continuation between GW theory of the quintic threefold X_5 and FJRW theory of the affine LG model (W_5, μ_5). Such an analytic continuation was established in [**CR10**] in a different basis. In [**CIR**] the analytic continuation was found to be consistent with Orlov equivalence. In the GLSM language hemisphere partition function provides the analytic continuation between $Z_w(\mathfrak{B}|_{K_{\mathbb{P}^4}})$ and $Z_w(\mathfrak{B}|_{[\mathbb{C}^5/\mu_5]})$, so Orlov equivalence is satisfied automatically.

5. Open/closed correspondence and extended LG/CY correspondence

The B-model disk potential satisfies the extended Picard-Fuchs equation (2.15). Correspondingly, it has the following Mellin-Barnes type integral representation [**Wa07**, Section 3]:

$$\mathcal{T}(q) = \frac{\pi^2}{2} \cdot \frac{1}{2\pi i} \oint_{i\mathbb{R}} \mathrm{d}\sigma \, \frac{\Gamma(-\sigma + 1/2)\Gamma(5\sigma + 1)\Gamma(\sigma + 1/2)}{\Gamma(\sigma + 1)^5} e^{i\pi(\sigma - 1/2)} q^{\sigma}, \tag{5.1}$$

where $\mathcal{T}(q) = \mathcal{T}^{\mathrm{CY}}(q)$ for the small q expansion. We rewrite this integral using reflection relation for the Gamma function:

$$\Gamma(x)\Gamma(1-x) = \frac{\pi}{\sin(\pi x)}, \tag{5.2}$$

and we also replace σ by $-\sigma$:

(5.3)

$$\mathcal{T}(q) = \frac{1}{64\pi^3} \times$$
$$\times \frac{1}{2\pi i} \oint_{i\mathbb{R}+\delta} \mathrm{d}\sigma \, \Gamma(\sigma)^5 \Gamma(1 - 5\sigma) \Gamma\big(\sigma + \frac{1}{2}\big) \Gamma\big(-\sigma + \frac{1}{2}\big) q^{-\sigma} e^{4\pi i \sigma} (1 - e^{-2\pi i \sigma})^5,$$

where $0 < \delta < 1/5$. We observe below that the second line of (5.3) is nothing but a quasimap central charge for a particular (abelian, Calabi-Yau) GLSM (see Section 3.7–3.8). In this paper we call this particular GLSM the extended GLSM corresponding to the real quintic.

5.1. Extended GLSM corresponding to the real quintic. More precisely, the input data of the extended GLSM corresponding to the real quintic is

$$(V = \operatorname{Spec} \mathbb{C}[x_1, \ldots, x_5, p, u, v] \simeq \mathbb{C}^8, \; G = \mathbb{C}^* \times \mu_2, \; \mathbb{C}^*_R, \; W, \; \zeta)$$

where the dimension 8 of V is equal to the number of gamma functions on the right hand side of (5.3), $\mu_2 = \{\pm 1\}$, and $G \cap \mathbb{C}^*_R = \mu_2 \subset \mathbb{C}^*_R$. The weights of $G_0 = \mathbb{C}^*$ (the connected component of 1 of the gauge group G) and $\mathbb{C}^*_\omega = \mathbb{C}^*_R/\mu_2$ on V can be read from the Gamma function arguments; the weights of $G_0 = \mathbb{C}^*$, μ_2, $\mathbb{C}^*_R$, $\mathbb{C}^*_\omega$ are summarized in the following table:

	x_1	x_2	x_3	x_4	x_5	p	u	v
$G_0 = \mathbb{C}^*$	1	1	1	1	1	-5	1	-1
μ_2	$+$	$+$	$+$	$+$	$+$	$+$	$-$	$-$
$\mathbb{C}^*_\omega$	0	0	0	0	0	1	1/2	1/2
$\mathbb{C}^*_R$	0	0	0	0	0	2	1	1

where we distinguished the last two columns for later convenience. It is clear, that the above extended GLSM is obtained from the GLSM of the quintic threefold (see Section 4.1) by an addition of two coordinates u and v of weights 1 and -1 with respect to G. The new coordinates have weights 1/2 with respect to $\mathbb{C}^*_\omega$ and therefore have to be orbifolded by an additional μ_2-symmetry (see discussion below on how G is related to Γ).

Definition of the superpotential W is ambiguous, but we can take a simple family of polynomials such as:

$$W_t(x, p, u, v) = p \sum_{i=1}^{5} x_i^5 + tuv, \quad t \in \mathbb{C}. \tag{5.4}$$

Let Γ be the subgroup of $\mathrm{GL}(V) = \mathrm{GL}_8(\mathbb{C})$ generated by G and $\mathbb{C}^*_R$. Then $\Gamma \simeq \mathbb{C}^* \times \mathbb{C}^*_R$ acts on V by

$$(t_1, t_2) \cdot (x, p, u, v) = (t_1 x, t_1^{-5} t_2^2 p, t_1 t_2 u, t_1^{-1} t_2 v).$$

For any $t \in \mathbb{C}$,

$$(t_1, t_2) \cdot W_t(x, p, u, v) = (t_2)^2 W_t(x, p, u, v).$$

So $\chi : \Gamma = (\mathbb{C}^*)^2 \to \mathbb{C}^*_\omega$ is given by $\chi(t_1, t_2) = t_2^2$, and

$$G = \mathrm{Ker}(\chi) = \{(t_1, t_2) \in (\mathbb{C}^*)^2 : (t_2)^2 = 1\} \simeq \mathbb{C}^* \times \{\pm 1\}.$$

The stability condition is a rational G character $\zeta \in \widehat{G} \otimes \mathbb{Q} \simeq \mathbb{Q}$.

5.2. Central charges in the positive phase $\zeta > 0$. The GIT quotient stack $\mathcal{X}_\zeta = [V /\!\!/_\zeta G]$ for the positive choice of the stability parameter ζ is

$$\mathcal{X}_\zeta = \left[\mathrm{tot}\left(\mathcal{O}_{\mathbb{P}^5}(-5) \oplus \mathcal{O}_{\mathbb{P}^5}(-1)\right)/\mu_2\right]$$

where -1 acts on $\mathrm{tot}(\mathcal{O}_{\mathbb{P}^5}(-5) \oplus \mathcal{O}_{\mathbb{P}^5}(-1))$ by

$$-1 \cdot [x, p, u, v] = [x, p, -u, -v] = [-x, -p, u, v], \quad x = (x_1, \ldots, x_5). \tag{5.5}$$

CASE 1 ($t \neq 0$). For $t \neq 0$ the critical locus of the supertotential (5.4) is

$$\mathrm{Crit}(w_\zeta) = X_5 \times B\mu_2 \overset{W_5(x)=0}{\subset} \mathbb{P}^4 \times B\mu_2 \overset{u=0}{\subset} [\mathbb{P}^5/\mu_2]$$
$$\overset{p=v=0}{\subset} \left[\mathrm{tot}\left(\mathcal{O}_{\mathbb{P}^5}(-5) \oplus \mathcal{O}_{\mathbb{P}^5}(-1)\right)/\mu_2\right].$$

Therefore, in this case there are no additional central charges compared to the quintic example. On the level of integral representations, for $t \neq 0$ all the branes

$$\mathfrak{B} \in \mathrm{MF}([V/G], w) = \mathrm{MF}_G(V, W_t = pW_5(x) + tuv)$$

contain the matrix factorization of uv. In particular, the Chern character contains a factor $(1+e^{2\pi\sqrt{-1}\sigma})$ that cancels all the poles of the two additional gamma functions

$$\Gamma(\sigma + \frac{1}{2})\Gamma(-\sigma + \frac{1}{2}) = \frac{\pi}{\cos(\pi\sigma)}.$$

CASE 2 ($t = 0$). For $t = 0$, the critical locus becomes larger. Let

$$CX_5 = \{W_5(x) = 0\} \subset \mathbb{P}^5$$

be the cone over the Fermat quintic threefold $X_5 \subset \mathbb{P}^4$, and let $\mathcal{O}_{CX_5}(-1)$ be the restriction of $\mathcal{O}_{\mathbb{P}^5}(-1)$ to CX_5. For $t = 0$ the critical locus of the superpotential is singular and is the union of two irreducible components:

$$\mathrm{Crit}(w_\zeta) = \{W_5(x) = px_i^4 = 0\} = C_1 \cup C_2$$

where

$$C_1 = \left[\mathrm{tot}\left(\mathcal{O}_{CX_5}(-1)\right)/\mu_2\right] \overset{W_5=p=0}{\subset} \mathcal{X}_\zeta = \left[\mathrm{tot}\left(\mathcal{O}_{\mathbb{P}^5}(-5) \oplus \mathcal{O}_{\mathbb{P}^5}(-1)\right)/\mu_2\right].$$

is a smooth closed substack of codimension 2, and

$$C_2 = \{x_1^4 = \cdots = x_5^4 = 0\} \subset \mathcal{X}_\zeta$$

is a non-reduced closed substack of codimension 5.

The disk potential $\mathcal{T}^{\mathrm{CY}}(q)$ appears as an additional central charge in this case for a particular choice of a brane

$$\mathfrak{B} \in \mathrm{MF}(\mathcal{X}_\zeta, w_\zeta) = \mathrm{MF}_G(V_G^{ss}(\zeta), W_0 = pW_5(x)).$$

Let $\mathcal{D}_i \subset \mathcal{X}_\zeta$ be the toric divisor defined by $x_i = 0$, and let

$$U = \bigoplus_{i=1}^{5} \mathcal{O}_{\mathcal{X}_\zeta}(\mathcal{D}_i).$$

The closed substack $\bigcap_{i=1}^{5} \mathcal{D}_i = (C_2)_{\text{red}} \subset \mathcal{X}_\zeta$ is the zero locus of the regular section

$$b = (x_1, \ldots, x_5) \in H^0(\mathcal{X}_\zeta, U).$$

Let

$$a = (px_1^4, \ldots, px_5^4) \in H^0(\mathcal{X}_\zeta, U^\vee).$$

Then

$$\langle a, b\rangle = p \sum_{i=1}^{5} x_i^5 = W_0.$$

The brane $\mathfrak{B} = \mathcal{L} \otimes \mathfrak{B}_0$ is a tensor product of a line bundle $\mathcal{L} = \mathcal{O}_{\mathcal{X}_\zeta}(\mathcal{D}_1)^{\otimes 2}$ with the Koszul matrix factorization

$$\mathfrak{B}_0 = \{a, b\} = \Big[\textstyle\bigoplus_i \Lambda^{2i} U^\vee \underset{\partial}{\overset{\partial = i_b + a\wedge}{\rightleftarrows}} \textstyle\bigoplus_i \Lambda^{2i+1} U^\vee \Big] \tag{5.6}$$

where i_b is the interior product with the section b of U and $a\wedge$ is the wedge product with the section a of $U^\vee$. The matrix factorization (5.6) is a deformation of the Koszul resolution of the closed substack $\bigcap_{i=1}^{5} \mathcal{D}_i \subset \mathcal{X}_\zeta$. The Chern character of $\mathfrak{forget}[\mathfrak{B}] \in K(\mathcal{X}_\zeta)$ is equal to

$$\operatorname{Ch}(\mathfrak{forget}[\mathfrak{B}]) = e^{4\pi\sqrt{-1}\sigma}(1 - e^{-2\pi\sqrt{-1}\sigma})^5. \tag{5.7}$$

We have the following open/closed correspondence relating disk invariants $\mathcal{T}^{\text{CY}}(q)$ of $(X_5, \mathbb{R}X_5)$ to closed string invariants $Z(\mathfrak{B})$ of the extended GLSM.

PROPOSITION 5.1 (open/closed correspondence for $(X_5, \mathbb{R}X_5)$ at $\epsilon = 0^+$).

$$Z_{D^2}(\mathfrak{B}) = 64\pi^3 \mathcal{T}^{\text{CY}}(q)\Big|_{q=e^{-\theta}}. \tag{5.8}$$

PROOF. This follows from (3.14), (5.3), and (5.7). □

5.3. Central charges in the negative phase $\zeta < 0$. The brane $\mathfrak{B}$ satisfies the grade restriction rule with respect to the unique wall separating the chambers $\zeta > 0$ and $\zeta < 0$. Indeed, the grade restriction rule takes the form:

$$|B + \mathbf{t}| < 3.$$

The brane $\mathfrak{B}$ consists of $\mathbf{t} \in \{-3, -2, -1, 0, 1, 2\}$, so for $B \in (0, 1)$ the brane satisfies the grade restriction rule. Then $Z_{D_2}(\mathfrak{B})$ realizes the analytic continuation between $Z_w(\mathfrak{B}_{\zeta>0})$ and $Z_w(\mathfrak{B}_{\zeta<0})$. Notice that the analogous brane for the original quintic GLSM would restrict to zero in the geometric phase and, in particular, would be not grade restricted.

The wall-crossing of the extended model provides GLSM viewpoint on the LG/CY correspondence (2.16) in the open string sector.

To compute the decomposition of $Z_{D_2}(\mathfrak{B})$ into $\mathcal{T}^{\text{LG}}$ and a closed period of the FJRW model we can decompose the class of the brane $\mathfrak{forget}[\mathfrak{B}] \in K([V/G])$. Let $t \in \widehat{G}$ be the fundamental character such that $\mathfrak{forget}[\mathfrak{B}] = t^2(1 - t^{-1})^5$.

From the formula (5.3) it is clear, that $\mathcal{T}^{\mathrm{LG}}$ must come from a part of the brane whose class is proportional to $1-t^{-5}$ to cancel the poles of $\Gamma(1-5\sigma)$ and the closed period $\mathcal{T}_c$ must come from the part proportional to $1+t^{-1}$ to cancel the poles of $\Gamma(1/2-\sigma)$. Therefore, to find representation of the form

$$\mathcal{T}^{\mathrm{CY}} = \mathcal{T}^{\mathrm{LG}} + \mathcal{T}_c$$

we need to decompose:

$$t^2(1-t^{-1})^5 = f(1+t^{-1}) + g(1-t^{-5}).$$

Such a decomposition is easy to find:

$$t^2(1-t^{-1})^5 = (-15t^2+10t-10t^{-1}+15t^{-2})(1+t^{-1}) + 16t^2(1-t^{-5}).$$

Then, using (5.8) we have

$$\begin{aligned}
\mathcal{T}^{\mathrm{LG}} &= \frac{\sqrt{-1}}{2\pi^2}\frac{1}{2\pi\sqrt{-1}}\int_{\mathbb{R}+\delta} d\sigma\,\frac{\Gamma(\sigma)^5}{\Gamma(5\sigma)}\Gamma(1/2-\sigma)\Gamma(1/2+\sigma)q^{-\sigma}e^{-\pi\sqrt{-1}\sigma}\\
\mathcal{T}_c &= \frac{1}{32\pi^2}\frac{1}{2\pi\sqrt{-1}}\int_{\mathbb{R}+\delta} d\sigma\,\Gamma(\sigma)^5\Gamma(1-5\sigma)q^{-\sigma}e^{-\pi\sqrt{-1}\sigma}\times\\
&\times(-15e^{4\pi\sqrt{-1}\sigma}+10e^{2\pi\sqrt{-1}\sigma}-10e^{-2\pi\sqrt{-1}\sigma}+15e^{-4\pi\sqrt{-1}\sigma}).
\end{aligned}$$

Notice, that the second line is (up to a factor and coordinate change $q\to -q$) the integral representation of the quintic hemisphere partition function (4.1) for the brane $\mathfrak{B}_c$ with $\mathrm{Ch}(\mathfrak{forget}[\mathfrak{B}_c]) = -15e^{4\pi\sqrt{-1}\sigma}+10e^{2\pi\sqrt{-1}\sigma}-10e^{-2\pi\sqrt{-1}\sigma}+15e^{-4\pi\sqrt{-1}\sigma}$.

5.4. The fan and minimal anticones. The coarse moduli space X_ς of the GIT quotient stack $\mathcal{X}_\varsigma$ is a semi-projective simplicial toric variety defined by a fan. Choice of the fan is non-canonical and depends on the choice of coordinates on the torus. For the extended model, the fan in the positive and negative phases have the same 1-dimensional cones whose generators $v_1, \dots, v_8 \in \mathbb{Z}^7$ can be chosen to be the column vectors of the following 7×8 matrix:

$$v = \begin{pmatrix} 1&1&1&1&1&1&1&1\\ 1&0&0&0&-1&0&0&0\\ 0&1&0&0&-1&0&0&0\\ 0&0&1&0&-1&0&0&0\\ 0&0&0&1&-1&0&0&0\\ 0&0&0&0&0&0&1&1\\ 0&0&0&0&2&0&-1&1 \end{pmatrix} \tag{5.9}$$

The 5×6 submatrix in the upper-left corner is the standard choice of 1-dimensional cones for the $K_{\mathbb{P}^4}$. We modify this fan by adding 2 vectors and 2 coordinates. Resulting 8 vectors satisfy the relation $\sum_{i=1}^5 v_i - 5v_6 + v_7 - v_8 = 0$. In addition, the coefficients are chosen to represent the correct μ_2 action.

The set of minimal anticones is

$$\mathcal{A}_\varsigma^{\min} = \begin{cases} \{\{1\},\{2\},\{3\},\{4\},\{5\},\{7\}\}, & \varsigma>0;\\ \{\{6\},\{8\}\}, & \varsigma<0. \end{cases}$$

Minimal anti-cones are in one-to-one correspondence with torus fixed points in X_ς.

5.5. I-functions in the positive phase $\zeta > 0$.

$$\mathrm{Box}_\zeta = \left\{ b_0 = 0,\ b_1 = \frac{1}{2}(v_7 + v_8) = \begin{pmatrix}1\\0\\0\\0\\0\\1\\0\end{pmatrix},\ b_3 = \frac{1}{2}\sum_{i=1}^{6} v_i = \begin{pmatrix}3\\0\\0\\0\\0\\0\\1\end{pmatrix} \right\} \tag{5.10}$$

where $\mathrm{age}(b_k) = k$. The three elements $b_0, b_1, b_3 \in \mathrm{Box}_\zeta$ correspond to the following three elements in $G \subset (\mathbb{C}^*)^8$:

$$\begin{aligned} g(b_0) &= (1,1,1,1,1,1,1,1) \quad \text{(identity)} \\ g(b_1) &= (1,1,1,1,1,1,-1,-1), \\ g(b_3) &= (-1,-1,-1,-1,-1,-1,1,1). \end{aligned} \tag{5.11}$$

The inertia stack of $\mathcal{X}_\zeta$ is a disjoint union of three connected components:

$$I\mathcal{X}_\zeta = \mathcal{X}_{\zeta,0} \cup \mathcal{X}_{\zeta,1} \cup \mathcal{X}_{\zeta,3}$$

where $\mathcal{X}_{\zeta,0} = \mathcal{X}_\zeta$ is the untwisted sector, and

$$\begin{aligned} \mathcal{X}_{\zeta,1} &:= \mathcal{X}_{\zeta,b_1} = \{u = v = 0\} \simeq K_{\mathbb{P}^4} \times B\mu_2, \\ \mathcal{X}_{\zeta,3} &:= \mathcal{X}_{\zeta,b_3} = \{x_i = p = 0\} \simeq B\mu_2 \times \mathbb{C}. \end{aligned}$$

As a graded complex vector space, the orbifold Chen-Ruan cohomology of $\mathcal{X}_\zeta$ (which is the state space of orbifold quasimap theory to the GIT quotient stack $\mathcal{X}_\zeta = [V /\!\!/_\zeta G]$) is

$$H^*_{\mathrm{CR}}(\mathcal{X}_\zeta) = H^*(\mathcal{X}_{\zeta,0})\mathbf{1}_0 \oplus H^*(\mathcal{X}_{\zeta,1})\mathbf{1}_1 \oplus H^*(\mathcal{X}_{\zeta,3})\mathbf{1}_3,$$

where $\deg \mathbf{1}_k = 2k$, and

$$H^*(\mathcal{X}_{\zeta,0}) \simeq H^*(\mathbb{P}^5), \quad H^*(\mathcal{X}_{\zeta,1}) \simeq H^*(\mathbb{P}^4), \quad H^*(\mathcal{X}_{\zeta,3}) = H^*(\bullet),$$

Therefore, the Poincaré polynomial is

$$P_t\left(H^*_{\mathrm{CR}}(\mathcal{X}_\zeta)\right) := \sum_k \dim_{\mathbb{C}} t^k H^k_{\mathrm{CR}}(\mathcal{X}_\zeta) = 1 + 2t^2 + 2t^4 + 3t^6 + 2t^8 + 2t^{10}. \tag{5.12}$$

As a graded complex vector space, the A-model state space of the extended GLSM in the positive phase is

$$\mathcal{H}_W = \mathcal{H}_{W,0}\mathbf{1}_0 \oplus \mathcal{H}_{W,1}\mathbf{1}_1 \oplus \mathcal{H}_{W,3}\mathbf{1}_3$$

where $\deg \mathbf{1}_k = 2k - 4$, and

$$\mathcal{H}_{W,k} = \mathbb{H}^*\left(\mathcal{X}_{\zeta,k}, (\Omega^*_{\mathcal{X}_{\zeta,k}}, dw_{\zeta,k}\wedge)\right).$$

Let $\mathcal{Z}^w_{\zeta,k} = \mathrm{Crit}(w_{\zeta,k}) \subset \mathcal{X}_{\zeta,k}$. We consider the following cases:

	$\mathcal{Z}^w_{\zeta,0}$	$\mathcal{Z}^w_{\zeta,1}$	$\mathcal{Z}^w_{\zeta,3}$
(1) $W = 0$	$\mathcal{X}_{\zeta,0} = \mathcal{X}_\zeta$	$\mathcal{X}_{\zeta,1} = K_{\mathbb{P}^4} \times B\mu_2$	$\mathcal{X}_{\zeta,3} = B\mu_2 \times \mathbb{C}$
(2) $W = pW_5(x)$	$C_1 \cup C_2$	$X_5 \times B\mu_2$	$B\mu_2 \times \mathbb{C}$
(3) $W = uv$	$K_{\mathbb{P}^4} \times B\mu_2$	$K_{\mathbb{P}^4} \times B\mu_2$	$\emptyset$
(4) $W = pW_5(x) + uv$	$X_5 \times B\mu_2$	$X_5 \times B\mu_2$	$\emptyset$

where $\emptyset$ denotes the empty set. In cases (1)–(3), $\mathcal{Z}^w_{\varsigma,0}$ is not proper and $\mathcal{H}_{W,0}$ is infinite dimensional. In case (4), or more generally, when $W = pW_5(x) + tuv$ where $t \neq 0$,

$$\mathcal{H}_W \simeq H^*(X_5) \oplus H^*(X_5),$$

so Poincaré polynomial is

$$P_t(\mathcal{H}_W) := \sum_k t^k \dim_{\mathbb{C}} \mathcal{H}^k_W = 2P_t(H^*(X_5)) = 2 + 2t^2 + 408t^3 + 2t^4 + 2t^6. \tag{5.13}$$

The effective classes are $\mathrm{Eff}_\varsigma = \{d \in \frac{1}{2}\mathbb{Z} \mid d \geq 0\}$. Let

$$\begin{aligned} V_d &= H^0(\mathbb{P}^1, \mathcal{O}_{\mathbb{P}^1}(d))^{\oplus 5} \oplus H^0(\mathbb{P}^1, \mathcal{O}_{\mathbb{P}^1}(d - \frac{1}{2})) \\ W_d &= H^1(\mathbb{P}^1, \mathcal{O}_{\mathbb{P}^1}(-5d-1) \oplus H^1(\mathbb{P}^1, \mathcal{O}_{\mathbb{P}^1}(-d - \frac{1}{2})) \end{aligned}$$

Note that $V_0 = \mathbb{C}^5$ and $W_0 = 0$, so

$$L_0 = E_0 = \mathbb{P}^4 \times B\mu_2.$$

The degree d LG loop space is

$$\begin{aligned} L_d &= [(V_d - 0)/G] \\ &= \begin{cases} [\mathbb{P}^{6d+4}/\mu_2] \text{ where -1 acts on the last } d \text{ coordinates}, & d \in \mathbb{Z}_{>0}; \\ [\mathbb{P}^{6d+1}/\mu_2] \text{ where -1 acts on the last } d - \frac{1}{2} \text{ coordinates}, & d \in \frac{1}{2} + \mathbb{Z}_{\geq 0}. \end{cases} \end{aligned}$$

The degree d obstruction bundle is

$$\begin{aligned} E_d &= [(V_d - 0) \times W_d)/\mathbb{C}^*] \\ &= \begin{cases} [\mathcal{O}_{\mathbb{P}^{6d+4}}(-5)^{\oplus 5d} \oplus \mathcal{O}_{\mathbb{P}^{6d+4}}(-1)^{\oplus d}/\mu_2], & d \in \mathbb{Z}_{>0}; \\ [\mathcal{O}_{\mathbb{P}^{6d+1}}(-5)^{\oplus(5d+1/2)} \oplus \mathcal{O}_{\mathbb{P}^{6d+1}}(-1)^{\oplus(d-1/2)}/\mu_2], & d \in \frac{1}{2} + \mathbb{Z}_{\geq 0}. \end{cases} \end{aligned}$$

$$\dim L_d - \operatorname{rank} E_d = \begin{cases} 4, & d \in \mathbb{Z}_{\geq 0}; \\ 1, & d \in \frac{1}{2} + \mathbb{Z}_{\geq 0} \end{cases}.$$

We have

$$\mathcal{F}_d = \begin{cases} \mathbb{P}^4 \times B\mu_2 \overset{p=0}{\subset} \mathcal{X}_{\varsigma,1} = \mathrm{tot}(\mathcal{O}_{\mathbb{P}^4}(-5)) \times B\mu_2, & d \in \mathbb{Z}_{>0}; \\ B\mu_2 \overset{v=0}{\subset} \mathcal{X}_{\varsigma,3} = B\mu_2 \times \mathbb{C}, & d \in \frac{1}{2} + \mathbb{Z}_{\geq 0}. \end{cases}$$

For any superpotential W,

$$z^{-1} I_w = I_{w,0} + I_{w,1} + I_{w,3}$$

where $I_{w,0} = 0$. In cases (3) and (4), $I_{w,3} = 0$.

(1) If $W = 0$ then

$$I_{w,1}(q,z) = e^{\log qH/z} \cdot \sum_{d=0}^{\infty} q^d \frac{\prod_{m=1}^{5d}(-5H - mz)\prod_{m=1}^{d}(-H-(m-\frac{1}{2})z)}{\prod_{m=1}^{d}(H+mz)^5 \prod_{m=1}^{d}(H+(m-1/2)z)} \mathbf{1}_1$$
$$\cdot \operatorname{tdch}_{\mathbb{P}^4 \times B\mu_2}^{K_{\mathbb{P}^4} \times B\mu_2}\{0,p\}$$
$$= e^{\log qH/z} \cdot \sum_{d=0}^{\infty} q^d \frac{\prod_{m=1}^{5d}(5H+mz)}{\prod_{m=1}^{d}(H+mz)^5} \mathbf{1}_1 \cdot \operatorname{tdch}_{\mathbb{P}^4 \times B\mu_2}^{K_{\mathbb{P}^4} \times B\mu_2}\{0,p\},$$

$$I_{w,3}(q,z) = \sum_{k=0}^{\infty} q^{k+\frac{1}{2}} \frac{\prod_{m=0}^{5k+2}(-(m+\frac{1}{2})z)\prod_{m=1}^{k}(-mz)}{\prod_{m=0}^{k}((m+\frac{1}{2})z)^5 \prod_{m=1}^{k}(mz)} \mathbf{1}_3 \cdot \operatorname{tdch}_{B\mu_2}^{B\mu_2 \times \mathbb{C}}\{0,v\}$$
$$= -\sum_{k=0}^{\infty} q^{k+\frac{1}{2}} \frac{\prod_{m=0}^{5k+2}((m+\frac{1}{2})z)}{\prod_{m=0}^{k}((m+\frac{1}{2})z)^5} \mathbf{1}_3 \cdot \operatorname{tdch}_{B\mu_2}^{B\mu_2 \times \mathbb{C}}\{0,v\}$$
$$= -4 \sum_{\substack{d>0 \\ d \text{ odd}}} q^d \frac{(5d)!!}{(d!!)^5} \frac{\mathbf{1}_3}{z^2} \operatorname{tdch}_{B\mu_2}^{B\mu_2 \times \mathbb{C}}\{0,v\}.$$

(2) If $W = pW_5(x)$ then

$$I_{w,1}(q,z) = e^{\log qH/z} \cdot \sum_{d=0}^{\infty} q^d \frac{\prod_{m=1}^{5d}(5H+mz)}{\prod_{m=1}^{d}(H+mz)^5} \mathbf{1}_1 \cdot \operatorname{tdch}_{X_5 \times B\mu_2}^{K_{\mathbb{P}^4} \times B\mu_2}\{W_5(x),p\},$$

$$I_{w,3}(q,z) = -4 \sum_{\substack{d>0 \\ d \text{ odd}}} q^d \frac{(5d)!!}{(d!!)^5} \frac{\mathbf{1}_3}{z^2} \operatorname{tdch}_{B\mu_2}^{B\mu_2 \times \mathbb{C}}\{0,v\}.$$

Let $\mathfrak{B}$ be defined as in Section 5.2. Then

$$Z_w(\mathfrak{B}) = \langle I_w(q,-1), \hat{\Gamma}\operatorname{ch}_w([\mathfrak{B}])\rangle$$
$$= 8 \sum_{\substack{d>0 \\ d \text{ odd}}} q^d \frac{(5d)!!}{(d!!)^5} \Gamma(\frac{1}{2})^6 2^5 \int_{B\mu_2} 1$$
$$= 128\pi^3 \sum_{\substack{d>0 \\ d \text{ odd}}} q^d \frac{(5d)!!}{(d!!)^5}$$
$$= 64\pi^3 \mathcal{T}^{\mathrm{CY}}(q).$$

By (5.8),

$$Z_{D^2}(\mathfrak{B}) = 64\pi^3 \mathcal{T}^{CY}(q)\Big|_{q=e^{-\theta}}.$$

This is consistent with Higgs-Coulomb correspondence (Theorem 3.7).

(3) If $W = uv$ then

$$I_{w,1}(q,z) = e^{\log qH/z} \cdot \sum_{d=0}^{\infty} q^d \frac{\prod_{m=1}^{5d}(5H+mz)}{\prod_{m=1}^{d}(H+mz)^5} \mathbf{1}_1 \cdot \operatorname{tdch}_{\mathbb{P}^4 \times B\mu_2}^{K_{\mathbb{P}^4} \times B\mu_2}\{0,p\},$$

$$I_{w,3}(q,z) = -4 \sum_{\substack{d>0 \\ d \text{ odd}}} q^d \frac{(5d)!!}{(d!!)^5} \frac{\mathbf{1}_3}{z^2} \operatorname{tdch}_{\emptyset}^{B\mu_2 \times \mathbb{C}}\{u,v\} = 0.$$

(4) If $W = pW_5(x) + uv$ (or more generally $W = pW_5(x) + tuv$ where $t \neq 0$) then

$$I_{w,1}(q,z) = e^{\log qH/z} \cdot \sum_{d=0}^{\infty} q^d \frac{\prod_{m=1}^{5d}(5H+mz)}{\prod_{m=1}^{d}(H+mz)^5} \mathbf{1}_1 \cdot \mathrm{tdch}_{X_5\times B\mu_2}^{K_{\mathbb{P}^4}\times B\mu_2}\{W_5(x), p\},$$

$$I_{w,3}(q,z) = 0.$$

5.6. A mirror conjecture in the positive phase. We now specialize to the case (2): $W = pW_5(x)$. Define

$$\mathbf{e} := \mathrm{tdch}_{X_5\times B\mu_2}^{K_{\mathbb{P}^4}\times B\mu_2}\{W_5(x), p\}\mathbf{1}_1,$$
$$\mathbf{p} := \mathrm{tdch}_{B\mu_2}^{B\mu_2\times\mathbb{C}}\{0, v\}\mathbf{1}_3.$$

Then

$$(\mathbf{p}, \mathbf{1}_3) = \int_{B\mu_2\times\mathbb{C}} \mathrm{tdch}_{B\mu_2}^{B\mu_2\times\mathbb{C}}\{0, v\} = \int_{B\mu_2} 1 = \frac{1}{2}.$$

Then the GLSM I-function can be rewritten as

$$I_w^+(q,z) := I_w(q,z) = z\left(\sum_{k=0}^{3} I_k^{\mathrm{CY}}(q)\frac{H^k}{z^k}\mathbf{e} - 2\mathcal{T}^{\mathrm{CY}}(q)\frac{\mathbf{p}}{z^2}\right)$$

Let $J_w^+(Q,z)$ be the (yet-to-be-defined) GLSM (small) J-function of the extended GLSM in the positive phase $\zeta > 0$.

CONJECTURE 5.2 (mirror conjecture for the extended GLSM in the positive phase).

$$J_w^+(Q,z) = \frac{I_w^+(q,z)}{I_0^{\mathrm{CY}}(q,z)} \quad \textit{under the mirror map} \quad \log Q = \frac{I_1^{\mathrm{CY}}(q)}{I_0^{\mathrm{CY}}(q)}. \tag{5.14}$$

Define a generating function of genus-zero primary GLSM invariants:

$$F^+(Q) := z(J_w^+(Q,z), \mathbf{1}_3) = \sum_{\substack{d\in\mathbb{Z}_{>0}\\ d \text{ odd}}} \langle \mathbf{1}_3\rangle_{\frac{d}{2}} Q^{\frac{d}{2}} \tag{5.15}$$

Conjecture 5.2 implies

$$F^+(Q) = -\frac{\mathcal{T}^{\mathrm{CY}}(q)}{I_0^{\mathrm{CY}}(q)} \quad \text{under the mirror map} \quad \log Q = \frac{I_1^{\mathrm{CY}}(q)}{I_0^{\mathrm{CY}}(q)}. \tag{5.16}$$

(5.15), (5.16) and Theorem 2.3 (open mirror theorem in the Calabi-Yau phase) imply

$$F^+(Q) = -F_{0,1}^{\mathrm{CY}}(Q)$$

which is equivalent to the following.

COROLLARY 5.3 (numerical open/closed correspondence in CY phase). *For all positive odd number d,*

$$\langle \mathbf{1}_3\rangle_{\frac{d}{2}} = -N_d^{disk}.$$

where $-N_d^{disk}$ is the disk invariants defined by the non-trivial spin structure on the real quintic $\mathbb{R}X_5 \simeq \mathbb{RP}^3$.

It will be interesting to have a direct A-model proof of the above numerical open/closed correspondence, as in [**LY22, LY**].

The GLSM J-function J_w^+ should be defined in terms of genus-zero ∞-stable LG quasimaps to the extended GLSM. The definition of ϵ-stability in [**FJR18, FK**] requires a good lift of ζ, which does not exist for the extended GLSM. We leave the precise definition of J_w^+, as well as the proofs of Conjecture 5.2 and Corollary 5.3, to future work.

5.7. I-functions in the negative phase $\zeta < 0$. In the negative phase $\zeta < 0$,

$$\mathcal{X}_\zeta = \left[\operatorname{tot}\left(\mathcal{O}_{\mathbb{P}[5,1]}(-1)^{\oplus 6}\right)/\mu_2\right].$$
$$\mathrm{Box}_\zeta = \{b_0, b_1, b_3\} \cup \{b_{j,k} : j \in \{1,2,3,4\}, k \in \{0,1\}\}$$

where b_0, b_1, b_3 are given in Equation (5.10),

$$b_{j,0} = \frac{j}{5}\left(\sum_{i=1}^{5} v_i + v_7\right) + \left(1 - \frac{j}{5}\right)v_8 = \begin{pmatrix} j+1 \\ 0 \\ 0 \\ 0 \\ 0 \\ 1 \\ 1 \end{pmatrix}$$

and

$$b_{j,1} = \frac{j}{5}\sum_{i=1}^{5} v_i + \left\{\frac{1}{2} + \frac{j}{5}\right\}v_7 + \left\{\frac{1}{2} - \frac{j}{5}\right\}v_8 = \begin{pmatrix} j+1 \\ 0 \\ 0 \\ 0 \\ 0 \\ 1 \\ 1+\epsilon_j \end{pmatrix}, \quad \epsilon_j = \begin{cases} -1, & j = 1,2 \\ 1, & j = 3,4. \end{cases}$$

We have $\mathrm{age}(b_{j,k}) = j+1$ for $j \in \{1,2,3,4\}$ and $k \in \{0,1\}$.

For $k \in \{0,1,3\}$, $g(b_k) \in G$ is given in Equation (5.11). Let $\zeta_5 = e^{2\pi\sqrt{-1}/5}$. For $j \in \{1,2,3,4\}$ and $k \in \{0,1\}$,

$$g(b_{j,k}) = \left(\xi_5^j, \xi_5^j, \xi_5^j, \xi_5^j, \xi_5^j, 1, (-1)^k \xi_5^j, (-1)^k \xi_5^{-j}\right) \in G. \tag{5.17}$$

The inertia stack of $\mathcal{X}_\zeta$ is a disjoint union of eleven connected components:

$$I\mathcal{X}_\zeta = \mathcal{X}_0 \cup \mathcal{X}_1 \cup \mathcal{X}_3 \cup \bigcup_{\substack{j \in \{1,2,3,4\} \\ k \in \{0,1\}}} \mathcal{X}_{j,k}$$

where

$$\begin{aligned}
\mathcal{X}_0 = \mathcal{X}_\zeta &= \left[\operatorname{tot}\left(\mathcal{O}_{\mathbb{P}[5,1]}(-1)^{\oplus 6}\right)/\mu_2\right], \\
\mathcal{X}_1 := \mathcal{X}_{\zeta, b_1} &= \{u = v = 0\} = [\mathbb{C}^5/\mu_5] \times B\mu_2, \\
\mathcal{X}_3 := \mathcal{X}_{\zeta, b_3} &= \{x_i = p = 0\} = B\mu_2 \times \mathbb{C}, \\
\mathcal{X}_{j,k} := \mathcal{X}_{\zeta, b_{j,k}} &= \{x_i = u = v = 0\} = B(\mu_5 \times \mu_2).
\end{aligned}$$

As a graded complex vector space, the Chen-Ruan orbifold cohomology of $\mathcal{X}_\varsigma$ is

$$H^*_{\mathrm{CR}}(\mathcal{X}_\varsigma) = \bigoplus_{k\in\{0,1,3\}} H^*(\mathcal{X}_k)\mathbf{1}_k \oplus \bigoplus_{\substack{j=\{1,2,3,4\}\\ k\in\{0,1\}}} H^*(\mathcal{X}_{j,k})\mathbf{1}_{j,k}$$

where $\deg \mathbf{1}_k = 2k$, $\deg \mathbf{1}_{j,k} = 2j+2$, and

$$H^*(\mathcal{X}_0) \simeq H^*(\mathbb{P}[5,1]) = \mathbb{C}1 \oplus \mathbb{C}p, \quad H^*(\mathcal{X}_1) \simeq H^*(\mathcal{X}_3) \simeq H^*(\mathcal{X}_{j,k}) \simeq H^*(\bullet).$$

The Poincaré polynomial is

$$(5.18) \quad P_t(H^*_{\mathrm{CR}}(\mathcal{X}_\varsigma)) := \sum_k t^k \dim_{\mathbb{C}} H^k_{\mathrm{CR}}(\mathcal{X}_\varsigma) = 1 + 2t^2 + 2t^4 + 3t^6 + 2t^8 + 2t^{10}$$

which agrees with (5.12) in the positive phase, as expected.

As a graded complex vector space, the A-model state space of the extended GLSM in the negative phase is

$$\mathcal{H}_W = \bigoplus_{k\in\{0,1,3\}} \mathcal{H}_{W,k}\mathbf{1}_k \oplus \bigoplus_{\substack{j=\{1,2,3,4\}\\ k\in\{0,1\}}} \mathcal{H}_{W,j,k}\mathbf{1}_{j,k}$$

where $\deg \mathbf{1}_k = 2k-4$, $\deg \mathbf{1}_{j,k} = 2j-2$, and

$$\mathcal{H}_{W,k} = \mathbb{H}^*\left(\mathcal{X}_k, (\Omega^*_{\mathcal{X}_k}, dw_k\wedge)\right), \quad \mathcal{H}_{W,j,k} = \mathbb{H}^*\left(\mathcal{X}_{j,k}, (\Omega^*_{\mathcal{X}_{j,k}}, dw_{j,k}\wedge)\right).$$

Let $\mathcal{Z}^w_k = \mathrm{Crit}(w_k) \subset \mathcal{X}_k$ and let $\mathcal{Z}^w_{j,k} = \mathrm{Crit}(w_{j,k}) \subset \mathcal{X}_{j,k}$. Let

$$X_0 = \{W_5(x) = 0\} \subset \mathbb{C}^5, \quad \mathbf{o} = \{x_i^4 = 0\} \subset \mathbb{C}^5,$$

$$C_1^- = \{W_5(x) = p = 0\} \simeq [X_0/\mu_5] \times \mathbb{C} \times B\mu_5 \subset \mathcal{X}_\varsigma, \quad C_2^- = \{x_i^4 = 0\} \subset \mathcal{X}_\varsigma,$$

	$\mathcal{Z}^w_0$	$\mathcal{Z}^w_1$	$\mathcal{Z}^w_3$	$\mathcal{Z}^w_{j,k}$
(1) $W = 0$	$\mathcal{X}_\varsigma$	$[\mathbb{C}^5/\mu_5] \times B\mu_2$	$B\mu_2 \times \mathbb{C}$	$B(\mu_5 \times \mu_2)$
(2) $W = pW_5(x)$	$C_1^- \cup C_2^-$	$[\mathbf{o}/\mu_5] \times B\mu_2$	$B\mu_2 \times \mathbb{C}$	$B(\mu_5 \times \mu_2)$
(3) $W = uv$	$[\mathbb{C}^5/\mu_2] \times B\mu_2$	$[\mathbb{C}^5/\mu_5] \times B\mu_2$	$\emptyset$	$B(\mu_5 \times \mu_2)$
(4) $W = pW_5(x) + uv$	$[\mathbf{o}/\mu_5] \times B\mu_2$	$[\mathbf{o}/\mu_5] \times B\mu_2$	$\emptyset$	$B(\mu_5 \times \mu_2)$

In cases (1)–(3), $\mathcal{Z}^w_0$ is not proper and $\mathcal{H}_{W,0}$ is infinite dimensional. In case (4), or more generally, when $W = pX_5(x) + tuv$ where $t \neq 0$,

$$\mathcal{H}_{W,0} = \mathcal{H}^7_{W,0} = \mathbb{C}^{204}, \quad \mathcal{H}_{W,1} = \mathcal{H}^5_{W,1} = \mathbb{C}^{204}, \quad \mathcal{H}_{W,3} = 0, \quad \mathcal{H}_{W,j,k} \simeq H^*(\bullet),$$

so the Poincaré polynomial is

$$(5.19) \qquad P_t(\mathcal{H}_W) := \sum_k t^k \dim_{\mathbb{C}} \mathcal{H}^k_W = 2 + 2t^2 + 408t^3 + 2t^4 + 2t^6$$

which agrees with (5.13) in the positive phase, as expected.

The effective classes are

$$\begin{aligned}\mathrm{Eff}_\varsigma &= \{d : -5d - 1 \in \mathbb{Z}_{\geq 0}\} \cup \{d : -d - \frac{1}{2} \in \mathbb{Z}_{\geq 0}\} \\ &= \{-\frac{k+1}{5} : k \in \mathbb{Z}_{\geq 0}\} \cup \{-\frac{1}{2} - k : k \in \mathbb{Z}_{\geq 0}\}.\end{aligned}$$

Let $k \in \mathbb{Z}_{\geq 0}$.

$$V_{-\frac{k+1}{5}} = H^0(\mathbb{P}^1, \mathcal{O}_{\mathbb{P}^1}(k)) \oplus H^0\Big(\mathbb{P}^1, \mathcal{O}_{\mathbb{P}^1}(\frac{2k-3}{10})\Big)$$
$$W_{-\frac{k+1}{5}} = H^1\Big(\mathbb{P}^1, \mathcal{O}_{\mathbb{P}^1}(-\frac{k+1}{5})\Big)^{\oplus 5} \oplus H^1(\mathbb{P}^1, \frac{-2k-7}{10})$$

The degree $-(k+1)/5$ LG loop space is

$$L_{-\frac{k+1}{5}} = \Big[\mathbb{P}[5^{k+1}, 1^{\lfloor \frac{2k-3}{10} \rfloor + 1}]/\mu_2\Big]$$

where -1 acts on the last $\lfloor \frac{2k-3}{10} \rfloor + 1$ coordinates. The obstruction bundle is

$$E_{-\frac{k+1}{5}} = \Big[\mathrm{tot}\left(\mathcal{O}_{\mathbb{P}}(-1)^{\oplus 5(\lceil \frac{k+1}{5} \rceil - 1) + \lceil \frac{2k+7}{10} \rceil - 1}\right)/\mu_2\Big]$$

where $\mathbb{P} = \mathbb{P}\big[5^{k+1}, 1^{\lfloor \frac{2k-3}{10} \rfloor + 1}\big]$.

$$\mathcal{F}_{-\frac{k+1}{5}} = B(\mu_5 \times \mu_2) \subset \begin{cases} \mathcal{X}_1 = [\mathbb{C}^5/\mu_5] \times B\mu_5, & k \in 5\mathbb{Z}+4, \\ \mathcal{X}_{5\{\frac{k+1}{5}\},1} = B(\mu_5 \times \mu_2), & k \notin 5\mathbb{Z}+4. \end{cases}$$

$$V_{-\frac{1}{2}-k} = H^0(\mathbb{P}^1, \mathcal{O}_{\mathbb{P}^1}(5k + \frac{3}{2})) \oplus H^0\big(\mathbb{P}^1, \mathcal{O}_{\mathbb{P}^1}(k)\big)$$
$$W_{-\frac{1}{2}-k} = H^1\Big(\mathbb{P}^1, \mathcal{O}_{\mathbb{P}^1}(-\frac{1}{2} - k)\Big)^{\oplus 5} \oplus H^1(\mathbb{P}^1, \mathcal{O}(-1-k))$$

The degree $-1/2 - k$ LG loop space is

$$L_{-\frac{1}{2}-k} = \big[\mathbb{P}[5^{5k+2}, 1^{k+1}]/\mu_2\big]$$

where -1 acts on the last $k+1$ coordinates. The obstruction bundle is

$$E_{-\frac{1}{2}-k} = \Big[\mathrm{tot}\left(\mathcal{O}_{\mathbb{P}}(-1)^{\oplus 5(\lceil \frac{1}{2} + k \rceil - 1) + k}\right)/\mu_2\Big]$$

where $\mathbb{P} = \mathbb{P}\big[5^{5k+2}, 1^{k+1}\big]$.

$$\mathcal{F}_{-\frac{1}{2}-k} = B\mu_2 \overset{u=0}{\subset} \mathcal{X}_3 = B\mu_2 \times \mathbb{C}.$$

For any superpotential W,

$$z^{-1} I_w = I_{w,0} + I_{w,1} + I_{w,3} + \sum_{j=1}^{4} I_{w,j,0} + \sum_{j=1}^{4} I_{w,j,1}$$

where

$$I_{w,0} = I_{w,1} = I_{w,j,0} = 0,$$

and

$$\begin{aligned}\sum_{j=1}^{4} I_{w,j,1} &= \sum_{\substack{k \in \mathbb{Z}_{\geq 0} \\ k \notin 4+5\mathbb{Z}}} q^{-(k+1)/5} \frac{\Gamma(1 - \{\frac{k+1}{5}\})^5 \Gamma(1 - \{\frac{2k+7}{10}\}) \Gamma(1 - \{-\frac{2k-3}{10}\})}{\Gamma(1 - \frac{k+1}{5})^5 \Gamma(1 - \frac{2k+7}{10}) \Gamma(1 + \frac{2k-3}{10})} \\ &\quad \cdot \frac{1}{k!} \frac{\mathbf{1}_{5\{\frac{k+1}{5}\},1}}{z^{5\{\frac{k+1}{5}\}-1}} \\ &= \sum_{\substack{k \in \mathbb{Z}_{\geq 0} \\ k \notin 4+5\mathbb{Z}}} q^{-(k+1)/5} \frac{\Gamma(1 - \{\frac{k+1}{5}\})^5}{\Gamma(1 - \frac{k+1}{5})^5} \frac{(-1)^{\lfloor \frac{2k+7}{10} \rfloor}}{k!} \frac{\mathbf{1}_{5\{\frac{k+1}{5}\},1}}{z^{5\{\frac{k+1}{5}\}-1}}.\end{aligned}$$

(1) If $W = 0$ then

$$I_{w,3} = \sum_{k=0}^{\infty} q^{-\frac{1}{2}-k} \frac{\prod_{m=0}^{k-1}(-(m+\frac{1}{2})z)^5 \prod_{m=1}^{k}(-mz)}{\prod_{m=0}^{5k+1}((m+\frac{1}{2})z)\prod_{m=1}^{k}(mz)} \mathbf{1}_3 \cdot \mathrm{tdch}_{B\mu_2}^{B\mu_2\times\mathbb{C}}\{0,u\}$$

$$= 4\sum_{k=0}^{\infty} q^{-\frac{1}{2}-k} \frac{((2k+1)!!)^5}{(2k+1)^5(10k+3)!!} \frac{\mathbf{1}_3}{z^2} \cdot \mathrm{tdch}_{B\mu_2}^{B\mu_2\times\mathbb{C}}\{0,u\}$$

$$= 4 \sum_{d>0, d \text{ odd}} q^{-\frac{d}{2}} \frac{(d!!)^5}{d^5(5d-2)!!} \frac{\mathbf{1}_3}{z^2} \cdot \mathrm{tdch}_{B\mu_2}^{B\mu_2\times\mathbb{C}}\{0,u\}.$$

(2) If $W = pW_5(x)$ then

$$I_{w,3} = 4 \sum_{d>0, d \text{ odd}} q^{-\frac{d}{2}} \frac{(d!!)^5}{d^5(5d-2)!!} \frac{\mathbf{1}_3}{z^2} \mathrm{tdch}_{B\mu_2}^{B\mu_2\times\mathbb{C}}\{0,u\}.$$

Let $\mathfrak{B} \in \mathrm{MF}(\mathcal{X}_\varsigma, w_\varsigma)$ such that $\mathfrak{forget}([\mathfrak{B}]) = 16t^2(1-t^{-5})$, where $t = \mathcal{O}_{\mathcal{X}_\varsigma}(\mathcal{D}_1)$ (see Section 5.3). Then

$$\begin{aligned} Z_w(\mathfrak{B}) &= \langle I_w(q,-1), \hat{\Gamma}\,\mathrm{ch}_w([\mathfrak{B}])\rangle \\ &= -8 \sum_{\substack{d>0 \\ d \text{ odd}}} q^{-\frac{d}{2}} \frac{(d!!)^5}{d^5(5d-2)!!} \Gamma(\frac{1}{2})^6 32 \int_{B\mu_2} 1 \\ &= -128\pi^3 \sum_{\substack{d>0 \\ d \text{ odd}}} q^{-\frac{d}{2}} \frac{(d!!)^5}{d^5(5d-2)!!} \\ &= 64\pi^3 \mathcal{T}^{\mathrm{LG}}(q). \end{aligned}$$

(3) If $W = uv$ then

$$I_{w,3} = 4 \sum_{d>0, d \text{ odd}} q^{-\frac{d}{2}} \frac{(d!!)^5}{d^5(5d-2)!!} \frac{\mathbf{1}_3}{z^2} \mathrm{tdch}_{\emptyset}^{B\mu_2\times\mathbb{C}}\{v,u\} = 0.$$

(4) If $W = pW_5(x) + uv$, or more generally $W = pW_5(x) + tuv$ where $t \neq 0$, then

$$I_{w,3} = 0.$$

5.8. A mirror conjecture in the negative phase. We now specialize to the case (2): $W = pW_5(x)$. Define

$$\widetilde{\phi}_k := \cdot \mathbf{1}_{5\{\frac{k+1}{5}\},1}, \quad k = 0,1,2,3,$$

and

$$\mathbf{p}_- := \mathrm{tdch}_{B\mu_2}^{B\mu_2\times\mathbb{C}}\{0,u\}\mathbf{1}_3.$$

Then

$$(\mathbf{p}_-, \mathbf{1}_3) = \int_{B\mu_2\times\mathbb{C}} \mathrm{tdch}_{B\mu_2}^{B\mu_2\times\mathbb{C}}\{0,u\} = \int_{B\mu_2} 1 = \frac{1}{2}.$$

Then GLSM I-function can be rewritten as (recall that $q = t^{-5}$)

$$I_w^-(q,z) := I_w(q,z) = z\left(\sum_{k=0}^{3} \epsilon_k I_k^{\mathrm{LG}}(t)\frac{\widetilde{\phi}_k}{z^k} - 2\mathcal{T}^{\mathrm{LG}}(t)\frac{\mathbf{p}_-}{z^2}\right)$$

where $\epsilon_0 = \epsilon_1 = 1$ and $\epsilon_2 = \epsilon_3 = -1$.

Let $J_w^-(\tau, z)$ be the (yet-to-be-defined) GLSM (small) J-function of the extended GLSM in the negative phase $\zeta < 0$.

CONJECTURE 5.4 (mirror conjecture for the extended GLSM in the negative phase).

$$J_w^-(\tau, z) = \frac{I_w^-(t,z)}{I_0^{\mathrm{LG}}(t,z)} \quad \textit{under the mirror map} \quad \tau = \frac{I_1^{\mathrm{LG}}(t)}{I_0^{\mathrm{LG}}(t)}. \tag{5.20}$$

We leave the precise definition of J_w^- and the proof of Conjecture 5.4 to future work.

Define a generating function of genus-zero primary GLSM invariants:

$$F^-(\tau) := z(J_w^-(\tau, z), \mathbf{1}_3). \tag{5.21}$$

Conjecture 5.4 implies

$$F^-(\tau) = -\frac{\mathcal{T}^{\mathrm{LG}}(t)}{I_0^{\mathrm{LG}}(t)} \quad \text{under the mirror map} \quad \tau = \frac{I_1^{\mathrm{LG}}(t)}{I_0^{\mathrm{LG}}(t)}. \tag{5.22}$$

(5.21), (5.22) and Conjecture 2.4 (open mirror conjecture for (W_5, μ_5)) imply the following open/closed correspondence:

$$F^-(\tau) = -F_{0,1}^{\mathrm{LG}}(\tau).$$

6. B-model

Mirror theorem [**Gi96, LLY97**] for the quintic threefold in particular implies that central charges of the quintic are equal to the period integrals of the mirror quintic. More precisely, consider the following family $\mathcal{Q}$ of quintic threefolds in $\mathbb{P}^4$ over $\mathbb{C}_t$:

$$W_t(x) = \sum_{i=1}^{5} x_i^5 + t\prod_{i=1}^{5} x_i = 0. \tag{6.1}$$

This family is invariant with respect to $(\mu_5)^3 \subset \mathbb{P}^4$ that preserves the product $\prod_i x_i$. Mirror quintic is the family $[\mathcal{Q}/\mu_5^3]$. The mirror family admits a holomorphic volume form Ω. Period integrals of Ω over cycles $\gamma \in H_3(\mathcal{Q}, \mathbb{Z})$ satisfy the Picard-Fuchs equation and are equal to the central charges of the quintic for a particular choice of branes.

$$\omega_\gamma(t) = \int_\gamma \Omega = \frac{1}{(2\pi\sqrt{-1})^2} \int_{T(\gamma)} \frac{\mathrm{d}^5 x}{W_t(x)}, \tag{6.2}$$

where $T(\gamma) \in H_5(\mathbb{C}^5 \backslash \{W_t = 0\})$ is the Hopf fibration lift of the tubular neighborhood of $\gamma \in H_3(\mathcal{Q}_t, \mathbb{Z})$. Here we lift a 3-cycle γ from a given fiber to a Gauss-Manin flat family of 3-cycles in $\mathcal{Q}$.

Another well-known representation for the periods of the quintic is the LG representation by oscillatory integrals:

$$\int_\Gamma e^{-W_t(x)} \mathrm{d}^5 x, \quad \Gamma \in H_5(\mathbb{C}^5, \Re(W) \gg 0). \tag{6.3}$$

6.1. Open B-model. There exist a B-model representation for the disk potential τ [**MW09**]. Let

$$C_\pm := \{x_1 + x_2 = 0,\ x_3 + x_4 = 0,\ x_5^2 \pm \sqrt{t}x_1x_3\} \subset \mathcal{Q} \tag{6.4}$$

be two families of curves in $\mathcal{Q}$. They descend to the mirror family $[\mathcal{Q}/\mu_5^3]$. Now, consider a family C of 3-chains $C_t \in C_3(\mathcal{Q}_t, \mathbb{Z})$ such that $\partial C_t = [C_{+,t}] - [C_{-,t}]$. It was shown that the partial periods:

$$\int_C \Omega \tag{6.5}$$

satisfy the extended Picard-Fuchs equation (2.15).

Here we show that the open-closed correspondence of this paper holds true on the B-side as well, namely the partial periods are equal to actual (oscillatory) periods associated to the extended GLSM of Section 5.

6.2. Extended B-model. In this and the next two subsections, we show how to reproduce the disk potentials in CY (6.15) and LG (6.21) phases using mirror symmetry for the extended GLSM.

We use a version of Givental-Hori-Vafa mirror construction for $\mathcal{X}_\varsigma = [\mathbb{C}^8 /\!/_\varsigma G]$ where $G = \mathbb{C}^* \times \mu_2$. Recall from Section 5.4 that the generators of 1-dimensional cones are the column vectors $v_1, \ldots, v_8$ of the matrix

$$v = \begin{pmatrix} 1 & 1 & 1 & 1 & 1 & 1 & 1 & 1 \\ 1 & 0 & 0 & 0 & -1 & 0 & 0 & 0 \\ 0 & 1 & 0 & 0 & -1 & 0 & 0 & 0 \\ 0 & 0 & 1 & 0 & -1 & 0 & 0 & 0 \\ 0 & 0 & 0 & 1 & -1 & 0 & 0 & 0 \\ 0 & 0 & 0 & 0 & 0 & 0 & 1 & 1 \\ 0 & 0 & 0 & 0 & 2 & 0 & -1 & 1 \end{pmatrix} \tag{6.6}$$

The mirror superpotential is given by the formula

$$F(y, u, v) = \sum_i q_i y^{v_i} = y_0 f(y, u, v)$$

where

$$f(y, u, v) = \sum_{i \le 4} y_i + \frac{v^2}{\prod_i y_i} + t + u/v + uv. \tag{6.7}$$

The superpotential gets deformed according to the R-charges:

$$F_q = F + \sum_{i=1}^{8} \frac{R_i}{2} \log(y^{v_i})$$

or $\log(y_0) + 1/2\log(y_0uv) + 1/2\log(y_0u/v) = 2\log(y_0) + \log(u)$. Thus, the period integrals can be written as

$$I_\Gamma = \int_\Gamma e^{-y_0 f} y_0 u \, \mathrm{d}y_0 \, \omega, \tag{6.8}$$

where

$$\omega = \frac{\mathrm{d}u}{u} \frac{\mathrm{d}v}{v} \prod_{i=1}^{4} \frac{\mathrm{d}y_i}{y_i} \tag{6.9}$$

is the canonical volume form on the torus $(\mathbb{C}^*)^6_{u,v,y}$ and Γ is any contour that makes the integral convergent. The integral depends only on the class of Γ in $H_7((\mathbb{C}^*)^7, \Re(F_q) \gg 0)$.

6.3. Extended period in the positive phase $\zeta > 0$. In order to evaluate the integral we perform the following (multivalued) coordinate change:

$$x_i := y_0 y_i, \quad 1 \le i \le 4, \quad x_5 := \frac{v^2 y_0}{\prod_i y_i}, \tag{6.10}$$
$$p_1 := y_0 t, \quad p_2 := y_0 uv.$$

Then

$$F = \sum_{i=1}^{5} x_i + p_1 + p_2 + t^{-5} p_1^5 p_2 / \prod_{i=1}^{5} x_i,$$
$$y_0^2 u = \prod_{i=1}^{5} x_i^{-1/2} p_1^{7/2} p_2 t^{-7/2}, \tag{6.11}$$
$$\frac{dy_0}{y_0}\omega = \frac{1}{2}\prod_{i=1}^{5}\frac{dx_i}{x_i}\prod_{a=1}^{2}\frac{dp_a}{p_a}.$$

Therefore,

$$I_\Gamma = \frac{t^{-1}}{2}\sum_{m\ge 0}(-1)^m \frac{t^{-5(m+1/2)}}{m!}\int_\Gamma e^{-\sum_i x_i - \sum_a p_a} p_1^{5(m+1/2)} p_2^m \cdot \prod_i x_i^{-m-3/2}\prod_i dx_i \prod_a dp_a. \tag{6.12}$$

Integration contour. We can choose a particular integration contour that transforms the integral in the product of 7 Gamma function integrals. We define the integration contour Γ_{geom} parametrically, using the Hankel contour $\mathcal{C}$. The Hankel contour $\mathcal{C} \subset \mathbb{C}$ is a contour that goes from infinity to 0 along the real axis, goes around the origin on the counter-clockwise direction and then continues along the positive real axis to positive infinity, $\log \mathcal{C} = (+\infty, 0] \cup [0, 2\pi i] \cup [2\pi i, +\infty + 2\pi i)$.

To define Γ_{geom} we let $y_0 \in (0,\infty)$, $y_1, \ldots, y_4 \in y_0^{-1}\mathcal{C}$, $v^2 \in \prod_{i=1}^4 y_i/y_0\mathcal{C}$, $u \in y_0^{-1}v^{-1}(0,+\infty)$.

The contour separates the variables in the integral reduces to a product of 7 integrals:

$$\int_\Gamma e^{-\sum_i x_i - \sum_a p_a} p_1^{5(m+1/2)} p_2^m \prod_i x_i^{-m-3/2}\prod_i dx_i \prod_a dp_a$$
$$= \prod_{i=1}^{5}\int_{\mathcal{C}} e^{-x_i} x_i^{-m-3/2} dx_i \cdot \int_0^\infty e^{-p_1} p_1^{5(m+1/2)} dp_1 \cdot \int_0^\infty e^{-p_2} p_2^m dp_2 \tag{6.13}$$
$$= 32\Gamma(-m-1/2)^5\Gamma(5m+7/2)m!.$$

Plugging it back to the sum we get

$$
\begin{aligned}
I_\Gamma &= t^{-1} 16 \sum_{m\geq} \sum_{m\geq 0} (-1)^m t^{-5(m+1/2)} \Gamma(5m+7/2)\Gamma(-m-1/2)^5 \\
&= t^{-1} 16 \Gamma(7/2)\Gamma(-1/2)^5 t^{-5/2} + \cdots \\
&= -32\pi^3 t^{-1}(30t^{-5/2} + \cdots).
\end{aligned} \tag{6.14}
$$

Then, after the coordinate change $t \to q = t^{-5}$ we get:

$$
tI_\Gamma = -32\pi^3 \mathcal{T}^{\mathrm{CY}}(q). \tag{6.15}
$$

6.4. Extended mirror period in the negative phase $\zeta < 0$. In the negative phase we expand the integral (6.8) differently. Here we perform the coordinate change:

$$
\begin{aligned}
&x_i := y_0 y_i, \quad 1 \leq i \leq 4, \quad x_5 := \frac{v^2 y_0}{\prod_i y_i}, \\
&\qquad\qquad p_1 := y_0 t, \quad p_2 := y_0 u/v.
\end{aligned} \tag{6.16}
$$

So that

$$
\begin{aligned}
F &= \sum_{i=1}^5 x_i + p_1 + p_2 + t^5 p_1^{-5} p_2 \prod_{i=1}^5 x_i, \\
y_0^2 u &= \prod_{i=1}^5 x_i^{1/2} p_1^{-3/2} p_2 t^{3/2}, \\
\frac{\mathrm{d}y_0}{y_0}\omega &= \frac{1}{2} \prod_{i=1}^5 \frac{\mathrm{d}x_i}{x_i} \prod_{a=1}^2 \frac{\mathrm{d}p_a}{p_a}.
\end{aligned} \tag{6.17}
$$

Expanding the exponential in the integral we compute

$$
\begin{aligned}
I_{\Gamma_{LG}} = \frac{t^{-1}}{2} &\sum_{m\geq 0} \frac{(-1)^m}{m!} t^{5(m+1/2)} \\
&\cdot \int_{\Gamma_{LG}} e^{-\sum_i x_i - p_1 - p_2} \prod_i x_i^{m-1/2} p_1^{-5(m+1/2)} p_2^m \mathrm{d}p_1 \mathrm{d}p_2 \mathrm{d}^5 x.
\end{aligned} \tag{6.18}
$$

Contour. We construct the integration contour $\Gamma_{LG} \simeq \mathcal{C}^6 \times \mathbb{R}$ parametrically in a similar fashion to the geometric positive case. We let $y_0 \in \mathcal{C}$, $y_i \in y_0^{-1}\mathcal{C}$ for $1 \leq i \leq 4$. Then $v^2 \in y_0^{-1} \prod_{i=1}^4 y_i \mathcal{C}$ and $u \in v y_0^{-1}(0,\infty)$

The integral factorizes into a product of 7 gamma functions:

$$
\begin{aligned}
\int_{\Gamma_{LG}} e^{-\sum_i x_i - p_1 - p_2} \prod_i x_i^{m-1/2} p_1^{-5(m+1/2)} p_2^m \mathrm{d}p_1 \mathrm{d}p_2 \mathrm{d}^5 x = \\
= 64\Gamma(m+1/2)^5 \Gamma(-5m-3/2) m!.
\end{aligned} \tag{6.19}
$$

Plugging it back into the expression for $I_{\Gamma_{LG}}$ we obtain:

$$
\begin{aligned}
I_{\Gamma_{LG}} = 32t^{-1} \sum_{m\geq 0} t^{5(m+1/2)} \Gamma(-5m-3/2)\Gamma(m+1/2)^5 = \\
= -64t^{-1}\pi^3(-2/3t^{5/2} + \cdots).
\end{aligned} \tag{6.20}
$$

We compute that up to a multiple and the coordinate change $q = t^{-5}$ this expression coincides with $\mathcal{T}^{\mathrm{LG}}(q)$:

$$tI_{\Gamma_{LG}}|_{t^{-5}=q} = -64\pi^3\mathcal{T}^{\mathrm{LG}}(q). \tag{6.21}$$

Acknowledgments

We thank Kentaro Hori, Johanna Knapp, Sebastian Nill, Renata Picciotto, and Johannes Walcher for helpful discussion. We thank Sebastian Nill for explaining his work. We thank Gang Tian, Johannes Walcher, Guangbo Xu, Song Yu, and the anonymous referee for their comments on early versions of this paper. The second named author wishes to thank Ezra Getzler for numerous helpful conversations on GW theory, open GW theory, and mirror symmetry, as well as his encouragement, help, and support, over the past twenty years.

References

[AL23] Konstantin Aleshkin and Chiu-Chu Melissa Liu, *Higgs-Coulomb correspondence and wall-crossing in abelian GLSMs*, arXiv:2301.01266, (2023).

[BCS] Lev A. Borisov, Linda Chen, and Gregory G. Smith, *The orbifold Chow ring of toric Deligne-Mumford stacks*, J. Amer. Math. Soc. **18** (2005), no. 1, 193–215, DOI 10.1090/S0894-0347-04-00471-0. MR2114820

[BH] Lev A. Borisov and R. Paul Horja, *On the K-theory of smooth toric DM stacks*, Snowbird lectures on string geometry, Contemp. Math., vol. 401, Amer. Math. Soc., Providence, RI, 2006, pp. 21–42, DOI 10.1090/conm/401/07551. MR2222527

[BCT19] Alexandr Buryak, Emily Clader, and Ran J. Tessler, *Closed extended r-spin theory and the Gelfand-Dickey wave function*, J. Geom. Phys. **137** (2019), 132–153, DOI 10.1016/j.geomphys.2018.11.007. MR3894288

[BCT22] Alexandr Buryak, Emily Clader, and Ran J. Tessler, *Open r-spin theory I: Foundations*, Int. Math. Res. Not. IMRN **14** (2022), 10458–10532, DOI 10.1093/imrn/rnaa345. MR4458550

[BCT24] Alexandr Buryak, Emily Clader, and Ran J. Tessler, *Open r-spin theory II: The analogue of Witten's conjecture for r-spin disks*, J. Differential Geom. **128** (2024), no. 1, 1–75, DOI 10.4310/jdg/1721075259. MR4773181

[CdGP] Philip Candelas, Xenia C. de la Ossa, Paul S. Green, and Linda Parkes, *A pair of Calabi-Yau manifolds as an exactly soluble superconformal theory*, Nuclear Phys. B **359** (1991), no. 1, 21–74, DOI 10.1016/0550-3213(91)90292-6. MR1115626

[CKL] Huai-Liang Chang, Young-Hoon Kiem, and Jun Li, *Algebraic virtual cycles for quantum singularity theories*, Comm. Anal. Geom. **29** (2021), no. 8, 1749–1774, DOI 10.4310/cag.2021.v29.n8.a1. MR4429243

[CL] Huai-Liang Chang and Jun Li, *Gromov-Witten invariants of stable maps with fields*, Int. Math. Res. Not. IMRN **18** (2012), 4163–4217, DOI 10.1093/imrn/rnr186. MR2975379

[CLL] Huai-Liang Chang, Jun Li, and Wei-Ping Li, *Witten's top Chern class via cosection localization*, Invent. Math. **200** (2015), no. 3, 1015–1063, DOI 10.1007/s00222-014-0549-5. MR3348143

[CCK] Daewoong Cheong, Ionuţ Ciocan-Fontanine, and Bumsig Kim, *Orbifold quasimap theory*, Math. Ann. **363** (2015), no. 3-4, 777–816, DOI 10.1007/s00208-015-1186-z. MR3412343

[CIR] Alessandro Chiodo, Hiroshi Iritani, and Yongbin Ruan, *Landau-Ginzburg/Calabi-Yau correspondence, global mirror symmetry and Orlov equivalence*, Publ. Math. Inst. Hautes Études Sci. **119** (2014), 127–216, DOI 10.1007/s10240-013-0056-z. MR3210178

[CR10] Alessandro Chiodo and Yongbin Ruan, *Landau-Ginzburg/Calabi-Yau correspondence for quintic three-folds via symplectic transformations*, Invent. Math. **182** (2010), no. 1, 117–165, DOI 10.1007/s00222-010-0260-0. MR2672282

[Ch08] Cheol-Hyun Cho, *Counting real J-holomorphic discs and spheres in dimension four and six*, J. Korean Math. Soc. **45** (2008), no. 5, 1427–1442, DOI 10.4134/JKMS.2008.45.5.1427. MR2439744

[CKS] Dongwook Choa, Bumsig Kim, and Bhamidi Sreedhar, *Riemann-Roch for stacky matrix factorizations*, Forum Math. Sigma **10** (2022), Paper No. e108, 29, DOI 10.1017/fms.2022.99. MR4519062

[CK14] Ionuţ Ciocan-Fontanine and Bumsig Kim, *Wall-crossing in genus zero quasimap theory and mirror maps*, Algebr. Geom. **1** (2014), no. 4, 400–448, DOI 10.14231/AG-2014-019. MR3272909

[CK20] Ionuţ Ciocan-Fontanine and Bumsig Kim, *Quasimap wall-crossings and mirror symmetry*, Publ. Math. Inst. Hautes Études Sci. **131** (2020), 201–260, DOI 10.1007/s10240-020-00114-0. MR4106795

[CKM] Ionuţ Ciocan-Fontanine, Bumsig Kim, and Davesh Maulik, *Stable quasimaps to GIT quotients*, J. Geom. Phys. **75** (2014), 17–47, DOI 10.1016/j.geomphys.2013.08.019. MR3126932

[CFGKS] Ionut Ciocan-Fontanine, David Favero, Jérémy Guéré, Bumsig Kim, and Mark Shoemaker, *Fundamental factorization of a GLSM Part I: Construction*, Mem. Amer. Math. Soc. **289** (2023), no. 1435, iv+96, DOI 10.1090/memo/1435. MR4632308

[Fa20] Bohan Fang, *Central charges of T-dual branes for toric varieties*, Trans. Amer. Math. Soc. **373** (2020), no. 6, 3829–3851, DOI 10.1090/tran/7734. MR4105511

[FL13] Bohan Fang and Chiu-Chu Melissa Liu, *Open Gromov-Witten invariants of toric Calabi-Yau 3-folds*, Comm. Math. Phys. **323** (2013), no. 1, 285–328, DOI 10.1007/s00220-013-1771-5. MR3085667

[FLT22] Bohan Fang, Chiu-Chu Melissa Liu, and Hsian-Hua Tseng, *Open-closed Gromov-Witten invariants of 3-dimensional Calabi-Yau smooth toric DM stacks*, Forum Math. Sigma **10** (2022), Paper No. e58, 56, DOI 10.1017/fms.2022.57. MR4458543

[FK] David Favero and Bumsig Kim, *General GLSM Invariants and Their Cohomological Field Theory*, `arXiv:2006.12182`, (2021).

[FJR07] Huijin Fan, Tyler J. Jarvis, and Yongbin Ruan, *The Witten equation and its virtual fundamental cycle*, `arXiv:0712.4025`, (2011).

[FJR18] Huijun Fan, Tyler J. Jarvis, and Yongbin Ruan, *A mathematical theory of the gauged linear sigma model*, Geom. Topol. **22** (2018), no. 1, 235–303, DOI 10.2140/gt.2018.22.235. MR3720344

[Gi96] Alexander B. Givental, *Equivariant Gromov-Witten invariants*, Internat. Math. Res. Notices **13** (1996), 613–663, DOI 10.1155/S1073792896000414. MR1408320

[GKT] Mark Gross, Tyler L. Kelly, and Ran J. Tessler, *Open FJRW Theory and Mirror Symmetry*, `arXiv:2203.02435`, (2022).

[Ho00] Shinobu Hosono, *Local mirror symmetry and type IIA monodromy of Calabi-Yau manifolds*, Adv. Theor. Math. Phys. **4** (2000), no. 2, 335–376, DOI 10.4310/ATMP.2000.v4.n2.a5. MR1838444

[HR13] Kentaro Hori and Mauricio Romo, *Exact Results In Two-Dimensional (2,2) Supersymmetric Gauge Theories With Boundary*, `arXiv:1308.2438`, (2013).

[Ir09] Hiroshi Iritani, *An integral structure in quantum cohomology and mirror symmetry for toric orbifolds*, Adv. Math. **222** (2009), no. 3, 1016–1079, DOI 10.1016/j.aim.2009.05.016. MR2553377

[KL13] Young-Hoon Kiem and Jun Li, *Localizing virtual cycles by cosections*, J. Amer. Math. Soc. **26** (2013), no. 4, 1025–1050, DOI 10.1090/S0894-0347-2013-00768-7. MR3073883

[KL20] Young-Hoon Kiem and Jun Li, *Quantum singularity theory via cosection localization*, J. Reine Angew. Math. **766** (2020), 73–107, DOI 10.1515/crelle-2019-0018. MR4145203

[LM] Wolfgang Lerche, Peter Mayr, *On $N = 1$ mirror symmetry for open type II strings*, `arXiv:hep-th/0111113`, (2002).

[LLY97] Bong H. Lian, Kefeng Liu, and Shing-Tung Yau, *Mirror principle. I*, Asian J. Math. **1** (1997), no. 4, 729–763, DOI 10.4310/AJM.1997.v1.n4.a5. MR1621573

[LY22] Chiu-Chu Melissa Liu and Song Yu, *Open/closed correspondence via relative/local correspondence.* part A, Adv. Math. **410** (2022), no. part A, Paper No. 108696, 43, DOI 10.1016/j.aim.2022.108696. MR4487974

[LY] Chiu-Chu Melissa Liu and Song Yu, *Orbifold open/closed correspondence and mirror symmetry*, `arXiv:2210.11721`, (2025).

[Ma] Peter Mayr, $\mathcal{N} = 1$ *mirror symmetry and open/closed string duality*, Adv. Theor. Math. Phys. **5** (2001), no. 2, 213–242, DOI 10.4310/ATMP.2001.v5.n2.a1. MR1909365

[MW09] David R. Morrison and Johannes Walcher, *D-branes and normal functions*, Adv. Theor. Math. Phys. **13** (2009), no. 2, 553–598, DOI 10.4310/atmp.2009.v13.n2.a5. MR2481273

[Ni] Sebastian Nill, *Extended FJRW theory of the quintic threefold in genus zero*, talk in Séminaire de Géométrie Enumérative, 2 December 2021.

[PSW08] Rahul Pandharipande, Jake P. Solomon, and Johannes Walcher, *Disk enumeration on the quintic 3-fold*, J. Amer. Math. Soc. **21** (2008), no. 4, 1169–1209, DOI 10.1090/S0894-0347-08-00597-3. MR2425184

[PV11] Alexander Polishchuk and Arkady Vaintrob, *Matrix factorizations and singularity categories for stacks* (English, with English and French summaries), Ann. Inst. Fourier (Grenoble) **61** (2011), no. 7, 2609–2642, DOI 10.5802/aif.2788. MR3112502

[PV16] Alexander Polishchuk and Arkady Vaintrob, *Matrix factorizations and cohomological field theories*, J. Reine Angew. Math. **714** (2016), 1–122, DOI 10.1515/crelle-2014-0024. MR3491884

[RR17] Dustin Ross and Yongbin Ruan, *Wall-crossing in genus zero Landau-Ginzburg theory*, J. Reine Angew. Math. **733** (2017), 183–201, DOI 10.1515/crelle-2015-0005. MR3731328

[So] Jake P. Solomon, *Intersection theory on the moduli space of holomorphic curves with Lagrangian boundary conditions*, ProQuest LLC, Ann Arbor, MI, 2006. Thesis (Ph.D.)–Massachusetts Institute of Technology. MR2717339

[TX18] Gang Tian and Guangbo Xu, *Analysis of gauged Witten equation*, J. Reine Angew. Math. **740** (2018), 187–274, DOI 10.1515/crelle-2015-0066. MR3824787

[TX21] Gang Tian and Guangbo Xu, *Virtual cycles of gauged Witten equation*, J. Reine Angew. Math. **771** (2021), 1–64, DOI 10.1515/crelle-2020-0022. MR4234101

[TX16] Gang Tian and GuangBo Xu, *Correlation functions of gauged linear σ-model*, Sci. China Math. **59** (2016), no. 5, 823–838, DOI 10.1007/s11425-016-5134-5. MR3484484

[TX20] Gang Tian and Guangbo Xu, *A wall-crossing formula and the invariance of GLSM correlation functions*, Peking Math. J. **3** (2020), no. 2, 235–291, DOI 10.1007/s42543-019-00019-w. MR4171914

[TX17] Gang Tian and Guangbo Xu, *The symplectic approach of gauged linear σ-model*, Proceedings of the Gökova Geometry-Topology Conference 2016, Gökova Geometry/Topology Conference (GGT), Gökova, 2017, pp. 86–111. MR3676084

[TX] Gang Tian and Guangbo Xu, *Gauged Linear Sigma Model in Geometric Phases. I*, `arXiv:1809.00424`, (2024); *Gauged linear sigma model in geometric phases. II. the virtual cycle*, `arXiv:2407.14545`, (2024).

[TZ1] Ran J. Tessler and Yizhen Zhao, *The point insertion technique and open r-spin theories I: moduli and orientation*, `arXiv:2310.13185`, (2026).

[TZ2] Ran J. Tessler and Yizhen Zhao, *The point insertion technique and open r-spin theories II: intersection theories in genus-zero*, `arXiv:2311.11779`, (2023).

[Wa07] Johannes Walcher, *Opening mirror symmetry on the quintic*, Comm. Math. Phys. **276** (2007), no. 3, 671–689, DOI 10.1007/s00220-007-0354-8. MR2350434

[Wa08] Johannes Walcher, *Open strings and extended mirror symmetry*, Modular forms and string duality, Fields Inst. Commun., vol. 54, Amer. Math. Soc., Providence, RI, 2008, pp. 279–297, DOI 10.1090/fic/054/13. MR2454328

[Wa09] Johannes Walcher, *Evidence for tadpole cancellation in the topological string*, Commun. Number Theory Phys. **3** (2009), no. 1, 111–172, DOI 10.4310/CNTP.2009.v3.n1.a3. MR2504755

[We05] Jean-Yves Welschinger, *Invariants of real symplectic 4-manifolds and lower bounds in real enumerative geometry*, Invent. Math. **162** (2005), no. 1, 195–234, DOI 10.1007/s00222-005-0445-0. MR2198329

[We] Jean-Yves Welschinger, *Spinor states of real rational curves in real algebraic convex 3-manifolds and enumerative invariants*, Duke Math. J. **127** (2005), no. 1, 89–121, DOI 10.1215/S0012-7094-04-12713-7. MR2126497

[Wi93] Edward Witten, *Phases of $N = 2$ theories in two dimensions*, Nuclear Phys. B **403** (1993), no. 1-2, 159–222, DOI 10.1016/0550-3213(93)90033-L. MR1232617

Department of Mathematics, Columbia University, 2990 Broadway, New York, New York 10027

Current address: Kavli Institute for the Physics and Mathematics of the Universe (WPI), The University of Tokyo Institutes for Advanced Study, The University of Tokyo, Kashiwa, Chiba 277-8583, Japan

Email address: konstantin.aleshkin@ipmu.jp

Department of Mathematics, Columbia University, 2990 Broadway, New York, New York 10027

Email address: ccliu@math.columbia.edu

Contemporary Mathematics
Volume **841**, 2026
https://doi.org/10.1090/conm/841/16833

A small catalogue of E_n-operads

André Beuckelmann and Ieke Moerdijk

To Ezra Getzler, on the occasion of his 60th birthday.

Abstract. In this largely expository paper, we will present a list of E_n-operads and give complete, and in some cases new, proofs of the equivalences between these operads.

1. Introduction

The theory of operads started in topology with the analysis of (iterated) loop spaces by Stasheff [**Sta63**], Boardman-Vogt [**BV73**] and May [**May72**], in terms of actions on spaces by the little n-cubes operads $\mathscr{C}_n$ for $1 < n < \infty$. An E_n-operad is by definition an operad that is (weakly) equivalent to $\mathscr{C}_n$. These E_n-operads and their associated homology have played a central role in many parts of algebra, geometry and mathematical physics ever since, and Getzler has played an important part in the development of the theory (see for example, Getzer-Jones [**GJ94**], Getzler-Kapranov [**GK95**] and [**GK98**], Getzler's Batalin-Vilkovisky paper [**Get94**]). We refer to [**MSS02**] for a brief early history of the theory of operads. More recently new E_n-operads of a more combinatorial nature have been constructed, for example in the context of the Deligne-conjecture [**MS04**] and of formality problems [**Tam98**].

The goal of this paper is to give an exposition of a series of E_n-operads, and to provide complete proofs of the equivalence between all these. It is not our intention to give a complete list, but instead focus on the following operads:

(1) The classical little n-cubes operad $\mathscr{C}_n$ introduced in [**May72**] and [**BV73**].
(2) The complete graphs operad $\mathscr{G}_n$ introduced by Berger [**Ber97**], and its extension $\mathscr{G}_n^{\text{ext}}$ introduced by Brun, Fiedorowicz and Vogt [**BFV07**].
(3) The n-th filtration of the Barratt-Eccles operad [**BE74**] introduced by Smith [**Smi89**], who conjectured it to be an E_n-operad, and studied by Kashiwabara [**Kas93**]. It will be denoted Γ_n in this paper.
(4) The Fulton-MacPherson compactification FM_n of the space of configurations of finitely many distinct points in $\mathbb{R}^n$.
(5) The operad $\mathscr{M}_n$ of Balteanu et al. defined in terms of n-fold monoidal categories, see [**BFSV03**].
(6) The McClure-Smith operad introduced in [**MS04**] and used in their proof of the Deligne conjecture. This operad will be denoted MS_n below.

2020 *Mathematics Subject Classification.* Primary 55P48.

©2026 American Mathematical Society

All these operads will be defined in detail below, and no prior knowledge of them is expected of the reader.

It is generally believed to be known that these operads are all equivalent, and hence all deserve the attribute "E_n-operad". However, the literature contains some proofs that seem to need some further clarification or contain gaps. In particular, there seems to be no correct proof in the literature that $\mathscr{G}_n$ is actually an E_n-operad. We will be more explicit about this below (see Appendix B). But to give just one example, in [**Ber97**], Berger argues that $\mathscr{G}_n$ is an E_n-operad, and [**BFV07**] remark that $\mathscr{G}_n$ should really be replaced by what we denote $\mathscr{G}_n^{\mathrm{ext}}$, but we have been unable to find a proof in the literature that the inclusion from $\mathscr{G}_n$ into $\mathscr{G}_n^{\mathrm{ext}}$ is an equivalence of operads. We will present a direct combinatorial proof of such an equivalence, which is one of the new elements in this paper. Besides this, we have tried to streamline some arguments in the literature, or replaced them by simpler ones. As another example, Salvatore [**Sal01**] has proved that the Fulton-MacPherson operad FM_n is an E_n-operad, by constructing a map into it from the Boardman-Vogt resolution of the little n-cubes operad. Instead, we will define a map from the Fulton-MacPherson operad into the Boardman-Vogt resolution $W\mathscr{G}_n$ of the complete graph operad $\mathscr{G}_n$, which provides a different and perhaps somewhat simpler proof that FM_n is an E_n-operad.

The contents of our paper, then, is as follows. After recalling some basic definitions in Section 2, we explain in the next section a general method for comparing operads in different categories, notably in spaces and in posets. Our method can be viewed as using a stratification of topological operads, rather than cellular decomposition as in [**Ber97**] or [**BFV07**]. In Section 4 we introduce the complete graphs operads $\mathscr{G}_n$ and $\mathscr{G}_n^{\mathrm{ext}}$ and prove their equivalence. (These are operads in posets, and we denote their geometric realisations by absolute value signs.) This equivalence is then used to relate the little n-cubes operad $\mathscr{C}_n$ to the graphs in the following way: we will construct a commutative diagram (of symmetric sequences of spaces)

$$\begin{array}{ccc} \mathsf{Conf}_n & \xrightarrow{\ \sim\ } & \mathscr{C}_n \\ {\scriptstyle \wr}\downarrow{\scriptstyle \psi} & & \downarrow{\scriptstyle \varphi} \\ |\mathscr{G}_n| & \xhookrightarrow{\ \sim\ } & |\mathscr{G}_n^{\mathrm{ext}}| \end{array}$$

The top map in this diagram is the evident homotopy equivalence relating the configurations of cubes to the configurations of their centres. The objects $\mathscr{C}_n$, $|\mathscr{G}_n|$ and $|\mathscr{G}_n^{\mathrm{ext}}|$ are operads and φ is a (lax) morphism of operads, but Conf_n is not. We will show that the map $\psi : \mathsf{Conf}_n \to |\mathscr{G}_n|$ is a weak homotopy equivalence between symmetric sequences of spaces, cf. Theorem 5.6 below. Further, the inclusion of $\mathscr{G}_n$ into $\mathscr{G}_n^{\mathrm{ext}}$ has been shown to be an equivalence, as well (Theorem 4.5), which shows that $\mathscr{C}_n \to |\mathscr{G}_n^{\mathrm{ext}}|$ is an equivalence. In particular, this proves that $\mathscr{G}_n$ and $\mathscr{G}_n^{\mathrm{ext}}$ are E_n-operads. Earlier attempts to prove this make use of a smaller symmetric sequence $\mathbb{P}_n \subseteq \mathscr{G}_n$ of posets, consisting of so called *proper* graphs (cf. [**Ber97**] and [**Kas93**]). We comment on the role of these proper graphs in the second part of the section.

In Section 6, we introduce the Barratt-Eccles operad Γ and its filtration by smaller operads Γ_n. There is a (lax) morphism of operads $\gamma : \Gamma \to \mathscr{G}$ which maps

Γ_n into $\mathscr{G}_n$, and it turns out to be surprisingly easy to show that the latter maps are homotopy equivalences. Having established that $\mathscr{G}_n$ is an E_n-operad, this gives an easy (and new) proof of the conjecture by Smith mentioned above.

In a next section, we consider the Fulton-McPherson operad FM_n. The spaces in this operad carry a stratification by trees, of exactly the same shape as the stratification of the Boardman-Vogt resolution $W\mathscr{G}_n$ of the complete graph operad $\mathscr{G}_n$. This results in a map $\mathsf{FM}_n \to W(\mathscr{G}_n)$, and the equivalence $\psi : \mathsf{Conf}_n \to \mathscr{G}_n$ shows that this map is an equivalence on strata. The fact that FM_n is equivalent to $\mathscr{G}_n$ (Theorem 9.1 below) then follows easily.

There are two more operads on our list above. The operad $\mathscr{M}_n$ coming from n-fold monoidal categories and the operad MS_n of McClure and Smith. Following [**BFSV03**] we will show that $\mathscr{M}_n$ is an E_n-operad by relating it to the suboperad of $\mathscr{C}_n$ given by the decomposable (see [**Dun88**]) little n-cubes. We conclude our paper with a presentation of the McClure-Smith operad in Section 12 and the construction of an equivalence to the complete graphs operad $\mathscr{G}_n$, showing that it is indeed an E_n-operad. The ideas for this last proof are mostly contained in [**MS04**], although we hope to have simplified the presentation somewhat, for example by making optimal use of the formalism of multisimplicial sets.

The paper has two appendices: one explaining some notations concerning the latter formalism, and one in which we explain in more detail the relation of our proofs to the arguments occurring in the literature.

2. Operads

DEFINITION 2.1. We fix a concrete category $\mathscr{E}$. In this paper, we will only use the categories of sets, simplicial sets, a convenient category of spaces, posets, categories, or diagrams in one of these categories. An *operad* P in $\mathscr{E}$ assigns to each finite set A an object $P(A)$ in $\mathscr{E}$ of "operations taking an A-indexed family of inputs". Further, we require structure maps

- $1_a \in P(\{a\})$, the *unit*.
- For every $\sigma : B \xrightarrow{\sim} A$ a map $\sigma_* : P(B) \to P(A)$
- For every A, B, and $a \in A$ a *composition* map

$$\circ_a : P(A) \times P(B) \to P(A[B/a])$$

where $A[B/a] := A \setminus \{a\} \amalg B$ (A with a replaced by B).

These should satisfy the natural compatibility conditions for symmetries, units, and associativity. For a detailed description, see [**May72**].

REMARK 2.2. Since the collection of all finite sets forms a class, we will technically have to work with only the finite sets contained in some convenient universe/level of the von Neumann hierarchy. We will disregard this point in the rest of the paper.

EXAMPLE 2.3.

(1) The *commutative operad* Com is given by $\mathsf{Com}(A) = \{*\}$ for every A.
(2) The *associative operad* Ass is given by $\mathsf{Ass}(A) = \Sigma_A := \left\{\text{bijections } A \xrightarrow{\sim} A\right\}$.

REMARK 2.4.

(1) Every finite set is isomorphic to a set of the form $n = \{0, \ldots, n-1\}$ for $n \in \mathbb{N}$. With this, it is enough to specify the $P(n)$, as is usually done in the literature. The definition above is the "coordinate free" version.
(2) Instead of the $\circ_a$ composition, we could equivalently specify composition maps for functions $f : B \to A$ between finite sets:

$$P(A) \times \prod_{a \in A} P\left(f^{-1}(a)\right) \to P(B)$$

EXAMPLE 2.5 (Little n-Cubes). The operad $\mathscr{C}_n$ of little n-cubes has as space of operations $\mathscr{C}_n(A)$ the space of injections

$$\coprod_{a \in A} (0,1)^n \xhookrightarrow{\coprod_{a \in A} c_a} (0,1)^n$$

such that the component maps c_a are rectilinear (i.e. maps of the form $c_a(x_1, \ldots, x_n) = (u_1 x_1 + v_1, \ldots, u_n x_n + v_n) = u \odot x + v$ where we are using $\odot$ for the pointwise multiplication of vectors and where all the u_i are positive). As for the operad structure, the unit is the identity embedding, the action is by permuting the c_a (i.e. for $\sigma : B \xrightarrow{\sim} A$ and $c \in \mathscr{C}_n(B)$, we set $\sigma_* c_a = c_{\sigma^{-1}(a)}$), and the composition is defined by composing embeddings

$$\begin{aligned} c^A \circ_a c^B : \coprod_{x \in A[B/a]} (0,1)^n &= \coprod_{x \in A \setminus \{a\}} (0,1)^n \amalg \coprod_{b \in B} (0,1)^n \\ &\xhookrightarrow{\mathrm{id} \amalg c^B} \coprod_{x \in A \setminus \{a\}} (0,1)^n \amalg (0,1)^n \\ &\cong \coprod_{x \in A} (0,1)^n \overset{c^A}{\hookrightarrow} (0,1)^n \end{aligned}$$

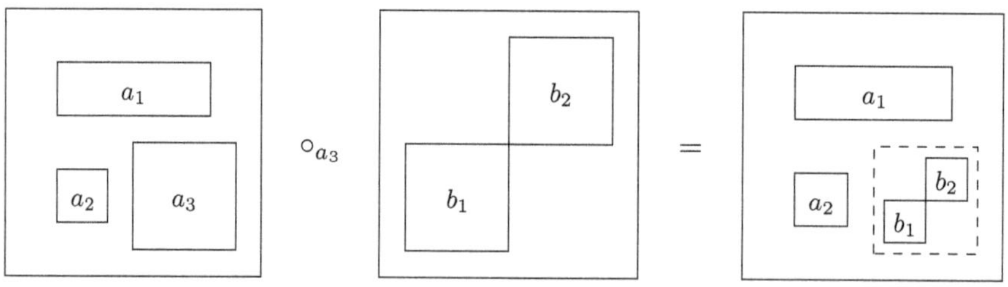

DEFINITION 2.6. The *configuration space* $\mathsf{Conf}_n(A)$ is the space of injections $A \to (0,1)^n$ where we are viewing A as a discrete space — equivalently, this is the space of configurations of A-indexed families of distinct points in $(0,1)^n$. For one such injection $x : A \to (0,1)^n$, we will often write x_a for the image of $a \in A$ under x.

REMARK 2.7. We can define a homotopy equivalence $\gamma : \mathscr{C}_n(A) \to \mathsf{Conf}_n(A)$ by sending each cube to its centre. A homotopy inverse is given by replacing each point of a configuration with a sufficiently small cube (e.g. the cube whose side lengths are half the distance to the closest other point or boundary of the cube $(0,1)^n$). The configuration spaces do not form an operad. They are, however, what is sometimes called a pre-operad: a contravariant functor from the category of finite, non-empty sets and injections to the category of spaces.

3. Equivalences

DEFINITION 3.1. Let P and Q be operads in some category $\mathscr{E}$. A morphism $\varphi : P \to Q$ of operads is given by a family of morphisms $\varphi_A : P(A) \to Q(A)$, one for each finite set, which together respect the operad structures of P and Q. For example, we want the maps to be compatible with the composition in the sense that the diagram

$$\begin{array}{ccc} P(A) \times P(B) & \xrightarrow{\varphi_A \times \varphi_B} & Q(A) \times Q(B) \\ {\scriptstyle \circ_a}\downarrow & & \downarrow{\scriptstyle \circ_a} \\ P(A[B/a]) & \xrightarrow[\varphi_{A[B/a]}]{} & Q(A[B/a]) \end{array}$$

should commute. For more details, see [**May72**].

In the case of operads in spaces, we call such a morphism an *equivalence*, if each $\varphi_A : P(A) \to Q(A)$ is a weak homotopy equivalence, and we write $\varphi : P \xrightarrow{\sim} Q$ in this case. Two operads in spaces are said to be equivalent if they can be related by a zigzag of equivalences

$$P \xleftarrow{\sim} P_1 \xrightarrow{\sim} P_2 \xleftarrow{\sim} \dots \xrightarrow{\sim} Q$$

In fact, it suffices to consider zigzags of the form $P \xleftarrow{\sim} P_1 \xrightarrow{\sim} Q$. In this case, we will write $P \simeq Q$.

DEFINITION 3.2 (E_n-operad). An operad P in spaces is said to be an *E_n-operad* if it is equivalent to the little n-cubes operad and, further, the Σ_n-action on $P(n)$ is free.

We wish to extend these definitions to operads in other categories.

DEFINITION 3.3. We call a morphism $P \to Q$ between operads in simplicial sets, posets, or categories an *equivalence*, if the induced map $\varphi : |P| \to |Q|$ between the geometric realisations is an equivalence of operads in spaces. Similarly, we will call an operad P in one of these categories an *E_n-operad*, if $|P|$ is.

In showing that certain operads are E_n-operads, we will often need a weaker version of morphisms between them.

DEFINITION 3.4. Let P and Q be operads in posets. A *lax morphism* $\varphi : P \to Q$ is given by a family $\varphi_A : P(A) \to Q(A)$ of maps of posets, compatible with the symmetries, but preserving composition only in the weak sense that

$$\varphi(p \circ_a p') \leq \varphi(p) \circ_a \varphi(p')$$

for any $p \in P(A), p' \in P(B)$, and $a \in A$. We say that a lax morphism φ is a *lax equivalence*, if each $\varphi_A : P(A) \to Q(A)$ induces a weak equivalence of classifying spaces $|P(A)| \to |Q(A)|$.

REMARK 3.5. Everything we will say about lax morphisms also holds for morphisms that are, instead, lax in the opposite sense that

$$\varphi(p) \circ_a \varphi(p') \leq \varphi(p \circ_a p')$$

This can be seen by simply replacing P and Q by the operads P^{op} and Q^{op} all of whose posets $P^{\mathrm{op}}(A)$ are simply the opposite of the posets $P(A)$ (and analogously for Q).

LEMMA 3.6. *If $\varphi : P \to Q$ is a lax equivalence between operads in posets, then P and Q are equivalent operads.*

PROOF. Define a new operad R in posets by

$$R = \{(p, q) \in P(A) \times Q(A) \mid \varphi(p) \leq q\}$$

with the pointwise ordering. Then R has the structure of an operad making the two projections $P \leftarrow R \to Q$ morphisms of operads (this uses that φ is a lax morphism). It suffices to see that each of these is an equivalence. To this end, we consider for each set A the diagram of posets

$$\begin{array}{ccc} P(A) & \overset{\pi}{\underset{\iota}{\leftrightarrows}} & R(A) \\ & \searrow^{\varphi_A} & \downarrow \lambda \\ & & Q(A) \end{array}$$

where π and λ are the projections and $\iota(p) = (p, \varphi(p))$. Then π and ι are an adjunction of posets and, hence, induce a weak equivalence. Since $\varphi_A = \lambda \circ \iota$ is a weak equivalence by assumption, λ is one as well. □

We will need a variation of this to relate operads in different categories.

DEFINITION 3.7. Let P be an operad in spaces and Q one in posets. A *lax morphism* $\varphi : P \to Q$ is a lax morphism of operads in posets, where we view P as an operad in posets by forgetting the topology and giving it the discrete partial order.

From such a lax morphism, we wish to construct an actual zigzag of morphisms between operads by applying the following procedure to each component map:

Let $\varphi : X \to \mathbb{P}$ be a map (of sets) from a topological space to a poset. We define $X_{\mathbb{P}}$ to be the topological poset of pairs (x, p) with $x \in X, p \in \mathbb{P}$, and $\varphi(x) \leq p$, ordered by $(x, p) \leq (y, q)$ if and only if $x = y$ and $p \leq q$. We write $|X_{\mathbb{P}}|$ for the geometric realisation of this poset. The projections $X \overset{\mu}{\leftarrow} X_{\mathbb{P}} \overset{\pi}{\to} \mathbb{P}$ then induce maps of spaces $X \overset{\mu}{\leftarrow} |X_{\mathbb{P}}| \overset{\pi}{\to} |\mathbb{P}|$. Further, for each $p \in \mathbb{P}$, we write $X_{\downarrow p}$ for the subspace of X given by $\{x \in X \mid \varphi(x) \leq p\}$.

LEMMA 3.8. *For $\varphi : X \to \mathbb{P}$ and $X_{\mathbb{P}}$ as above*

(a) *The map $\mu : |X_{\mathbb{P}}| \to X$ is a weak homotopy equivalence if one of the following holds*

  (i) *X is a smooth manifold and each $X_{\downarrow p}$ for $p \in \mathbb{P}$ is open in X.*

  (ii) *$X = \operatorname{colim}_{p \in \mathbb{P}} X_{\downarrow p}$, the poset $\mathbb{P}$ is well-founded (i.e. if there are no infinite descending chains) and for each p the map $\operatorname{colim}_{q<p} X_{\downarrow q} \to X_{\downarrow p}$ is a cofibration.*

(b) *If each subspace $X_{\downarrow p}$ of X is a retract of a CW-complex, then $|X_{\mathbb{P}}|$ is (a model for) the homotopy colimit of the functor $\mathbb{P} \to \mathsf{Top}$ sending p to $X_{\downarrow p}$.*

(c) *If, further, each of the $X_{\downarrow p}$ is weakly contractible, then $\pi : |X_{\mathbb{P}}| \to |\mathbb{P}|$ is a weak homotopy equivalence, as well.*

PROOF. Recall first that if each $X_{\downarrow p}$ is cofibrant in the Kan-Quillen model structure on spaces, then $|X_{\mathbb{P}}|$ is a model (the classical Bousfield-Kan model, see e.g. [**HM22**], Example 10.34) for the homotopy colimit of the diagram $p \mapsto X_{\downarrow p}$ of spaces over $\mathbb{P}$. This shows part (b). Part (c) is a direct consequence of this, as the pointwise weak equivalence of diagrams $X_{\downarrow p} \to *$ induces a weak equivalence $\operatorname{hocolim}_{\mathbb{P}} X_{\downarrow \bullet} \to \operatorname{hocolim}_{\mathbb{P}} *$, i.e. $|X_{\mathbb{P}}| \xrightarrow{\sim} |\mathbb{P}|$, proving the lemma.

For part (a)(ii), we proceed by a similar argument: if $\mathbb{P}$ is well-founded, then the generalised Reedy model structure on functors $\mathbb{P} \to \mathsf{Top}$ exists, where we take all morphisms in $\mathbb{P}$ to be positive and define the degree of p via transfinite recursion as $\deg(p) = \sup_{q<p}(\deg(q)+1)$. As this makes all the matching objects trivial, it is the same as the projective model structure. Further, the condition in point (ii) is precisely that for a functor to be cofibrant in this model structure, hence its colimit $\operatorname{colim}_{p} X_{\downarrow p} = X$ is a model for its homotopy colimit, but so is $|X_{\mathbb{P}}|$.

Finally, for part (a)(i), we argue as follows: we use the projective model structure on the category $\mathcal{P}(X)$ of simplicial presheaves on X. We will think of the objects of this category as simplicial spaces over X and note that we obtain a left Quillen functor $|\bullet| : \mathcal{P}(X) \to \mathsf{Top}/X$ given by geometric realisation (since each open subset of X can be given a CW-structure and, so, is cofibrant). This model structure on $\mathcal{P}(X)$ admits a Bousfield localisation $\mathcal{P}'(X)$ by hypercovers where the weak equivalences are stalkwise weak equivalences of simplicial sets (see Jardine [**Jar87**]) and $|\bullet|$ factors as a left Quillen functor $\mathcal{P}'(X) \to \mathsf{Top}/X$ because it sends each hypercover to a weak equivalence, see Dugger and Isaksen [**DI04**]. Now suppose that each $X_{\downarrow p}$ is open and consider the two objects Y and Z of $\mathcal{P}(X)$ defined by

$$Y_n = \coprod_{p_0 \le \cdots \le p_n} X_{\downarrow p_0}, \qquad Z_n = \coprod_{q_0,\dots,q_n} X_{\downarrow q_0} \cap \ldots \cap X_{\downarrow q_n}$$

There is an evident inclusion $Y_\bullet \hookrightarrow Z_\bullet$ and $Z_\bullet$ is a standard hypercover, so $|Z| \to X$ is a weak equivalence by [**DI04**]. The map $Y_\bullet \to Z_\bullet$ is also a weak equivalence, as the stalk of $Y_\bullet$ at a point x is the nerve of $\varphi(x)/\mathbb{P}$, hence also a contractible. Thus, we obtain a weak equivalence $|Y_\bullet| \to X$. □

DEFINITION 3.9. We will call a map $X \to \mathbb{P}$ from a topological space to a poset *sufficiently cofibrant* if it fulfills one of the conditions of part (a) of the previous lemma.

REMARK 3.10. If X is the geometric realisation of a simplicial set and each $X_{\downarrow p}$ is the realisation of a simplicial subset, then condition (a)(ii) is automatically satisfied, i.e. the map $X \to \mathbb{P}$ is sufficiently cofibrant.

We can now apply this lemma to lax morphisms as defined in Definition 3.7.

PROPOSITION 3.11. *Let $\varphi\colon P \to Q$ be a lax morphism from an operad P in spaces to an operad Q in posets. Then φ induces a zigzag*

$$P \xleftarrow{\mu} |R| \xrightarrow{\pi} |Q|$$

of operads in spaces. If $\varphi\colon P(A) \to Q(A)$ is sufficiently cofibrant for each finite set A, then the morphism μ is an equivalence. If for each finite set A and each $q \in Q(A)$ the space $P(A)_{\downarrow q} := \{p \in P(A) \mid \varphi(p) \le q\}$ is a weakly contractible retract of a CW-complex, then $\pi\colon |R| \to |Q|$ is a weak equivalence.

PROOF. This follows by applying the previous lemma to each of the maps $P(A) \to Q(A)$ — the intermediate operad $|R|$ is the realisation of the operad in simplicial spaces given by

$$R_n(A) = \{(p, q_1, \dots, q_n) \in P(A) \times Q(A)^n \mid \varphi(p) \le q_1 \le \cdots \le q_n\} \qquad \square$$

Summarising, we obtain the following result.

COROLLARY 3.12. *Let $\varphi : P \to Q$ be a lax morphism from an operad P in spaces to an operad Q in posets. If for each A and each $q \in Q(A)$ the space $\{p \in P(A) \mid \varphi(p) \leq q\}$ is weakly contractible and the map $P(A) \to Q(A)$ is sufficiently cofibrant, then φ induces an equivalence $P \simeq |Q|$.*

4. The complete graphs operad

DEFINITION 4.1. A directed graph is called *complete*, if the underlying undirected graph is. In graph theory, complete directed graphs are sometimes referred to as *tournaments*. Let W be a poset whose elements we will call *weights*. A graph is said to be *weighted* by W if to each edge a weight $w \in W$ is assigned. We will consider weighted directed complete graphs g.

If $a \to b$ is a directed edge in g, we write $w(a,b)$ for the weight and write $a \xrightarrow{w} b$, if $w = w(a,b)$. If g and h are graphs on the same set A of vertices, we say that $g \leq h$, if, for any $a, b \in A$, if $a \xrightarrow{v} b$ in g, then either $a \xrightarrow{w} b$ in h with $v \leq w$, or $b \xrightarrow{w} a$ in h with $v < w$ strictly. In other words, given a graph h, one can get a smaller graph by reducing weights, and possibly reversing arrows if the weight has been reduced strictly. In this way, directed weighted complete graphs on a finite set of vertices form a poset.

DEFINITION 4.2 (Complete graphs operad). The *complete graphs operad* is the operad, in posets, where $\mathcal{G}_n(A)$ is the poset of *cycle-free* or *acyclic* directed weighted complete graphs on the set A of vertices, weighted by the set $\{1, \ldots, n\}$ with the natural order. The condition that the graph is acyclic means that the directions of the graph define a linear order on the set A of vertices.

The action $\sigma_* : \mathcal{G}_n(B) \to \mathcal{G}_n(A)$ for $\sigma : B \xrightarrow{\sim} A$ is the obvious one. The operad composition is defined as follows: if $g \in \mathcal{G}_n(A)$ and $h \in \mathcal{G}_n(B)$, then, for $a \in A$, the graph $g \circ_a h$ on the set $A[B/a]$ is obtained by replacing the vertex a in g by the graph h and equipping the resulting graph with the natural weights and directions. Explicitly, in $g \circ_a h$, we have:

$$\begin{cases} a' \xrightarrow{w} a'', & a', a'' \in A \setminus \{a\} \text{ and } a' \xrightarrow{w} a'' \text{ in } g \\ b \xrightarrow{w} b', & b, b' \in B \text{ and } b \xrightarrow{w} b' \text{ in } h \\ a' \xrightarrow{w} b, & a' \in A \setminus \{a\}, b \in B \text{ and } a' \xrightarrow{w} a \text{ in } g \\ b \xrightarrow{w} a', & a' \in A \setminus \{a\}, b \in B \text{ and } a \xrightarrow{w} a' \text{ in } g \end{cases}$$

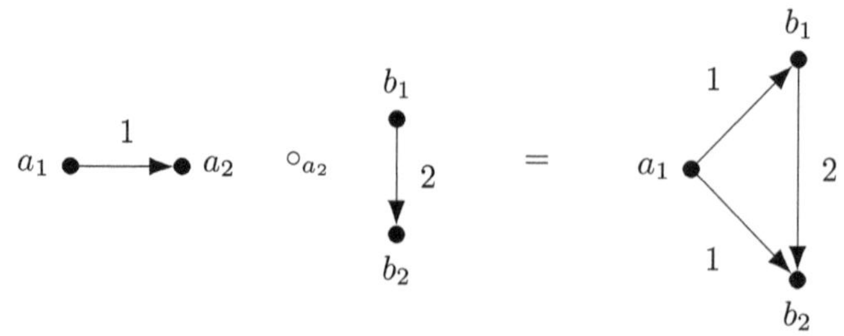

giving a well-defined element in $\mathcal{G}_n(A[B/a])$. This composition is clearly order-preserving, so gives a composition operation $\mathcal{G}_n(B) \times \mathcal{G}_n(A) \to \mathcal{G}_n(A[B/a])$.

DEFINITION 4.3 (Variation). The *extended* complete graphs operad $\mathscr{G}_n^{\text{ext}}$ is defined in the same way, except that the acyclicity condition is weakened to the condition that there are no cycles of *uniform weight*.

REMARK 4.4. We will write $|\mathscr{G}_n|$ and $|\mathscr{G}_n^{\text{ext}}|$ for the operad in topological spaces obtained by taking the classifying spaces of the posets involved (viewed as categories).

THEOREM 4.5. *The inclusion $\mathscr{G}_n \hookrightarrow \mathscr{G}_n^{\text{ext}}$ induces an equivalence of topological operads $|\mathscr{G}_n| \hookrightarrow |\mathscr{G}_n^{\text{ext}}|$.*

PROOF. Let us write $\varphi : \mathscr{G}_n \hookrightarrow \mathscr{G}_n^{\text{ext}}$ for the inclusion. Fix a set A. By Quillen's Theorem A [**Qui73**], it suffices to show that, for each $g \in \mathscr{G}_n^{\text{ext}}(A)$, the space $|\varphi/g|$ is contractible (in particular, it is also non-empty). Of course, this is clear if $g \in \mathscr{G}_n$ (and, hence, for all minimal g).

We proceed by induction on the order on $\mathscr{G}_n^{\text{ext}}(A)$. Suppose that $g \in \mathscr{G}_n^{\text{ext}}(A)$ does not belong to $\mathscr{G}_n(A)$, and that we have proved that φ/h is contractible for all $h \in \mathscr{G}_n^{\text{ext}}(A)$ with $h < g$.

By assumption, g contains cycles, although none of uniform weight. Let us order the cycles as follows: for a cycle c in g, write $\mu_i(c)$ for the number of edges of weight i in c and $\mu(c) := (\mu_n(c), \mu_{n-1}(c), \ldots, \mu_1(c))$. We write $c \leq d$, if $\mu(c) \leq \mu(d)$ in the lexicographical order on $\mathbb{N}^n$. Now, fix a minimal cycle c in g in this ordering, say $c = \left(a_0 \overset{w_1}{\to} a_1 \overset{w_2}{\to} \ldots \overset{w_{k-1}}{\to} a_{k-1} \overset{w_k}{\to} a_k = a_0\right)$. Let us consider the poset φ/g. If $h \leq g$ belongs to $\mathscr{G}_n(A)$, then h contains no cycles, so h must have reversed at least one edge in c (necessarily of weight > 1). Let us consider finite subsets $\emptyset \neq U \subseteq \{1, \ldots, k\}$ for which the weight w_i for $i \in U$ on $a_{i-1} \overset{w_i}{\to} a_i$ in c is strictly larger than 1, and write g_U for the graph obtained from c by reversing the edge $a_{i-1} \overset{w_i}{\to} a_i$ in c and replacing the weight w_i by $w_i - 1$ for all $i \in U$. Then $g_U < g$ and

$$\varphi/g = \bigcup_U \varphi/g_U$$

The intersection of φ/g_U and φ/g_V for two such subsets is again of the form φ/g_W, as one easily checks. Thus, to show that $|\varphi/g|$ is contractible, it suffices to show that each g_U belongs to $\mathscr{G}_n^{\text{ext}}(A)$, because then the induction hypothesis will give that each $|\varphi/g_U|$ is contractible, hence so is $|\varphi/g|$ (being covered by contractible subspaces with contractible and, in particular, non-empty intersections). We formulate this as the following lemma, which then finishes the proof. □

LEMMA 4.6. *Each graph of the form $g_U < g$ belongs to $\mathscr{G}_n^{\text{ext}}(A)$.*

PROOF. We must show that g_U does not contain cycles of uniform weight. Suppose to the contrary that d is a cycle of uniform weight in g_U. By replacing U by a smaller subset of $\{1, \ldots, k\}$, we may assume that d passes through each of the reversed edges $a_i \to a_{i-1}$ of c for $i \in U$. Pick one such $i \in U$ and let j be the index of the next reversed edge $a_j \to a_{j-1}$ on the cycle d. (So if $i < j$ then no number between i and j belongs to U, and if $i > j$ no number larger than j or smaller than i belongs to U, and if U contains just one element, then $i = j$). But then the segments $a_j \to \cdots \to a_{i-1}$ of c and $a_{i-1} \to \cdots \to a_j$ of d together form a cycle e in the original graph g with $\mu(e) < \mu(c)$, contradicting the minimality of c.

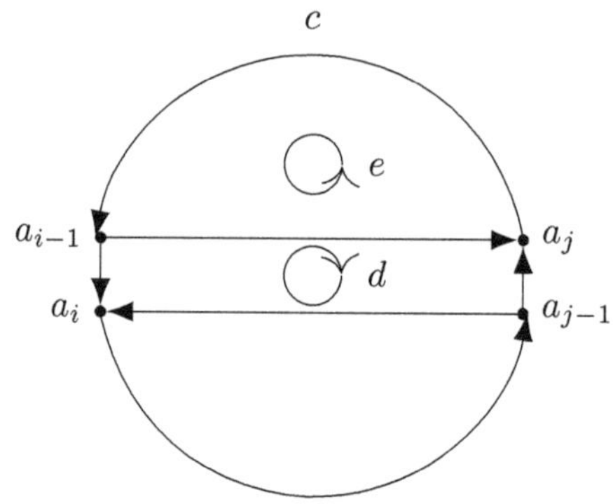

□

5. Complete graphs and little cubes

In this section, we wish to prove that the complete graphs operad is, indeed, an E_n-operad. For this, we will first construct a (lax) morphism

$$\varphi : \mathscr{C}_n \to \mathscr{G}_n^{\text{ext}}$$

from the little n-cubes operad to the extended complete graphs operad.

DEFINITION 5.1. Let c be a point in $\mathscr{C}_n(A)$, given by some embedding of cubes c_a for each $a \in A$. Further, let $i \in \{1, \ldots, n\}$. We say that two cubes c_a and c_b are *i-separated* if they lie on either side of the hyperplane $\{x \in (0,1)^n \mid x_i = t\}$ for some $t \in (0,1)$. In this case, we will say c_a is *i-below* c_b if $c_a \subseteq \{x \in (0,1)^n \mid x_i < t\}$. Since c_a and c_b are disjoint, they must be i-separated for at least one i.

Now, let $\varphi(c)$ be the graph on the set A of vertices whose edges are given as follows: we assign the weight i to the edge between a and b if i is the *smallest* number such that c_a and c_b are i-separated. Further, we direct the edge from a to b if c_a is i-below c_b, and from b to a, otherwise. This construction defines a directed weighted complete graph which does not contain a cycle of uniform weight: indeed, such a circle would give us numbers $t_0, \ldots, t_k \in (0,1)$ with $t_0 < t_1 < \cdots < t_k < t_0$, which is impossible.

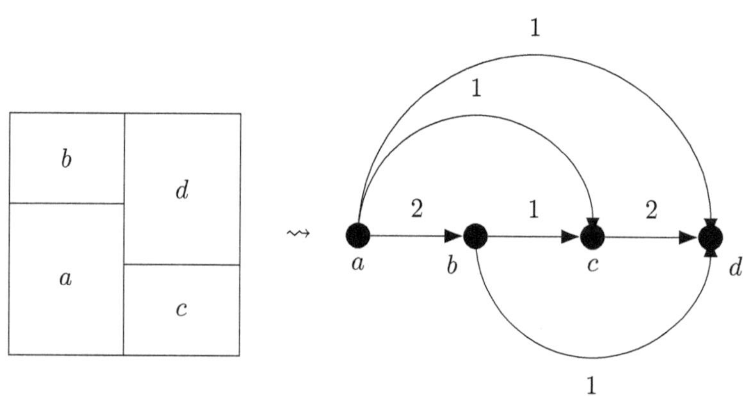

Thus, φ gives a well-defined family of maps $\mathscr{C}_n(A) \to \mathscr{G}_n^{\text{ext}}(A)$.

LEMMA 5.2. *The map φ is a lax morphism of operads.*

PROOF. Let $c \in \mathscr{C}_n(A), d \in \mathscr{C}_n(B)$, and $a \in A$. Consider $c \circ_a d \in \mathscr{C}(A[B/a])$. If c_a and $c_{a'}$ are i-separated, then d_b and $c_{a'}$ are i-separated in $c \circ_a d$ for each $b \in B$ as the image of d_b is contained in that of c_a. □

REMARK 5.3. (1) Note that strict equality does not necessarily hold, as can be seen by the example

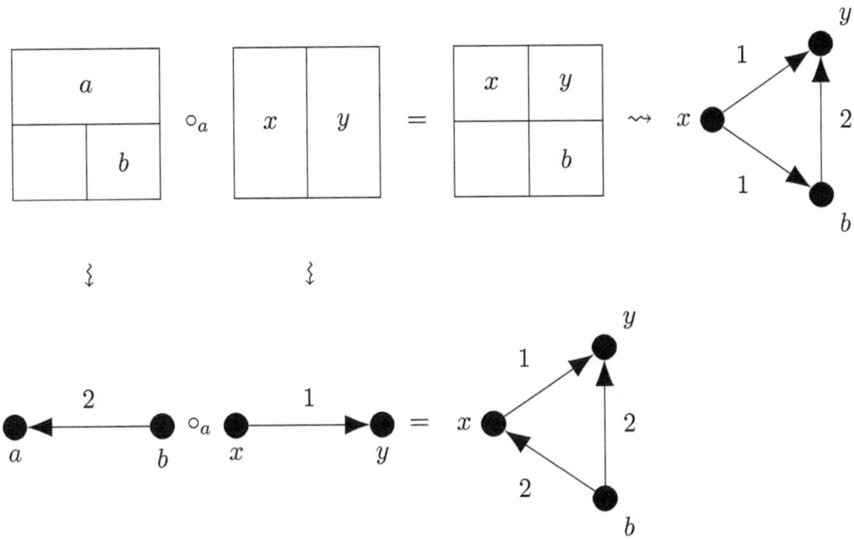

(2) Here, it is indeed necessary to use the extended complete graphs operad rather than the complete graphs operad, as can be seen by the example

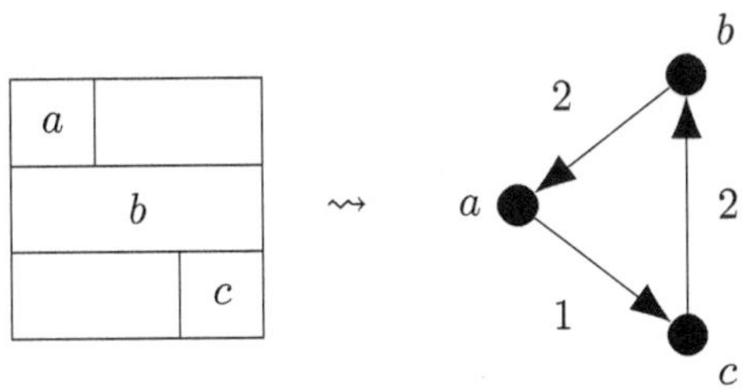

LEMMA 5.4. *The sets $\varphi^{-1}(\downarrow g)$ are CW-complexes.*

PROOF. For a point $c \in \mathscr{C}_n(A)$ and $a \in A$ given by $c_a = u(a) \odot x + v(a)$ with the pointwise multiplication $\odot$, we will define $w(a) := u(a) + v(a)$ — informally, $v(a)$ is the lower left and $w(a)$ the upper right corner of the cube. Using this notation, c_a is i-below c_b if and only if $w(a)_i \leq v(b)_i$. Thus, if $a \xrightarrow{i} b$ in g and $\varphi(c) \leq g$, then either $w(a)_j \leq v(b)_j$ for $j \leq i$ or $w(b)_j \leq v(a)_j$ for some $j < i$ strictly. Thus, if we identify $\mathscr{C}_n(A)$ with a subspace of $[0,1]^{A+A}$ via $c \mapsto (v(a), w(a))_{a \in A}$ then

$$\varphi^{-1}(\downarrow g) = \bigcap_{a \xrightarrow{i} b \text{ in } g} \left(\bigcup_{j \leq i} \left\{ w(a)_j \leq v(b)_j \right\} \cup \bigcup_{j<i} \left\{ w(b)_j \leq v(a)_j \right\} \right)$$

If all the sets of the form $\{\cdots \leq \ldots\}$ on the right were to be interpreted in $[0,1]^{A+A}$ instead, then this would be a union of simplices intersecting along their faces — hence, a CW-complex. For those points coming $\mathscr{C}_n(A)$, however, we have the additional requirement that $v(a) < w(a)$, pointwise. This means that $\varphi^{-1}(\downarrow g)$ is obtained by omitting faces of the form $v(a)_i = w(a)_i$, but such a space is still a CW-complex.

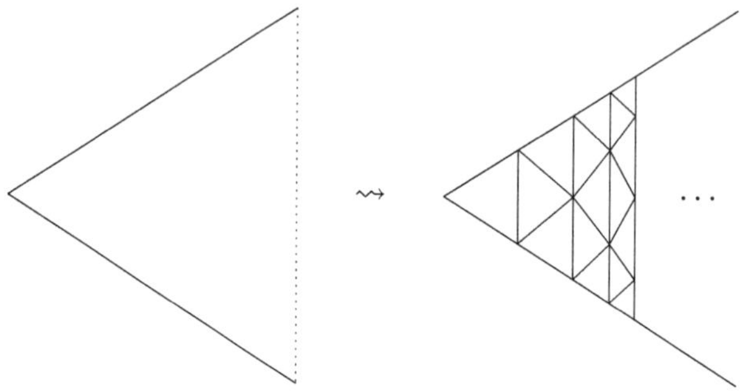

□

REMARK 5.5. Analogously to the definition of φ, we can define a map $\psi : \mathsf{Conf}_n \to \mathscr{G}_n^{ext}$ by sending a configuration x on A to the graph on the vertices A with the edge between a and b having the weight i, if i is the smallest number such that $x(a)_i \neq x(b)_i$. Further, we direct it from a to b, if $x(a)_i < x(b)_i$, and from b to a, otherwise. One can see that the homotopy equivalence $\gamma : \mathscr{C}_n \to \mathsf{Conf}_n$ restricts to a homotopy equivalence $\varphi^{-1}(\downarrow g) \to \psi^{-1}(\downarrow g)$ — the homotopy inverse is defined similarly to that of γ, but one needs to be a bit more careful about how one chooses the size of the cubes.

The image of this map, in fact, lands in $\mathscr{G}_n$: in $\psi(x)$, we will have an edge from $a \to b$ if $x(a) < x(b)$ in the lexicographical order of the coordinates — this is a linear order and so there cannot be any cycles.

The following proof is inspired by the argument of Kashiwabara in [**Kas93**], where he shows the equivalence between the filtration stage Γ_n of the Barratt-Eccles operad and the configuration spaces Conf_n.

THEOREM 5.6. *The map* $\psi : \mathsf{Conf}_n \to \mathscr{G}_n$ *induces an equivalence of preoperads* $\mathsf{Conf}_n \simeq |\mathscr{G}_n|$.

PROOF. Let us fix some finite set A, an element $a \in A$, and write $B := A \setminus \{a\}$. Further, let $p := |B|$. Then there exists an obvious restriction map $\rho : \mathscr{G}_n(A) \to \mathscr{G}_n(B)$. Now, for a fixed element $g \in \mathscr{G}_n(B)$, we wish to study the fibre of ρ over it. Note that g defines some linear order $b_1 \to \cdots \to b_p$ on B — any extension of g to A then places a either to the very front of this linear order, or just behind some b_i for $i \in \{1, \ldots, p\}$ (we will say that a is in the 0-th position, if it is placed at the very front, and in the i-th position, if it is placed right behind b_i). For $i \in \{0, \ldots, p\}$, let us denote by g_i the graph that places a into the i-th position and labels every edge into or out of it by n.

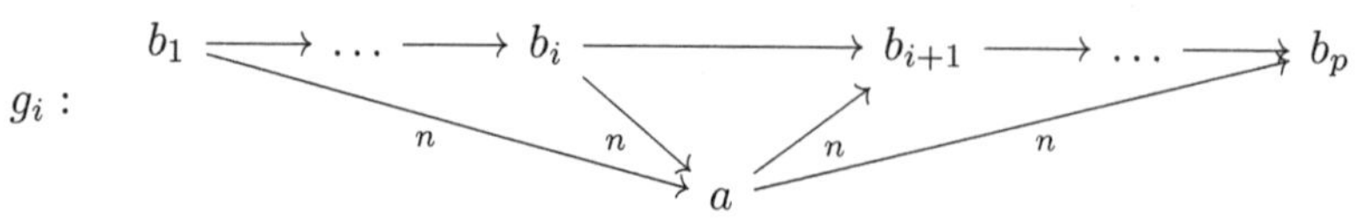

Further, for $k \in \{1, \ldots, n\}$ and $i \in \{0, \ldots, p-1\}$, we will write $h_i^+[k]$ for the graph obtained from g_{i+1} by changing the label of the edge $b_{i+1} \to a$ to k, and

$h_i^-[k]$ for the graph obtained from g_i by changing the label of the edge $a \to b_{i+1}$ to k.

$h_i^-[k]$:

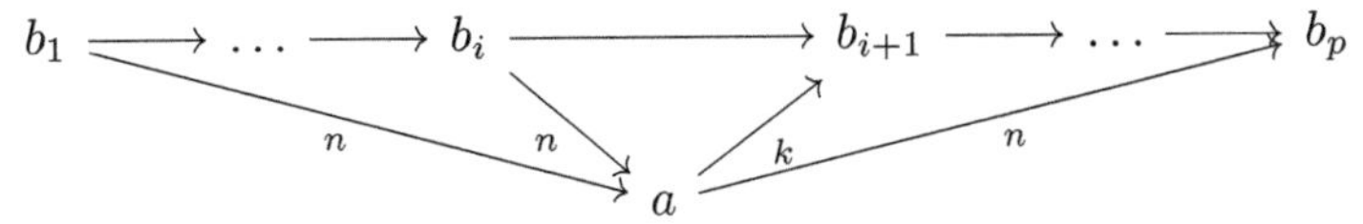

$h_i^+[k]$:

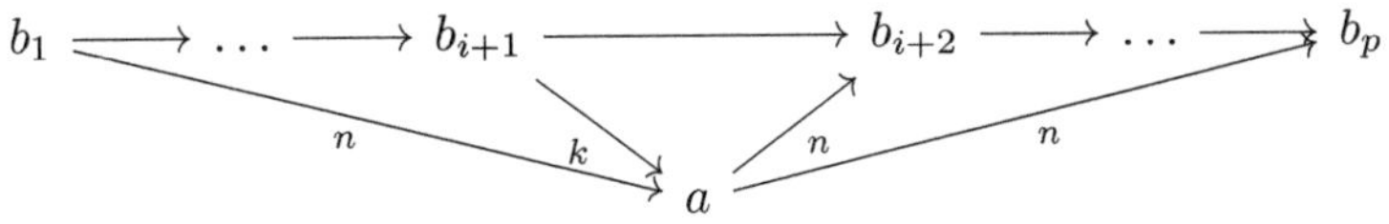

With these definitions, we note the following properties:

(1) $\rho/g = \bigcup_i \downarrow g_i$.
(2) $\downarrow g_s \cap \downarrow g_{t+1} \subseteq \downarrow g_t \cap \downarrow g_{t+1}$ for $s \leq t$.
(3) $\downarrow g_t \cap \downarrow g_{t+1} = \downarrow h_t^-[n-1] \cup \downarrow h_t^+[n-1]$.
(4) $\downarrow h_t^-[1] \cap \downarrow h_t^+[1] = \emptyset$ and $\downarrow h_t^-[k] \cap \downarrow h_t^+[k] = \downarrow h_t^-[k-1] \cup \downarrow h_t^+[k-1]$ for $k > 1$.

This means that the classifying space of $\downarrow g_t \cup \downarrow g_{t+1}$ is an $(n-1)$-sphere and the sequence $\downarrow g_0 \subseteq \downarrow g_0 \cup \downarrow g_1 \subseteq \cdots \subseteq \bigcup_i \downarrow g_i$ corresponds to a sequence of attaching spheres. Thus, we have

$$\left| \bigcup_{i=0}^{p} \downarrow g_i \right| \simeq \bigvee_{i=1}^{p} S^{n-1} = \bigvee_B S^{n-1}$$

Also, note that the attachment of spheres depends only on the order, but not on the weights of g, where the order corresponds to the order in which the copies of S^{n-1} are attached. Thus, for $g \leq h$, we obtain that the map $|\rho/g| \to |\rho/h|$ is given by some permutation of the copies of the spheres.

$$\begin{array}{ccc} |\rho/g| & \longrightarrow & |\rho/h| \\ \wr\downarrow & & \downarrow\wr \\ \bigvee_B S^{n-1} & \xrightarrow[\tau]{} & \bigvee_B S^{n-1} \end{array}$$

By Quillen's theorem B [**Qui73**], this means that the homotopy fibre of ρ is $\bigvee_B S^{n-1}$.

Let $c \in \mathsf{Conf}_n(B)$ be some configuration with $\psi(c) \leq g$. Consider the diagram

$$\begin{array}{ccc} \mathbb{R}^n \setminus c(B) & \xrightarrow{\psi_c} & \rho/g \\ \downarrow & & \downarrow \\ \mathsf{Conf}_n(A) & \xrightarrow{\psi_A} & \mathcal{G}_n(A) \\ \downarrow & & \downarrow \\ \mathsf{Conf}_n(B) & \xrightarrow{\psi_B} & \mathcal{G}_n(B) \end{array}$$

The maps ψ_A and ψ_B are the map ψ defined earlier, and the map ψ_c is defined by sending the point x to the image under ψ_A of the configuration given by extending c via $c(a) = x$. This makes the diagram commute.

The column on the left consists of topological spaces, the one on the right of posets and the horizontal maps are maps as discussed in Section 3. By Lemma 3.8 part (a), we obtain a diagram

$$\begin{array}{ccccc}
\mathbb{R}^n \setminus c(B) & \xleftarrow{\sim} & \left|(\mathbb{R}^n \setminus c(B))_{\rho/g}\right| & \longrightarrow & |\rho/g| \\
\downarrow & & \downarrow & & \downarrow \\
\mathsf{Conf}_n(A) & \xleftarrow{\sim} & \left|\mathsf{Conf}_n(A)_{\mathscr{G}_n(A)}\right| & \longrightarrow & |\mathscr{G}_n(A)| \\
\downarrow & & \downarrow & & \downarrow \\
\mathsf{Conf}_n(B) & \xleftarrow{\sim} & \left|\mathsf{Conf}_n(B)_{\mathscr{G}_n(B)}\right| & \longrightarrow & |\mathscr{G}_n(B)|
\end{array}$$

where the horizontal maps on the left are equivalences, as $\mathbb{R}^n \setminus c(B)$ and $\mathsf{Conf}_n(A)$ are manifolds and the preimages of the downsets are open, as they are defined by strict inequalities. In this diagram, the left hand column is the Fadell-Neuwirth fibration [**FN62**], so the middle column is a fibration up to homotopy. By Quillen's theorem B, the right hand column is as well, as we have just seen. To finish the proof, it now suffices to show that the horizontal maps on the right are weak equivalences, as well. Since this is obviously the case if A is a single point, this follows by induction on the size of A if we can show that the top right horizontal map is a weak equivalence. Thus, the following lemma completes the proof.

LEMMA 5.7. *The map ψ_c induces an equivalence $\left|(\mathbb{R}^n \setminus c(B))_{\rho/g}\right| \to |\rho/g|$.*

PROOF. Since we have just shown that the homotopy type of ρ/g depends only on the ordering of g, but not the particular g, it suffices to prove the claim for one particular g. Thus, let us choose g to be the graph $b_1 \to \cdots \to b_p$ with all edges labelled by 1 and let c be the configuration with the points b_i arranged from left to right on a line along the first coordinate axis (say, we place b_i at $(i, 0, \dots, 0)$).

By our analysis of the poset ρ/g at the beginning of the proof of the theorem, it suffices to prove that the spaces $\psi_c^{-1}(\downarrow h)$ are contractible CW-complexes, if h is either one of the g_i or one of the $h_i^-[k], h_i^+[k]$.

- For one of the g_i, the space $\psi_c^{-1}(\downarrow g_i)$ is just $\mathbb{R}^n$ without the line segments of the form $(j, 0, \dots, 0, t)$ where $j \in \{1, \dots, p\}$, $t \geq 0$, if $j > i$ and $t \leq 0$, if $j \leq i$. By shrinking those line segments towards $\pm\infty$, it can be seen that this space is homotopic to $\mathbb{R}^n$ and, thus, contractible.
- For one of the $h_i^+[k]$, the space $\psi_c^{-1}\left(\downarrow h_i^+[k]\right)$ is $\mathbb{R}^n$ without the line segments of the form $(j, 0, \dots, 0, t)$ where $j \in \{1, \dots, p\}$, $t \geq 0$, if $j > i+1$ and $t \leq 0$, if $j < i+1$, as well as the points of the form $(i+1, 0, \dots, 0, t, *, \dots, *)$ with $t \leq 0$ where t is the k-th entry. The space of points of the latter form can similarly be shrunk towards $(i+1, 0, \dots, 0, -\infty, *, \dots, *)$. Thus, this space is, equally, contractible.
- For one of the $h_i^-[k]$, we get the analogous space to h_i^- where we, instead, have the points with $t \leq 0$. $\square$

$\square$

COROLLARY 5.8. *The map $\varphi : \mathscr{C}_n \to \mathscr{G}_n^{\text{ext}}$ induces an equivalence of operads between $\mathscr{C}_n$ and $|\mathscr{G}_n^{\text{ext}}|$.*

PROOF. This is clear from the diagram

$$\begin{array}{ccc}
\mathsf{Conf}_n(A) & \overset{\sim}{\hookrightarrow} & \mathscr{C}_n(A) \\
\wr\downarrow & & \downarrow \\
\mathscr{G}_n(A) & \underset{\sim}{\hookrightarrow} & \mathscr{G}_n^{\mathrm{ext}}
\end{array}$$

where the map on the left induces an equivalence, as we have just proved, while the map on the bottom is an equivalence by Theorem 4.5. □

In the literature, configuration spaces and the little cubes operad have been related to a smaller collection $\mathbb{P}_n$ of proper graphs; see Appendix B for precise references. We begin by recalling the definitions and deduce from Theorem 5.6 that the inclusion $\mathbb{P}_n \hookrightarrow \mathscr{G}_n$ is an equivalence, as well — a question left open in the literature.

DEFINITION 5.9. For a graph $g \in \mathscr{G}_n^{\mathrm{ext}}$, we will call $\psi^{-1}(g)$ the *cell corresponding to* g. Further, we will call the subset $\psi^{-1}(g) \setminus \bigcup_{h<g} \psi^{-1}(h)$ its *interior*.

We will call a graph $g \in \mathscr{G}_n^{\mathrm{ext}}$ *proper* if its corresponding cell has a non-empty interior, i.e. if the inclusion

$$\bigcup_{h<g} \psi^{-1}(h) \subseteq \psi^{-1}(g)$$

is proper. We will write $\mathbb{P}_n(A)$ for the poset of proper graphs on A. The following lemma gives a criterion for a graph to be proper. The proper graphs do not form a suboperad of the complete graphs operad.

LEMMA 5.10. *The poset* $\mathbb{P}_n(A)$ *consists precisely of those (acyclic) graphs* g *for which, for any path* $v_0 \to \cdots \to v_k$*, the weight of the edge* $v_0 \to v_k$ *is the minimum of the weights of the* $v_i \to v_{i+1}$.

PROOF. If we have some configuration x of points in the interior of the cell corresponding to g, then, in particular, we must have $\psi(x) = g$. Now, if $v_0 \to \cdots \to v_k$ is a path in g and l is the lowest weight on any of the edges, say $v_i \xrightarrow{l} v_{i+1}$, then, by definition of ψ, we must have $x(v_p)_j = x(v_{p+1})_j$ for all $j < l$ and $x(v_p)_l \leq x(v_{p+1})_l$ for all $p \in \{0, \dots, k-1\}$ with at least one inequality holding strictly. In particular, we have $x(v_0)_j = x(v_k)_j$ for $j < l$ and $x(v_0)_l < x(v_k)_l$. Thus, the edge $v_0 \to v_k$ has weight l in $\psi(x) = g$ and the condition is necessary.

To see that it is sufficient, let g be a graph satisfying it. Without loss of generality, we assume $A = \{0, \dots, m\}$ and inductively define a configuration on it. We set $x(0) = 0$ and, if $x(i)$ has already been defined and the edge $i \to i+1$ has weight l in g, then we set $x(i+1) := x(i) + (0, \dots, 1, \dots, 0)$ with the 1 in the l-th position. It is clear that $\psi(x)$ agrees with g on edges of the form $i \to i+1$ and, by the property, it also does on all others. □

THEOREM 5.11. *The map* $\psi : \mathsf{Conf}_n \to \mathbb{P}_n$ *induces an equivalence* $\mathsf{Conf}_n \simeq |\mathbb{P}_n|$.

PROOF. We will apply Corollary 3.12 to the map ψ. Again, Conf_n is a manifold and the preimages of the downsets are open. Thus, it remains to be shown that the $\psi^{-1}(\downarrow g)$ are also contractible for all proper g. The CW-part is analogous to the argument in Lemma 5.4, only that all inequalities are strict, i.e. we omit all faces. For contractibility, let g be a proper graph and $x \in \mathsf{Conf}_n(A)$ be a point with $\psi(x) = g$.

Since $\psi(x) = g$, we have that, for every edge $a \xrightarrow{i} b$ in g, $x(a)_j = x(b)_j$ for $j < i$ and $x(a)_i < x(b)_i$. We set

$$Z_k := \left\{ y \in \psi^{-1}(\downarrow g) \mid y(a)_i = x(a)_i \text{ for all } a \in A \text{ and } i > k \right\}$$

In particular, we have $Z_n = \psi^{-1}(\downarrow g)$ and $Z_0 = \{x\}$. To show that $\psi^{-1}(\downarrow g)$ is contractible, it suffices to see that we can retract Z_k to Z_{k-1}.

We define a retraction as follows: for $t \in [0,1]$, we map a point $y \in Z_k$ to the point $y^{(t)}$ given by

$$y^{(t)}(a)_i := \begin{cases} y(a)_i, & i \neq k \\ (1-t)\,y(a)_k + t x(a)_k, & i = k \end{cases}$$

It remains to be seen that this is, indeed, a well-defined retraction. In other words, we need to see that $y^{(t)}$ is an element of $\psi^{-1}(\downarrow g)$ for all t. Suppose we have some edge $a \xrightarrow{i} b$ in g. If $i > k$, then $y^{(t)}(a)_i = x(a)_i$ and, since $g = \psi(x)$, also $y^{(t)}(a)_i < y^{(t)}(b)_i$. If $i < k$, then $y^{(t)}(a)_i = y(a)_i$ and $y \in \psi^{-1}(\downarrow g)$. Lastly, we need to check the case $i = k$. If $y(a)_j \neq y(b)_j$ for some $j < i$, then also $y^{(t)}(a)_j \neq y^{(t)}(b)_j$. Otherwise, we must have that $y(a)_k < y(b)_k$, since $y \in \psi^{-1}(\downarrow g)$, but x also satisfies $x(a)_k \neq x(b)_k$ and therefore we also have $y^{(t)}(a)_k \neq y^{(t)}(b)_k$.

Thus, we have $y^{(t)} \in \psi^{-1}(\downarrow g)$. □

REMARK 5.12. In [**Ber97**], it has been claimed that the inclusion $\mathbb{P}(A)/\Sigma_A \leftrightarrows \mathscr{G}(A)/\Sigma_A$ has a right adjoint, which would make this into a direct proof of the equivalence between $\mathscr{C}_n$ and $\mathscr{G}_n$. Unfortunately, such an adjunction cannot exist. If we do not quotient by Σ_A, the following gives a counterexample:

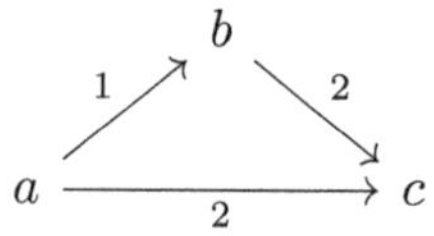

then the poset $\mathbb{P}_2/g$ is given by

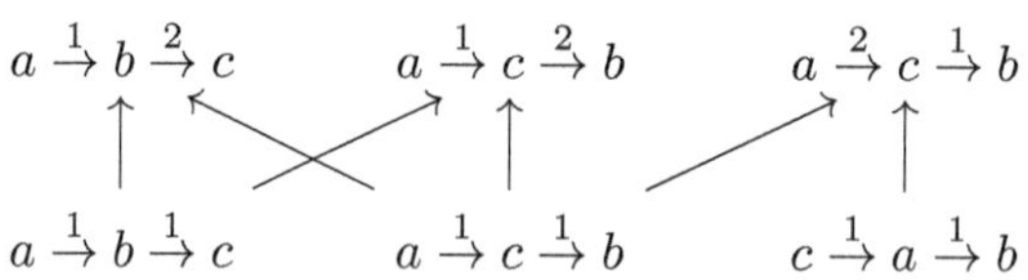

and, hence, not contractible (it is homotopy equivalent to S^1). So $\mathbb{P}(A) \hookrightarrow \mathscr{G}(A)$ cannot have a right adjoint. The same conclusion then follows for $\mathbb{P}(A)/\Sigma_A \leftrightarrows \mathscr{G}(A)/\Sigma_A$. (Indeed, for a general category $\mathscr{E}$ with a free left action by a group H, the quotient category $\mathrm{Ar}(\mathscr{E})/H \rightrightarrows \mathrm{Ob}(\mathscr{E})/H$ is equivalent to the Grothendieck construction $\int_H \mathscr{E}$. Its slice $\left(\int_H \mathscr{E}\right)/x$ over an object x has the slice $\mathscr{E}/x$ as a deformation retract, so it has the same homotopy type.)

A left adjoint cannot exist, either. For this, consider the graph

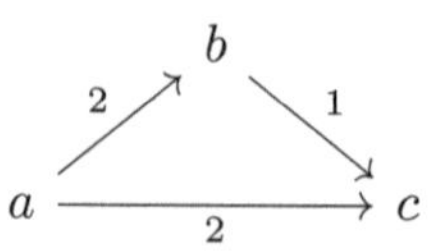

then the poset $g/\mathbb{P}_2$ consists of only to incomparable elements, $a \xrightarrow{2} b \xrightarrow{2} c$ and $a \xrightarrow{2} c \xrightarrow{2} b$. Therefore, it is not connected and, in particular, not contractible.

COROLLARY 5.13. *The inclusion* $\mathbb{P}_n \to \mathscr{G}_n$ *induces an equivalence* $|\mathbb{P}_n| \hookrightarrow |\mathscr{G}_n|$ *of preoperads in spaces.*

PROOF. This follows from 5.11 and 5.6. □

6. The Barratt-Eccles operad

DEFINITION 6.1. For a finite set A, we let $\mathsf{Lin}(A)$ denote the category of *linear orders on* A. More precisely, we define it to be the category whose objects consist of pairs (L, l) with L a linearly ordered set and $l : A \to L$ a bijection. Further, we define a morphism $(L, l) \to (M, m)$ to consist of a function $f : L \to M$ (not necessarily respecting the orders on L and M), such that the triangle

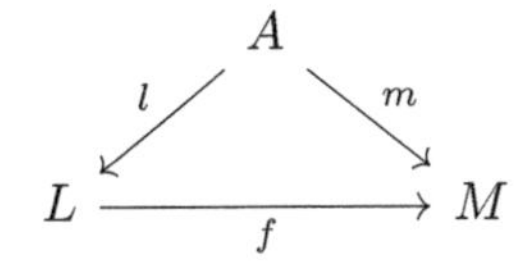

commutes. Note that any such morphism is necessarily a bijection. Moreover, between any two objects, there is exactly one morphism.

Given a bijection $\sigma : B \xrightarrow{\sim} A$, we obtain a functor $\mathsf{Lin}(B) \to \mathsf{Lin}(A)$ given by sending an object (L, l) to the object $(L, l \circ \sigma^{-1})$. Further, we can define a composition for these categories via

$$\begin{aligned}\circ_a : \mathsf{Lin}(A) \times \mathsf{Lin}(B) &\to \mathsf{Lin}(A[B/a]) \\ ((L, l), (M, m)) &\mapsto (L[M/l(a)], l[m/a])\end{aligned}$$

where $L[M/l(a)]$ is the linear order whose underlying set is $L[M/l(a)]$ and whose order is defined by

$$x < y :\Leftrightarrow \begin{cases} x <_L y, & x, y \in L \\ x <_M y, & x, y \in M \\ x <_L l(a), & x \in L, y \in M \\ l(a) <_L y, & x \in M, y \in L \end{cases}$$

and $l[m/a] : A[B/a] \to L[M/l(a)]$ is defined via $l[m/a](x) = l(x)$ for $x \notin B$ and $l[m/a](y) = m(y)$ for $y \in B$. This gives Lin the structure of an operad.

Taking the nerves of these categories, for all finite sets A, gives the *Barratt-Eccles* operad Γ in sSet. The spaces $|\Gamma(A)|$ have a free Σ_A action and are contractible — in other words, $|\Gamma|$ defines an E_∞-operad.

REMARK 6.2. The classical definition of the Barratt-Eccles operad is slightly different, using the categories whose objects are the permutations on A, instead of $\mathsf{Lin}(A)$. If we choose some linear order on A, it is clear that these two categories are equivalent.

DEFINITION 6.3. For a k-simplex x of $\Gamma(A)$, i.e. a diagram of the form

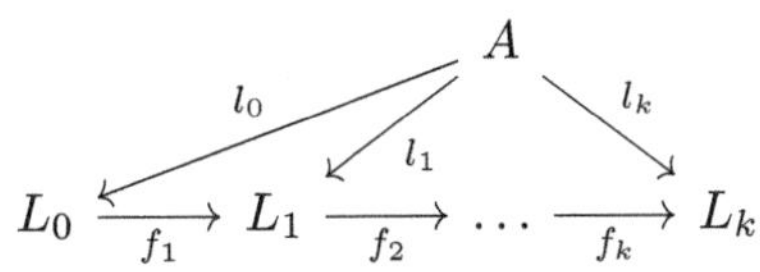

and two elements $a, b \in A$, we define the *weight* $w_x(a,b)$ as 1 plus the number of f_i that switch a and b, or, more precisely, it is 1 plus the number of indices i such that the order of $l_{i-1}(a)$ and $l_{i-1}(b)$ is different from that of $f_i(l_{i-1}(a)) = l_i(a)$ and $l_i(b)$. For a face map d_i or a degeneracy map s_j, we have $w_{d_i x}(a,b) \leq w_x(a,b)$ and $w_{s_j x}(a,b) = w_x(a,b)$, so we can define a simplicial subset

$$\Gamma_n(A) := \{x \in \Gamma(A) \mid w_x(a,b) \leq n \ \forall a, b \in A\}$$

These spaces give a suboperad Γ_n of Γ. We will return to the exact properties of these suboperads in the next subsection.

EXAMPLE 6.4. For the simplex x given by

$$\begin{array}{c} \cdots < a < \cdots < b < \cdots \\ \downarrow \\ \cdots < a < \cdots < b < \cdots \\ \downarrow \\ \cdots < b < \cdots < a < \cdots \\ \downarrow \\ \cdots < a < \cdots < b < \cdots \end{array}$$

we have $w_x(a,b) = 3$.

6.1. Relation to complete graphs. In this section, we will prove the equivalence between the filtrations of the complete graphs and the Barratt-Eccles operad by constructing a map between the two.

DEFINITION 6.5. For an m-simplex $x \in \Gamma(A)$ and $a \neq b \in A$, recall the definition of the weights $w_x(a,b)$ from Definition 6.3. Assume x is of the form

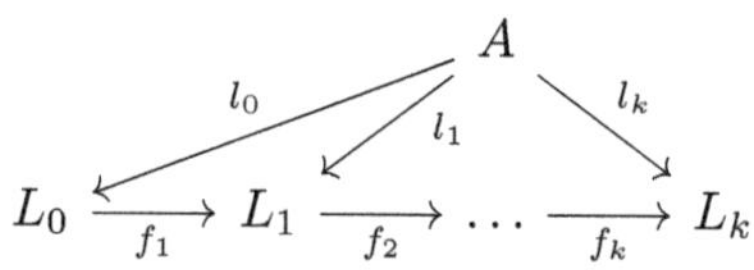

then we will associate to it the graph $\gamma(x)$ defined by an edge $a \xrightarrow{i} b$ for any two $a \neq b \in A$, if $w_x(a,b) = i$ and $l_0(a) < l_0(b)$. In other words, the underlying linear order of $\gamma(x)$ is L_0 and the edges are weighted by the weight in x between their endpoints.

This defines a function

$$\gamma : \Gamma(A)_k \to \mathscr{G}(A)$$

Further, by definition of the suboperads Γ_n, it is clear that this restricts to a map

$$\gamma : \Gamma_n(A) \to \mathscr{G}_n(A)$$

for any n.

LEMMA 6.6. *The map γ gives a morphism of operads $|\Gamma_n| \to \mathscr{G}_n$ as in Definition 3.7 — in fact, we have equalities instead of inequalities.*

PROOF. It suffices to prove that γ respects the operad structure as a map from Γ_n (as an operad in simplicial sets), as the geometric realisation is functorial. For A, B and $a \in A$, we need to consider the functors

$$\circ_a : \mathsf{Lin}(A) \times \mathsf{Lin}(B) \to \mathsf{Lin}(A[B/a])$$

If f is a morphism in $\mathsf{Lin}(A)$, and g is a morphism in $\mathsf{Lin}(B)$, then we will write $f \circ_a g$ for the corresponding morphism given by applying the functor, i.e.

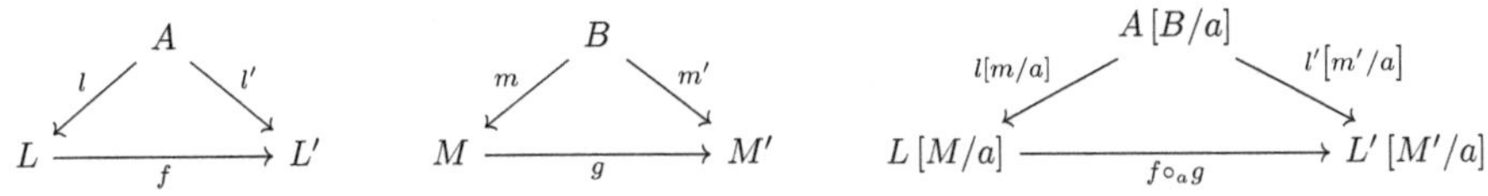

For $x, y \in A \setminus \{a\}$ and $z, w \in B$, note the following:

(1) f switches x and y, if and only if $f \circ_a g$ does.
(2) g switches z and w, if and only if $f \circ_a g$ does.
(3) f switches a and x, if and only if $f \circ_a g$ switches x and z (or w).

Applying this observation to the strings of morphisms of two k-simplices, $x \in \Gamma(A)_k$ and $y \in \Gamma(B)_k$, we see that $\gamma(x \circ_a y) = \gamma(x) \circ_a \gamma(y)$. $\square$

DEFINITION 6.7. For a graph $g \in \mathscr{G}_n(A)$, we will write $\Gamma_g \subseteq \Gamma_n(A)$ for the simplicial set

$$\Gamma_g := \{x \in \Gamma(A) \mid \gamma(x) \leq g\}$$

LEMMA 6.8. *For any $g \in \mathscr{G}_n(A)$, the space $|\Gamma_g|$ is contractible (and a CW-complex).*

PROOF. Let L_g be the set A equipped with the linear order given by g. Then $(L_g, \mathrm{id} : A \to L_g)$ (which we, by abuse of notation, will also denote by L_g) defines a vertex of Γ_g. We claim that Γ_g is a retract of the simplicial set L_g/Γ_g whose k-simplices are given by

$$(L_g/\Gamma_g)_k := \left\{x = (L_0 \to \cdots \to L_{k+1}) \in (\Gamma_g)_{k+1} \mid L_0 = L_g\right\}$$

Note that this simplicial set is contractible (there is an obvious contraction to the vertex $L_g \overset{\mathrm{id}}{\to} L_g$).

To prove the claim, we wish to define a section σ of the map $\pi : L_g/\Gamma_g \to \Gamma_g$ sending a k-simplex x (i.e. a certain $k+1$-simplex of Γ_g) to the simplex given by forgetting the first vertex (viewing $(L_g/\Gamma_g)_k$ as a subset of $(\Gamma_g)_{k+1}$, this is simply the face map d_0 of Γ_g (but not the face map of the simplicial set L_g/Γ_g)).

To do so, suppose we have a k-simplex $x \in \Gamma_g$ of the form

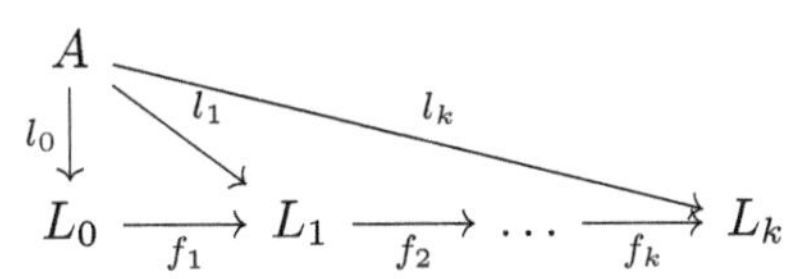

Then we define a $k+1$-simplex $\sigma(x) \in \Gamma_g$ by

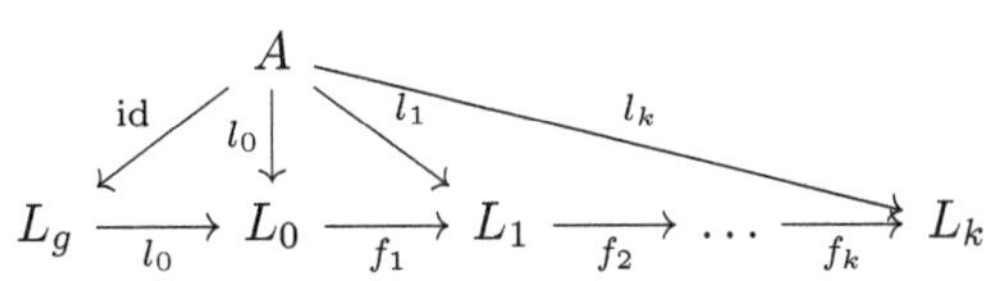

If this indeed defines a k-simplex of L_g/Γ_g, then it is clear that σ gives a section of π, so it remains to check just that. To this end, let $a, b \in A$, let x be a k-simplex of Γ_g (i.e. $\gamma(x) \leq g$) and suppose $a \xrightarrow{i} b$ in g (in particular, we have $a < b$ in L_g). If $l_0(a) < l_0(b)$, as well, then $a \xrightarrow{j} b$ in $\gamma(x)$ for some $j \leq i$, and the same holds in $\gamma(\sigma(x))$. Otherwise, if $l_0(b) < l_0(a)$, then we must have $b \xrightarrow{j} a$ in $\gamma(x)$ for some j strictly less than i, by definition of γ and the ordering on $\mathscr{G}(A)$. This means that, for $\gamma(\sigma(x))$, we will have $a \xrightarrow{j+1} b$, and $j+1 \leq i$. Thus, $\gamma(\sigma(x)) \leq g$, as required. □

THEOREM 6.9. *The operad $|\Gamma_n|$ is equivalent to $\mathscr{G}_n$.*

PROOF. This follows by Corollary 3.12 using the previous two lemmas, as $|\Gamma_n|$ is the geometric realisation of a simplicial set and each $|\Gamma_g|$ is the geometric realisation of a simplicial subset. □

7. The Fulton-MacPherson compactification

We begin with a quick review of the construction of this compactification, and refer to [**Kon99**] or [**GJ94**] for details.

DEFINITION 7.1. Recall the definition of the configuration space from 2.6. The group $\mathbb{R}_{>0} \ltimes \mathbb{R}^n$ acts on $\mathsf{Conf}_n(A)$ via scaling and translating. As this action is free, we obtain a bundle $\mathsf{Conf}_n(A) \to \mathsf{Conf}_n(A)/\mathbb{R}_{>0} \ltimes \mathbb{R}^n$. This bundle has a section given by scaling an equivalence class of configurations in $\mathsf{Conf}_n(A)/\mathbb{R}_{>0} \ltimes \mathbb{R}^n$ so that the maximum distance between two points is equal to 1 and translating it such that the barycentre is the origin. Thus, the bundle is trivial. Writing $\mathsf{NConf}_n(A)$ for the image of this section (we will refer to its image as *normalised* configurations), we obtain a diagram

$$\begin{array}{ccc} \mathsf{Conf}_n(A) & \twoheadrightarrow & \mathsf{Conf}_n(A)/\mathbb{R}_{>0} \ltimes \mathbb{R}^n \\ \uparrow & \nearrow_{\sim} & \\ \mathsf{NConf}_n(A) & & \end{array}$$

There exists an embedding of $\mathsf{NConf}_n(A)$ into the space $\prod_{a,b,c \in A, a \neq b \neq c \neq a} \mathbb{R}_{>0} \times \prod_{a,b \in A, a \neq b} S^{n-1}$ given by sending a configuration x to the point whose coordinate in $\mathbb{R}_{>0}$ for $a, b, c \in A$ is $\left|\frac{x(a)-x(c)}{x(b)-x(c)}\right|$ and whose coordinate in S^{n-1} for $a, b \in A$ is $\frac{x(a)-x(b)}{|x(a)-x(b)|}$. We define $\mathsf{FM}_n(A)$ to be the compactification of the image of $\mathsf{NConf}_n(A)$ in $\prod_{a,b,c \in A, a \neq b \neq c \neq a} [0, \infty] \times \prod_{a,b \in A, a \neq b} S^{n-1}$.

REMARK 7.2. The space $\mathsf{FM}_n(A)$ has a stratification indexed by rooted trees whose set of leaves is A and which do not have any vertices with only one input edge. For two such trees, we will write $S \rightarrowtail T$, if S can be obtained from T by contracting some number of inner edges.

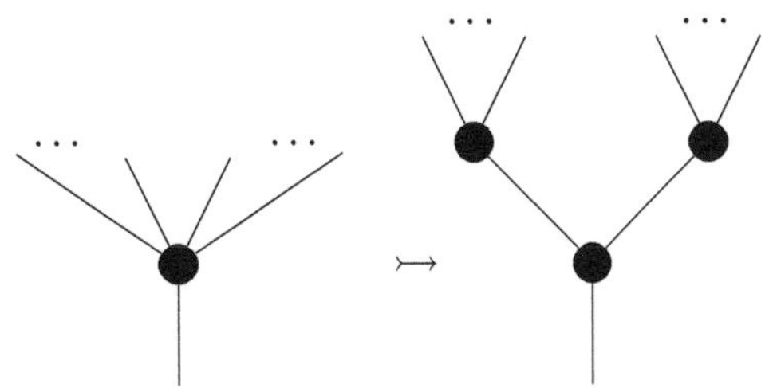

The stratification has the following properties:

(1) $\mathsf{FM}_n(A)$ is partitioned into subspaces $\mathsf{FM}_n(T)$ indexed by such trees.
(2) $\mathsf{FM}_n(T) \subseteq \overline{\mathsf{FM}_n(S)}$, if $S \rightarrowtail T$.
(3) If C_A denotes the corolla whose set of leaves is A, then $\mathsf{FM}_n(A) \supseteq \mathsf{NConf}_n(A) = \mathsf{FM}_n(C_A)$.
(4) $\mathsf{FM}_n(A)$ is a manifold with corners, its interior is given by $\mathsf{FM}_n(C_A)$, and the boundary is the union of the $\mathsf{FM}_n(T)$ taken over trees T with at least one inner edge.
(5) For each tree T, the stratum $\mathsf{FM}_n(T)$ is canonically homeomorphic to a product of normalised configuration spaces

$$\mathsf{FM}_n(T) \cong \prod_{v \in T} \mathsf{NConf}_n(\mathrm{in}(v))$$

where the product is taken over the vertices v of T and $\mathrm{in}(v)$ denotes the set of input edges of v.

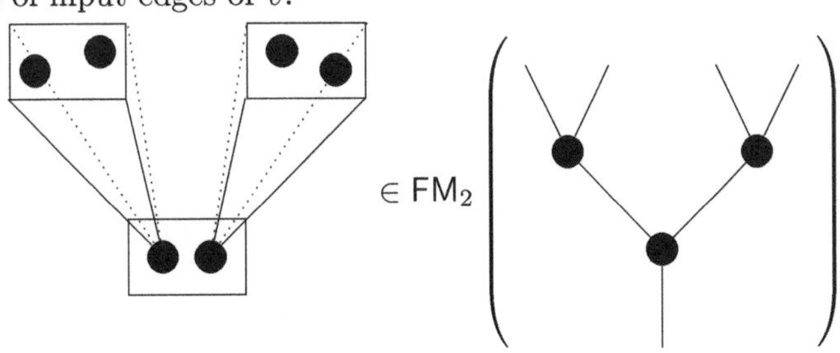

EXAMPLE 7.3. For a set A with $|A| \leq 2$, the space $\mathsf{NConf}_n(A)$ is already compact — in other words, we have $\mathsf{FM}_n(A) = \mathsf{NConf}_n(A)$ and the stratification consists only of the single tree C_A.

DEFINITION 7.4. The spaces $\mathsf{FM}_n(A)$ have the structure of an operad called the *Fulton-MacPherson operad.* The composition operation is given as follows:

For finite sets A and B, $a \in A$, and trees S and T whose sets of leaves are A and B, respectively, as well as elements

$$(y_v)_{v \in S} \in \prod_{v \in S} \mathsf{NConf}_n(\mathrm{in}(v)) = \mathsf{FM}_n(S)$$

and

$$(x_v)_{v \in T} \in \prod_{v \in T} \mathsf{NConf}_n(\mathrm{in}(v)) = \mathsf{FM}_n(T)$$

we can construct a new tree $S \circ_a T$ by grafting T onto the leaf a of S. The set of leaves of $S \circ_a T$ is $A[B/a]$, its set of vertices is the union of the sets of vertices of S and T, and the input edges of some vertex are the same as the input edges of the corresponding vertex of S or T. Thus, the family $(z_v)_{v\in S\circ_a T}$ with $z_v = x_v$ for $v \in T$ and $z_v = y_v$ for $v \in S$ defines a point of

$$\prod_{v\in S\circ_a T} \mathsf{NConf}_n(\mathrm{in}(v)) = \mathsf{FM}_n(S\circ_a T) \subseteq \mathsf{FM}_n(A[B/a])$$

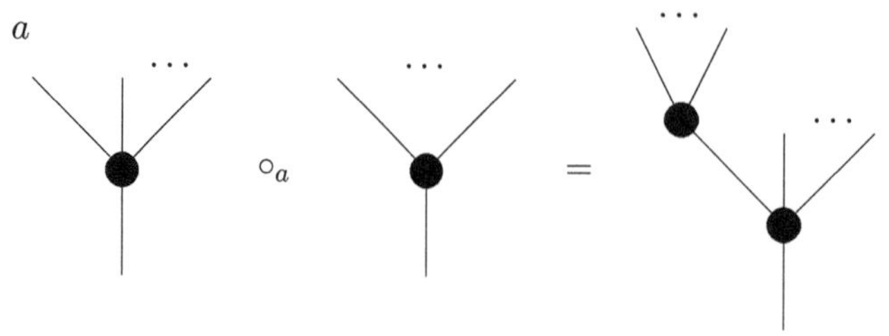

8. The Boardman-Vogt resolution

In this section, we recall the Boardman-Vogt resolution [**BV73**] in spaces and in posets.

DEFINITION 8.1. If P is a topological operad, its *Boardman-Vogt resolution* is an equivalence of operads $W(P) \to P$ where $W(P)(A)$, for some finite set A, is defined as follows: first, for a rooted tree T with A as its set of leaves, we consider the space of all labellings of the vertices v of T by points in $x(v) \in P(\mathrm{in}(v))$, and all of its inner edges e by a *length* $\lambda(e) \in [0,1]$. The space $W(P)(A)$ is obtained by quotienting the union of those spaces for all such trees

$$\coprod_T \left(\prod_v P(\mathrm{in}(v)) \times \prod_e [0,1]\right)$$

by the equivalence relation generated by isomorphisms of trees, as well as

(1) A point with an edge e between two vertices v_1 and v_2 labelled 0 is equivalent to the point of the tree obtained by contracting this edge given by the same labels on any edges besides e and all vertices besides v_1 and v_2 and whose label on the vertex given by the contraction of v_1 and v_2 is given by $x(v_1) \circ_e x(v_2)$ where $x(v_1)$ and $x(v_2)$ are the labels of v_1 and v_2 in the original point:

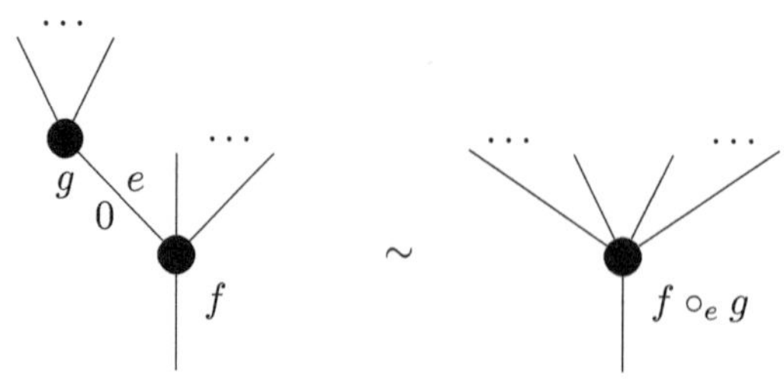

(2) A point of a tree T containing a vertex v with only a single input edge labelled by the unit id of the operad P is equivalent to the point in the tree obtained from T by removing v and linking the other vertex of its input edge to the other vertex of its output edge given by the same labels on all other edges and vertices and whose label on the new edge is given

by the maximum of the labels on the input and the output edge of v:

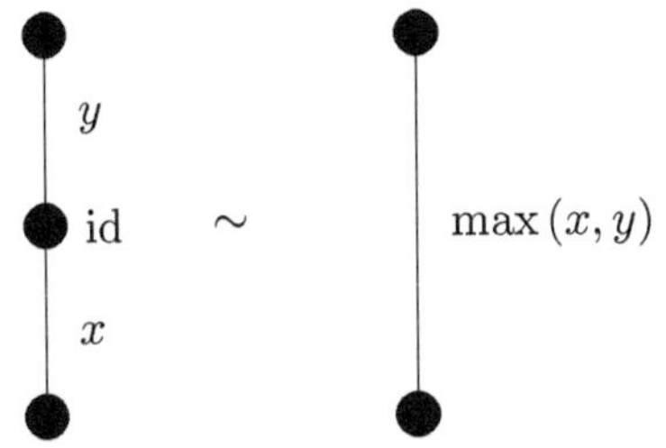

The spaces $W(P)(A)$ have the structure of an operad where the composition is given by grafting the trees and labelling the new inner edge by 1. The map $W(P) \to P$ is given as follows: for a point x given by some labelling of some tree T, we first map it to the point x' by changing the labels on all edges to zero. This point is equivalent to a labelling of C_A, the corolla on A, but such points are completely determined by an element of $P(A)$, that we can map x to. Informally, we might think of this map as just forgetting all edge labels and composing all labels on vertices.

REMARK 8.2. The Boardman-Vogt resolution also makes sense in categories other than spaces by replacing $[0,1]$ by some appropriate interval object. In posets, for example, we can take the linear order $[1] = \{0 \leq 1\}$ and define $W(P)$, accordingly. In this case, as any edge labelled 0 can be contracted, any point of $W(P)(A)$ is represented by a labelling that assigns 1 to all inner edges (in other words, we may disregard this labelling). Thus, we may represent the elements simply by some tree T all of whose vertices have labels by elements of P other than the unit. These labelled trees are ordered by $(S,p) \leq (T,q)$, if $S \rightarrowtail T$ and the labelling p is less than or equal to the one given by composing the elements in the labelling q of T along the contracted edges.

The corresponding map $\pi : W(P) \to P$ gives an equivalence of operads — its inverse is given by $\iota : P \to W(P)$ sending an element of $P(A)$ to the corolla C_A with the element as the label of the unique vertex (note that ι is only a map of symmetric sequences, not of operads). For a fixed A and a tree T with A as its set of leaves, we have $C_A \rightarrowtail T$, so that, by definition of the map π, we have $\iota \circ \pi \leq \mathrm{id}$ and, further, clearly, $\pi \circ \iota = \mathrm{id}$.

The poset $W(P)(A)$ also possesses a stratification indexed by trees with $W(P)(T)$ consisting of the labellings of T, and we have

$$W(P)(T) = \prod_{v \in T} P(\mathrm{in}(v))$$

9. Relating FM_n to $\mathscr{G}_n$

Recall the definition of the maps $\psi : \mathsf{NConf}_n(A) \subseteq \mathsf{Conf}_n(A) \to \mathscr{G}_n(A) \subseteq \mathscr{G}_n^{ext}(A)$ from Remark 5.5. Clearly, these extend to maps

$$\mathsf{FM}_n(T) = \prod_{v \in T} \mathsf{NConf}_n(A) \to \prod_{v \in T} \mathscr{G}_n(A) = W(\mathscr{G}_n)(T)$$

and they fit together to define a map $\psi : \mathsf{FM}_n(A) \to W(\mathscr{G}_n)(A)$ — this gives a map of operads.

THEOREM 9.1. *The map* $\mathsf{FM}_n \to W(\mathscr{G}_n)$ *gives rise to an equivalence of operads between* FM_n *and* $\mathscr{G}_n$.

PROOF. We have already seen that $W(\mathscr{G}_n) \to \mathscr{G}_n$ is an equivalence, so it suffices to see that we obtain an equivalence between FM_n and $W(\mathscr{G}_n)$. For this, consider the diagram

$$\begin{array}{ccccc}
\mathsf{FM}_n(A) & \longleftarrow & \left|\mathsf{FM}_n(A)_{W(\mathscr{G}_n)(A)}\right| & \longrightarrow & |W(\mathscr{G}_n)(A)| \\
\uparrow & & \uparrow & & \uparrow \\
\mathsf{NConf}_n(A) & \longleftarrow & \left|\mathsf{NConf}_n(A)_{\mathscr{G}_n(A)}\right| & \longrightarrow & |\mathscr{G}_n(A)| \\
\downarrow & & \downarrow & & \| \\
\mathsf{Conf}_n(A) & \longleftarrow & \left|\mathsf{Conf}_n(A)_{\mathscr{G}_n(A)}\right| & \longrightarrow & |\mathscr{G}_n(A)|
\end{array}$$

Here, all the maps in the first and last column are equivalences:

- The map $\mathsf{NConf}_n(A) \to \mathsf{FM}_n(A)$ is an equivalence as, by Remark 7.2, it is the inclusion of the interior of a manifold with corners.
- The map $\mathsf{NConf}_n(A) \to \mathsf{Conf}_n(A)$ is the equivalence we remarked on in Definition 7.1.
- The map $|\mathscr{G}_n(A)| \to |W(\mathscr{G}_n)(A)|$ is an equivalence, as it is the geometric realisation of the equivalence from Remark 8.2.
- The map $|\mathscr{G}_n(A)| \to |\mathscr{G}_n(A)|$ is just the identity.

Further, note that the bottom row is an equivalence by Theorem 5.6. Similarly, or via diagram chase, we obtain that the middle row is one, and so is the top-left map. Thus, we obtain the diagram

$$\begin{array}{ccccc}
\mathsf{FM}_n(A) & \xleftarrow{\sim} & \left|\mathsf{FM}_n(A)_{W(\mathscr{G}_n)(A)}\right| & \longrightarrow & |W(\mathscr{G}_n)(A)| \\
\wr\uparrow & & \uparrow & & \uparrow\wr \\
\mathsf{NConf}_n(A) & \xleftarrow{\sim} & \left|\mathsf{NConf}_n(A)_{\mathscr{G}_n(A)}\right| & \xrightarrow{\sim} & |\mathscr{G}_n(A)| \\
\wr\downarrow & & \downarrow & & \| \\
\mathsf{Conf}_n(A) & \xleftarrow{\sim} & \left|\mathsf{Conf}_n(A)_{\mathscr{G}_n(A)}\right| & \xrightarrow{\sim} & |\mathscr{G}_n(A)|
\end{array}$$

and by its commutativity, all other maps must be equivalences as well, proving the theorem. □

10. Operads from n-fold monoidal categories

In this section, we will define the operad $\mathscr{M}_n$ based on n-fold monoidal categories introduced by Balteanu, Fiedorowicz, Schwänzl and Vogt in [**BFSV03**] and give a direct proof that it is an E_n operad. Our formalism will, however, be quite different from that in [**BFSV03**].

DEFINITION 10.1 (The operad $\mathscr{M}_n$). For $n \geq 1$, we define the operad $\mathscr{M}_n$ in posets as follows: for a finite set A, we let $\mathscr{M}_n(A)$ be the equivalence class of planar rooted trees whose set of leaves is A and whose vertices are labelled by the numbers $1, \ldots, n$. The equivalence relation is generated by

(1) Any two stumps are equivalent.

(2) Any stump, other than one directly attached to the root edge, can be removed.

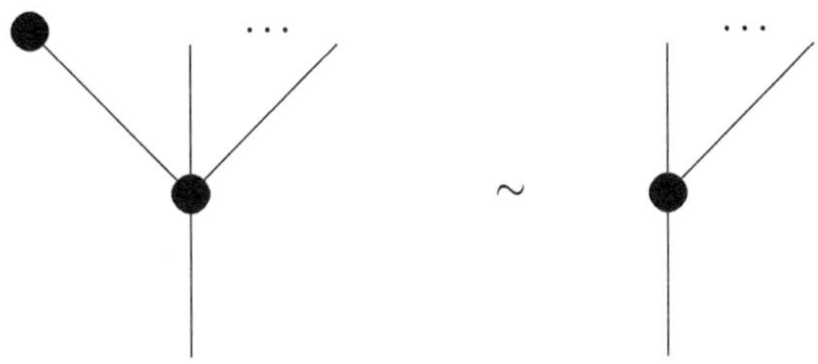

(3) Edges between vertices with the same label can be contracted.

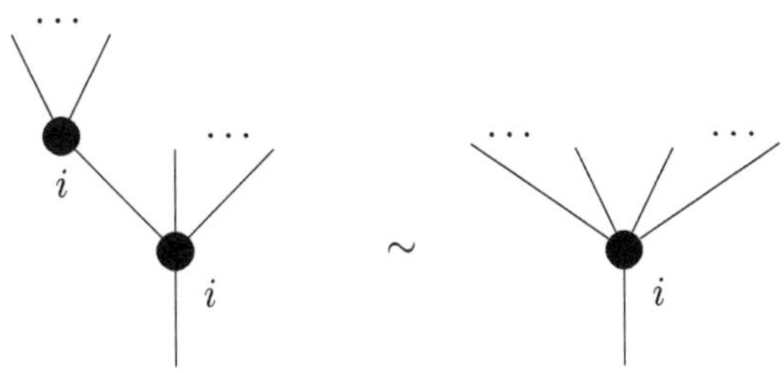

(4) Vertices with only one input edge can be removed.

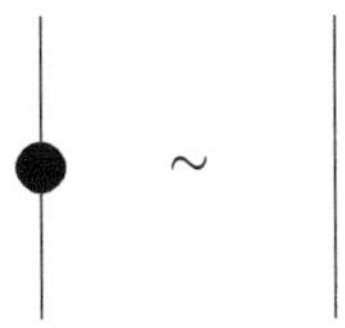

Further, for two such trees T_1 and T_2, we define the order by setting $T_1 \leq T_2$, if, for any two $a, b \in A$, the label on the lowest vertex of the path from a to b in T_1 is less than or equal to that in T_2, and the inequality is strict, if the order of a and b (with respect to the planar structure) is different in T_1 and T_2. Equivalently, this is the first vertex in the intersection between the paths from a and, respectively, b

to the root.

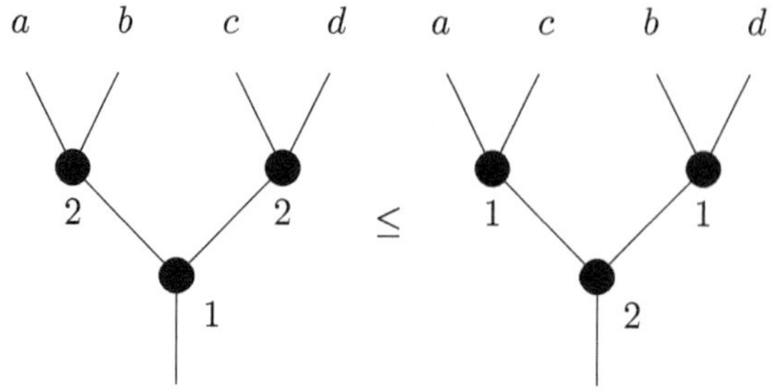

The operad structure is defined as follows: the composition is given by grafting trees and bijections $B \xrightarrow{\sim} A$ act by relabelling the vertices. The unit, for $\{a\} = A$, is given by the tree consisting of only an edge labelled a. This makes $\mathscr{M}_n$ into an operad in posets.

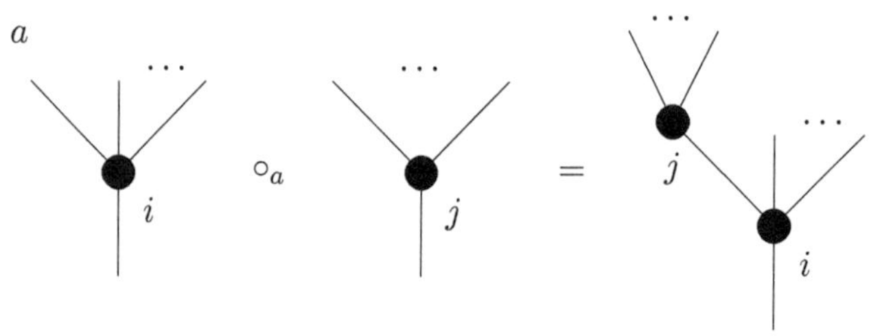

We will write $|\mathscr{M}_n|$ for the topological operad obtained by taking the classifying spaces of the posets. We will call this the *iterated monoidal operad.*

REMARK 10.2. Any such equivalence class of trees is canonically represented by a tree that contains no edges between vertices of the same label, vertices with only one input edge, or stumps (unless the tree itself is only a single stump, in which case we also have to choose a canonical label for this stump, say, 1). We will usually identify the equivalence class with its canonical representative.

REMARK 10.3. Such a labelled tree corresponds to an object of the n-fold monoidal category generated by one object for each element of A: inductively, we define the tree consisting of only a leaf (which is, at the same time, the root) labelled a (with $\{a\} = A$) to correspond to the object a. Now, for a tree whose root vertex is labelled l and whose inputs (viewed, again, as rooted trees themselves) correspond to the objects $o_1, \ldots, o_k$ (in the order corresponding to the planar structure), we define the corresponding object to be $o_1 \Box_l \ldots \Box_l o_k$. In this context, the first two relations on our trees correspond to the fact that we have a common unit, and the last two correspond to (strict) associativity and unitality of the monoidal products.

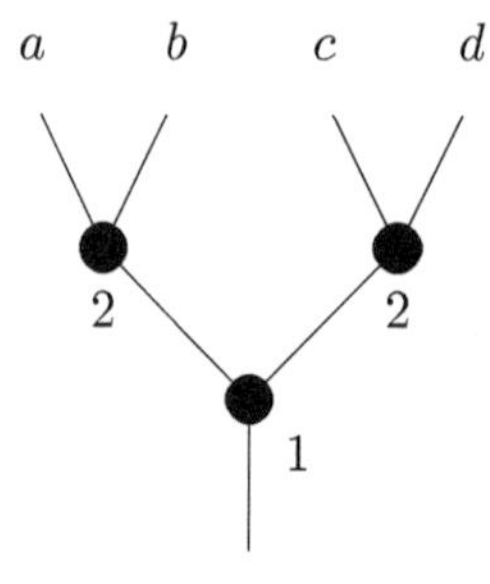

$\sim \quad (a\Box_2 b)\Box_1 (c\Box_2 d)$

This was the original definition used in [**BFSV03**]. They set $\mathscr{M}_n(k)$ to be the free n-fold monoidal category on the objects $1, \ldots, k$ whose morphisms are generated by the interchange maps of the form

$$(a_1\Box_i \ldots \Box_i a_m)\,\Box_j\,(b_1\Box_i \ldots \Box_i b_m) \to (a_1\Box_j b_1)\,\Box_i \ldots \Box_i\,(a_m\Box_j b_m)$$

for $j < i$ (in the case $m = 2$, this interchange map is simply the example on the previous page). They have shown that this makes the $\mathscr{M}_n(k)$ into a poset whose order is the same as the one we defined on the trees.

Now, we will relate this operad to the little cubes.

REMARK 10.4. It is well known that the little cubes operad $\mathscr{C}_n$ is equivalent to that of decomposable configurations (recall that a configuration c of cubes is decomposable, if it consists either of only a single cube, or there is some hyperplane, orthogonal to one of the axes, dividing the configuration into two (non-empty) decomposable configurations). Thus, we may instead work to relate the operad $\mathscr{D}_n$ of decomposable configurations of little cubes to $\mathscr{M}_n$. See, for example, Dunn [**Dun88**] for reference.

DEFINITION 10.5. We will associate to each tree $T \in \mathscr{M}_n(A)$ a subspace $G(T) \subseteq \mathscr{D}_n(A)$ of those configurations possessing a decomposition resembling T. More precisely, we set $G(T)$ to be the set of such configurations c, such that, for each $a, b \in A$ with a to the left of b in the planar structure on T and with the lowest vertex in the path from a to b labelled i, we have that i-th coordinates of the points of the cube c_a are lower than those of the i-th coordinates of the points of the cube c_b. Equivalently, $G(T)$ is recursively defined as follows: $G(\{a\}) = \mathscr{D}_n(\{a\})$ for any singleton $\{a\}$. Further, for some tree T whose root vertex is labelled i and has, as inputs, the trees $T_1, \ldots, T_k$ (in that order according to the planar structure), the set $G(T)$ consists of those configurations c such that there are $k-1$ hyperplanes orthogonal to the i-th axis dividing the configuration into k smaller configurations $c_1, \ldots, c_k$ (in order along the i-th axis) with $c_j \in G(T_j)$.

Next, we define spaces $F(T)$, for each $T \in \mathscr{M}_n(A)$ via $F(T) := \bigcup_{T' \leq T} G(T')$. We will also define a map $\varphi : \mathscr{D}_n(A) \to \mathscr{M}_n(A)$ by mapping c to the smallest tree T with $c \in F(T)$ — this gives a lax morphism of operads. Note that $F(T) = \varphi^{-1}(\downarrow T)$, but we do not necessarily have $G(T) = \varphi^{-1}(T)$.

The following proof is mostly a direct translation of that of [**BFSV03**] to our setting.

PROPOSITION 10.6. *The map $\varphi : \mathscr{D}_n \to \mathscr{M}_n$ induces an equivalence. In particular, the operad $\mathscr{M}_n$ is an E_n-operad.*

PROOF. As remarked in [**BFSV03**],[1] the map φ is sufficiently cofibrant, since the inclusion of a finite union of convex subspaces of $\mathbb{R}^N$ into a larger such union is a cofibration. By Corollary 3.12, it thus suffices to show that each $F(T)$ is a retract of a CW-complex, and that it is contractible. The first point is clear, as the $G(T)$ can be defined in terms of simple, linear inequalities. The second point will be a consequence of the following remark and lemma:

[1] Note added in proof: although the conclusion of this proposition holds, this cofibrancy property does not, as has been discovered by [**Sch25**].

REMARK 10.7. Any equivalence class of trees with more than one leaf can also be represented by a full binary tree, i.e. one where all vertices have exactly two input edges (note that this representation is not unique, but it will suffice for our purposes). Further, our ordering on trees, as well as the subspaces G and F depend only on the equivalence class of the trees. In particular, we have

$$F(T) = \bigcup_{T' \leq T, T' \text{ is full binary}} G(T')$$

For two full binary trees T_1 and T_2, as well as $i \in \mathbb{N}$, write $T_1 \sim_t T_2$, if the first t levels of T_1 and T_2 (starting from the root) coincide. More formally, let $T_1^{(t)}$ be the subtree of T_1 containing all those vertices, edges, and leaves reachable from the root by a path containing at most t edges and define $T_2^{(t)}$ similarly, then we ask for the two to coincide.

LEMMA 10.8. *For every full binary tree T we have that the inclusion of spaces*

$$\bigcup_{T' \leq T, T' \sim_{t+1} T} G(T') \subseteq \bigcup_{T' \leq T, T' \sim_t T} G(T')$$

where the union is taken only over full binary trees is a deformation retract.

PROOF. Our ordering $\leq$ of trees is preserved by grafting trees (i.e. if $T_1 \leq T_1'$, $\ldots$, $T_k \leq T_k'$ and T is some tree, then we have $T \circ (T_1, \ldots, T_k) \leq T \circ (T_1', \ldots, T_k')$). Therefore, it suffices to show that $\bigcup_{T' \leq T, T' \sim_1 T} G(T') \subseteq F(T)$ is a retraction: if T is given by grafting some trees T_1, $\ldots$, T_k onto $T^{(t)}$, then $\bigcup_{T' \leq T, T' \sim_t T} G(T')$ is a subspace of $F(T_1) \times \cdots \times F(T_k) \times G\left(T^{(t)}\right)$, each of the factors of this product but the last retracts to $\bigcup_{T_j' \leq T_j, T_i' \sim_1 T_j} G\left(T_j'\right)$, the combined retraction stays within the subspace, and ends up in the space $\bigcup_{T' \leq T, T' \sim_{t+1} T} G(T')$ which we view as a subspace of $\bigcup_{T_1' \leq T_1, T_1' \sim_1 T_1} G(T_1') \times \cdots \times \bigcup_{T_k' \leq T_k, T_k' \sim_1 T_k} G(T_k') \times G\left(T^{(t)}\right)$.

Now, let $T \in \mathscr{M}_n(A)$, and let the root vertex of T be labelled i and have inputs T_1, T_2. Further, let $T' \leq T$. If $T' \sim_1 T$, there is nothing for us to do, so suppose the root vertex of T' is instead labelled j for some $j < i$ (here, it is important that we use full binary trees, for otherwise it might also be the case that the root vertices have the same label but a different number of inputs). Further, let A_1, A_2 be the labels of the leaves of T_1, T_2, respectively, and let T_1', T_2' be the inputs of the root vertex of T'. Next, for $l \in \{1, 2\}$, let $h_l : [0,1] \times [0,1] \to [0,1]$ be some homotopy from the identity to the obvious affine map $[0,1] \to \left[\frac{l-1}{2}, \frac{l}{2}\right]$. We can then define a homotopy h from the inclusion $G(T') \to F(T)$ to a map whose image ends up in $\bigcup_{\tilde{T} \leq T, \tilde{T} \sim_1 T} G\left(\tilde{T}\right)$ as follows:

We let $h(t, c)$ be the configuration obtained by applying $h_l(t, \bullet)$ to the i-th coordinate of those cubes c_a with $a \in A_l$. This is well defined, as the cubes are separated along the j-th coordinate, which is unchanged by h, and, at $t = 1$, the i-th coordinate of the points of c_a with A_l are in $\left[\frac{l-1}{2}, \frac{l}{2}\right]$, so we obtain a desired decomposition by cutting along the family of hyperplanes defined by their i-th coordinate being one of the $\frac{l}{2}$.

Stitching those homotopies together gives the desired retraction. □

We proceed with the proof of Proposition 10.6.

Note that for any two trees $T_1, T_2 \in \mathscr{M}_n(A)$, we have $T_1 \sim_0 T_2$. Moreover, if h is the height of T_1, we have $T_1 \sim_h T_2$, if and only if $T_1 = T_2$. Thus, iterating the previous lemma, we see that, for $T \in \mathscr{M}_n$, the space $G(T)$ is a deformation retract of $F(T)$. But $G(T)$ is defined in terms of linear inequalities so it is convex. Therefore, the only thing left to show is that each $G(T)$ is non-empty, which we want to see by induction on the number of leaves of T.

For a tree T with only a single leaf, this is clear. Now, suppose T has a root vertex v with input trees $T_1, \ldots, T_k$ (in order of the planar structure) and label l, and suppose that $c_1, \ldots, c_k$ are configurations with $c_i \in G(T_i)$. Further, we define c_i' to be the configuration given by scaling c_i along the l-th coordinate to lie in $\left[\frac{i-1}{k}, \frac{i}{k}\right]$ and set c to be the union of the c_i', then this configuration c is as required. Therefore, we see that $F(T)$ is also contractible which completes the proof. □

EXAMPLE 10.9. For the tree T given by

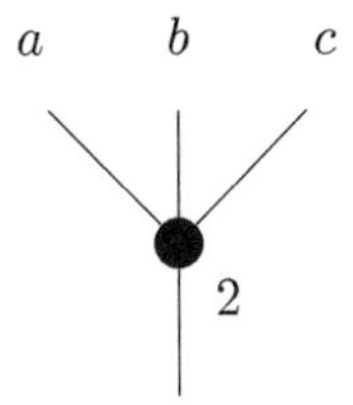

the retraction $F(T) \to \bigcup_{T' \leq T, T' \sim_1 T} G(T') = G(T)$ can be pictured as follows:

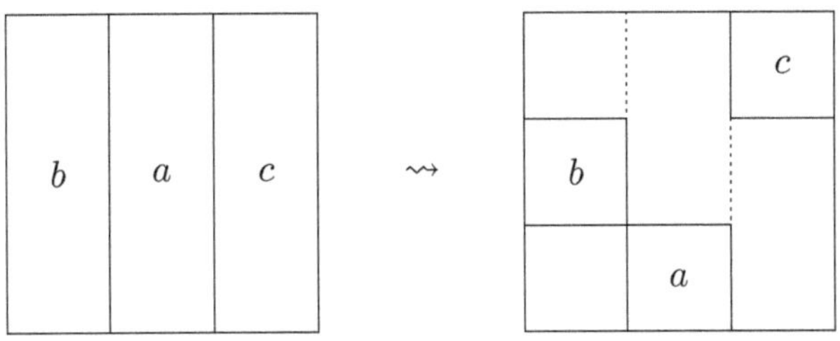

10.1. Relation to complete graphs. From the previous result, we obtain a diagram

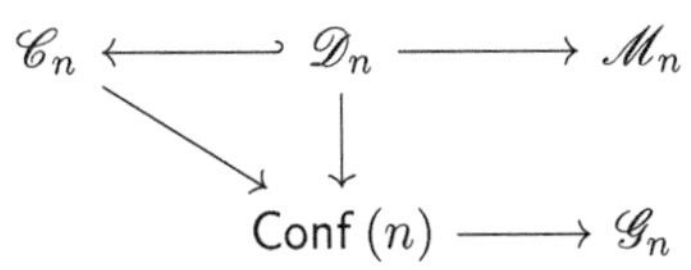

where all arrows induce equivalences. In particular, it follows that $\mathscr{M}_n$ and $\mathscr{G}_n$ are (abstractly) equivalent as operads. However, it would be convenient to find a way to complete this diagram with a map

$$\mu : \mathscr{M}_n \to \mathscr{G}_n$$

This, we will seek to construct, as follows.

Given some tree $T \in \mathscr{M}_n(A)$, we can construct a graph on A by defining the edge between a and b to be directed from a to b, if a precedes b in T with respect to the planar structure. Further, we label this edge by i, if i is the label on the lowest vertex of the path from a to b in T. Given the description of the order on both $\mathscr{M}_n$ and $\mathscr{G}_n$, it is easy to see that this is a morphism of operads.

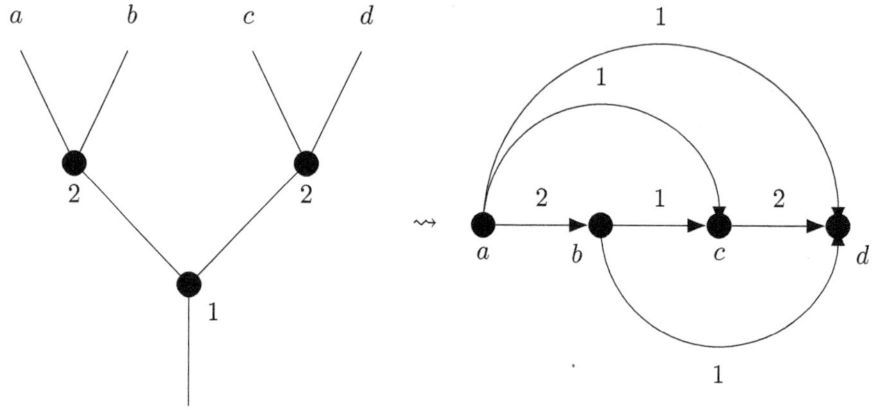

This map does not make the above diagram commute on the nose, as the configuration of the barycentres might enjoy better separation properties than the configuration of the cubes, themselves. However, we do obtain commutativity up to homotopy (or, rather, the ordering of $\mathscr{G}_n$). From that we get the following result.

COROLLARY 10.10. *The map μ induces an equivalence.*

We hope that there is some combinatorial way to prove this, as well. However, we have not been able to find one (a direct proof via Quillen's theorem A is not possible due to similar counterexamples as for the map $\mathbb{P}_n \to \mathscr{G}_n$ above, which can be seen using the next proposition).

For completeness sake, we provide a nice characterisation of the image of μ — in particular, this gives an alternative characterisation of the operad $\mathscr{M}_n$. We will, however, not use this in the remainder of the paper.

PROPOSITION 10.11. *Under μ, the elements of $\mathscr{M}_n$ correspond to those graphs $g \in \mathscr{G}$ which are decomposable in the following way:*

For all $a, b \in g$ and any path $a =: a_0 \to a_1 \to \cdots \to a_k := b$, there is an index i such that all edges $a_p \to a_q$ with $p \leq i < q$ have the same weight.

PROOF. First, we note that the property is equivalent to the following:

If the underlying linear order of g is $x_0 \to \cdots \to x_l$, then g is decomposable in the sense that there exists one edge $x_i \to x_{i+1}$, such that any edge from an element x_j, $j \leq i$, to an element x_k, $k \geq i+1$, has the same label as the edge $x_i \to x_{i+1}$, and the subgraphs on the vertices $x_0, \ldots, x_i$ and $x_{i+1}, \ldots, x_l$ are both decomposable (technically, this is a recursive definition where we have to add the base case that any graph on a single vertex is decomposable).

To see this, first assume that g is decomposable and let $a_0 \to a_1 \to \cdots \to a_k$ be any path, then, by iterating the assumption that g is decomposable, we obtain vertices $y_0 \to \cdots \to y_k$ such that all of the a_i are among these vertices, and such that there is one edge $y_j \to y_{j+1}$ such that any edge from a vertex to the left of or equal to y_j to one to the right of or equal to y_{j+1} has the same weight as the edge $y_j \to y_{j+1}$. In particular, if we let i be the largest index such that a_i is to the left of or equal to y_j, then a_{i+1} must be to the right of or equal to y_{j+1}, and the same must be true for all a_p and a_q with $p \leq i < q$, respectively. In particular, the edge from a_p to a_q must have the same weight as that $y_j \to y_{j+1}$, so i is an index, as required. The converse follows by repeatedly applying the property to paths of the form $x_p \to x_{p+1} \to \cdots \to x_q$ where the underlying linear order of g is $x_0 \to \cdots \to x_l$.

Now, we first show that $\mu(T)$ is decomposable for any tree T. We do this by induction. The base case is clear, so we may move straight to the induction step. For this, let us assume that the root vertex v of T has only two input edges connected to subtrees T_1 and T_2 where T_1 is to the left of T_2 in the planar structure. Now, the underlying linear order of $\mu(T)$ is of the form $x_0 \to \cdots \to x_k \to y_0 \to \cdots \to y_l$ where all the x_i are leaves of T_1 and all the y_j are leaves of T_2. We claim that the edge $x_k \to y_0$ is as required: any path from x_i to y_j must make use of the root vertex v and (as this is the root vertex) v is the lowest vertex on this path. Thus, the weight on any edge $x_i \to y_j$ must be the label of this root vertex — in particular, the weights are all equal to that of $x_k \to y_0$, as required. Now, the induction hypothesis applied to T_1 and T_2 gives that the subgraphs on the $x_0, \ldots, x_k$ and $y_0, \ldots, y_l$, respectively, are decomposable, finishing the proof of the claim.

For the converse, suppose we are given a decomposable graph g, then we need to find a tree T with $g = \mu(T)$. We do this by induction on the number of vertices of g. Again, the base case is clear, so we will move to the induction step. Assume the underlying linear order of g is $x_0 \to \cdots \to x_l$ and let $x_i \to x_{i+1}$ be an edge as in the definition of decomposable and let k be its weight. By induction hypothesis, we find trees T_1 and T_2, such that $\mu(T_1)$ is equal to the subgraph of g on the vertices $x_0, \ldots, x_i$ and $\mu(T_2)$ is that on $x_{i+1}, \ldots, x_l$. We define T to be the planar graph with a root vertex v labelled k having two inputs: one to the left connected to T_1 and one to the right connected to T_2. We claim that $\mu(T) = g$. For this, we only have to see that both agree on edges of the form $x_p \to x_q$ with $p \leq i < q$: in $\mu(T)$, all such edges are labelled k, as the path between the corresponding leaves must go through the root vertex, which is labelled k, but this is also their weight in g, as, by the definition of decomposability, their weight must be equal to that of the edge $x_i \to x_{i+1}$, which is how we defined k. □

11. The McClure-Smith operad

The operad we define in this section is originally due to McClure and Smith [**MS04**].

DEFINITION 11.1 (Lattice paths). For a finite set A and a natural number n, let $\mathscr{L}(A; n)$ be the set of isomorphism classes of diagrams of the form

$$A \overset{f}{\leftarrow} L \overset{\alpha}{\to} [n]$$

where L is a finite linear order, f is a surjective map of sets, and α is a map of linear orders into $[n] = \{0 < 1 < \cdots < n\}$. We will refer to such diagrams as *lattice paths*. Isomorphisms between two diagrams $A \overset{f}{\leftarrow} L \overset{\alpha}{\to} [n]$ and $A \overset{g}{\leftarrow} M \overset{\beta}{\to} [n]$ consist of isomorphisms of linear orders $\theta : L \to M$, such that the diagram

$$\begin{array}{ccc} L & \xrightarrow{\alpha} & [n] \\ {\scriptstyle f}\downarrow & \overset{\theta}{\underset{\sim}{\searrow}} & \uparrow{\scriptstyle \beta} \\ A & \xleftarrow[g]{} & M \end{array}$$

commutes.

DEFINITION 11.2. Later on, it will be convenient to also have a restriction operation for lattice paths: if we are given some subset $U \subseteq A$, then we set

$$\bullet|_U : \mathscr{L}(A;n) \to \mathscr{L}(U;n)$$

to be the map defined by sending some diagram $A \xleftarrow{f} L \xrightarrow{\alpha} [n]$ to the diagram

$$U \xleftarrow{f|_{f^{-1}(U)}} f^{-1}(U) \xrightarrow{\alpha|_{f^{-1}(U)}} [n]$$

REMARK 11.3. (1) For set theoretic reasons, we have to restrict this definition to linear orders L coming from some convenient universe/level of the von Neumann hierarchy.
(2) If two lattice paths are isomorphic, then the isomorphism between them is unique.

REMARK 11.4. The name lattice path comes from the following observation: if we are given a point $A \xleftarrow{f} L \xrightarrow{\alpha} [n]$ with $L = \{l_0 < \cdots < l_k\}$, and if the smallest element of L in the preimage of i under α is $l_{k(i)}$, then we can identify this with a subdivided string of symbols

$$f(l_0)\dots f(l_{k(1)-1})\,|f(l_{k(1)})\dots f(l_{k(2)-1})\,| \quad \dots \quad |f(l_{k(n)})\dots f(l_k)$$

Further, if $A = \{a_1, \dots, a_m\}$, then each such string corresponds to a path through the m-dimensional lattice of size $|f^{-1}(a_1)| \times \cdots \times |f^{-1}(a_m)|$ subdivided into $n+1$ pieces: we interpret each occurrence of a_i in the string of symbols as taking one step in the i-th direction.

$ab|a|bb \quad \rightsquigarrow$

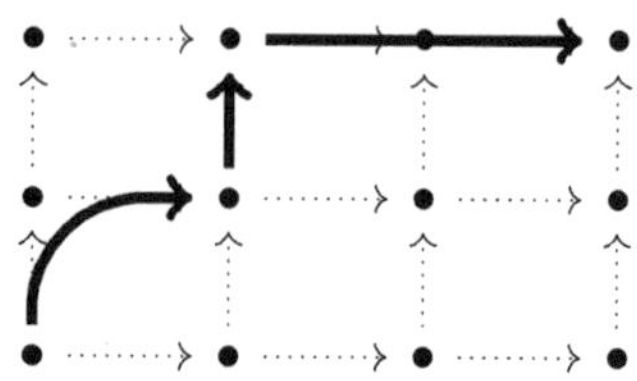

The idea of considering these lattice paths and using them to construct the McClure-Smith operad comes from Batanin and Berger [**BB09**].

REMARK 11.5. $\mathscr{L}(A;\bullet)$ has an obvious cosimplicial structure. It also has the structure of an A-fold simplicial object, for which we will need some additional notation.

DEFINITION 11.6. For $p : A \to \mathbb{N}$, we let $\mathscr{L}(A;n)_p$ be the set of diagrams $A \xleftarrow{f} L \xrightarrow{\alpha} [n]$ such that $f^{-1}(a) \cong [p(a)]$ for each $a \in A$ as a linear order. As this isomorphism is unique, we will identify $f^{-1}(a)$ with $[p(a)]$. Of course every element of $\mathscr{L}(A;n)$ lies in $\mathscr{L}(A;n)_p$ for some p.

Further, suppose we are given some $a \in A$, as well as $m \in \mathbb{N}$, then we will write $\mathscr{L}(A;n)_{a\mapsto m}$ for the union of the $\mathscr{L}(A;n)_p$ taken over all p with $p(a) = m$ — this is just the set of those diagrams $A \xleftarrow{f} L \xrightarrow{\alpha} [n]$ with $f^{-1}(a) \cong [m]$. We will refer to its elements as *m-simplices in the a-direction.* Now, if we are given a map $\rho : [k] \to [m]$, then we can define a map

$$\rho^* : \mathscr{L}(A;n)_{a\mapsto m} \to \mathscr{L}(A;n)_{a\mapsto k}$$

as follows: Given some diagram $A \overset{f}{\leftarrow} L \overset{\alpha}{\rightarrow} [n]$ in $\mathscr{L}(A;n)_p$ with $p(a) = m$ (here, we will view $[m]$ as a subset of L via the isomorphism $[m] \cong f^{-1}(a)$), we define a new diagram

$$\rho^* \left(A \overset{f}{\leftarrow} L \overset{\alpha}{\rightarrow} [n]\right) = A \xleftarrow{f[f\rho/[m]]} L\left[\,[k]\,/\,[m]\right] \xrightarrow{\alpha[\alpha\rho/[m]]} [n]$$

given by replacing $[m]$ by $[k]$ according to ρ. More precisely, $L\left[\,[k]\,/\,[m]\right]$ has as its underlying set the set $L \setminus [m] \amalg [k]$, ordered by

$$x \leq y :\Leftrightarrow \begin{cases} x \leq y, & x, y \in L \setminus [m] \\ x \leq y, & x, y \in [k] \\ x \leq \rho(y), & x \in L \setminus [m], y \in [k] \\ \rho(x) \leq y, & x \in [k], y \in L \setminus [m] \end{cases}$$

and, further, the maps $f\left[f\rho/\left[m\right]\right]$ and $\alpha\left[\alpha\rho/\left[m\right]\right]$ are defined via

$$f\left[f\rho/\left[m\right]\right](x) = f(x), \qquad \alpha\left[\alpha\rho/\left[m\right]\right](x) = \alpha(x)$$

for $x \in L \setminus [m]$ and

$$f\left[f\rho/\left[m\right]\right](y) = f(\rho(y)), \qquad \alpha\left[\alpha\rho/\left[m\right]\right](y) = \alpha(\rho(y))$$

for $y \in [k]$.

This gives the sets $\left(\mathscr{L}(A;n)_{a\mapsto m}\right)_{m\in\mathbb{N}}$ the structure of a simplicial set, which we will denote by $\mathscr{L}(A;n)_{a\mapsto\bullet}$ and refer to as the *simplicial set in the a-direction.*

REMARK 11.7. A diagram of the form $A \leftarrow L \rightarrow [n] \in \mathscr{L}(A;n)_{a\mapsto m}$ is a face (more precisely: iterated composition of faces) of another diagram $A \leftarrow M \rightarrow [n]$ in the a-direction, if there is a commutative diagram of the form

$$\begin{array}{ccc} L & \xrightarrow{\alpha} & [n] \\ {\scriptstyle f}\downarrow & \overset{\theta}{\searrow} & \uparrow{\scriptstyle \beta} \\ A & \underset{g}{\leftarrow} & M \end{array}$$

with θ injective and an isomorphism outside of $f^{-1}(a)$. Further, if there exists a diagram $A \leftarrow M \rightarrow [n]$ and a commutative diagram of the form

$$\begin{array}{ccc} L & \xrightarrow{\alpha} & [n] \\ {\scriptstyle f}\downarrow & \overset{\theta}{\searrow} & \uparrow{\scriptstyle \beta} \\ A & \underset{g}{\leftarrow} & M \end{array}$$

with θ surjective and an isomorphism outside of $f^{-1}(a)$, then $A \leftarrow L \rightarrow [n]$ is degenerate in the a-direction.

We now begin to use this multi-simplicial and cosimplicial structure to construct an operad in spaces from the $\mathscr{L}(A;n)$. For this, some of the details of the constructions can be found in the appendix.

DEFINITION 11.8. By geometrically realising $\mathscr{L}(A;n)$ in all simplicial directions (i.e. $\mathscr{L}(A;n)_{a\mapsto\bullet}$ for all a), we obtain a space that we will denote by $|\mathscr{L}(A;n)|$ and, taken together for all $n \in \mathbb{N}$, this gives a cosimplicial space $|\mathscr{L}(A;\bullet)|$. We can further take the *totalisation* of this cosimplicial space to obtain a single, topological space

$$\mathrm{Tot}\left(\left|\mathscr{L}\left(A;\bullet\right)\right|\right) := \mathsf{Top}^{\Delta}\left(\left|\Delta^{\bullet}\right|, \left|\mathscr{L}\left(A;\bullet\right)\right|\right)$$

Suggestively, we will already denote these spaces by $\mathsf{MS}\left(A\right)$.

REMARK 11.9. A point in the geometric realisation $\left|\mathscr{L}\left(A;n\right)\right|$ is represented by a pair

$$\left(A \overset{f}{\leftarrow} L \overset{\alpha}{\rightarrow} [n]\right) \otimes \left(t\left(a\right)\right)_{a\in A}$$

where $t\left(a\right) \in \Delta^{\left|f^{-1}(a)\right|-1}$. We will often identify the $\left(t\left(a\right)\right)_{a\in A}$ with a family $\left(t_l\right)_{l\in L}$ with $0 \leq t_l \leq 1$ such that if $f^{-1}\left(a\right) = \{l_0, \ldots, l_k\}$ and $l_0 < \cdots < l_k$, then $t\left(a\right) = \left(t_{l_0}, \ldots, t_{l_k}\right)$. In particular, we have $\sum_{i=0}^{k} t_{l_i} = 1$. This representation is unique, if we take $A \leftarrow L \rightarrow [n]$ to be non-degenerate and each $t\left(a\right)$ to be an interior point. If $A \leftarrow L \rightarrow [n]$ is degenerate, i.e. we have a diagram

$$\begin{array}{ccc} L & \xrightarrow{\alpha} & [n] \\ {\scriptstyle f}\downarrow & \overset{\theta}{\searrow} & \uparrow{\scriptstyle \beta} \\ A & \xleftarrow[g]{} & M \end{array}$$

then the point $\left(A \leftarrow L \rightarrow [n]\right) \otimes \left(t_l\right)_{l\in L}$ is the same as the point $\left(A \leftarrow M \rightarrow [n]\right) \otimes \left(s_m\right)_{m\in M}$ with $s_m := \sum_{l:\theta(l)=m} t_l$.

DEFINITION 11.10. We can define a substitution operation for lattice paths. If we are given two elements $A \overset{f}{\leftarrow} L \overset{\alpha}{\rightarrow} [n]$ and $B \overset{g}{\leftarrow} M \overset{\beta}{\rightarrow} [m]$, and if $[m] \cong f^{-1}\left(a\right)$ for some $a \in A$, then we can replace each $l \in f^{-1}\left(a\right)$ by the corresponding fibre of β to obtain a new linear order $L \circ_a M$. Further, there are maps

$$\begin{array}{ccccc} L \circ_a M & \longrightarrow & L & \longrightarrow & [n] \\ {\scriptstyle f\circ_a g}\downarrow & & \downarrow & & \\ A\left[B/a\right] & \longrightarrow & A & & \end{array}$$

making $A\left[B/a\right] \leftarrow L \circ_a M \rightarrow [n]$ an element of $\mathscr{L}\left(A\left[B/a\right];[n]\right)$. In the cosimplicial structure, this element has the same degree as $A \leftarrow L \rightarrow [n]$. In the simplicial structure in the x-direction for $x \in A\backslash\{a\}$, it has the same degree as $A \leftarrow L \rightarrow [n]$, and in the b-direction for $b \in B$, it has the same degree as $B \leftarrow M \rightarrow [m]$.

REMARK 11.11. Let $A \overset{f}{\leftarrow} L \overset{\alpha}{\rightarrow} [n]$, $a \in A$, and $B \overset{g}{\leftarrow} M \overset{\beta}{\rightarrow} [m]$ with $[m] \cong f^{-1}\left(a\right)$ be as above, and assume we are given a map $\rho : [m'] \rightarrow [m]$ in Δ. The compositions

$$\rho^*\left(A \overset{f}{\leftarrow} L \overset{\alpha}{\rightarrow} [n]\right) \circ_a \left(B \overset{g}{\leftarrow} M \overset{\beta}{\rightarrow} [m']\right) \text{ and } \left(A \overset{f}{\leftarrow} L \overset{\alpha}{\rightarrow} [n]\right) \circ_a \rho_*\left(B \overset{g}{\leftarrow} M \overset{\beta}{\rightarrow} [m']\right)$$

where ρ^* is the simplicial, and ρ_* the cosimplicial action, coincide.

REMARK 11.12. The substitution operations induce maps

$$\circ_a : \mathsf{MS}\left(A\right) \times \mathsf{MS}\left(B\right) \rightarrow \mathsf{MS}\left(A\left[B/a\right]\right)$$

Given maps $\varphi : \Delta^{\bullet} \rightarrow \left|\mathscr{L}\left(A;\bullet\right)\right|$ and $\psi : \Delta^{\bullet} \rightarrow \left|\mathscr{L}\left(B;\bullet\right)\right|$, we can construct a map $\varphi \circ_a \psi : \Delta^{\bullet} \rightarrow \left|\mathscr{L}\left(A\left[B/a\right]\right)\right|$ as follows: suppose $t = \left(t_0, \ldots, t_n\right) \in \Delta^n$ and $\varphi\left(t\right)$ is represented by $\left(A \overset{f}{\leftarrow} L \rightarrow [n]\right) \otimes \left(s_l\right)_{l\in L}$. Then in particular $s_a =$

$(s_l)_{l\in f^{-1}(a)}$ is a point in Δ^m where $[m]\cong f^{-1}(a)$, so suppose $\psi(s_a)$ is represented by $(B\leftarrow M\rightarrow[m])\otimes(r_x)_{x\in M}$. Then we let $(\varphi\circ_a\psi)(t)$ be represented by

$$((A\leftarrow L\rightarrow[n])\circ_a(B\leftarrow M\rightarrow[m]))\otimes(s\circ_a r)$$

with $s\circ_a r$ defined by $(s\circ_a r)_l=s_l$ for $l\in L\setminus\{a\}$ and $(s\circ_a r)_x=r_x$ for $x\in M$. This makes the $\mathsf{MS}(A)$ into the McClure-Smith operad introduced in [**MS04**].

DEFINITION 11.13. For a lattice path $x=\left(A\xleftarrow{f}L\rightarrow[n]\right)$, as well as $a\neq b\in A$, we define the *weight of x between a and b*, $w_x(a,b)$, as follows: let $L|_{\{a,b\}}:=f^{-1}(a)\cup f^{-1}(b)$ and define an equivalence relation $\sim$ on $L|_{\{a,b\}}$ via $l\sim l'$, for $l<l'$, if f is constant on $\{m\in L|_{\{a,b\}}\mid l\le m\le l'\}$ and set $w_x(a,b):=\left|L|_{\{a,b\}}/\sim\right|-1$. In other words, if we imagine x to be represented by a (subdivided) string of symbols in A, and let x' be the string obtained from x by deleting all symbols other than a and b, then $w_x(a,b)$ is the number of times that the string x' switches between a and b.

Further, we define the *weight $w(x)$ of x* to be the maximum of the weights between any a and b, $w(x):=\max\limits_{a,b}w_x(a,b)$.

For each finite set A and each $n,m\in\mathbb{N}$ this allows us to define a subset

$$\mathscr{L}_m(A;n):=\{x\in\mathscr{L}(A;n)\mid w(x)\le m\}$$

These sets are clearly compatible with the cosimplicial operations (the maps $\alpha:L\rightarrow M$ do not even appear in the definition of the weights) and the simplicial operations (they can only decrease the number of equivalence classes of $L|_{\{a,b\}}$) and so we get subspaces

$$\mathsf{MS}_m(A)=\mathsf{Top}^{\Delta}(\Delta^\bullet,|\mathscr{L}_m(A;\bullet)|)\subseteq\mathsf{MS}(A)$$

It is, easy to verify that these subspaces are closed under the substitution operations. Thus, this defines a suboperad MS_m of MS.

12. The McClure-Smith operad and complete graphs

In this section, we wish to relate the McClure-Smith operad to the complete graphs operad. Before we do so, we need to study some technical details about the geometric realisation and totalisation.

The following lemma is due to McClure and Smith [**MS04**].

LEMMA 12.1. *For each finite set A, there is an isomorphism of cosimplicial spaces*

$$|\mathscr{L}(A;\bullet)|\cong|\mathscr{L}(A;0)|\times\Delta^\bullet,$$

i.e., for each n, we have homeomorphisms

$$|\mathscr{L}(A;n)|\cong|\mathscr{L}(A;0)|\times\Delta^n$$

compatible with the cosimplicial structure.

PROOF. For a fixed A, the simplex Δ^n is homeomorphic to the blown up simplex

$$\tilde{\Delta}^n:=\left\{(s_0,\ldots,s_n)\;\middle|\;\sum_i s_i=|A|\right\}$$

and it will be more convenient to work with these spaces, instead.

Recall that a point of $|\mathscr{L}(A;n)|$ is given by an equivalence class of objects of the form $\left(A \xleftarrow{f} L \xrightarrow{\alpha} [n]\right) \otimes (t_l)_{l\in L}$. Now, we can obtain a point in $\tilde{\Delta}^n$ from the t_l by summing over the fibres of α. More precisely, we set $s_i := \sum_{l\in\alpha^{-1}(i)} t_l$ to get the point $(s_0, \ldots, s_n)$. This defines a map

$$\begin{aligned}&|\mathscr{L}(A;n)| \to |\mathscr{L}(A;0)| \times \tilde{\Delta}^n\\ &\left(A \xleftarrow{f} L \xrightarrow{\alpha} [n]\right) \otimes (t_l)_{l\in L} \mapsto \left(\left(A \xleftarrow{f} L \to [0]\right) \otimes (t_l)_{l\in L}, (s_i)_{i\in[n]}\right)\end{aligned}$$

To see that this defines a homeomorphism, we wish to construct an inverse.

Suppose we are given a point $\left(\left(A \xleftarrow{f} L \to [0]\right) \otimes (t_l)_{l\in L}, (s_i)_{i\in[n]}\right)$. We wish to interpret the $(s_i)_{i\in[n]}$ as encoding a map $L \to [n]$. That is, we would like to define a map $\alpha : L \to [n]$, such that $s_i = \sum_{l\in\alpha^{-1}(i)} t_l$. Unfortunately, this is not always possible — however, we can find a point in the same equivalence class for which this works: as $\sum_l t_l = |A| = \sum_i s_i$, there will be unique $l_i \in L$, such that

$$\sum_{l<l_i} t_l < \sum_{j\le i} s_i \le \sum_{l\le l_i} t_l$$

for each $i \in [n]$.

Now, let L' be the linear order on the set

$$L \setminus \{l_i \mid i \in [n]\} \cup \{l_i^- \mid i \in [n]\} \cup \{l_i^+ \mid i \in [n]\}$$

ordered via

$$l < m :\Leftrightarrow \begin{cases} l < m, & l, m \in L \\ l < l_i, & l \in L, m = l_i^+ \text{ or } m = l_i^- \\ l_i < m, & m \in L, l = l_i^+ \text{ or } l = l_i^- \\ l_i < l_j, & l = l_i^+ \text{ or } l = l_i^- \text{ and } m = l_j^+ \text{ or } m = l_j^- \\ l = l_i^- \text{ and } m = l_i^+ \end{cases}$$

In other words, we split each l_i into two elements, l_i^- and l_i^+ with $l_i^- < l_i^+$ and leave the ordering between all other elements unchanged. There is an obvious map $\theta : L' \to L$ sending l_i^- and l_i^+ to l_i and any other elements l of L' to $l \in L$, and we set $f' := f \circ \theta$. Further, we can define a point $(t'_l)_{l\in L'}$ via $t'_l := t_l$ for $l \notin \{l_i^-, l_i^+ \mid i \in [n]\}$, $t'_{l_i^-} := \sum_{j\le i} s_j - \sum_{l<l_i} t_l$, and $t'_{l_i^+} := \sum_{l\le l_i} t_l - \sum_{j\le i} s_j$. Then, $(A \leftarrow L' \to [0]) \otimes (t'_l)_{l\in L'}$ represents the same point as $(A \leftarrow L \to [0]) \otimes (t_l)_{l\in L}$ as is witnessed by the diagram

$$\begin{array}{ccc} L' & \longrightarrow & [0] \\ {\scriptstyle f'}\big\downarrow & \overset{\theta}{\searrow} & \big\uparrow \\ A & \xleftarrow{\;f\;} & L \end{array}$$

Now, if we set $\alpha : L' \to [n]$ to be the map $\alpha(l) = i$ if $\sum_{j<i} s_j < \sum_{m\le l} t_m \le \sum_{j\le i} s_j$, then we have $s_i = \sum_{l\in\alpha^{-1}(i)} t'_l$ and so

$$\left(A \xleftarrow{f'} L' \xrightarrow{\alpha} [n]\right) \otimes (t'_l)_{l\in L'}$$

is the desired pre-image. □

DEFINITION 12.2. Recall the definition of the weights $w_x(a,b)$ from Definition 11.13. We can use these to define a morphism

$$\gamma : \mathscr{L}_m(A;n) \to \mathscr{G}_m(A)$$

by sending a lattice path

$$x = \left(A \overset{f}{\leftarrow} L \overset{\alpha}{\to} [n]\right)$$

to the graph on A having an edge $a \overset{i}{\to} b$, if the smallest l with $f(l) = a$ is smaller than the smallest m with $f(m) = b$ and if $w_x(a,b) = i$ — if we think of x as a string of symbols in A, the first part simply asks that a occurs before b does.

We already observed that the definition of the weights is compatible with the simplicial and cosimplicial structure, and so γ induces a map

$$\gamma : \mathsf{MS}_m(A) \to \mathscr{G}_m(A)$$

REMARK 12.3. γ defines a lax morphism of operads: if we have some $a \in A$ and composable lattice paths $A \overset{f}{\leftarrow} L \overset{\alpha}{\to} [n]$ and $B \overset{g}{\leftarrow} M \overset{\beta}{\to} [m]$, and if $x, y \in A \setminus \{a\}$, then the weights between x and y in the composition is equal to that in the first lattice path, as $L \circ_a M|_{\{x,y\}}$ depends only on L. Similarly, if $x, y \in B$, then the weight between x and y in the composition is equal to that in the second lattice path, as $L \circ_a M|_{\{x,y\}}$ depends only on M. Finally, for $x \in A \setminus \{a\}$ and $y \in B$, the weight between x and y in the composition is at most that between x and a in the first lattice path, as there is an obvious injection from the equivalence classes of $L \circ_a M|_{\{x,y\}}$ to those of $L|_{\{x,a\}}$. Further, the only way that x can occur before y in the composition, if a occurs before x in the first lattice path, is if this injection is *not* also a bijection — in other words, the direction of the edge between x and y in γ applied to the composition can only differ from that between x and a in γ applied to the first lattice path, if the weight between the former is strictly smaller than the weight between the latter.

PROPOSITION 12.4. *For each $g \in \mathscr{G}_n(A)$, the space $|\mathscr{L}_g(A;n)| := \left|\gamma^{-1}(\downarrow g)\right|$ is contractible.*

PROOF. First, note that it is non-empty: if the underlying linear order of g is $a_0 < \cdots < a_k$, then the string $a_0 a_1 \ldots a_k \in \mathscr{L}_n(A;0)$, together with some coordinates, gives a point of it.

Now, by Lemma 12.1, it suffices to see that the preimage of g in $|\mathscr{L}_n(A;0)|$ is contractible. For convenience, we will drop the maps $L \to [0]$ when denoting elements of $\mathscr{L}_g(A;0)$. We will proceed by induction on $|A|$. If $A = \{a\}$, then the claim is obvious. Now, for the induction step, let us fix g and let a_0 be the smallest element of A according to the underlying order of g. First of all, we note that it suffices to show that the simplicial set obtained from $\mathscr{L}_g(A;0)$ by fixing the simplicial degree in all directions other than a_0 is contractible, that is, the simplicial set

$$X_y := \left\{x = \left(A \overset{f}{\leftarrow} L\right) \in \mathscr{L}_g(A;0) \;\middle|\; x|_{A\setminus\{a_0\}} = y\right\}$$

for some fixed $y \in \mathscr{L}(A \setminus \{a_0\};0)$ with $\gamma(y) \leq g|_{A\setminus\{a_0\}}$. To see this, note that this will give us that the space $|\mathscr{L}_g(A;0)|$ is homotopy equivalent to the space

$$\left|\mathscr{L}_{g|_{A\setminus\{a_0\}}}(A \setminus \{a_0\};0)\right|$$

and we may proceed with the induction hypothesis.

Let us fix one such diagram $y = A \overset{f}{\leftarrow} L \in \mathscr{L}(A \setminus \{a_0\}; 0)$. A k-simplex $M \overset{e}{\to} A$ of X_y gives rise to, and can be identified with, a pullback diagram

$$\begin{array}{ccc} L & \hookrightarrow & M \\ {\scriptstyle f}\downarrow & & \downarrow{\scriptstyle e} \\ A \setminus \{a_0\} & \hookrightarrow & A \end{array}$$

such that $e^{-1}(a_0) \cong [k]$ and $\gamma(e) \leq g$. We can identify the map e with a monotone map $\tilde{e} : [k] \to \{-\infty\} \amalg L$ denoting that the $(i+1)$-st element of $e^{-1}(a_0)$ in M occurs just after $\tilde{e}(i)$. More precisely, we set

$$\tilde{e}(i) := \max \left\{ l \in L \mid \left| e^{-1}(a_0) \cap \{m \in M \mid m \leq l\} \right| \leq i \right\}$$

where we use the convention that $\max \emptyset = -\infty$. The inverse of this operation is given by setting $M := L \amalg [k]$ using the linear ordering with $l \leq m$, for $l \in L$ and $m \in [k]$, if $l \leq \tilde{e}(m)$. The simplicial set X_y has a canonical vertex v given by the map $\tilde{e} : [0] \to \{-\infty\} \amalg L$ with $\tilde{e}(0) = -\infty$. Again (as in the proof of Lemma 6.8), we consider the simplicial set v/X_y whose k-simplices are given by

$$(v/X_y)_k := \left\{ x \in [X_y]_{k+1} \mid x_0 = v \right\}$$

This simplicial set is clearly contractible. Thus, it suffices to see that the projection $\pi : v/X_y \to X_y$ sending a k-simplex x (i.e. a certain $k+1$-simplex of X_y) to the simplex given by forgetting the first vertex has some section σ. Indeed, if this is the case, then X_y is a retract of the contractible space v/X_y and, hence, contractible. If we are given a k-simplex

$$x = \left(A \overset{e}{\leftarrow} M \right)$$

in X_y, then we set

$$\sigma(x) := \left(A \xleftarrow{a_0 \amalg e} * \amalg M \right)$$

where $* \amalg M$ is the linear order given by adding a new smallest element $*$ to M and $a_0 \amalg e$ is the map defined by sending $*$ to a_0 and any element m of M to $e(m)$. Assuming that σ is well-defined, this clearly gives a section, so it remains to check just that. In other words, we need to see that, if $\gamma(x) \leq g$, then $\gamma(\sigma(x)) \leq g$, so let us fix some x. For any edge in $\gamma(\sigma(x))$ between vertices other than a_0, this graph coincides with $\gamma(x)$. Thus, it suffices to consider the edges between a_0 and some $a \in A \setminus \{a_0\}$. If $a_0 \overset{i}{\to} a$ in $\gamma(x)$, then the same is true in $\gamma(\sigma(x))$. To see this, note that $a_0 \to a$ in $\gamma(x)$ gives that the string obtained from x by removing all symbols besides a_0 and a begins with some string of a_0's, the corresponding string for $\sigma(x)$ also begins with a (one symbol longer) string of a_0's and the weight between the two elements is unchanged. If, on the other hand, $a \overset{i}{\to} a_0$ in $\gamma(x)$, then we have $a_0 \overset{i+1}{\to} a$ in $\gamma(\sigma(x_0))$ — a_0 now occurs before a in the string corresponding to $\sigma(x_0)$ and the symbols switch one additional time. Further, as a_0 is the smallest element of g, we must have $a_0 \overset{j}{\to} a$ in g for some $j > i$ and, so, $i + 1 \leq j$ and $\gamma(\sigma(x)) \leq g$, as well. □

THEOREM 12.5. *The map γ gives rise to an equivalence of operads between $\mathsf{MS}_m(A)$ and $|\mathscr{G}_m(A)|$.*

PROOF. The space $\mathsf{MS}_m(A)$ is the totalisation of a geometric realisation of a multisimplicial set and each $|\mathscr{L}_g(A;n)|$ is the totalisation of the realisation of a multisimplicial subset, hence γ is sufficiently cofibrant and so this follows from Corollary 3.12 using the previous proposition. □

Appendix A. Some (co)simplicial generalities

For the precise procedure of turning a multisimplicial cosimplicial set into a topological space, we will recall some formalities about tensors and cotensors between (co)simplicial objects.

(1) If X is a simplicial set and Y a cosimplicial one, the set $X \otimes Y$ is the set of equivalence classes of pairs $(x,y) \in X_m \times Y_m$ where, for each morphism $\theta : [n] \to [m]$ in Δ, $x \in X_m$ and $y \in Y_n$, one identifies the pairs $(\theta^* x, y)$ and $(x, \theta_* y)$. We will write $x \otimes y$ for the equivalence class of (x,y) — this allows the equivalence relation to be expressed in the familiar form $x\theta \otimes y = x \otimes \theta y$ where we write the simplicial and cosimplicial structures as right and left actions. The set $X \otimes Y$ is a coend and also written as $\int^n X_n \times Y^n$, $\int^\Delta X_n \times Y^n \mathrm{d}n$, or $X \otimes_n Y$.

More generally, if X is a simplicial object and Y is a cosimplicial object in a cocomplete monoidal category, we can construct $X \otimes Y$ as the coequaliser

$$\coprod_{f\colon m \to n \in \mathrm{Mor}(\Delta)} X_n \otimes Y^m \rightrightarrows \coprod_{m \in \Delta} X_m \otimes Y^m \to X \otimes Y$$

where the first map is given by the simplicial action of X to map $X_n \otimes Y^m$ to $X_m \otimes Y^m$ and the second map is given by the cosimplicial action of Y to map $X_n \otimes Y^m$ to $X_n \otimes Y^n$.

(2) If X is a simplicial set and Y is a cosimplicial *space* then $X \otimes Y$ inherits its topology as a quotient of a sum of copies of the spaces Y^n. In particular, if Y is the cosimplicial space of standard simplices $|\Delta^\bullet|$, then $X \otimes Y = |X|$ is the geometric realisation of X. More generally, one can think of $X \otimes Y$ as a geometric realisation of X with respect to Y and write $|X|_Y$ for $X \otimes Y$ in this case.

(3) The tensor product is associative in the following sense: if X is a simplicial set, $Y$is a simplicial cosimplicial set, and Z is a cosimplicial set, then

$$(X \otimes Y) \otimes Z \cong X \otimes (Y \otimes Z)$$

with the isomorphism natural in all three arguments. More explicitly, this means

$$\int^p \left(\int^n X_n \times Y_p^n \right) \times Z^p \cong \int^n X_n \times \left(\int^p Y_p^n \times Z^p \right)$$

For example, if X is a simplicial set and Y is a cosimplicial simplicial set, then the geometric realisation $|Y|$ is a cosimplicial space and the realisation of X with respect to $|Y|$ is the same as the realisation of $X \otimes Y$, i.e.,

$$|X|_{|Y|} = |X \times Y|$$

as the left hand side is $X \otimes (Y \otimes |\Delta^\bullet|)$ and the right is $(X \otimes Y) \otimes |\Delta^\bullet|$.

(4) In the case of multi-simplicial and multi-cosimplicial objects, one can integrate over each variable, separately. This means that, if X is an A-fold simplicial set and Y is a B-fold simplicial cosimplicial set, then, for any $a \in A$, the tensor

$$X \otimes_a Y = \int^a X_a \times Y^a$$

is an $A[B/a]$-fold simplicial set.
(5) For example the substitution of lattice paths from Definition 11.10 is a map $\mathscr{L}(A;n) \otimes_a \mathscr{L}(B;\bullet) \to \mathscr{L}(A[B/a];n)$. The fact that this is well-defined is simply the Remark 11.11.

In fact, this map is even an isomorphism: as the substitution does not affect the part of the lattice path over A outside of a, we may assume that $A = \{a\}$, but in this case $\mathscr{L}(A;n) \cong \Delta(n)$ is the standard simplicial n-simplex and $\mathscr{L}(A[B/a];n) \cong \mathscr{L}(B;n)$ with the isomorphism induced by the substitution. Since $\Delta^\bullet$ is the unit for the tensor (by the Yoneda lemma), this just means that our map is given by $\mathscr{L}(A;n) \otimes_a \mathscr{L}(B;\bullet) = \Delta^n \otimes_a \mathscr{L}(B;\bullet) \cong \mathscr{L}(B;n) \xrightarrow{\sim} \mathscr{L}(A[B/a];n)$.
(6) If X is a cosimplicial space, then its totalisation $\mathrm{Tot}(X)$ (for which we introduced the alternative notation $-X-$) is the space of natural transformations from the standard cosimplicial space $|\Delta^\bullet|$ into X

$$\mathrm{Tot}(X) = \mathsf{Top}^\Delta(|\Delta^\bullet|, X^\bullet)$$

It has a natural topology as a subspace of the product of the spaces $(X^n)^{\Delta^n}$ of maps $\Delta^n \to X^n$.
(7) There is also a more general way to interpret the totalisation construction: if we are given two cosimplicial spaces X and Y, then the mapping space $(X^\bullet)^{Y^\bullet}$ is both simplicial and cosimplicial, allowing us to take the end over the category Δ to obtain a space $\int_n (X_n)^{Y^n}$ — this happens to be the space of natural transformations $X^\bullet \to Y^\bullet$. In particular, if we choose $Y = |\Delta^\bullet|$, this recovers the notion of the totalisation. Thus, we may think of totalising as a sort of dual construction to geometrically realising.

LEMMA A.1. *If X and Y are simplicial cosimplicial spaces, then there is a natural map*

$$\mathrm{Tot}(|X|) \times \mathrm{Tot}(|Y|) \to \mathrm{Tot}(|X \otimes Y|)$$

PROOF. First, note that, for a simplicial cosimplicial space X, the geometric realisation $|X^\bullet_\bullet|$ is again a cosimplicial space. On the other hand, $X \otimes Y = \left(\int^n X^p_n \times Y^n_m\right)_{p,m}$ is a cosimplicial simplicial space, so the space $\mathrm{Tot}(|X \otimes Y|)$ is, indeed, well-defined. Now, given natural transformations $\varphi : |\Delta^\bullet| \to |X|$ and $\psi : |\Delta^\bullet| \to |Y|$, the latter defines a natural transformation of cosimplicial spaces

$$\psi_* : |X^p| = X^p \otimes |\Delta^\bullet| \to X^p \otimes |Y|$$

but the realisation $X^p \otimes |Y|$ of X with respect to $|Y|$ is $|X^p \otimes Y|$, as seen previously, so, in fact, $\psi_* : |X| \to |X \otimes Y|$. Composing this map with φ gives a map $|\Delta^\bullet| \to |X \otimes Y|$. Thus, we obtain a natural map

$$\mathrm{Tot}(|X|) \times \mathrm{Tot}(|Y|) \to \mathrm{Tot}(|X \otimes Y|)$$
$$(\varphi, \psi) \mapsto \psi_* \circ \varphi$$

□

Appendix B. Some remarks on the literature

As mentioned in the introduction, the definition of the little n-cubes operads goes back to Boardman-Vogt [**BV73**] and May [**May72**], who also introduced the notion of an E_n-operad. For $n = \infty$, Barratt and Eccles introduced their operad in [**BE74**]; this operad is essentially our operad Γ given by the categories Lin. The suboperads Γ_n were introduced by Jeff Smith in [**Smi89**], and conjectured to be E_n-operads in this paper.

Kashiwabara [**Kas93**] was probably the first person to publish about this conjecture. He already used graphs, and wrote a directed graph on $A = \{1,\dots,p\}$ with weights 1, ..., n as a function $f : \mathrm{Ind}(p) \to \mathrm{J}(n)$ where $\mathrm{Ind}(p)$ is the set of two-element subsets of A and $\mathrm{J}(n)$ is the product of $\{+,-\}$ (for the directions in the graph) and $\{1,\dots,n\}$ (for the weights). He uses the terms coherent for what we (following Berger) call complete, and non-degenerate for what we call proper. Furthermore, he writes $C_n\Sigma_p$ for the simplicial set $\Gamma_n(A)$ (where $A = \{1,\dots,p\}$ again), and $\mathrm{C}(n,p)$ for what we call $\mathbb{P}_n(A)$ (the proper graphs), cf. [**Kas93**, Definition 3.7]. Using F. Cohen's description of the homology of configuration spaces [**Coh76**] Kashiwabara proceeds to prove a homology equivalence between (in our notation) the configuration space $\mathsf{Conf}_n(A)$ and the geometric realization $|\mathbb{P}_n(A)|$ of the proper graphs (cf. [**Kas93**, 3.10]), as well as a homotopy equivalence $|\mathbb{P}_n(A)| \xrightarrow{\sim} |\Gamma_n(A)|$ [**Kas93**, 6.1], and concludes in a somewhat indirect way that $|\Gamma_n(A)|$ is homotopy equivalent to $\mathsf{Conf}_n(A)$ [**Kas93**, 6.2].

The decomposition of Conf_n into proper cells ($\psi^{-1}(g)$ in our Definition 5.9) goes back to Fox and Neuwirth [**FoN62**] for $n = 2$, and has been discussed extensively in the literature. For example, Getzler and Jones [**GJ94**] defined the cells in terms of nested preorders $\pi_1 \leq \cdots \leq \pi_n$ on A, which are related to the proper graphs g by defining $a \leq b$ in π_i if and only if $a \xrightarrow{j} b$ in g for some $j \leq i$. For further discussion of Fox-Neuwirth cells and applications, we refer to Giusti-Sinha [**GS12**], Ayala-Hepworth [**AH14**], and the references cited there.

In a subsequent paper [**Ber97**], Berger introduces the complete graph operad $\mathscr{G}_n$, and shows that it is equivalent to Γ_n as an operad, as stated above Theorem 6.9. Another proof of this fact appears in [**BFSV03**]. Berger next argues that $\mathscr{C}_n$ is equivalent to $\mathscr{G}_n$ by proving an equivalence between Conf_n and $\mathbb{P}_n$ (like [**Kas93**] does) and by arguing that $\mathbb{P}_n$ is equivalent to $\mathscr{G}_n$ by claiming that the inclusion of $\mathbb{P}_n$ into $\mathscr{G}_n$ has an adjoint, after quotienting by the action of $\Sigma_\bullet$ — unfortunately, there was a slight mistake in the proof and such an adjoint cannot exist, as mentioned above. Moreover, Berger's construction really lands in the larger operad $\mathscr{G}_n^{\mathrm{ext}}$, as observed in Brun et.al. [**BFV07**], who claim without further explanation that one can prove that $\mathscr{C}_n$ is equivalent to $\mathscr{G}_n^{\mathrm{ext}}$ by the same arguments as used in [**Ber97**]. A slightly more detailed version of Berger's arguments can also be found in [**Ber96**].

We hope to have established in a more rigorous way that $\mathbb{P}_n$ is equivalent to $\mathscr{G}_n$ as symmetric sequence of posets, and that $\mathscr{G}_n$ and $\mathscr{G}_n^{\mathrm{ext}}$ are indeed equivalent to each other and to $\mathscr{C}_n$ as operads. Our argument that Γ_n is an E_n-operad could therefore be based on a map of operads from Γ_n to $\mathscr{G}_n$ as we have seen. Our proof that $\mathscr{M}_n$ is an E_n-operad largely follows [**BFSV03**], as already mentioned.

The fact that the Fulton-MacPherson operad FM_n is an E_n-operad was established by Salvatore, who constructed an equivalence of operads $W\mathscr{C}_n \to \mathsf{FM}_n$, where W denotes the Boardman-Vogt resolution. For us, having established that $\mathscr{G}_n$ is an E_n-operad via a map $\mathsf{Conf}_n \to \mathscr{G}_n$ (our Proposition 5.6) the fact that FM_n is also equivalent to $\mathscr{G}_n$ follows rather easily, as we have seen. Our proof that the McClure-Smith operad MS_n is an E_n-operad closely follows the same ideas as in [**MS04**], although we hope to have clarified and simplified a few points. The observation that elements in $\mathscr{L}(A;n)$ can be pictured as paths in an A-dimensional lattice with n "stops", i.e. as paths subdivided in $n+1$ segments, comes from [**BB09**]. They show that these lattice paths form a coloured operad from which the McClure-Smith operad can be constructed using an operation they call condensation.

Acknowledgments

This paper grew out of several activities on the topic: a seminar at Utrecht, the master's thesis by one of the authors under supervision of the other, and a crash course by the second author at Sheffield. We would like to thank Lennart Meier for useful comments during the seminar, the active audience at Sheffield, and Clemens Berger for helpful correspondence. Further, we would like to thank Vladimir Dotsenko for providing helpful feedback on the preprint of this paper. Finally, we would like to express our thanks to Paolo Salvatore for helping us improve the exposition, and in particular for helping us to correct the formulation of one of the statements in an earlier version of the paper.

References

[AH14] David Ayala and Richard Hepworth, *Configuration spaces and* Θ_n, Proc. Amer. Math. Soc. **142** (2014), no. 7, 2243–2254, DOI 10.1090/S0002-9939-2014-11946-0. MR3195750

[BB09] M. A. Batanin and C. Berger, *The lattice path operad and Hochschild cochains*, Alpine perspectives on algebraic topology, Contemp. Math., vol. 504, Amer. Math. Soc., Providence, RI, 2009, pp. 23–52, DOI 10.1090/conm/504/09874. MR2581904

[BE74] M. G. Barratt and Peter J. Eccles, Γ^+*-structures. I. A free group functor for stable homotopy theory*, Topology **13** (1974), 25–45, DOI 10.1016/0040-9383(74)90036-6. MR348737

[Ber96] Clemens Berger, *Opérades cellulaires et espaces de lacets itérés* (French, with English and French summaries), Ann. Inst. Fourier (Grenoble) **46** (1996), no. 4, 1125–1157, DOI 10.5802/aif.1543. MR1415960

[Ber97] Clemens Berger, *Combinatorial models for real configuration spaces and* E_n*-operads*, Operads: Proceedings of Renaissance Conferences (Hartford, CT/Luminy, 1995), Contemp. Math., vol. 202, Amer. Math. Soc., Providence, RI, 1997, pp. 37–52, DOI 10.1090/conm/202/02582. MR1436916

[BFSV03] C. Balteanu, Z. Fiedorowicz, R. Schwänzl, and R. Vogt, *Iterated monoidal categories*, Adv. Math. **176** (2003), no. 2, 277–349, DOI 10.1016/S0001-8708(03)00065-3. MR1982884

[BFV07] Morten Brun, Zbigniew Fiedorowicz, and Rainer M. Vogt, *On the multiplicative structure of topological Hochschild homology*, Algebr. Geom. Topol. **7** (2007), 1633–1650, DOI 10.2140/agt.2007.7.1633. MR2366174

[BV73] J. M. Boardman and R. M. Vogt, *Homotopy invariant algebraic structures on topological spaces*, Lecture Notes in Mathematics, Vol. 347, Springer-Verlag, Berlin-New York, 1973. MR420609

[Coh76] F. Cohen, *The homology of SF(n+1)*, pp. 352–398, Springer Berlin Heidelberg, Berlin, Heidelberg, 1976.

[DI04] Daniel Dugger and Daniel C. Isaksen, *Topological hypercovers and* $\mathbb{A}^1$*-realizations*, Math. Z. **246** (2004), no. 4, 667–689, DOI 10.1007/s00209-003-0607-y. MR2045835

[Dun88] Gerald Dunn, *Tensor product of operads and iterated loop spaces*, J. Pure Appl. Algebra **50** (1988), no. 3, 237–258, DOI 10.1016/0022-4049(88)90103-X. MR938617

[FN62] Edward Fadell and Lee Neuwirth, *Configuration spaces*, Math. Scand. **10** (1962), 111–118, DOI 10.7146/math.scand.a-10517. MR141126

[FoN62] R. Fox and L. Neuwirth, *The braid groups*, Math. Scand. **10** (1962), 119–126, DOI 10.7146/math.scand.a-10518. MR150755

[Get94] E. Getzler, *Batalin-Vilkovisky algebras and two-dimensional topological field theories*, Comm. Math. Phys. **159** (1994), no. 2, 265–285. MR1256989

[GJ94] E. Getzler and J.D.S. Jones, *Operads, homotopy algebra and iterated integrals for double loop spaces*, 1994.

[GK95] E. Getzler and M. M. Kapranov, *Cyclic operads and cyclic homology*, Geometry, topology, & physics, Conf. Proc. Lecture Notes Geom. Topology, IV, Int. Press, Cambridge, MA, 1995, pp. 167–201. MR1358617

[GK98] E. Getzler and M. M. Kapranov, *Modular operads*, Compositio Math. **110** (1998), no. 1, 65–126, DOI 10.1023/A:1000245600345. MR1601666

[GS12] Chad Giusti and Dev Sinha, *Fox-Neuwirth cell structures and the cohomology of symmetric groups*, Configuration spaces, CRM Series, vol. 14, Ed. Norm., Pisa, 2012, pp. 273–298, DOI 10.1007/978-88-7642-431-1_12. MR3203643

[HM22] Gijs Heuts and Ieke Moerdijk, *Simplicial and dendroidal homotopy theory*, Ergebnisse der Mathematik und ihrer Grenzgebiete. 3. Folge. A Series of Modern Surveys in Mathematics [Results in Mathematics and Related Areas. 3rd Series. A Series of Modern Surveys in Mathematics], vol. 75, Springer, Cham, [2022] ©2022, DOI 10.1007/978-3-031-10447-3. MR4485749

[Jar87] J. F. Jardine, *Simplicial presheaves*, J. Pure Appl. Algebra **47** (1987), no. 1, 35–87, DOI 10.1016/0022-4049(87)90100-9. MR906403

[Kas93] Takuji Kashiwabara, *On the homotopy type of configuration complexes*, Algebraic topology (Oaxtepec, 1991), Contemp. Math., vol. 146, Amer. Math. Soc., Providence, RI, 1993, pp. 159–170, DOI 10.1090/conm/146/01221. MR1224913

[Kon99] Maxim Kontsevich, *Operads and motives in deformation quantization*, Lett. Math. Phys. **48** (1999), no. 1, 35–72, DOI 10.1023/A:1007555725247. Moshé Flato (1937–1998). MR1718044

[May72] J. P. May, *The geometry of iterated loop spaces*, Lecture Notes in Mathematics, Vol. 271, Springer-Verlag, Berlin-New York, 1972. MR420610

[MS04] James E. McClure and Jeffrey H. Smith, *Cosimplicial objects and little n-cubes. I*, Amer. J. Math. **126** (2004), no. 5, 1109–1153. MR2089084

[MSS02] Martin Markl, Steve Shnider, and Jim Stasheff, *Operads in algebra, topology and physics*, Mathematical Surveys and Monographs, vol. 96, American Mathematical Society, Providence, RI, 2002, DOI 10.1090/surv/096. MR1898414

[Qui73] Daniel Quillen, *Higher algebraic K-theory. I*, Algebraic K-theory, I: Higher K-theories (Proc. Conf., Battelle Memorial Inst., Seattle, Wash., 1972), Lecture Notes in Math., vol. Vol. 341, Springer, Berlin-New York, 1973, pp. 85–147. MR0338129

[Sal01] Paolo Salvatore, *Configuration spaces with summable labels*, Cohomological methods in homotopy theory (Bellaterra, 1998), Progr. Math., vol. 196, Birkhäuser, Basel, 2001, pp. 375–395. MR1851264

[Sch25] C. Schlichtkrull, *Combinatorial and homotopical aspects of E_n-operads*, arXiv preprint, `arXiv:2408.17236`, 2025.

[Smi89] Jeffrey Henderson Smith, *Simplicial group models for $\Omega^n S^n X$*, Israel J. Math. **66** (1989), no. 1-3, 330–350, DOI 10.1007/BF02765902. MR1017171

[Sta63] James Dillon Stasheff, *Homotopy associativity of H-spaces. I, II*, Trans. Amer. Math. Soc. **108** (1963), 293–312, DOI 10.1090/s0002-9947-1963-0158400-5. **108** (1963), 275-292; ibid. MR158400

[Tam98] D. E. Tamarkin, *Formality of chain operad of small squares*, 1998.

Department of Mathematics, Universität Hamburg, Mittelweg 177, 20148 Hamburg, Germany

Email address: `andre.beuckelmann@uni-hamburg.de`

Department of Mathermatics, Universiteit Utrecht, 3584 CS Utrecht, The Netherlands

Email address: `i.moerdijk@uu.nl`

Contemporary Mathematics
Volume **841**, 2026
https://doi.org/10.1090/conm/841/16834

Characteristic forms for holomorphic families of local systems

Indranil Biswas and Eduard Looijenga

To Ezra, on the occasion of his 60th birthday.

Abstract. Let Γ be a finitely generated group and G a complex Lie group with Lie algebra $\mathfrak{g}$. The complex variety of group homomorphisms $\mathrm{Hom}(\Gamma, G)$ is equipped with an action of G by inner automorphism and thus it defines a complex-analytic stack. We associate to each $\Phi \in (\mathrm{Sym}^n \mathfrak{g})^G$ a closed holomorphic n-form on this stack with values in $H^n(\Gamma, \mathbb{C})$. This construction is multiplicative so that we get a graded algebra homomorphism. We obtain this as a universal refinement of a Chern-Weil homomorphism.

If we take $n = 2$ and let Γ be the fundamental group of a closed orientable surface, then we recover a classical construction due to Goldman.

Introduction

Let Γ be the fundamental group of closed connected orientable surface of genus > 1, G a Lie group and σ a nondegenerate bilinear form on its Lie algebra $\mathfrak{g}$ that is invariant under the adjoint representation. A well-known theorem of Goldman [**Go**] asserts that σ determines on the space of group homomorphisms $\Gamma \longrightarrow G$ given up to G-conjugacy (understood in the sense of geometric invariant theory) a symplectic form. One of the main goals of this paper is to show that this is a special case of what one might call a *universal relative Chern-Weil homomorphism.*

In more detail, let us first recall the Chern-Weil recipe for extracting topological invariants from a principal G-bundle $\mathcal{P} \longrightarrow M$ over manifold a M. One begins with choosing a connection ∇ on the bundle $\mathcal{P}$ over M. Its curvature form $R(\nabla)$ is then a 2-form on M that takes values in an associated bundle of Lie algebras $\mathrm{ad}(\mathcal{P}) = \mathcal{P} \times^G \mathfrak{g}$. A fiber of $\mathrm{ad}(\mathcal{P})$ is isomorphic to the Lie algebra $\mathfrak{g}$ of G with the isomorphism given up to the adjoint action of G on $\mathfrak{g}$ and so it makes sense to evaluate on $R(\nabla)$ a homogeneous polynomial function on $\mathfrak{g}$ that is invariant under this adjoint action of G, the result then being a differential form on M of degree twice that of the polynomial. The vector bundle $\mathrm{ad}(\mathcal{P})$ inherits a connection from $\mathcal{P}$ and the Bianchi identity, which says that $R(\nabla)$ is closed with respect to the inherited connection, implies that such a differential form is closed. A simple homotopy argument shows that its de Rham cohomology class is independent of

2020 *Mathematics Subject Classification.* Primary 14J42, 53C07, 53B10, 57R20.

Key words and phrases. Holomorphic connection, characteristic form, projective structure, symplectic form.

©2026 American Mathematical Society

the choice of ∇ and we thus obtain the Chern-Weil homomorphism $\mathbb{C}[\mathfrak{g}]^G \longrightarrow H^{\text{even}}(M; \mathbb{C})$.

The relative version we alluded to begins (in the complex-analytic setting, say) with a proper holomorphic submersion $f : \mathcal{X} \longrightarrow S$ of complex manifolds and takes for G a complex linear algebraic group (so with a complex Lie algebra $\mathfrak{g}$), a holomorphic principal G-bundle $\mathcal{P}$ over $\mathcal{X}$ and a holomorphic connection ∇ on $\mathcal{P}$ *relative to* f (it is given by a holomorphic splitting of the relative Atiyah exact sequence, which means that only covariant derivation along the fibers of f is defined). We assume that ∇ is flat and so this can be thought of as a holomorphically varying family of compact complex manifolds endowed with a holomorphic principal G-bundle with flat connection. We show that a refinement of the Chern-Weil homomorphism yields a graded algebra homomorphism

$$\Psi_{(f,\mathcal{P},\nabla)} : \mathbb{C}[\mathfrak{g}]^G \longrightarrow \bigoplus_{n\geq 0} H^0(S, \Omega^n_{S,cl} \otimes R^n f_* \mathbb{C}),$$

where $\Omega^n_{S,cl}$ stands for the sheaf of closed holomorphic n-forms on S.

Our generalization of Goldman's construction is in a sense a universal homomorphism of this type. It is stated in terms of the *character space* that is associated with a finitely generated group Γ (that takes the role of the fundamental group of a fiber of f) and a complex Lie group G as above: this is the complex-analytic space of group homomorphisms $\operatorname{Hom}(\Gamma, G)$. It comes with an evident action of $\operatorname{Aut}(\Gamma) \times \operatorname{Aut}(G)$.The following theorem (stated here only in the holomorphic setting) is, we believe, the most natural generalization of the main result of [**Go**].

THEOREM 0.1. *There is a natural graded* $\mathbb{C}$*-algebra homomorphism* $\Psi_{\Gamma,G}$ *from* $\mathbb{C}[\mathfrak{g}]^G$ *to the algebra of* $\operatorname{Aut}(\Gamma)$*-invariant sections of* $\bigoplus_{n\geq 0} \Omega^n_{\operatorname{Hom}(\Gamma,G),cl} \otimes H^n(\Gamma, \mathbb{C})$ *that are killed by contraction with the vector fields defined by* $\mathfrak{g}$. *It is universal as a (refined) Chern-Weil homomorphism in the sense that is explained below.*

Let us first comment on the equivariance properties of $\Psi_{\Gamma,G}$. Naturality of $\Psi_{\Gamma,G}$ is here understood as contravariance in G and covariance in Γ. Since $\operatorname{Aut}(\Gamma)$ and the inner automorphisms of G act trivially on $\mathbb{C}[\mathfrak{g}]^G$, this implies that $\Psi_{\Gamma,G}$ takes values in forms that are $\operatorname{Aut}(\Gamma) \times G$-invariant (the $\Gamma \times G$-invariance also easily seen directly). So we may just as well say that $\Psi_{\Gamma,G}$ is $\operatorname{Out}(G)$-equivariant and takes values in $\operatorname{Out}(\Gamma)$-invariant forms.

In order to explain its universal nature, it is best to reformulate the first assertion in terms of the *character stack* $\underline{\chi}_G(\Gamma)$ defined by the G-action on $\operatorname{Hom}(\Gamma, G)$. A *regular form* on this stack $\underline{\chi}_G(\Gamma)$ is by definition a regular form on the space of group homomorphisms $\operatorname{Hom}(\Gamma, G)$ (which comes with the structure of a complex-analytic space) that is *basic* with respect to the G-action. This means that both the form and its exterior derivative are killed by $\mathfrak{g}$. So in the above theorem we can replace $\Omega^n_{\operatorname{Hom}(\Gamma,G)}$ by $\Omega^n_{\underline{\chi}_G(\Gamma)}$ and regard $\Psi_{\Gamma,G}$ as a graded, $\operatorname{Out}(G)$-equivariant algebra homomorphism

$$\underline{\Psi}_{\Gamma,G} : \mathbb{C}[\mathfrak{g}]^G \longrightarrow \Big(\bigoplus_{n\geq 0} H^0(\underline{\chi}_G(\Gamma), \Omega^n_{\underline{\chi}_G(\Gamma),cl}) \otimes H^n(\Gamma, \mathbb{C})\Big)^{\operatorname{Out}(\Gamma)}.$$

Now suppose that we are given a triple $(f : \mathcal{X} \to S, \mathcal{P}, \nabla)$ as above with connected fibers. Since our interest is focused on a property that is local on S, we shall assume that S is simply connected and f admits a section $x : S \longrightarrow \mathcal{X}$. This implies that for each $s \in S$, the inclusion map $X_s \subset \mathcal{X}$ induces an isomorphism on fundamental groups. Let us write Γ for $\pi_1(\mathcal{X}, x)$, where we think of x as a (relative) base point. We then have a natural homotopy class of maps κ from $\mathcal{X}$ to a classifying space $B\Gamma$ of Γ. The associated map

$$\kappa^* : H^\bullet(\Gamma, \mathbb{C}) \cong H^\bullet(B\Gamma; \mathbb{C}) \longrightarrow H^\bullet(\mathcal{X}, \mathbb{C}) \longrightarrow H^0(S, R^n f_*\mathbb{C})$$

is independent of x. Each fiber X_s carries a flat principal G-bundle. So if we choose a lift of x to $\mathcal{P}$, then the monodromy representations define a morphism $\rho : S \longrightarrow \mathrm{Hom}(\Gamma, G)$. But if we pass to the stack, then the resulting morphism $\underline{\rho} : S \longrightarrow \underline{\chi_G}(\Gamma)$ no longer depends on x and its lift to $\mathcal{P}$. The last sentence of Theorem 0.1 is meant to say that $\Psi_{(f,\mathcal{P},\nabla)}$ is equal to the composite

$$\mathbb{C}[\mathfrak{g}]^G \xrightarrow{\Psi_{\Gamma,G}} \bigoplus_{n\geq 0} H^0(S, \Omega^n_{\underline{\chi}_G(\Gamma),cl}) \otimes H^n(\Gamma, \mathbb{C}) \xrightarrow{\underline{\rho}^*\otimes\kappa^*} \bigoplus_{n\geq 0} H^0(S, \Omega^n_{S,cl} \otimes R^n f_*\mathbb{C}).$$

This makes it apparent that the cohomology of the fibers only involves classes coming from their fundamental group (as is to be expected).

Goldman is concerned with G-invariant quadratic forms B on $\mathfrak{g}$ (so $n = 2$) and takes for Γ the fundamental group of a closed orientable genus g surface. There is no group cohomology in his formulation, as implicit in his discussion is an identification $H^2(\Gamma) \cong \mathbb{Z}$. It should however not be left out if one demands (as he does) $\mathrm{Out}(\Gamma)$-invariance. Indeed, an orientation reversing automorphism of the closed surface induces minus the identity in $H^2(\Gamma)$. Our proof, when restricted to this special case seems different from his. We should however add that Goldman also shows that if B is nondegenerate, then so is the associated closed 2-form on $\underline{\chi_G}(\Gamma)$, so that the latter becomes what one might call a *holomorphically symplectic stack* (he works in the differentiable category and avoids stacks, however this difference is merely formal).

We note here that Guruprasad, Huebschmann, Jeffrey, and Weinstein [**GHJW**] obtained an extension of Goldman's theorem to a situation where the fibers of f are compact (oriented) surfaces with boundary (but where the G-invariant is still a quadratic form on its Lie algebra) and where one might want to fix the holonomy on each boundary component. Here the universal situation not only involves the fundamental group of the surface, but also the fundamental groups of the boundary components and the way these groups are related in a groupoid fashion. In our set-up, this means that instead of requiring f to be proper, we only demand that it be topologically locally trivial. It is likely that our approach lends itself to an extension which incorporates that case (we note that the formula (1.5) only requires f to be smooth) but we have not pursued this.

There is presumably also a link with work of Bloch and Esnault. Given a family of connections with certain properties, there are secondary cohomology invariants associated with it which are known as Cheeger–Chern–Simons classes [**CherS**], [**CheeS**]. In [**BE**], refinements of these are defined in an algebro-geometric set-up (see [**BE**, Theorem 3.7] and [**BE**, Definition 3.10]). Their constructions use the Atiyah class (which is also the key ingredient here), but their invariants take values in a different complex of forms: their focus is on vector bundles rather than on principal bundles (in other words, G is then a general linear group) and their invariants live on the associated projectivized bundle. It is at present not clear to us whether in this case the closed (group cohomology valued) forms that we find can be extracted from theirs.

Goldman's theorem was tailored to be applied to the moduli space of projective structures on a fixed closed connected orientable surface Σ of genus $g \geq 2$ (these are projective structures given up to a diffeomorphism of Σ isotopic to the identity map) and indeed, this is what he did in the same paper. He thus obtained on that moduli space a natural holomorphic symplectic form. A projective structure on Σ determines a complex-analytic structure and hence this moduli space maps to the Teichmüller space of Σ in a manner that is equivariant for the action of the mapping class group of Σ. It is however not clear a priori that this can be done is a purely algebraic setting, where we work with schemes (or Deligne-Mumford stacks) of finite type over an algebraically closed field of characteristic zero. At the outset one needs an algebraic definition of a projective structure. Indeed, to do this as well as to show that it leads to symplectic form was our original motivation. In a subsequent paper we will show that our set-up allows us to accomplish this in a straightforward manner.

We happily dedicate this paper to Ezra Getzler on the occasion of his 60th birthday.

1. Relative versions of the Chern-Weil homomorphism

1.1. Chern-Weil homomorphism for a relative connection. We fix a complex Lie group G. Let $f : \mathcal{X} \longrightarrow S$ be a holomorphic submersion of complex manifolds, and let $\pi : \mathcal{P} \longrightarrow \mathcal{X}$ be a holomorphic principal G-bundle. We assume that $\mathcal{P}$ is endowed with a *holomorphic connection* ∇ *relative to* f. Let us first spell out what this exactly means. First take the direct image under π of the short exact sequence

$$0 \longrightarrow \theta_{\mathcal{P}/\mathcal{X}} \longrightarrow \theta_{\mathcal{P}} \longrightarrow \pi^*\theta_{\mathcal{X}} \longrightarrow 0$$

of the sheaf of holomorphic vector fields; it is still exact and that remains so if we subsequently pass to G-invariants. So letting $\mathrm{ad}(\mathcal{P})$ stand for the $\mathcal{O}_{\mathcal{X}}$-module of Lie algebras $\pi_*\theta^G_{\mathcal{P}/\mathcal{X}}$, we have a short exact sequence on $\mathcal{X}$

$$0 \longrightarrow \mathrm{ad}(\mathcal{P}) \longrightarrow \pi_*(\theta_{\mathcal{P}})^G \longrightarrow \theta_{\mathcal{X}} \longrightarrow 0. \tag{1.1}$$

A *holomorphic connection* ∇ *on* $\mathcal{P}$ *relative to* f is by definition a holomorphic splitting of this sequence over the subsheaf $\theta_{\mathcal{X}/S} \subset \theta_{\mathcal{X}}$. One might picture the triple $(\mathcal{X}/S, \mathcal{P}, \nabla)$ as a holomorphic family of compact complex manifolds, each of which is endowed with a holomorphic principal G-bundle together with a holomorphic connection on it, that is parametrized by S. Given these data, consider the commutative diagram of short exact sequences, one of them coming with a section

(1.2)
$$\begin{array}{ccccccccc}
 & & & & 0 & & 0 & & \\
 & & & & \downarrow & \overset{\nabla}{\swarrow\!\!\frown} & \downarrow & & \\
0 & \longrightarrow & \mathrm{ad}(\mathcal{P}) & \longrightarrow & \pi_*\theta_{\mathcal{P}/S} & \longrightarrow & \theta_{\mathcal{X}/S} & \longrightarrow & 0 \\
 & & \| & & \downarrow & & \downarrow & & \\
0 & \longrightarrow & \mathrm{ad}(\mathcal{P}) & \longrightarrow & \pi_*\theta_{\mathcal{P}} & \longrightarrow & \theta_{\mathcal{X}} & \longrightarrow & 0 \\
 & & & & \downarrow & & \downarrow & & \\
 & & & & f^*\theta_S & = & f^*\theta_S & & \\
 & & & & \downarrow & & \downarrow & & \\
 & & & & 0 & & 0 & &
\end{array}$$

The relative holomorphic connection ∇ lifts locally on $\mathcal{X}$ to an absolute holomorphic connection (i.e., lifts to local holomorphic homomorphisms from $\theta_{\mathcal{X}}$ to $\pi_*\theta_{\mathcal{P}}$). The obstruction against being able to do this globally defines the Atiyah class

(1.3) $$\mathfrak{at}_{\mathcal{X}}(\mathcal{P}) \in H^1(\mathcal{X}, \mathcal{H}om(\theta_{\mathcal{X}}, \mathrm{ad}(\mathcal{P})) \cong H^1(\mathcal{X}, \Omega_{\mathcal{X}} \otimes \mathrm{ad}(\mathcal{P})).$$

It is by definition the extension class defined by the second row in (1.2). If we work relative to S, then this evidently produces an element $\mathfrak{at}_{\mathcal{X}/S}(\mathcal{P}) \in R^1 f_*(\Omega_{\mathcal{X}} \otimes \mathrm{ad}(\mathcal{P}))$. But then the given relative connection ∇ lifts this to

(1.4)
$$\mathfrak{at}^{\nabla}_{\mathcal{X}/S}(\mathcal{P}) \in H^0(S, R^1 f_*(f^{-1}\Omega_S \otimes_{f^{-1}\mathcal{O}_S} \mathrm{ad}(\mathcal{P}))) \cong H^0(S, \Omega_S \otimes R^1 f_* \,\mathrm{ad}(\mathcal{P})).$$

Indeed, the cokernel of $\theta_{\mathcal{X}/S} \xrightarrow{\nabla} \pi_*\theta_{\mathcal{P}/S} \subset \pi_*\theta_{\mathcal{P}}$ is an extension of $f^*\theta_S$ by $\mathrm{ad}(\mathcal{P})$ and $\mathfrak{at}^{\nabla}_{\mathcal{X}/S}(\mathcal{P})$ in (1.4) is its extension class. This *relative Atiyah class* in (1.4) can be explicated as follows. Write $\mathcal{E}^{p,q}_{\mathcal{X}}$ for the (flabby) sheaf of C^∞-forms on $\mathcal{X}$ of type (p, q). The relative connection ∇ lifts locally on $\mathcal{X}$ to an absolute one and a partition of unity then allows us do this globally with C^∞ coefficients so that we obtain a C^∞ connection

$$\widetilde{\nabla} : \theta_{\mathcal{X}} \longrightarrow \pi_*\theta_{\mathcal{P}}.$$

Its curvature form $R(\widetilde{\nabla})$ will be a global section of

$$(\mathcal{E}^{2,0}_{\mathcal{X}} \oplus \mathcal{E}^{1,1}_{\mathcal{X}}) \otimes \mathrm{ad}(\mathcal{P}) = (\mathcal{E}^{1,0}_{\mathcal{X}} \oplus \mathcal{E}^{0,1}_{\mathcal{X}}) \wedge \Omega^1_{\mathcal{X}} \otimes \mathrm{ad}(\mathcal{P}).$$

Its restriction to every fiber of f is holomorphic and hence the component of type $(1, 1)$ must be a section of the subsheaf $\mathcal{E}^{0,1}_{\mathcal{X}} \wedge f^*\Omega^1_S \otimes \mathrm{ad}(\mathcal{P}) \subset \mathcal{E}^{0,1}_{\mathcal{X}} \wedge \Omega^1_{\mathcal{X}} \otimes \mathrm{ad}(\mathcal{P})$. The Bianchi identity says that $R(\widetilde{\nabla})$ is $\widetilde{\nabla}$-closed and this implies (see for example [**MS**]) that the above component of type $(1, 1)$ must be $\overline{\partial}$-closed. It thus defines a section of

$$R^1 f_*(f^*\Omega^1_S \otimes \mathrm{ad}(\mathcal{P})) \cong \Omega_S \otimes R^1 f_*(\mathrm{ad}(\mathcal{P}))$$

and this is indeed our $\mathfrak{at}^{\nabla}_{\mathcal{X}/S}(\mathcal{P})$.

Associated Chern-Weil homomorphism. Denote by $\mathfrak{g}$ the Lie algebra of G, and let

$$\Phi \in \mathbb{C}[\mathfrak{g}]^G$$

be a G-invariant homogeneous polynomial of degree n. We can evaluate Φ on $\mathfrak{at}_{\mathcal{X}}(\mathcal{P})$ in (1.3) and with the help of the cup product obtain an element of $H^n(\mathcal{X}, \Omega^n_{\mathcal{X}})$. When $\mathcal{X}$ is compact Kähler, we have $H^n(\mathcal{X}, \Omega^n_{\mathcal{X}}) \cong H^{n,n}(\mathcal{X})$ and according to Atiyah [**At**] this reproduces the cohomology class given by the image of Φ under the Chern-Weil homomorphism associated to $\mathcal{P}$.

The given relative connection ∇ allows us to construct the following refinement of the above construction: The evaluation of Φ on $\mathfrak{at}^{\nabla}_{\mathcal{X}/S}(\mathcal{P})$ yields

$$\Phi(\mathfrak{at}^{\nabla}_{\mathcal{X}/S}(\mathcal{P})) \in H^0(\Omega^n_S \otimes R^n f_* \mathcal{O}_{\mathcal{X}}).$$

We thus obtain a homomorphism of graded algebras

$$\mathbb{C}[\mathfrak{g}]^G \longrightarrow \bigoplus_{n\geq 0} H^0(S,\, \Omega^n_S \otimes R^n f_* \mathcal{O}_{\mathcal{X}}). \tag{1.5}$$

If we assume that f is proper, then by the Grauert direct image theorem [**Gr**], each $R^n f_* \mathcal{O}_{\mathcal{X}}$ is a coherent $\mathcal{O}_S$-module. The relative Chern-Weil homomorphism then commutes with base change: if $g : S' \longrightarrow S$ is a holomorphic map of complex manifolds and

$$\widetilde{g} : \mathcal{X}' := \mathcal{X} \times_S S' \longrightarrow \mathcal{X}$$

is the pulled back family, so that we have the pull-back $(\widetilde{g}^*\mathcal{P},\, \widetilde{g}^*\nabla)$ of $(\mathcal{P},\, \nabla)$ over $\mathcal{X}'/S'$, then

$$\Phi(\mathfrak{at}^{\widetilde{g}^*\nabla}_{\mathcal{X}'/S'}) = g^*\Phi(\mathfrak{at}^{\nabla}_{g^*\mathcal{X}/S'}).$$

If $\mathcal{X}$ is even compact and admits a Kähler metric, then by a classical result of Deligne [**De1**] the Leray spectral sequence for f applied to the constant sheaf $\underline{\mathbb{C}}_{\mathcal{X}}$ degenerates. If we combine this with the Hodge decomposition theorem we find that $H^0(S, \Omega^n_S \otimes R^n f_* \mathcal{O}_{\mathcal{X}})$ naturally sits inside the n-th Leray subquotient of $H^{2n}(\mathcal{X}; \underline{\mathbb{C}})$ and that this is also the image of the Chern-Weil class of $\mathcal{P}$ in $H^{2n}(\mathcal{X}; \underline{\mathbb{C}})$ defined by Φ.

1.2. Chern-Weil homomorphism for a flat relative connection. *We here assume that f is proper and that ∇ is* flat *as a relative holomorphic connection.* By the last property we mean that the associated section of $\pi_*\theta_{\mathcal{P}/S} \longrightarrow \theta_{\mathcal{X}/S}$ preserves the Lie bracket of vector fields. This is equivalent to the curvature form $R(\nabla)$ (which is here a section of $\Omega^2_{\mathcal{X}/S} \otimes \mathrm{ad}(\mathcal{P})$) being zero. The homomorphism $\mathrm{ad}(\mathcal{P}) \longrightarrow \Omega_{\mathcal{X}/S} \otimes \mathrm{ad}(\mathcal{P})$ given by ∇ will, for notational convenience, again be denoted by ∇. The kernel of the map $\nabla : \mathrm{ad}(\mathcal{P}) \longrightarrow \Omega_{\mathcal{X}/S} \otimes \mathrm{ad}(\mathcal{P})$ is then a subsheaf

$$\mathrm{ad}(\mathcal{P})^{\nabla} \subset \mathrm{ad}(\mathcal{P})$$

of locally free $f^{-1}\mathcal{O}_S$-modules with the property that

$$\mathcal{O}_{\mathcal{X}} \otimes_{f^{-1}\mathcal{O}_S} \mathrm{ad}(\mathcal{P})^{\nabla} \longrightarrow \mathrm{ad}(\mathcal{P})$$

is an isomorphism of $\mathcal{O}_{\mathcal{X}}$-modules. One might call this then a *local system relative to* f and regard $(\mathcal{X}/S,\, \mathcal{P},\, \nabla)$ as a family of compact complex manifolds, each of which is endowed with a holomorphic principal G-bundle and a holomorphic flat connection on it. The higher direct images $R^n f_*\, \mathrm{ad}(\mathcal{P})^{\nabla}$ can be computed with a relative de Rham resolution. Since f is proper, it then follows that these are in fact coherent $\mathcal{O}_S$-modules.

LEMMA 1.1. *In the above set-up the class $\mathfrak{at}^{\nabla}_{\mathcal{X}/S}(\mathcal{P}) \in H^0(S,\, \Omega_S \otimes R^1 f_*\, \mathrm{ad}(\mathcal{P}))$ has as a natural lift*

$$\mathcal{A}t^{\nabla}_{\mathcal{X}/S}(\mathcal{P}) \in H^0(S,\, \Omega_S \otimes R^1 f_*\, \mathrm{ad}(\mathcal{P})^{\nabla}).$$

Moreover, if $\widetilde{\nabla}$ is a C^∞-lift of ∇ to an absolute connection on $\mathrm{ad}(\mathcal{P})$*, then $\widetilde{\nabla}$ induces a connection on $f_*\,\mathrm{ad}(\mathcal{P})^\nabla$ as well as on the higher direct images, and $\mathcal{A}t^\nabla_{\mathcal{X}/S}(\mathcal{P})$ is closed for the resulting connection on $R^1 f_*\,\mathrm{ad}(\mathcal{P})^\nabla$.*

PROOF. If we go through the preceding procedure, we note that $R(\widetilde{\nabla})$ has now the additional property that its restriction to every fiber of f is zero. This means that $R(\widetilde{\nabla})$ is a section of

$$\mathcal{E}^1_{\mathcal{X}} \wedge f^*\Omega_S \otimes \mathrm{ad}(\mathcal{P}) = \mathcal{E}^1_{\mathcal{X}} \wedge_{f^{-1}\mathcal{O}_S} \left(f^{-1}\Omega_S \otimes_{f^{-1}\mathcal{O}_S} \mathrm{ad}(\mathcal{P})^\nabla\right), \tag{1.6}$$

where $\mathcal{E}^1_{\mathcal{X}} = \mathcal{E}^{1,0}_{\mathcal{X}} \oplus \mathcal{E}^{0,1}_{\mathcal{X}}$. The Bianchi identity says that $R(\widetilde{\nabla})$ is $\widetilde{\nabla}$-closed. Since on the right hand side of (1.6) we have already passed to the relative local system $\mathrm{ad}(\mathcal{P})^\nabla$, this implies that its image in $\mathcal{E}^1_{\mathcal{X}/S}\otimes_{f^{-1}\mathcal{O}_S}(f^{-1}\Omega_S\otimes_{f^{-1}\mathcal{O}_S}\mathrm{ad}(\mathcal{P})^\nabla)$ is $d_{\mathcal{X}/S}$-closed. Hence we now obtain a section $\mathcal{A}t^\nabla_{\mathcal{X}/S}(\mathcal{P})$ of $R^1 f_*(f^{-1}\Omega_S\otimes_{f^{-1}\mathcal{O}_S}\mathrm{ad}(\mathcal{P})^\nabla) = \Omega_S \otimes R^1 f_*\,\mathrm{ad}(\mathcal{P})^\nabla$.

The direct image $R^1 f_*\,\mathrm{ad}(\mathcal{P})^\nabla$ inherits from $\widetilde{\nabla}$ a connection: the covariant derivative with respect to a local C^∞ vector field on S is computed by choosing a lift of that vector field to $\mathcal{X}$ and then taking the covariant derivative with respect to it for the connection $\widetilde{\nabla}$. The induced action on the higher direct images of $\mathrm{ad}(\mathcal{P})$ is independent of the lift of the vector field because two such lifts differ by a vector field along the fibers of f and covariant derivation with respect to such a vector field acts trivially on $R^\bullet f_*\,\mathrm{ad}(\mathcal{P})^\nabla$. Since $\mathcal{A}t^\nabla_{\mathcal{X}/S}(\mathcal{P})$ represents the curvature form, the Bianchi identity implies that $\mathcal{A}t^\nabla_{\mathcal{X}/S}(\mathcal{P})$ is closed relative to the above connection on $R^1 f_*\,\mathrm{ad}(\mathcal{P})^\nabla$. □

We can understand $\mathcal{A}t^\nabla_{\mathcal{X}/S}(\mathcal{P})$ as a *Kodaira-Spencer class*: if we regard $(\mathcal{X}/S, \mathcal{P}, \nabla)$ as a family of holomorphic principal G-bundles with flat connections, then it is the first obstruction against this family being locally trivial over S. In this context we often write $\mathcal{A}t^\nabla_{\mathcal{X}/S}(\mathcal{P})$ as a homomorphism

$$\mathcal{A}t^\nabla_{\mathcal{X}/S}(\mathcal{P}) : \theta_S \longrightarrow R^1 f_*\,\mathrm{ad}(\mathcal{P})^\nabla. \tag{1.7}$$

In Section 2.2 we shall define a morphism from S to the so-called character stack and then interpret this homomorphism $\mathcal{A}t^\nabla_{\mathcal{X}/S}(\mathcal{P})$ as the derivative of this morphism.

For $\Phi \in \mathbb{C}[\mathfrak{g}]^G$ homogeneous of degree n, its evaluation on $\mathcal{A}t^\nabla_{\mathcal{X}/S}(\mathcal{P})$ now yields

$$\Phi(\mathcal{A}t^\nabla_{\mathcal{X}/S}(\mathcal{P})) \in H^0(S, \Omega^n_S \otimes R^n f_* f^{-1}\mathcal{O}_S) \cong H^0(S, \Omega^n_S \otimes R^n f_*\underline{\mathbb{C}}_{\mathcal{X}}).$$

So this is a twisted holomorphic n-form on S, to be precise, it is a holomorphic n-form with values in the local system $R^n f_*\underline{\mathbb{C}}_{\mathcal{X}}$. We thus obtain a homomorphism of graded algebras $\mathbb{C}[\mathfrak{g}]^G \longrightarrow \bigoplus_{n\geq 0} H^0(S, \Omega^n_S \otimes R^n f_*\underline{\mathbb{C}}_{\mathcal{X}})$, where the cup product furnishes the product structure on the right hand side.

COROLLARY 1.2. *In this situation, the relative Chern-Weil homomorphism takes its values in the closed forms: it is a graded algebra homomorphism*

$$\mathbb{C}[\mathfrak{g}]^G \longrightarrow \bigoplus_{n\geq 0} H^0(S, \Omega^n_{S,cl} \otimes R^n f_*\underline{\mathbb{C}}_{\mathcal{X}})$$

whose definition commutes with base change. In particular, if n is the real fiber dimension of f (so that n is even) and $\Phi \in \mathbb{C}[\mathfrak{g}]^G$ is homogeneous of degree n, *then the holomorphic n-form* $f_!\Phi(\mathfrak{at}^{\nabla}_{\mathcal{X}/S})$ *on* S *obtained by integration along fibers is closed, where* $\mathfrak{at}^{\nabla}_{\mathcal{X}/S}$ *is the section constructed in* (1.4).

PROOF. The first assertion follows from Lemma 1.1 and a standard argument in the Chern-Weil theory. The base change property is evident. □

2. The universal construction

2.1. The character stack. We first recall the notion of a characteristic space (for which we use Sikora [**Si**] as a general reference). Let Γ be a finitely generated group, admitting p generators say, so that there exists an epimorphism from the free group F_p on p generators to Γ. Then the group homomorphisms from Γ to G make up a complex space $\mathrm{Hom}(\Gamma, G)$ of finite type over $\mathbb{C}$. We note that it can be realized in an evident manner as a closed subspace of G^p: every element in the kernel of the projection $F_p \longrightarrow \Gamma$ defines a morphism $G^p \longrightarrow G$ for which the preimage of $1 \in G$ defines a closed subspace of G^p and $\mathrm{Hom}(\Gamma, G)$ is then simply the intersection of these subspaces. It is independent of the epimorphism, but it need not be reduced. Its definition is covariant in G and contravariant in Γ. In particular, $\mathrm{Aut}(\Gamma) \times \mathrm{Aut}(G)$ acts on $\mathrm{Hom}(\Gamma, G)$. Let us note in passing that the noetherian nature of complex-analytic geometry implies that the ideal defining $\mathrm{Hom}(\Gamma, G)$ in G^p is locally finitely generated and hence will involve only finitely many relations. So for local considerations there is no loss in generality in assuming that Γ is finitely presented.

Composition with inner automorphisms of G makes G act on $\mathrm{Hom}(\Gamma, G)$ in the category of complex spaces. The kernel of this action clearly contains the center $Z(G)$ of G and so the action of G is via $\overline{G} := G/Z(G)$. We could form the categorical quotient $\overline{G}\backslash \mathrm{Hom}(\Gamma, G)$ (or in case G is a linear algebraic group, $\mathrm{Spec}\,\mathbb{C}[\mathrm{Hom}(\Gamma, G)]^{\overline{G}}$), but we prefer to treat the quotient as a stack, denoting it by $\underline{\chi}_G(\Gamma)$. We shall not recall the formal definition as for our purpose it suffices to know what it means to have a stack morphism from a complex space S to $\underline{\chi}_G(\Gamma)$ (often called an S-valued point of $\underline{\chi}_G(\Gamma)$). In the algebraic setting (so where S is a complex scheme of finite type), it simply assigns to each closed point of S a G-orbit in $\mathrm{Hom}(\Gamma, G)$ in such a manner that this is locally (in some Grothendieck topology that has to be specified) presented as a scheme morphism from S to $\mathrm{Hom}(\Gamma, G)$. The same description continues to be valid in the context of other categories of ringed spaces such as the holomorphic and C^∞ categories (where G is a complex Lie group resp. just a Lie group). An advantage of this approach is that it allows us to avoid imposing irrelevant additional assumptions.

It is then clear that if $f : \mathcal{X} \longrightarrow S$ is in one of these categories, and f is topologically locally trivial with connected fiber and simply connected base S, then a principal G-bundle $\mathcal{P}$ over $\mathcal{X}$ endowed with a connection ∇ relative to f determines a morphism in this category from S to $\underline{\chi}_G(\Gamma)$, where Γ is the fundamental group of a fiber (so given up to an inner automorphism). Conversely, such a morphism determines the pair $(\mathcal{P}, \nabla)$ uniquely up to an obvious notion of isomorphism.

A *regular form* on $\underline{\chi}_G(\Gamma)$ is by definition a regular form ω on the complex space $\mathrm{Hom}(\Gamma, G)$ that is *basic* with respect to the G-action; this means that

$$i_{\mathsf{v}}\omega = 0 \quad \text{and} \quad i_{\mathsf{v}}d\omega = 0$$

for all $v \in \mathfrak{g}$, where $\mathbf{v}$ is the vector field on $\operatorname{Hom}(\Gamma, G)$ corresponding to v for the action of G on $\operatorname{Hom}(\Gamma, G)$. If ω is closed, then the second property is obviously automatic and hence such a form ω is basic if and only if its contraction with $\mathfrak{g}$ is basic.

Let us also note that the obvious action of $\operatorname{Aut}(\Gamma)$ on $\operatorname{Hom}(\Gamma, G)$ gives rise to an action of $\operatorname{Aut}(\Gamma)$ on the space of regular forms on $\underline{\chi}_G(\Gamma)$ and that (as one would expect) this action factors through the outer automorphism group $\operatorname{Out}(\Gamma)$.

We have a similar way of describing the Zariski tangent space of $\underline{\chi}_G(\Gamma)$ at a point, represented by $\rho \in \operatorname{Hom}(\Gamma, G)$, say. Recall that the space of 1-cycles $Z^1(\Gamma, \operatorname{Ad}\rho)$ consists of maps $\Gamma \longrightarrow \mathfrak{g}$ that are completely given by their values on a set of generators. For $Z^1(F_p, \operatorname{Ad}\rho)$ these values can be arbitrary so that $Z^1(\Gamma, \operatorname{Ad}\rho)$ becomes a subspace of

$$Z^1(F_p, \mathfrak{g}) \cong Z^1(F_p) \otimes \mathfrak{g} \cong \mathfrak{g}^{\oplus p}$$

(which is independent of ρ). When we regard ρ as an element of the algebraic group G^p, then the right translation with ρ identifies $Z^1(\Gamma, \operatorname{Ad}\rho)$ with the Zariski tangent space of $\operatorname{Hom}(\Gamma, G)$ at ρ [**Si**, Theorem 35]. To be precise: if $t \longmapsto \rho_t$ is a curvelet in $\operatorname{Hom}(\Gamma, G)$ given up to first order with $\rho_0 = \rho$ and we write $\rho_t \equiv e^{t\sigma}\rho_0 \pmod{t^2}$, then the condition $\rho_t(\gamma_1\gamma_2) = \rho_t(\gamma_1)\rho_t(\gamma_2)$ boils down to $\sigma(\gamma_1\gamma_2) = \sigma(\gamma_1) + \rho(\gamma_1)\sigma(\gamma_2)\rho(\gamma_1)^{-1}$, which indeed amounts to $\sigma \in Z^1(\Gamma, \mathfrak{g})$. Under this map, the space of 1-boundaries $B^1(\Gamma, \operatorname{Ad}\rho)$ is mapped on the tangent space of the G-orbit of ρ and therefore the group cohomology space $H^1(\Gamma, \operatorname{Ad}\rho)$ can be understood as the Zariski tangent space of $\underline{\chi}_G(\Gamma)$ at $\underline{\rho}$. There is a similar notion for the Zariski tangent space of an S-valued point (in a manner that is straightforward to spell out); this then appears as a coherent $\mathcal{O}_S$-module over S that we shall denote by $\theta_{\underline{\chi}_G(\Gamma),S}$.

The regular forms on $\underline{\chi}_G(\Gamma)$ of interest here take values in $H^\bullet(\Gamma, \mathbb{C})$ and are defined by the G-invariants in $\mathbb{C}[\mathfrak{g}]$: any $\Phi \in \mathbb{C}[\mathfrak{g}]^G$ that is homogeneous of degree n combines with the cup product to give a linear map

$$\eta(\rho, \Phi) : \wedge^n H^1(\Gamma, \operatorname{Ad}\rho) \longrightarrow H^n(\Gamma, (\operatorname{Ad}\rho)^{\otimes n}) \longrightarrow H^n(\Gamma, \mathbb{C}).$$

We can now establish part of our main Theorem 0.1.

PROPOSITION 2.1. *The linear maps $\eta(\rho, \Phi)$ define a $H^n(\Gamma, \mathbb{C})$-valued n-form η_Φ on* $\operatorname{Hom}(\Gamma, G)$ *that is basic relative to the G-action. We thus obtain a graded $\mathbb{C}$-algebra homomorphism*

$$\eta : \mathbb{C}[\mathfrak{g}]^G \longrightarrow \bigoplus_{n\geq 0} \left(H^0(\underline{\chi}_G(\Gamma), \Omega^n_{\underline{\chi}_G(\Gamma)}) \otimes H^n(\Gamma, \mathbb{C})\right)^{\operatorname{Aut}(\Gamma)},$$

where the degree of the summand indexed by n is $2n$.

PROOF. Consider the natural linear map

$$\wedge^n(Z^1(F_p) \otimes \mathfrak{g}) \longrightarrow \wedge^n Z^1(F_p) \otimes \operatorname{Sym}^n \mathfrak{g} \xrightarrow{1\otimes\Phi} \wedge^n Z^1(F_p, \mathbb{C}).$$

(It descends to a map $\wedge^n H^1(F_p, \operatorname{Ad}\rho) \longrightarrow H^n(F_p, \mathbb{C})$, but this is of little interest here, as $H^n(F_p, \mathbb{C}) = 0$ for all $n > 1$.) We identify $Z^1(F_p) \otimes \mathfrak{g} \cong \mathfrak{g}^{\oplus p}$ with the Lie algebra of $G^p \cong \operatorname{Hom}(F_p, G)$ and then use the left translation by G^p to make this an n-form $\tilde{\eta}_\Phi$ on $\operatorname{Hom}(F_p, G) = G^p$ with values in $\wedge^n Z^1(F_p, \mathbb{C})$.

This is, besides being invariant under the left G^p-action, also invariant under simultaneous conjugation by G (for Φ is G-invariant). The same is true for its exterior derivative and so $\widetilde{\eta}_\Phi$ is basic.

The restriction of $\widetilde{\eta}_\Phi$ to the closed subspace $\mathrm{Hom}(\Gamma, G)$ of $\mathrm{Hom}(F_p, G)$ is of course still basic. But it will now take its values in the subspace $\wedge^n Z^1(\Gamma) \subset \wedge^n Z^1(F_p, \mathbb{C})$ and η_Φ then appears after composition with the natural map

$$\wedge^n Z^1(\Gamma, \mathbb{C}) \longrightarrow \wedge^n H^1(\Gamma, \mathbb{C}) \longrightarrow H^n(\Gamma, \mathbb{C}).$$

This proves that η_Φ is basic.

The naturality of the construction implies that η_Φ is $\mathrm{Aut}(\Gamma)$-invariant. □

REMARK 2.2. We already noted that $\underline{\chi}_G(\Gamma)$ can be regarded as the moduli stack of G-local systems on a reasonable space with fundamental group isomorphic to Γ. In case that space is a connected compact C^∞-manifold M, there is a geometric interpretation of $H^1(\Gamma, \mathrm{Ad}\,\rho)$ in terms of flat connections that is rooted in Yang-Mills theory (as developed by Atiyah and Bott in [**AB**]) and which in the present context was first explicated by Goldman [**Go**]. For this, assume that we are given a universal cover $\widetilde{M} \longrightarrow M$ with group of deck transformations Γ (acting on the left). Every $\rho \in \mathrm{Hom}(\Gamma, G)$ gives rise to a principal G-bundle P_ρ on M with a flat connection; it is obtained as the quotient of $\widetilde{M} \times G$ by the diagonal action of Γ (acting on G by left translations via ρ). But if we vary ρ, then the underlying C^∞ principal bundle does not really change if we ignore the connection. In other words, if we fix a $\rho \in \mathrm{Hom}(\Gamma, G)$ and write (P, ∇) for P_ρ, then a variation of ρ can be implemented by a variation of ∇ as a flat connection. Any C^∞-connection ∇' on P is of the form $\nabla + A$ with A being a section of $\mathcal{E}^1_M \otimes \mathrm{ad}(P)$ and the connection $\nabla + A$ is flat if and only if $d_\nabla A + A \wedge A = 0$ (the exterior derivative being taken with respect to the connection ∇). For the first order deformations we can ignore the quadratic term $A \wedge A$ so that this becomes $d_\nabla A = 0$. Those that come from an infinitesimal automorphism (and hence 'do not count') are in the d_∇-image of a smooth section of $\mathrm{ad}(P)$ and so this realizes the space of genuine first order deformations of ∇ with the de Rham cohomology space $H^1(M, \mathrm{ad}(P, \nabla)) = H^1(M, \mathrm{ad}(P_\rho))$. The standard isomorphism

$$H^1(\Gamma, \mathrm{Ad}\,\rho) \xrightarrow{\cong} H^1(M, \mathrm{ad}(P_\rho)) \tag{2.1}$$

thus acquires a geometric interpretation, namely as the tangent map at ρ of the map which assigns to any $[\rho'] \in \underline{\chi}_G(\Gamma)$, the isomorphism type of $P_{\rho'}$. We will encounter an analogue of it for S-valued points of $\underline{\chi}_G(\Gamma)$ in the next subsection.

2.2. A comparison. We return to the situation of Section 1.2. Since the issue that we will address is local on S, we assume that S is a contractible Stein manifold, that $\mathcal{X}$ is connected and that f admits a holomorphic section $x : S \longrightarrow \mathcal{X}$ such that $x^*\mathcal{P}$ is trivial, i.e., $x^*\mathcal{P}$ admits a holomorphic section $\widetilde{x}$. Let $p : \widetilde{\mathcal{X}} \longrightarrow \mathcal{X}$ be the universal cover defined by x with group of deck transformations $\Gamma := \pi_1(\mathcal{X}, x)$. Since S is contractible, the map p defines a universal cover of every fiber of f. The pull-back $p^*\mathcal{P}$ comes with a connection relative to $f \circ p$ that is flat. Since the fibers of $f \circ p$ are simply connected, the global sections of $p^*\mathcal{P}$ that are relatively flat identify this pull-back with $\widetilde{\mathcal{X}} \times_S x^*\mathcal{P}$. We use the section $\widetilde{x}$ of $x^*\mathcal{P}$ to identify this even with $\widetilde{\mathcal{X}} \times G$. Via this isomorphism the Γ-action on $p^*\mathcal{P}$ then becomes one on

$\widetilde{\mathcal{X}} \times G$. The latter action is given by a holomorphic holonomy map

$$\rho : S \longrightarrow \operatorname{Hom}(\Gamma, G), \quad s \longmapsto \rho_s$$

in the sense that $\gamma \in \Gamma$ takes $(\widetilde{x}, g)$ to $(\gamma.\widetilde{x}, \rho_{f(p(\widetilde{x}))}(\gamma)g)$. A different choice of x or $\widetilde{x}$ alters ρ by conjugation with a holomorphic map $S \longrightarrow G$, so that the resulting holomorphic map $\underline{\rho} : S \longrightarrow \underline{\chi}_G(\Gamma)$ is intrinsic to the situation.

The Γ-cover $\widetilde{\mathcal{X}}$ of $\mathcal{X}$ defines a natural homotopy class κ of maps from $\mathcal{X}$ to a classifying space for Γ. It has its own characteristic algebra, given by the natural map:

$$\kappa^* : H^\bullet(\Gamma) \longrightarrow H^\bullet(\mathcal{X}).$$

Proposition 2.3 and Corollary 2.4 below imply the remaining assertions of Theorem 0.1.

PROPOSITION 2.3. *If we regard κ^* as a homomorphism of trivial local systems on S:*

$$\kappa^* : \underline{H^\bullet(\Gamma)}_S \longrightarrow R^\bullet f_* \underline{\mathbb{Z}}_{\mathcal{X}},$$

then $1 \otimes \kappa^*$ takes $\rho^* \eta_\Phi \in H^0(S, \Omega^n_S) \otimes H^n(\Gamma, \mathbb{C})$ to $\Phi(\mathcal{A}t^\nabla_{\mathcal{X}/S}(\mathcal{P}))$, where $\mathcal{A}t^\nabla_{\mathcal{X}/S}(\mathcal{P})$ and η are as in Lemma 1.1 *and Proposition* 2.1 *respectively.*

PROOF. For $s \in S$, put $X_s := f^{-1}(s)$ and let $(P_s/X_s, \nabla_s)$ be the restriction $(\mathcal{P}, \nabla)$ to X_s. We have the Kodaira-Spencer homomorphism

$$\mathcal{A}t^\nabla_{\mathcal{X}/S}(\mathcal{P}) : \theta_S \longrightarrow R^1 f_* \operatorname{ad}(\mathcal{P})^\nabla$$

in (1.7), whereas the variation of the local system defines a similar homomorphism from θ_S its Zariski tangent bundle $\theta_{\underline{\chi}_G(\Gamma), S}$ when considered as a S-valued point (for a closed point s this would be $H^1(\Gamma, \operatorname{Ad} \rho_s)$). As noted above, the isomorphism $H^1(\Gamma, \operatorname{Ad} \rho_s) \xrightarrow{\cong} H^1(X_s, \operatorname{ad}(P_s))$ has an interpretation as an isomorphism of tangent spaces. Over S this gives rise to an isomorphism $\theta_{\underline{\chi}_G(\Gamma), S} \longrightarrow R^1 f_* \operatorname{ad}(\mathcal{P})^\nabla$ and what remains to show is that this is compatible with the homomorphisms from θ_S to its source and target. The proof of this is left to the reader. □

COROLLARY 2.4. *The map η in Proposition* 2.1 *takes values in the* closed *$H^\bullet(\Gamma, \mathbb{C})$-valued forms on $\underline{\chi}_G(\Gamma)$.*

PROOF. We observed that the characteristic stack $\underline{\chi}_G(\gamma)$ is locally realized by a finitely presented group that has Γ as a quotient and so there is no loss in generality in assuming that Γ is already of that type. If Γ appears as the fundamental group of compact complex manifold, then Proposition 2.3 gives the corollary for such a Γ. At this point we can finish the proof by invoking a theorem of Taubes, [**Ta**, Theorem 1.1], which implies that any finitely presented group is isomorphic to the fundamental group a compact complex manifold of dimension three (which here appears as the twistor space of an anti-self-dual Riemannian 4-manifold). Instead of relying on such a deep theorem, one might alternatively observe that Proposition 2.3 and its proof are also valid in case $\mathcal{X} \longrightarrow S$ is a trivial family with fiber a compact complex manifold M with boundary. In that case it is clear that any finitely presented group can be realized as the fundamental group of such a manifold: just take a finite cell complex K with that fundamental group, embed K in $\mathbb{C}^3$ and take for M be a closed regular neighborhood of its image. □

REMARK 2.5. It would of course be more satisfying to have a direct proof of Corollary 2.4. For this it is perhaps useful to point out that the corollary does not hold without tensoring with $H^\bullet(\Gamma, \mathbb{C})$ (which can be zero). For example, if $\Gamma = F_p$ with $p \geq 2$, then the image of the Killing form on $\mathfrak{sl}(2, \mathbb{C})$ in $\mathrm{Hom}(\Gamma, \mathrm{PSL}_2(\mathbb{C}))$ will not be closed, but after tensoring with $H^2(F_p, \mathbb{C}) = 0$ it is of course trivially so.

References

[At] M. F. Atiyah, *Complex analytic connections in fibre bundles*, Trans. Amer. Math. Soc. **85** (1957), 181–207, DOI 10.2307/1992969. MR86359

[AB] M. F. Atiyah and R. Bott, *The Yang-Mills equations over Riemann surfaces*, Philos. Trans. Roy. Soc. London Ser. A **308** (1983), no. 1505, 523–615, DOI 10.1098/rsta.1983.0017. MR702806

[BE] Spencer Bloch and Hélène Esnault, *Relative algebraic differential characters*, Motives, polylogarithms and Hodge theory, Part I (Irvine, CA, 1998), Int. Press Lect. Ser., vol. 3, Int. Press, Somerville, MA, 2002, pp. 47–73. MR1977584

[CherS] Shiing Shen Chern and James Simons, *Characteristic forms and geometric invariants*, Ann. of Math. (2) **99** (1974), 48–69, DOI 10.2307/1971013. MR353327

[CheeS] Jeff Cheeger and James Simons, *Differential characters and geometric invariants*, Geometry and topology (College Park, Md., 1983/84), Lecture Notes in Math., vol. 1167, Springer, Berlin, 1985, pp. 50–80, DOI 10.1007/BFb0075216. MR827262

[De1] P. Deligne, *Théorème de Lefschetz et critères de dégénérescence de suites spectrales* (French), Inst. Hautes Études Sci. Publ. Math. **35** (1968), 259–278. MR244265

[De2] Pierre Deligne, *Équations différentielles à points singuliers réguliers* (French), Lecture Notes in Mathematics, Vol. 163, Springer-Verlag, Berlin-New York, 1970. MR417174

[Go] William M. Goldman, *The symplectic nature of fundamental groups of surfaces*, Adv. in Math. **54** (1984), no. 2, 200–225, DOI 10.1016/0001-8708(84)90040-9. MR762512

[Gr] Hans Grauert, *Ein Theorem der analytischen Garbentheorie und die Modulräume komplexer Strukturen* (German), Inst. Hautes Études Sci. Publ. Math. **5** (1960), 64. MR121814

[GHJW] K. Guruprasad, J. Huebschmann, L. Jeffrey, and A. Weinstein, *Group systems, groupoids, and moduli spaces of parabolic bundles*, Duke Math. J. **89** (1997), no. 2, 377–412, DOI 10.1215/S0012-7094-97-08917-1. MR1460627

[MS] John W. Milnor and James D. Stasheff, *Characteristic classes*, Annals of Mathematics Studies, No. 76, Princeton University Press, Princeton, NJ; University of Tokyo Press, Tokyo, 1974. MR440554

[Si] Adam S. Sikora, *Character varieties*, Trans. Amer. Math. Soc. **364** (2012), no. 10, 5173–5208, DOI 10.1090/S0002-9947-2012-05448-1. MR2931326

[Ta] Clifford Henry Taubes, *The existence of anti-self-dual conformal structures*, J. Differential Geom. **36** (1992), no. 1, 163–253. MR1168984

DEPARTMENT OF MATHEMATICS, SHIV NADAR UNIVERSITY, NH91, TEHSIL DADRI, GREATER NOIDA, UTTAR PRADESH 201314, INDIA

Email address: `indranil.biswas@snu.edu.in; indranil29@gmail.com`

MATHEMATISCH INSTITUUT UNIVERSITEIT UTRECHT, NETHERLANDS AND DEPT. OF MATHEMATICS, UNIVERSITY OF CHICAGO

Email address: `e.j.n.looijenga@uu.nl`

Contemporary Mathematics
Volume **841**, 2026
https://doi.org/10.1090/conm/841/16884

The 2-associahedra are Eulerian

Nathaniel Bottman and Dylan Mavrides

Abstract. The 2-associahedra $W_{\mathbf{n}}$ are a family of abstract polytopes, indexed by sequences of nonnegative integers of arbitrary length. They arise in symplectic geometry, where they govern the functoriality properties of the Fukaya category, but they have purely combinatorial and topological interpretations: for instance, $W_{\mathbf{n}}$ indexes the possible degenerations in a configuration space of marked vertical lines in $\mathbb{R}^2$, up to translations and dilations. In this paper, we exploit the recursive structure of the 2-associahedra in order to further explore their combinatorial properties. First, we define a family of generating functions that count elements of 2-associahedra, and establish a system of equations satisfied by these generating functions. Next, we use this computation to prove that the 2-associahedra are Eulerian.

1. Introduction

In [**Bot19a**], the first author introduced the *2-associahedra*, a family $(W_{\mathbf{n}})$ of abstract polytopes which forms the algebraic underpinnings for the naturality structure on the Fukaya category [**Bot19b,BC21,BW18,MWW18**]. While these posets originate in symplectic geometry, they are natural combinatorial objects, which can be thought of either as the posets of degenerations in the configuration spaces of marked vertical lines in $\mathbb{R}^2$ (up to translations and dilations), or as the posets that control the coherences in an $(A_\infty, 2)$-category.

As we describe in §1.1, the 2-associahedra have a recursive structure: each closed face of a 2-associahedron has a canonical decomposition as a product of fiber products of smaller 2-associahedra, where this fiber product is with respect to the forgetful map from 2-associahedra to associahedra. Just as one can use the analogous (but simpler) recursive structure for the associahedra K_r to derive an equation satisfied by the generating function that counts faces of K_r, we can use this recursive structure to derive equations satisfied by the generating functions that count faces of $W_{\mathbf{n}}$. Specifically, in §2, we define for any stable rooted ribbon tree $T \in K_r$ a function $F_T(t, \mathbf{x})$, where t is a single variable and $\mathbf{x} = (x_1, \ldots, x_n)$ has length r. The coefficient on the monomial $t^m\mathbf{x}^{\mathbf{n}}$ counts the number of faces in $W_{\mathbf{n}}$ of dimension m, that in addition lie over T under the forgetful map $W_{\mathbf{n}} \to K_r$. Our first result is a system of equations satisfied by this collection of generating functions.

2020 *Mathematics Subject Classification.* Primary 53D37, 05E45, 52B05.

©2026 American Mathematical Society

THEOREM 2.1, PART (B). *The collection* $\left(F_T(t,\mathbf{x})\right)_{\substack{r\geq 1,\\ T\in K_r}}$ *satisfies the following equations:*

(1)

$$F_\bullet = \frac{F_\bullet^2}{1-tF_\bullet} + x, \qquad F_T = \frac{F_T^2}{t^p - tF_T} + t^{p-1}\left(\frac{t^{p_1}}{t^{p_1} - tF_{T_1}} \cdots \frac{t^{p_k}}{t^{p_k} - tF_{T_k}} - 1\right),$$

In the first equation, $\bullet$ *is the tree with a single vertex and no edges, which by convention is the unique element of* K_1. *In the second equation, we have decomposed* $T \in K_r, r \geq 2$ *into a corolla with* k *edges, all incident to the root, together with the subtrees* $T_1, \ldots, T_k$ *that are attached to the non-root ends of these* k *edges. By* p *and* p_i *we denote the dimensions of* T *and* T_i*, when thought of as elements of associahedra.*

We expect that this result will be of interest independently of the application described in the next paragraph. For instance, just as the Catalan numbers can be realized as the number of vertices of the associahedra, one could study the "2-Catalan numbers", i.e. the number of vertices of the 2-associahedra. Extensive searches in the OEIS [**SI**] performed by the first author suggest that the 2-Catalan numbers do not form a known family of numbers, and using this theorem could shed light on the properties of these numbers.

In §3, we use Theorem 2.1 to prove that $\widehat{W_{\mathbf{n}}} := W_{\mathbf{n}} \cup \{F_{\mathbf{n}}^{\min}\}$ is Eulerian. This poset is a formal augmentation of $W_{\mathbf{n}}$; $F_{\mathbf{n}}^{\min}$ is a formal minimal element, of dimension -1. A ranked poset is *Eulerian* if every nontrivial interval has the same number of even-rank elements as it does odd-rank elements. The face lattice of a convex polytope is an example, and many results about the combinatorics of convex polytopes can be extended to Eulerian posets.

THEOREM 1.1. *For every* $r \geq 1$ *and* $\mathbf{n} \in \mathbb{Z}_{\geq 0}^r \setminus \{\mathbf{0}\}$*, the augmented 2-associahedron* $\widehat{W_{\mathbf{n}}}$ *is an Eulerian poset.*

One motivation for this theorem is that one can associate to an Eulerian poset an important invariant called the *cd-index*, which is a concise encoding of the flag f-vector [**Sta94**]. By using the recursive structure of the 2-associahedra as described in §2, we hope to compute the *cd*-indices of the 2-associahedra algorithmically in future work. Another motivation for Thm. 1.1 is that it provides further evidence that the 2-associahedra can be realized as the face lattices of convex polytopes.

1.1. An introduction to the 2-associahedra for combinatorialists. 2-associahedra are likely to be unfamiliar to most combinatorialists, because the 2-associahedra arose in symplectic geometry. In this subsection we give a brief overview of these objects. For more detail, we encourage the interested reader to consult [**Bot19a**].

Fix $r \geq 1$ and $\mathbf{n} \in \mathbb{Z}_{\geq 0}^r \setminus \{\mathbf{0}\}$. The 2-associahedron $W_{\mathbf{n}}$ is a poset, whose definition is motivated[1] by the compactified configuration space $\overline{2M}_{\mathbf{n}}$ of r vertical lines in $\mathbb{R}^2$ with $n_1, \ldots, n_r$ marked points, up to translations and positive dilations. Specifically, $W_{\mathbf{n}}$ is the poset of strata of $\overline{2M}_{\mathbf{n}}$. In [**Bot19a**], the first author defined two isomorphic presentations of $W_{\mathbf{n}}$, denoted $W_{\mathbf{n}}^{\text{tree}}$ and $W_{\mathbf{n}}^{\text{br}}$. Before we give the precise definition of $W_{\mathbf{n}}^{\text{tree}}$, we use a figure from [**Bot19a**] to illustrate $\overline{2M}_{\mathbf{n}}$, $W_{\mathbf{n}}^{\text{tree}}$, and $W_{\mathbf{n}}^{\text{br}}$ in the example of $r = 3$, $\mathbf{n} = (2,0,0)$.

[1] In fact, defining $\overline{2M}_{\mathbf{n}}$, as the first author did in [**Bot19b**], requires as input the poset $W_{\mathbf{n}}$.

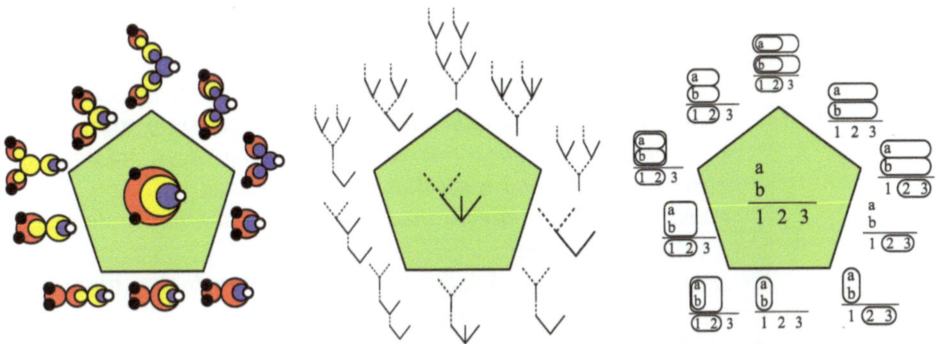

FIGURE 1. From left to right, the compactified moduli space $\overline{2M}_{200}$ and the isomorphic posets W_{200}^{tree} and W_{200}^{br}. (We depict $\overline{2M}_{200}$ in a slightly different, but equivalent, way, compared to its definition in the text: here, instead of three vertical lines in $\mathbb{R}^2$, we have a disk with two interior circles; the white marked point corresponds to the point at ∞.) Note that in W_{200}^{tree}, we have depicted only T_b, not the entire tree-pair $T_b \to T_s$, for reasons of space.

The following theorem, which is the main result in [**Bot19a**], records some important structural properties of the 2-associahedra.

THEOREM 4.1, [**Bot19a**]. *For any $r \geq 1$ and $\mathbf{n} \in \mathbb{Z}_{\geq 0}^r \backslash \{\mathbf{0}\}$, the 2-associahedron $W_{\mathbf{n}}$ is a poset, the collection of which satisfies the following properties:*

(abstract polytope) *$\widehat{W_{\mathbf{n}}} := W_{\mathbf{n}} \cup \{F_{-1}\}$ is an abstract polytope.*

(forgetful) *There are forgetful maps $\pi\colon W_{\mathbf{n}} \to K_r$ to the associahedra.*

(recursive) *The closure of any facet of $W_{\mathbf{n}}$ can be decomposed as a product of fiber products over the associahedra of lower-dimensional 2-associahedra.*

In addition, we can view the associahedra and multiplihedra as instances of 2-associahedra: as proven in [**Bot19a**, Lemma 3.9], there are poset isomorphisms

$$W_d \simeq K_d, \qquad W_{d,0} \simeq J_d \simeq W_{0,d}, \tag{2}$$

where J_d is the $(d-1)$-dimensional multiplihedron. It does not seem possible to cast the 2-associahedra as instances of known polytopes: for instance, they are not generalized permutohedra, because W_{21} is the face lattice of an octagon, but no generalized permutohedron is an octagon.

We conclude this subsection by recording the definition of $W_{\mathbf{n}}^{\text{tree}}$, which is the version of $W_{\mathbf{n}}$ that we will work with in this paper. This definition may not seem the most natural way of presenting these combinatorial data; the reason that the first author has chosen this phrasing is because of the relationship with the compactified moduli spaces $\overline{2M}_{\mathbf{n}}$. For a detailed explanation of this definition and how it arises naturally from $\overline{2M}_{\mathbf{n}}$, we direct the reader to [**Bot19a**, §1.1].

DEFINITION 3.1, [**Bot19a**]. *A* stable tree-pair of type $\mathbf{n}$ *is a datum $2T = T_b \xrightarrow{f} T_s$, with T_b, T_s, f described below:*

- *The* bubble tree T_b *is a rooted ribbon tree (RRT) whose edges are either solid or dashed, which must satisfy these properties:*

- *The vertices of* T_b *are partitioned as* $V(T_b) = V_{\mathrm{comp}} \sqcup V_{\mathrm{seam}} \sqcup V_{\mathrm{mark}}$, *where:*
  - $*$ *every* $\alpha \in V_{\mathrm{comp}}$ *has* ≥ 1 *solid incoming edge, no dashed incoming edges, and either a dashed or no outgoing edge;*
  - $*$ *every* $\alpha \in V_{\mathrm{seam}}$ *has* ≥ 0 *dashed incoming edges, no solid incoming edges, and a solid outgoing edge; and*
  - $*$ *every* $\alpha \in V_{\mathrm{mark}}$ *has no incoming edges and either a dashed or no outgoing edge.*

  We partition $V_{\mathrm{comp}} =: V^1_{\mathrm{comp}} \sqcup V^{\geq 2}_{\mathrm{comp}}$ *according to the number of incoming edges of a given vertex.*
- ***(stability)*** *If* α *is a vertex in* V^1_{comp} *and* β *is its incoming neighbor, then* $\#\mathrm{in}(\beta) \geq 2$*; if* α *is a vertex in* $V^{\geq 2}_{\mathrm{comp}}$ *and* $\beta_1, \ldots, \beta_\ell$ *are its incoming neighbors, then there exists* j *with* $\#\mathrm{in}(\beta_j) \geq 1$.

- *The* seam tree T_s *is an element of* K^{tree}_r.
- *The* coherence map *is a map* $f\colon T_b \to T_s$ *of sets having these properties:*
  - f *sends root to root, and if* $\beta \in \mathrm{in}(\alpha)$ *in* T_b*, then either* $f(\beta) \in \mathrm{in}(f(\alpha))$ *or* $f(\alpha) = f(\beta)$.
  - f *contracts all dashed edges, and every solid edge whose terminal vertex is in* V^1_{comp}.
  - *For any* $\alpha \in V^{\geq 2}_{\mathrm{comp}}$, f *maps the incoming edges of* α *bijectively onto the incoming edges of* $f(\alpha)$*, compatibly with* $<_\alpha$ *and* $<_{f(\alpha)}$.
  - f *sends every element of* V_{mark} *to a leaf of* T_s*, and if* $\lambda^{T_s}_i$ *is the i-th leaf of* T_s*, then* $f^{-1}\{\lambda^{T_s}_i\}$ *contains* n_i *elements of* V_{mark}*, which we denote by* $\mu^{T_b}_{i1}, \ldots, \mu^{T_b}_{in_i}$.

We denote by $W^{\mathrm{tree}}_{\mathbf{n}}$ *the set of isomorphism classes of stable tree-pairs of type* $\mathbf{n}$. *Here an isomorphism from* $T_b \xrightarrow{f} T_s$ *to* $T'_b \xrightarrow{f'} T'_s$ *is a pair of maps* $\varphi_b\colon T_b \to T'_b$ *and* $\varphi_s\colon T_s \to T'_s$ *that fit into a commutative square in the obvious way and that respect all the structure of the bubble trees and seam trees.*

For reasons of space, we omit a description of the partial order on $W^{\mathrm{tree}}_{\mathbf{n}}$. *(The poset structure will not be of significant importance to us in this paper.) We invite the interested reader to consult* [**Bot19a**, Definition 3.8]. △

As in [**Bot19a**, Definition 3.3], we equip $W^{\mathrm{tree}}_{\mathbf{n}}$ with the structure of a ranked poset by defining a *dimension function* d by

$$d(2T) := |\mathbf{n}| + r - \#V^1_{\mathrm{comp}}(T_b) - \#(T_s)_{\mathrm{int}} - 2. \tag{3}$$

This should be compared with the usual dimension function on K^{tree}_r, which is defined by

$$d(T) := \sum_{i=1}^{k} q_i - \#T_{\mathrm{int}} - 1. \tag{4}$$

In both formulas, T_{int} denotes the interior vertices of a rooted tree, i.e. all vertices except the leaves.

1.2. An overview of the arguments in this paper. We now describe the plan for our paper.

§2: In this section, we define a collection of generating functions F_T, where F_T counts the elements of the 2-associahedra that project to a given element T of an associahedron. The main result of §2 is Thm. 2.1, which reformulates the recursive structure of the 2-associahedra as a set of equations satisfied by the collection (F_T).

§3: We exploit Thm. 2.1 to prove Thm. 1.1. Specifically, we divide the proof of Thm. 1.1 among §§3.1–3.3 like so:

§3.1: In this subsection, we use Thm. 2.1 to show that the alternating sum over the completed 2-associahedron $\widehat{W_{\mathbf{n}}}$ is zero (Lemma 3.3). The key is to note that since the variable t in F_T tracks the dimension, we can compute the alternating sum by specializing to $t = -1$.

§3.2: Next, we use Lemma 3.3 to show that the completion of any fiber product $W_{\mathbf{m}_1} \times_{K_r} \cdots \times_{K_r} W_{\mathbf{m}_k}$ is balanced (Lemma 3.4), where we are using the forgetful map $W_{\mathbf{m}_i} \to K_r$ defined in [**Bot19a**, Thm. 4.1]. This, plus the recursive structure of 2-associahedra (see (32)), allows us to show that the alternating sum over any interval in $\widehat{W_{\mathbf{n}}}$ of the form $[F_{\mathbf{n}}^{\min}, 2T]$ is zero (Lemma 3.5).

§3.3: Finally, we show in Lemma 3.8 that any interval in $W_{\mathbf{n}}$ has alternating sum zero. Together with Lemma 3.5, this completes the proof of Thm. 1.1. The proof of Lemma 3.5 is somewhat involved, but is conceptually simpler than that of Lemma 3.5 because an interval in $W_{\mathbf{n}}$ (as opposed to an interval in $\widehat{W_{\mathbf{n}}}$ with lower bound $F_{\mathbf{n}}^{\min}$) essentially decomposes into a product, with one term for each bracket.

2. Generating functions associated to 2-associahedra

In this section, we will study the following generating functions:

$$f(t, \mathbf{x}) := \sum_{m \geq 0, r \geq 1} a_{mr} t^m x^r, \quad a_{mr} := \#\{T \in K_r \mid d(T) = m\} =: \#K_{m,r}, \tag{5}$$

$$F_T(t, \mathbf{x}) := \sum_{\substack{m \geq 0, \\ \mathbf{n} \in \mathbb{Z}_{\geq 0}^r \setminus \{\mathbf{0}\}}} A_{T,m,\mathbf{n}} t^m \mathbf{x}^{\mathbf{n}}, \tag{6}$$

$$A_{T,m,\mathbf{n}} := \#\left\{2T \in W_{\mathbf{n}} \;\middle|\; \substack{\pi(2T)=T, \\ d(2T)=m}\right\} =: \#W_{T,m,\mathbf{n}},$$

where $T \in K_r$ is a stable rooted ribbon tree (RRT). That is, f counts faces of the associahedra, and F_T counts faces of the 2-associahedra whose image under the forgetful map is T. For instance, K_4 is CW-isomorphic to a pentagon, so f should include the terms $5x^4$, $5tx^4$, and t^2x^4.

Our main result in this section is the following characterization of f and F_T, which will be the key input to our proof that $W_{\mathbf{n}}$ is Eulerian.

THEOREM 2.1.

(a) *f is characterized uniquely by the following equations:*

$$f = \frac{f^2}{1 - tf} + x, \quad \frac{\partial f}{\partial x}(0, 0) = 1. \tag{7}$$

(b) *The collection* $\big(F_T(t,\mathbf{x})\big)_{\substack{r\geq 1,\\ T\in K_r}}$ *satisfies the following equations:*

$$F_\cdot = \frac{F_\cdot^2}{1-tF_\cdot} + x, \tag{8}$$

$$F_T = \frac{F_T^2}{t^p - tF_T} + t^{p-1}\left(\frac{t^{p_1}}{t^{p_1}-tF_{T_1}}\cdots\frac{t^{p_k}}{t^{p_k}-tF_{T_k}} - 1\right), \quad T = C(T_1,\ldots,T_k),$$

where in the latter equation, we have set $p := d(T)$, $p_i := d(T_i)$.

2.1. Counting faces in K_r. In this subsection we will prove part (a) of Thm. 2.1. We begin by defining the operation of concatenation. Set $T_r^{\max}$ to be the unique element of K_r with $d(T_{\max}^r) = r-2$. (That is, $T_{\max}^1$ is a single vertex, and for $r \geq 2$, $T_{\max}^r$ is a root vertex with r incoming neighbors.)

Definition-Proposition 2.2. *Fix $k \geq 1$ and a k-tuple $(T_i \in K_{p_i,q_i})_{1\leq i\leq k}$ of stable RRTs. Then the* concatenation *of this data is the stable RRT $C(T_1,\ldots,T_r)$ defined by attaching T_1, $\ldots$, T_r to the r leaves of $T_{\max}^r$. The concatenated tree has $\sum_{i=1}^k q_i$ leaves and the following dimension:*

$$d\big(C(T_1,\ldots,T_r)\big) = \begin{cases} p_1, & k=1, \\ \sum_{i=1}^k p_i + k - 2, & k\geq 2. \end{cases} \tag{9}$$

Proof. The $k=1$ case is trivial, so from now on we assume $k \geq 2$.

Set $T := C(T_1,\ldots,T_r)$. We must establish the formula $d(T) = \sum_{i=1}^k p_i + k - 2$ and show that T has $\sum_{i=1}^k q_i$ leaves. The latter equality follows immediately from the definition of T. The former equality follows from a calculation:

$$\begin{aligned} d(T) &= \sum_{i=1}^k q_i - \#T_{\text{int}} - 1 \\ &= \sum_{i=1}^k q_i - \left(\sum_{i=1}^k \#(T_i)_{\text{int}} + 1\right) - 1 \\ &= \sum_{i=1}^k (q_i - \#(T_i)_{\text{int}} - 1) + k - 2 \\ &= \sum_{i=1}^k p_i + k - 2. \end{aligned}$$

□

We are now ready to prove part (a) of Thm. 2.1.

Proof of Theorem 2.1(a).

Step 1. *f satisfies* (7).

For $m \geq 0, r\geq 2$, there is a bijective correspondence between the following two sets:

$$K_{m,r} \longleftrightarrow \bigsqcup_{k=2}^{\infty} \bigsqcup_{\substack{p_1+\cdots+p_k=m-k+2\\ q_1+\cdots+q_k=r\\ p_i\geq 0,\, q_i\geq 1}} K_{p_1,q_1}\times\cdots\times K_{p_k,q_k} \tag{10}$$

Indeed, the right-to-left direction of this correspondence is defined by concatenation, as in Def.-Prop. 2.2, and the left-to-right direction follows from the observation that concatenation is reversible.

For $m \geq 0, r \geq 2$, the following recursion is an immediate consequence of (10):

$$a_{mr} = \sum_{k=2}^{\infty} \sum_{\substack{p_1+\cdots+p_k=m-k+2 \\ q_1+\cdots+q_k=r \\ p_i \geq 0,\, q_i \geq 1}} a_{p_1q_1} \cdots a_{p_kq_k}. \tag{11}$$

Moreover, we have $a_{01} = 1$ and $a_{m1} = 0$ for $m \geq 1$. These equations in (a_{mr}) imply the following equation in f:

$$f = \sum_{k=2}^{\infty} t^{k-2} f^k + x, \tag{12}$$

which can be rewritten as

$$f = \frac{f^2}{1-tf} + x. \tag{13}$$

The equation $a_{01} = 1$ is equivalent to $\frac{\partial f}{\partial x}(0,0) = 1$, so f satisfies (7).

STEP 2. *(7) has only one solution.*

Next, we show that (7) has only one solution. The first equation in (7) has two solutions:

$$f_\pm := \frac{1 + tx \pm \sqrt{1 - 4x - 2tx + t^2x^2}}{2(1+t)}. \tag{14}$$

Then $f_\pm$ has $\frac{\partial f_\pm}{\partial x}(0,0) = \mp 1$, so $f_- = f$ is the unique solution of (7). □

2.2. Counting faces in $W_{\mathbf{n}}$. As in the previous section, we will now characterize the family $\big(F_T(t,\mathbf{x})\big)_{\substack{r\geq 1, \\ T\in K_r}}$ in terms of equations that the functions in this family satisfy. We begin with a concatenation operation on tree-pairs. For any $r \geq 0$ and $\mathbf{n} \in \mathbb{Z}^r_{\geq 0} \setminus \{\mathbf{0}\}$, let $2T^{\max}_{\mathbf{n}}$ denote the unique element of $W_{\mathbf{n}}$ with $d(2T^{\mathbf{n}}_{\max}) = |\mathbf{n}| + r - 3$.

DEFINITION-PROPOSITION 2.3. *Fix $k \geq 1$, $\mathbf{a} \in \mathbb{Z}^k_{\geq 0} \setminus \{\mathbf{0}\}$, a k-tuple of stable RRTs $(T_i \in K_{p_i,q_i})_i$, and for every $i \in [1,r]$ an a_i-tuple of stable tree-pairs $(2T_{ij} \in W_{T_i,P_i,\mathbf{Q}_{ij}})_j$. Then the* concatenation *of this data is the stable tree-pair $2C\big((T_i),(2T_{ij})\big) = \big(T_b \xrightarrow{\pi} T_s\big)$ defined in the $k = 1, a_1 = 1$ case to be $2T_{11}$, and in the remaining cases like so:*

- *T_s is the concatenation $C\big((T_i)\big)$ as in Def.-Prop. 2.2.*
- *We define T_b by attaching the root of $2T_{ij}$ to $\mu_{ij} \in V_{\text{mark}}(2T^{\max}_{\mathbf{a}})$.*

In the case $\mathbf{a} \neq (1)$, $2C\big((T_i),(2T_{ij})\big)$ is a stable tree-pair of type $\big(\sum_{j=1}^{a_1} \mathbf{Q}_{1j}, \ldots, \sum_{j=1}^{a_k} \mathbf{Q}_{kj}\big)$, and has the following dimension:

$$d\Big(2C\big((T_i),(2T_{ij})\big)\Big) = \sum_{i=1,j=1}^{k,a_i} P_{ij} - \sum_{i=1}^{k}(a_i - 1)p_i + |\mathbf{a}| + k - 3. \tag{15}$$

PROOF. The only thing we need to check is the dimension formula (15).

$$\begin{aligned}
d\Big(2C\big((T_i),(T_{ij})\big)\Big) &= \sum_{i=1,j=1}^{k,a_i} |\mathbf{Q}_{ij}| + \sum_{i=1}^{k} q_i - \#V^1_{\text{comp}}(T_b) - \#(T_s)_{\text{int}} - 2 \\
&= \sum_{i=1,j=1}^{k,a_i} |\mathbf{Q}_{ij}| + \sum_{i=1}^{k} q_i - \Big(\sum_{i=1,j=1}^{k,a_i} \#V^1_{\text{comp}}(T_{b,ij}) + \delta_{k1}\Big) \\
&\qquad - \Big(\sum_{i=1}^{k} \#(T_{s,i})_{\text{int}} + (1-\delta_{k1})\Big) - 2 \\
&= \sum_{i=1,j=1}^{k,a_i} \Big(|\mathbf{Q}_{ij}| + q_i - \#V^1_{\text{comp}}(T_{b,ij}) - \#(T_{s,i})_{\text{int}} - 2\Big) \\
&\qquad - \sum_{i=1}^{k} (a_i - 1)\big(q_i - \#(T_{s,i})_{\text{int}} - 1\big) + \sum_{i=1}^{k} a_i + k - 3 \\
&= \sum_{i=1,j=1}^{k,a_i} P_{ij} - \sum_{i=1}^{k} (a_i - 1)p_i + |\mathbf{a}| + k - 3. \qquad \square
\end{aligned}$$

Now that we have defined concatenation for tree-pairs, we are ready to derive a collection of equations that $\big(F_T(t,\mathbf{x})\big)$ satisfies.

PROOF OF THEOREM 2.1(B). By [**Bot19a**, Lemma 3.9], the following sets are in bijection for $m \geq 0, n \geq 1$:

$$W_{\bullet,m,n} \simeq K_{m,n}. \tag{16}$$

It follows from this sublemma that $A_{\bullet,m,n} = a_{m,n}$, hence that $F_\bullet = f$, where f is the generating function defined in the last subsection. That $F_\bullet$ satisfies the first equation in (8) now follows from Thm. 2.1(a).

Next, fix $r \geq 2$ and $T \in K_r$, and write $T = C(T_1, \ldots, T_k)$. The following sets are then in bijection:

$$\begin{aligned}
W_{T,m,\mathbf{n}} \simeq \bigsqcup_{a \geq 2} \quad & \prod_{\substack{1 \leq j \leq a, \\ \sum m_j = m+(a-1)p-a+2 \\ \mathbf{n}=\sum \mathbf{Q}_j}} W_{T,m_j,\mathbf{Q}_j} \\
& + \bigsqcup_{\mathbf{a} \in \mathbb{Z}^k_{\geq 0} \setminus \{\mathbf{0}\}} \quad \prod_{\substack{1 \leq i \leq k,\, 1 \leq j \leq a_i, \\ (\sum \mathbf{Q}_{1j}, \ldots, \sum \mathbf{Q}_{kj}) = \mathbf{n}, \\ \sum m_{ij} = m + \sum a_i(p_i - 1) - \sum p_i - k + 3}} W_{T_i,m_{ij},\mathbf{Q}_{ij}},
\end{aligned}$$

where we have denoted $p := d(T)$. This follows from concatenation of tree-pairs, analogously with (10). Indeed, either the root of T_b has one solid incoming edge, which in turn has $\ell \geq 2$ dashed incoming edges; or it has $k \geq 2$ solid incoming edges, which in turn have $\ell_1, \ldots, \ell_k$ dashed incoming edges. Note that in the latter case, from any one of the dashed edges attached to the j-th solid incoming edge of the root we can extract a picture $T'_b \xrightarrow{f'} T_j$, where T_j is the j-th branch of T. (17)

implies the following equality on generating functions:

$$F_T = t^{-p+2}\sum_{a=2}^{\infty} t^{a(-p+1)}F_T^a + t^{\sum p_i+k-3}\left(\prod_{i=1}^{k}\sum_{a_i\geq 0} t^{a(-p_i+1)}F_{T_i}^a - 1\right), \tag{17}$$

which is equivalent to the equation

$$F_T = t^{p-2}\frac{t^{-2p+2}F_T^2}{1-t^{-p+1}F_T} + t^{\sum_i p_i+k-3}\left(\frac{1}{1-t^{-p_1+1}F_{T_1}}\cdots\frac{1}{1-t^{-p_k+1}F_{T_k}} - 1\right), \tag{18}$$

which simplifies to

$$F_T = \frac{F_T^2}{t^p - tF_T} + t^{p-1}\left(\frac{t^{p_1}}{t^{p_1}-tF_{T_1}}\cdots\frac{t^{p_k}}{t^{p_k}-tF_{T_k}} - 1\right)\square$$

3. The 2-associahedra are Eulerian

In this section, we will use the functional equations we derived in Thm. 2.1 to prove Thm. 1.1. We will split our proof into three parts:

(1) In Lemma 3.3, §3.1, we use Thm. 2.1 to show that $\widehat{W_{\mathbf{n}}}$ is balanced.
(2) In Lemma 3.4, §3.2, we use Lemma 3.3 and the recursive structure of 2-associahedra to show that any fiber product $W_{\mathbf{m}_1}\times_{K_r}\widehat{\cdots\times_{K_r}} W_{\mathbf{m}_k}$ has alternating sum 1. (See Definition 3.1 for the definition of the alternating sum of a ranked poset.)
(3) Finally, we show in Lemma 3.8, §3.3 that all remaining intervals (i.e. those intervals in $W_{\mathbf{n}}$ with lower bound not equal to $F_{\mathbf{n}}^{\min}$) are balanced.

Putting these results together, we now prove Thm. 1.1.

Proof of Thm. 1.1. Fix $F_1, F_2, \in W_{\mathbf{n}}$ with $F_1 < F_2$. If $F_1 = F_{\mathbf{n}}^{\min}$, then Lemma 3.4 implies that $[F_1, F_2]$ is balanced. Otherwise, it follows from Lemma 3.8 that $[F_1, F_2]$ is balanced. □

We will need the following basic notions in this section.

Definition 3.1. A *ranked poset* is a poset P together with a rank function $d\colon P \to \mathbb{Z}$, such that:

- $x < y$ implies $d(x) < d(y)$.
- If y covers x, then $d(y) = d(x) + 1$.

We define the *alternating sum* of the poset P to be

$$A(P) := \sum_{z\in P}(-1)^{d(z)}. \tag{19}$$

An interval $[x, y]$ is *balanced* if we have $A\big([x, y]\big) = 0$, and P is *Eulerian* if, for every $x < y$, $[x, y]$ is balanced. △

It is easy to see that A commutes with products, i.e.

$$A(P\times Q) = A(P)A(Q). \tag{20}$$

3.1. $\widehat{W_{\mathbf{n}}}$ is balanced. In this subsection, we will prove that $\widehat{W_{\mathbf{n}}}$ is balanced. We will derive this as a consequence of Thm. 2.1.

LEMMA 3.2. *For every $r \geq 1$ and $T \in K_r$, the generating function F_T satisfies the following identity when we evaluate at $t = -1$:*

$$F_T(-1, \mathbf{x}) = (-1)^p \left(\frac{1}{\prod_{i=1}^r (1 - x_i)} - 1 \right), \tag{21}$$

where we set $p := d(T)$.

PROOF.

STEP 1. *We establish the $r = 1$ case.*

By the $T = \bullet$ case of (8), $F_\bullet$ satisfies

$$F_\bullet(-1, x) = \frac{F_\bullet(-1, x)^2}{1 + F_\bullet(-1, x)} + x. \tag{22}$$

This yields

$$F_\bullet(-1, x) = \frac{1}{1 - x} - 1. \tag{23}$$

STEP 2. *We prove the general case by strong induction on r.*

Choose $T_i \in K_{p_i}$, $1 \leq i \leq k$, such that $T = C(T_1, \ldots, T_k)$. For convenience, we define and simplify a quantity X:

$$\begin{aligned}
X &= \prod_{i=1}^k \frac{(-1)^{p_i}}{(-1)^{p_i} + F_{T_i}} - 1 \\
&= \prod_{i=1}^k \frac{(-1)^{p_i}}{(-1)^{p_i} + \left(\frac{(-1)^{p_i}}{\prod_{j=1}^{\ell_i} (1 - y_{ij})} + (-1)^{p_i - 1} \right)} - 1 \\
&= \prod_{i=1}^k \prod_{j=1}^{\ell_i} (1 - y_{ij}) - 1 \\
&= \prod_{i=1}^r (1 - x_i) - 1,
\end{aligned} \tag{24}$$

where in the second equality we have used the inductive hypothesis. This allows us to deduce the inductive step of the current lemma from (8). Indeed, clearing denominators in that equation yields

$$(-1)^p F_T = -X + F_T (-1)^{p-1} X, \tag{25}$$

and solving this equation for F_T yields

$$\begin{aligned}
F_T &= \frac{-X}{(-1)^p (1 + X)} \\
&= \frac{(-1)^p + (-1)^{p-1} \prod_{i=1}^r (1 - x_i)}{\prod_{i=1}^r (1 - x_i)} \\
&= (-1)^p \left(\frac{1}{\prod_{i=1}^r (1 - x_i)} - 1 \right). \quad \square
\end{aligned} \tag{26}$$

LEMMA 3.3. *For any $r \geq 1$ and $\mathbf{n} \in \mathbb{Z}_{\geq 0}^r \setminus \{\mathbf{0}\}$, $\widehat{W_{\mathbf{n}}}$ is balanced.*

PROOF.

STEP 1. *We show that $\widehat{K_r}$ is balanced.*

By an argument identical to the derivation of (23), the generating function f associated to (K_r) satisfies the following identity:

$$f(-1,x) = \frac{1}{1-x} - 1 = x + x^2 + \cdots. \tag{27}$$

We have $A\left(\widehat{K_r}\right) = [f(-1,x)]_{x^r} - 1$, where the first term on the right-hand side indicates the coefficient on x^r in $f(-1,x)$ and where the second term comes from the minimal element in $\widehat{K_r}$, so the last displayed equation implies $A\left(K_r\right) = 0$.

STEP 2. *We prove the theorem.*

We interpret $A\left(\widehat{W_{\mathbf{n}}}\right)$ in terms of the generating functions F_T, then use Step 1 and Lemma 3.2 to prove the theorem:

$$\begin{aligned} A\left(\widehat{W_{\mathbf{n}}}\right) &= \sum_{T \in K_r} \left[F_T(-1,\mathbf{x})\right]_{\mathbf{x^n}} - 1 \\ &\overset{\text{Lem. 3.2}}{=} \sum_{T\in K_r} \left[(-1)^{d(T)}\left(\frac{1}{\prod_{i=1}^r(1-x_i)} - 1\right)\right]_{\mathbf{x}^n} - 1 \\ &= A(K_r)\left[\frac{1}{\prod_{i=1}^r(1-x_i) - 1} - 1\right]_{\mathbf{x}^n} - 1 \\ &= \left[(1+x_1+x_1^2+\cdots)\cdots(1+x_r+x_r^2+\cdots) - 1\right]_{\mathbf{x}^n} - 1 \\ &= 0. \quad \square \end{aligned} \tag{28}$$

3.2. Intervals of the form $[F_{\mathbf{n}}^{\min}, 2T]$ are balanced. In this subsection, we will generalize Lemma 3.3 to the statement that for every $2T \in W_{\mathbf{n}}$, the sublevel set $[F_{\mathbf{n}}^{\min}, 2T]$ is balanced. The key is the recursive structure of the 2-associahedra, which implies that $[F_{\mathbf{n}}^{\min}, 2T]$ decomposes as a completed fiber product of products of 2-associahedra.

LEMMA 3.4. *Fix $r \geq 0$ and $\mathbf{m}_1, \ldots, \mathbf{m}_k \in \mathbb{Z}_{\geq 0}^r \setminus \{\mathbf{0}\}$. Then the following identity holds, where the expression in the alternating sum denotes a fiber product with respect to the forgetful map $\pi\colon W_{\mathbf{m}_i} \to K_r$:*

$$A\left(\prod_{1\leq i\leq k}^{K_r} W_{\mathbf{m}_i}\right) = 1. \tag{29}$$

PROOF.

STEP 1. *We compute $A\left(\pi_{\mathbf{m}_1}^{-1}(T) \times \cdots \times \pi_{\mathbf{m}_k}^{-1}(T)\right)$, where T is any element of K_r and $\pi_{\mathbf{m}_i}\colon W_{\mathbf{m}_i} \to K_r$ is the forgetful map.*

$$
\begin{aligned}
A(\pi_{\mathbf{m}_1}^{-1}(T) \times \cdots \times \pi_{\mathbf{m}_k}^{-1}(T)) &= \sum_{\substack{F_i \in W_{\mathbf{m}_i},\\ 1\le i\le k}} (-1)^{d(F_1,\ldots,F_k)} \\
&= \sum_{\substack{F_i \in W_{\mathbf{m}_i},\\ 1\le i\le k}} (-1)^{d(F_1)+\cdots+d(F_k)-(k-1)d(T)} \\
&= (-1)^{-(k-1)d(T)} \prod_{i=1}^{k} \sum_{F_i\in\pi_{\mathbf{m}_i}^{-1}(T)} (-1)^{d(F_i)} \\
&= (-1)^{-(k-1)d(T)} \prod_{i=1}^{k} (-1)^{d(T)} A(W_{\mathbf{m}_i}) \\
&= (-1)^{d(T)}.
\end{aligned} \tag{30}
$$

STEP 2. *We prove the lemma.*

$$
\begin{aligned}
A\Big(\prod^{K_r}_{1\le i\le k} W_{\mathbf{m}_i}\Big) &= A\Big(\bigsqcup_{T\in K_r} (\pi_{\mathbf{m}_1}^{-1}(T) \times \cdots \times \pi_{\mathbf{m}_r}^{-1}(T))\Big) \\
&= \sum_{T\in K_r} A(\pi_{\mathbf{m}_1}^{-1}(T) \times \cdots \times \pi_{\mathbf{m}_k}^{-1}(T)) \\
&= \sum_{T\in K_r} (-1)^{d(T)} \\
&= 1. \quad \square
\end{aligned} \tag{31}
$$

LEMMA 3.5. *Fix* $r \geq 1$, $\mathbf{n} \in \mathbb{Z}^r_{\geq 0} \setminus \{\mathbf{0}\}$, *and* $2T \in W_{\mathbf{n}}$. *Then the sublevel set* $[F^{\min}_{\mathbf{n}}, 2T]$ *is balanced.*

PROOF. By the (RECURSIVE) part of [**Bot19a**, Thm. 4.1], the following posets are isomorphic:

$$
\begin{aligned}
&\prod_{\substack{\alpha\in V^1_{\mathrm{comp}}(T_b),\\ \mathrm{in}(\alpha)=(\beta)}} W^{\mathrm{tree}}_{\#\mathrm{in}(\beta)} \\
&\times \prod_{\rho\in V_{\mathrm{int}}(T_s)} \prod^{K_{\#\mathrm{in}(\rho)}}_{\substack{\alpha\in V^{\geq 2}_{\mathrm{comp}}(T_b)\cap\pi^{-1}\{\rho\},\\ \mathrm{in}(\alpha)=(\beta_1,\ldots,\beta_{\#\mathrm{in}(\rho)})}} W_{\#\mathrm{in}(\beta_1),\ldots,\#\mathrm{in}(\beta_{\#\mathrm{in}(\alpha)})} \simeq (F^{\min}_{\mathbf{n}}, 2T] \subset W_{\mathbf{n}}
\end{aligned} \tag{32}
$$

It now follows from (20) and Lemma 3.4 that $A\big((F^{\min}_{\mathbf{n}}, 2T]\big) = 1$, hence $A\big([F^{\min}_{\mathbf{n}}, 2T]\big) = 0$. $\square$

3.3. All remaining intervals are balanced.

DEFINITION 3.6. Suppose that $(\mathscr{B}, 2\mathscr{B})$ is a 2-bracket in $W^{\mathrm{br}}_{\mathbf{n}}$. We say that a bracket $B \in \mathscr{B}$ is *removable* if it is not a singleton and does not contain all elements of $(1,\ldots,r)$. We say that a 2-bracket $\mathbf{2B} \in 2\mathscr{B}$ is *removable* if it is not a singleton and is not the maximal 2-bracket $\mathbf{2B}^{\max}_{\mathbf{n}} := \{(i,j) \mid 1 \leq i \leq r, 1 \leq j \leq n_i\} \subset 2\mathscr{B}$.

LEMMA 3.7. *Fix* $r \geq 1$, $\mathbf{n} \in \mathbb{Z}^r_{\geq 0} \setminus \{\mathbf{0}\}$, *and* $(\mathscr{B}, 2\mathscr{B}) \in W_\mathbf{n} \setminus \{F^{\max}_\mathbf{n}\}$. *Then the interval*

$$I = \left[(\mathscr{B}, 2\mathscr{B}), F^{\max}_\mathbf{n}\right] \subset W^{\mathrm{br}}_\mathbf{n} \tag{33}$$

is balanced.

PROOF. We will induct on r, and subinduct on the number of 2-brackets in $2\mathscr{B}$. If $r = 1$, this lemma follows from the identification $W_n \simeq K_n$ made in [**Bot19a**, Lemma 3.9]. In the remainder of the proof, we fix $(\mathscr{B}, 2\mathscr{B}) \in W^{\mathrm{br}}_\mathbf{n}$ for $r \geq 2$, $\mathbf{n} \in \mathbb{Z}^r_{\geq 0}$, and use the inductive hypothesis to show that I is balanced.

STEP 1. *We prove that* I *is balanced in the case that* $\mathscr{B}$ *contains no removable brackets and every removable 2-bracket* $\mathbf{2B} \in 2\mathscr{B}$ *has* $\pi(\mathbf{2B}) = (1, \ldots, r)$.

We establish this step by strong induction on $d(F^{\max}_\mathbf{n}) - d(\mathscr{B}, 2\mathscr{B})$. This step is trivial when this difference in dimension is 1.

Next, consider the case when this difference is at least 2. There must then be a 2-bracket $\mathbf{2B} \in 2\mathscr{B}$ which properly contains another 2-bracket in $2\mathscr{B}$, and which is maximal in $2\mathscr{B} \setminus \{\mathbf{2B}^{\max}_\mathbf{n}\}$. Indeed, if there were no such 2-bracket, then every 2-bracket in $2\mathscr{B} \setminus \{\mathbf{2B}^{\max}_\mathbf{n}\}$ would be minimal, so by [**Bot19a**, Def. 3.12, (MARKED SEAMS ARE UNFUSED)], we would have $d(\mathscr{B}, 2\mathscr{B}) = d(F^{\max}_\mathbf{n}) - 1$.

Denote by $2\mathscr{B}'$ the 2-brackets that $\mathbf{2B}$ contains, and by $2\mathscr{B}''$ the 2-brackets other than $\mathbf{2B}^{\max}_\mathbf{n}$ that $\mathbf{2B}$ does not include. Then $(\mathscr{B}, 2\mathscr{B}') \in W^{\mathrm{br}}_{\mathbf{n}'}$, $(\mathscr{B}, 2\mathscr{B}'') \in W^{\mathrm{br}}_{\mathbf{n}''}$ are legal 2-bracketings. Using this notation, we can decompose decompose I according to whether an element of I does or does not contain $\mathbf{2B}$:

$$\begin{aligned} I &= \left[(\mathscr{B}, 2\mathscr{B}), F^{\max}_\mathbf{n}\right] \\ &\simeq \Big(\left[(\mathscr{B}, 2\mathscr{B}'), F^{\max}_{\mathbf{n}'}\right] \times \left(\left[(\mathscr{B}, 2\mathscr{B}''), F^{\max}_{\mathbf{n}''}\right] \setminus \{F^{\max}_{\mathbf{n}''}\}\right)\Big) \\ &\quad \sqcup \left[(\mathscr{B}, 2\mathscr{B} \setminus \{\mathbf{2B}\}), F^{\max}_\mathbf{n}\right]. \end{aligned} \tag{34}$$

Indeed, given an element of the product that appears on the left-hand side of the final expression, we can insert the 2-bracketing in $W^{\mathrm{br}}_{\mathbf{n}'}$ into $\mathbf{2B}$ and use the 2-bracketing in $W^{\mathrm{br}}_{\mathbf{n}''}$ to bracket the remaining elements. We do not allow the 2-bracketing in $W^{\mathrm{br}}_{\mathbf{n}''}$ to be the top element, because the presence of $\mathbf{2B}$ would then violate the (MARKED SEAMS ARE UNFUSED) property. On the other hand, an element in the right-hand part of the final expression in (34) is simply an element of I that does not include $\mathbf{2B}$.

It follows from (34) and the inductive hypothesis that I is balanced.

STEP 2. *We prove that* I *is balanced in the case that* $2\mathscr{B}$ *does not contains a 2-bracket of the form* $\mathbf{2B} = (B, (2B_i)) \neq \mathbf{2B}^{\max}_\mathbf{n}$ *with* $2B_i = \{(i,j) \mid 1 \leq j \leq n_i\}$ *for all* i.

Fix $B \in \mathscr{B}$, and set $s := \#B$ and $\mathbf{n}(B) := (n_i)_{i \in B} \in \mathbb{Z}^s_{\geq 0}$. We can then define a 2-bracketing $(\mathscr{B}(B), 2\mathscr{B}(B))$ like so:

$$\mathscr{B}(B) := \{(i)\}_{i \in B} \cup \{B\}, \quad 2\mathscr{B}(B) := \{(i,j)\}_{\substack{i \in B, \\ 1 \leq j \leq n_i}} \cup \left(2\mathscr{B} \cap \pi^{-1}\{B\}\right) \cup \{\mathbf{2B}^{\max}_{\mathbf{n}(B)}\}. \tag{35}$$

That is, $(\mathscr{B}(B), 2\mathscr{B}(B))$ contains only those nonremovable elements of $2\mathscr{B}$ that lie over B. Using this construction, we can decompose I in terms of the brackets in $\mathscr{B}$:

$$I \simeq \prod_{B \in \mathscr{B}} \left[\big(\mathscr{B}(B), 2\mathscr{B}(B)\big), F^{\max}_{\mathbf{n}(B)} \right]. \tag{36}$$

Indeed, given an element of this product, we can take the union of all the brackets and 2-brackets to obtain an element of I, and this correspondence is clearly bijective.

It follows from (35) and Step 1 that I is balanced.

STEP 3. *We prove the lemma.*

If I satisfies the hypothesis of Step 2, we are done. Otherwise, fix a 2-bracket $\mathbf{2B} \in 2\mathscr{B}$ of the sort excluded in Step 2. If $\mathbf{2B}$ does not properly contain any element of $2\mathscr{B} \cap \pi^{-1}\{B\}$, then we have

$$I \simeq \left[\big(\mathscr{B}, 2\mathscr{B} \setminus \{\mathbf{2B}\}\big), F^{\max}_{\mathbf{n}} \right], \tag{37}$$

so I is balanced by the inductive hypothesis.

Next, suppose that $\mathbf{2B}$ does properly contain an element of $2\mathscr{B} \cap \pi^{-1}\{B\}$. Similarly to (34), we can decompose I according to whether an element of I does or does not contain $\mathbf{2B}$:

$$I \simeq \left[\big(\mathscr{B}, 2\mathscr{B} \setminus \{\mathbf{2B}\}\big), F^{\max}_{\mathbf{n}} \right] \sqcup \left[\big(\mathscr{B}, 2\mathscr{B} \setminus \{\mathbf{2B}\}\big), F^{\max}_{\mathbf{n}} \right]. \tag{38}$$

Indeed, if we take an element of the first copy of $\left[\big(\mathscr{B}, 2\mathscr{B} \setminus \{\mathbf{2B}\}\big), F^{\max}_{\mathbf{n}} \right]$, we can add $\mathbf{2B}$ to produce a 2-bracketing in I that includes $\mathbf{2B}$. An element of the second copy of $\left[\big(\mathscr{B}, 2\mathscr{B} \setminus \{\mathbf{2B}\}\big), F^{\max}_{\mathbf{n}} \right]$ corresponds to a 2-bracketing in I that does not include $\mathbf{2B}$.

It follows from (38) and the inductive hypothesis that I is balanced. □

LEMMA 3.8. *Fix $r \geq 1$, $\mathbf{n} \in \mathbb{Z}^r_{\geq 0} \setminus \{\mathbf{0}\}$, and $\big(\mathscr{B}^{(i)}, 2\mathscr{B}^{(i)}\big) \in W^{\mathrm{br}}_{\mathbf{n}}$ for $i \in \{1,2\}$ with*

$$\big(\mathscr{B}^{(2)}, 2\mathscr{B}^{(2)}\big) < \big(\mathscr{B}^{(1)}, 2\mathscr{B}^{(1)}\big). \tag{39}$$

Then the interval $I = \left[\big(\mathscr{B}^{(2)}, 2\mathscr{B}^{(2)}\big), \big(\mathscr{B}^{(1)}, 2\mathscr{B}^{(1)}\big) \right]$ is balanced.

PROOF. By (32), it suffices to show that in any fiber product $W_{\mathbf{m}_1} \times_{K_r} \cdots \times_{K_r} W_{\mathbf{m}_k}$, any nontrivial interval with upper bound $\big(F^{\max}_{\mathbf{m}_1}, \ldots, F^{\max}_{\mathbf{m}_k}\big)$ is balanced. This follows from an argument very similar to our proof of Lemma 3.7. □

Acknowledgments

The first author thanks Richard Ehrenborg, Margaret Readdy, and Lauren Williams for useful conversations. The first author was supported by an NSF Mathematical Sciences Postdoctoral Research Fellowship and by an NSF Standard Grant (DMS-1906220).

References

[BC21] Nathaniel Bottman and Shachar Carmeli, *$(A_\infty, 2)$-categories and relative 2-operads*, High. Struct. **5** (2021), no. 1, 401–421. MR4367226

[Bot19a] Nathaniel Bottman, *2-associahedra*, Algebr. Geom. Topol. **19** (2019), no. 2, 743–806, DOI 10.2140/agt.2019.19.743. MR3924177

[Bot19b] Nathaniel Bottman, *Moduli spaces of witch curves topologically realize the 2-associahedra*, J. Symplectic Geom. **17** (2019), no. 6, 1649–1682, DOI 10.4310/JSG.2019.v17.n6.a3. MR4057724

[BW18] Nathaniel Bottman and Katrin Wehrheim, *Gromov compactness for squiggly strip shrinking in pseudoholomorphic quilts*, Selecta Math. (N.S.) **24** (2018), no. 4, 3381–3443, DOI 10.1007/s00029-018-0404-4. MR3848023

[MWW18] S. Ma'u, K. Wehrheim, and C. Woodward, *A_∞ functors for Lagrangian correspondences*, Selecta Math. (N.S.) **24** (2018), no. 3, 1913–2002, DOI 10.1007/s00029-018-0403-5. MR3816496

[SI] Neil J. A. Sloane and The OEIS Foundation Inc. The on-line encyclopedia of integer sequences.

[Sta94] Richard P. Stanley, *Flag f-vectors and the cd-index*, Math. Z. **216** (1994), no. 3, 483–499, DOI 10.1007/BF02572336. MR1283084

Max Planck Institute for Mathematics, Vivatsgasse 7, 53111 Bonn NRW, Germany
Email address: bottman@mpim-bonn.mpg.de

Department of Mathematics, Princeton University, Fine Hall, Washington Rd, Princeton, New Jersey 08544
Email address: mavrides@princeton.edu

Contemporary Mathematics
Volume **841**, 2026
https://doi.org/10.1090/conm/841/16835

A knot-theoretic approach to comparing the Grothendieck-Teichmüller and Kashiwara-Vergne groups

Zsuzsanna Dancso, Tamara Hogan, and Marcy Robertson

ABSTRACT. Homomorphic expansions are combinatorial invariants of knotted objects, which are universal in the sense that all finite-type (Vassiliev) invariants factor through them. Homomorphic expansions are also important as bridging objects between low-dimensional topology and quantum algebra. For example, homomorphic expansions of parenthesised braids are in one-to-one correspondence with Drinfel'd associators (Bar-Natan 1998), and homomorphic expansions of w-foams are in one-to-one correspondence with solutions to the Kashiwara-Vergne (KV) equations (Bar-Natan and the first author, 2017). The sets of Drinfel'd associators and KV solutions are both bi-torsors, with actions by the pro-unipotent Grothendieck-Teichmüller and Kashiwara-Vergne groups, respectively. The above correspondences are in fact maps of bi-torsors (Bar-Natan 1998, and the first and third authors with Halacheva 2022).

There is a deep relationship between Drinfel'd associators and KV equations—discovered by Alekseev, Enriquez and Torossian in the 2010s—including an explicit formula constructing KV solutions in terms of associators, and an injective map $\rho : \mathsf{GRT}_1 \to \mathsf{KRV}$. This paper is a topological/diagrammatic study of the image of the Grothendieck-Teichmüller groups in the Kashiwara-Vergne symmetry groups, using the fact that both parenthesised braids and w-foams admit respective finite presentations as an operad and as a tensor category (circuit algebra or prop).

1. Introduction

Given a class of knotted objects, $\mathcal{K}$, an *expansion*, or *universal finite type invariant*, is an assignment $Z : \widehat{\mathcal{K}} \to \mathcal{A}$ taking values in a vector space, $\mathcal{A}$, which is graded with respect to a *Vassiliev filtration* of $\mathcal{K}$, and satisfies a universality condition (Definition 2.2). Expansions are called *homomorphic* if they respect any operations defined on $\mathcal{K}$, such as braid composition, knot connected sum or strand doubling. One of the best known examples of a homomorphic expansion is the Kontsevich integral ([**Kon93**], [**CDM12**], [**Dan10**]) which is universal in the sense that any finite type invariant of oriented links factors through it.

In practice, constructing homomorphic expansions is difficult and, in many cases (e.g. for the planar algebra of tangles), they do not exist. The construction can be simplified, however, when $\mathcal{K}$ is finitely presented as an algebraic structure.

2020 *Mathematics Subject Classification.* Primary 18M85, 57K16; Secondary 18M15, 18M60.
The third author acknowledges the support of Australian Research Council Future Fellowship FT210100256.

©2026 by the authors

Examples of this include parenthesised braids viewed as an operad or prop (Section 2.1;[**BN98**]), welded tangles as a tensor category (Section 2.3; [**BND16**, Section 2]) and so forth. When this is the case, the Vassiliev filtration of $\mathcal{K}$ coincides with the I-adic filtration by powers of the augmentation ideal, and finding a homomorphic expansion is equivalent to solving a set of equations in the associated graded space $\mathcal{A}$, which admits a diagrammatic description as vector spaces spanned by Jacobi or Feynman diagrams (Section 2.2 and Definition 2.13).

Homomorphic expansions also have deep connections to quantum algebra. For example, using the description of parenthesised braids as a tensor category, Bar-Natan [**BN98**] showed that there is a bijection between Drinfel'd associators and homomorphic expansions of parenthesised braids. Similarly, work of Bar-Natan and the first author showed that homomorphic expansions of a four-dimensional class of knotted objects called *welded foams* are in bijection with solutions to the *Kashiwara-Vergne equations* (see Section 2.3 or [**BND17**, Theorems 4.8 and 4.23]).

This paper is part of an ongoing study of the relationship between the homomorphic expansions of parenthesised braids and welded foams, informed by equivalent phenomena in quantum algebra. Informally, the Kashiwara-Vergne (KV) problem ([**KV78**]) asks when one can simplify the expression:

$$e^x e^y = e^{x+y+\frac{1}{2}[x,y]+\frac{1}{12}[x,[x,y]]+\dots}, \tag{1.1}$$

where $[x,y] = xy - yx$. The Lie series in the exponent is called the Baker-Campbell-Hausdorff series [**Hal15**] of x and y, and is denoted by $\mathfrak{bch}(x,y)$. A solution to the Kashiwara-Vergne equations allows one to replace $\mathfrak{bch}(x,y)$ with a convergent power series and adjoint operations [**KV78**]. In practice, this problem transforms into a question about when *tangential*, or basis-conjugating, automorphisms of free Lie algebras (Section 2.4.1) satisfy two equations ([**AT12**, Section 5.3]).

Drinfel'd observed in [**Dri90**, Section 4] that elements of the Lie algebras of infinitesimal braids (also known as Drinfel'd-Kohno Lie algebras, Section 2.2) induce tangential derivations of the free Lie algebras (e.g. Example 2.10; Proposition 3.11 of [**AT12**]). In a series of papers ([**AT12**], [**AET10**]), Alekseev, Enriquez and Torossian exploited this observation to show that solutions of the Kashiwara-Vergne equations can be explicitly constructed from Drinfel'd associators.

In [**BND17**, Section 4], Bar-Natan and the first author showed that KV solutions are in bijection with homomorphic expansions of welded foams. The explicit construction of KV solutions from Drinfel'd associators will be carried out on the topological side by explicitly constructing homomorphic expansions of welded foams from expansions of parenthesised braids in the upcoming paper ([**BND24b**]). This paper accomplishes a similar goal for the symmetry groups: that is, an explicit topological construction of an element of KRV from an element of GRT_1, equivalent to the Alekseev-Torossian map. The main difficulty is to check that the result of the construction is, in fact, an element of KRV. This is a less complex problem than the corresponding task for KV solutions, as the first defining equation of the group KRV is simpler than the first Kashiwara–Vergne equation. Hence, a less involved topological construction – as presented in this paper – is sufficient.

The relationship between Drinfel'd associators and KV solutions described by Alekseev, Enriquez and Torossian extends to an intricate relationship between the symmetry groups of these objects: the prounipotent Grothendieck–Teichmüller groups (GT and GRT_1) and Kashiwara-Vergne groups (KV and KRV), respectively.

The sets of Drinfel'd associators and KV solutions are bi-torsors under the actions of these symmetry groups (Section 2.2 and Section 2.6, respectively). In particular, Alekseev and Torossian [**AT12**, Theorem 4.6] describe a map $\rho : \mathsf{GRT}_1 \to \mathsf{KRV}$. It is a still open conjecture to show that this map is nearly an isomorphism, i.e. that $\mathsf{KRV} \cong \mathsf{GRT}_1 \times \mathbb{K}$ ([**AT12**, Section 4]).

Several authors have demonstrated that the Grothendieck–Teichmüller groups, GT and GRT_1 are isomorphic to automorphism groups of (prounipotently completed) parenthesised braids and their associated graded, parenthesised *chord diagrams* ([**BN98, Tam98, Fre17, Wil15**]):

$$\mathrm{Aut}(\widehat{\mathsf{PaB}}) \cong \mathsf{GT} \quad \text{and} \quad \mathrm{Aut}(\mathsf{PaCD}) \overset{\tau}{\cong} \mathsf{GRT}_1.$$

Subsequently, the first and last authors, with Halacheva [**DHR23**], identified the Kashiwara-Vergne groups, KV and KRV, as automorphisms of (prounipotently completed) welded foams and their associated graded prop [**DHR21**] of *arrow diagrams*,

$$\mathrm{Aut}_v(\widehat{\mathsf{wF}}) \cong \mathsf{KV} \quad \text{and} \quad \mathrm{Aut}_v(\mathscr{A}) \overset{\Theta}{\cong} \mathsf{KRV}.$$

This paper builds on these results and those of [**BND17, DHR21, DHR23**] to construct an explicit topologically-based interpretation of the Alekseev-Torossian (AT) map ([**AT12**, Section 4]; Construction 3.4)

$$\rho : \mathsf{GRT}_1 \to \mathsf{KRV}.$$

Explicitly, we construct a map $\mathring{\rho} : \mathrm{Aut}(\mathsf{PaCD}) \to \mathrm{Aut}_v(\mathscr{A})$ which agrees with the Alekseev-Torossian map:

THEOREM (Theorem 3.7 and Corollary 3.14). *The following is a commutative diagram of prounipotent groups:*

$$\begin{array}{ccc} \mathrm{Aut}(\mathsf{PaCD}) & \xrightarrow{\ \ \mathring{\rho}\ \ } & \mathrm{Aut}_v(\mathscr{A}) \\ \cong \big\downarrow \tau & & \Theta \big\downarrow \cong \\ \mathsf{GRT}_1 & \xrightarrow[\ \ \rho\ \ ]{} & \mathsf{KRV}. \end{array}$$

In particular, we identify a subgroup of arrow diagram automorphisms isomorphic to GRT_1. This is a further step towards understanding the relationship between automorphisms of chord diagrams and arrow diagrams. Moreover, these automorphisms preserve the homomorphic expansions of w-foams which come from parenthesised braids.

The reader may ask, in light of this result, is an explicit 4-dimensional approach to the group homomorphism $\mathsf{GT} \to \mathsf{KV}$ now within reach? While results regarding the graded groups transfer formally to the filtered groups via the isomorphism induced by a choice of homomorphic expansion, explicitly describing the resulting map is much more difficult. The reason is that GT and KV are descibed as automorphisms of pro-unipotent completions of parenthesised braids, and welded foams, respectively, whose elements are much less tractable than chord- or arrow diagrams. This complexity is also reflected in the complexity of homomorphic expansions.

1.1. Structure and Conventions. Throughout this paper we assume that the reader is familiar with basic notations of category theory, in the form of linear tensor categories, as well as some basics around the theory of (coloured) operads. Some review of these details are provided in Appendix B as well as some basic definitions of (wheeled) props.

Sections 2.1 and 2.2 provide an expository review of homomorphic expansions of *parenthesised braids*,

$$\widehat{\mathbb{Q}[\mathsf{PaB}]} \xrightarrow{Z} \mathsf{PaCD},$$

and their relationship to the graded Grothendeick-Teichmüller group, GRT_1. Throughout, we use the fact that the class of parenthesised braids, PaB and their associated graded parenthesised chord diagrams PaCD, can be finitely presented as *tensor categories* or *props* as in [**BN98**], [**JS93**], [**AET10**] and [**Tam98**]. Readers conversant with this material may be more familiar with the presentation of parenthesised braids and chord diagrams as operads, as in [**Fre17**], [**Mer21**], etc., but the language of props leads to a more natural discussion of relationships between parenthesised braids and welded foams.

Section 2.3 introduces a class of four-dimensional tangles called *welded foams*. Welded foams, hereafter called *w-foams*, are a simultaneous generalisation of w-braids (aka loop braids, [**FRR97**], [**Dam17,BWC07**]) and classical tangles. Topologically, they are knotted ribbon tubes and ribbon tori in $\mathbb{R}^4$. In [**BND17**], Bar-Natan and the first author introduced the notion of a *circuit algebra* to encode the algebraic structure of w-foams and construct their homomorphic expansions. Later, in a paper with Halacheva, the first and third author showed that circuit algebras are themselves a type of tensor category called a *wheeled prop* [**DHR21**]. This algebraic structure is described in full detail in [**DHR21**, Sections 2 and 4], and also in abbreviated form in [**BND17,DHR23**]. While it provides foundations for this paper, it does not play a major role in the proof of our main theorem, and thus we push the details to Appendix B.

The construction of the map $\mathring{\rho} : \mathrm{Aut}(\mathsf{PaCD}) \to \mathrm{Aut}_v(\mathscr{A})$ appears in Section 3. We then go through the technical proof that this map is an injective group homomorphism and agrees with the Alekseev-Torossian map $\rho : \mathsf{GRT}_1 \to \mathsf{KRV}$. In particular, we conclude that $\mathrm{Aut}(\mathsf{PaCD})$ can be seen as a subgroup of $\mathrm{Aut}_v(\mathscr{A})$. Appendix A gives an alternative, more constructive, proof of the same result.

2. Preliminaries: Topology, Lie algebras, and expansions

In this section we give a short review of how invariants called *homomorphic expansions* provide a bridge between knot theory, quantum algebra and Lie theory. Expansions in knot theory are inspired by an idea from group theory, which aims to study a group ring through its associated graded ring. More explicitly, given an abstract group G, the complete *associated graded algebra* of the group ring $\mathbb{Q}[G]$ is the $\mathbb{Q}$-algebra

$$A(G) := \mathrm{gr}(\mathbb{Q}[G]) = \prod_{m \geq 0} \mathcal{I}^m / \mathcal{I}^{m+1},$$

where $\mathcal{I} = \ker\left(\mathbb{Q}[G] \xrightarrow{\epsilon} \mathbb{Q}\right)$ is the augmentation ideal.

DEFINITION 2.1. An *expansion* of the group G is a filtered algebra isomorphism

$$Z : \widehat{\mathbb{Q}[G]} \to A(G).$$

Here the hat notation denotes the pro-unipotent completion of the group-ring $\mathbb{Q}[G]$. An equivalent statement is to say that the induced map

$$\mathrm{gr}(Z) : \left(\mathrm{gr}(\widehat{\mathbb{Q}[G]}) = A(G)\right) \to (\mathrm{gr}(A(G)) = A(G))$$

is the identity of $A(G)$. The latter condition is saying that the degree m piece of Z, restricted to $\mathcal{I}^m$, is the projection onto $\mathcal{I}^m/\mathcal{I}^{m+1}$.

Replacing groups by classes of knotted objects, this idea extends to provide useful knot invariants. This is particularly true when the class of knotted objects $\mathcal{K}$ can be finitely presented as a tensor category. Examples include braids, viewed as a prop [**JS93, BN98**], (virtual) tangles, viewed as a ribbon category [**Bro19, JS93, Tur88, FY89**], and welded tangles, viewed as a wheeled prop [**BND17, DHR23**]. In the last example, the topological analogue of the group ring, $\mathbb{Q}\mathcal{K}$, is the linear tensor category whose morphism spaces are vector spaces spanned by elements of $\mathcal{K}$. The category $\mathbb{Q}\mathcal{K}$ is filtered by powers of the augmentation ideal $\mathcal{I} = \{\sum c_i K_i | c_i \in \mathbb{Q}, K_i \in \mathcal{K}, \sum c_i = 0\}$. As with the examples from group theory, the complete associated graded structure is the direct product $\mathcal{A} := \prod_{m\geq 0} \mathcal{I}^m/\mathcal{I}^{m+1}$. Further details on this type of filtration can be found in [**BN98**, Section 2.1.1] or [**DHR23**, Section 2.3].

Definition 2.2 ([**BND17**, Definition 2.5]). Let $\mathcal{K}$ denote a class of knotted objects finitely presented as a tensor category (possibly with extra structure). A *homomorphic expansion* of $\mathcal{K}$ is a filtered linear isomorphism[1] $Z : \widehat{\mathbb{Q}\mathcal{K}} \to \mathcal{A}$ which intertwines all operations on $\mathcal{K}$ with their associated graded counterparts on $\mathcal{A}$.

When a class of knotted objects $\mathcal{K}$ is finitely presented as a tensor category, defining a homomorphic expansion reduces to defining a functor at the values of the generators of $\mathcal{K}$ in $\mathcal{A}$. These values are subject to the relations of $\mathcal{K}$, which define a set of equations in $\mathcal{A}$. This seemingly simple exercise can build powerful bridges between knot invariants and algebra. We review two examples of interest:

(1) When $\mathcal{K}$ is *parenthesised braids*, the equations which determine a homomorphic expansion turn out to be the *pentagon* and *hexagon* equations which define *Drinfel'd associators*. Hence, homomorphic expansions of parenthesised braids are in bijection with Drinfel'd associators (Theorem 2.4). This bijection turns out to be an isomorphism of torsors: the expansion-preserving automorphism groups of (completed) parenthesised braids and their associated graded structure are identified with the symmetry groups of Drinfel'd associators, the Grothendieck-Teichmüller groups ([**BN97, BN98, Tam98, Fre17**], etc.).

(2) When $\mathcal{K}$ is the class of *welded foams*, the equations which determine homomorphic expansions are the *Kashiwara-Vergne* equations of Lie theory (Section 2.5). Thus, homomorphic expansions for welded foams are in bijection with Kashiwara-Vergne solutions ([**BND17**, Theorems 4.8, 4.23]). This bijection can also be promoted to an isomorphism of torsors: the expansion preserving automorphism groups of (completed) welded foams, and their associated graded structure, are isomorphic to the Kashiwara-Vergne symmetry groups [**DHR23**, Theorems 5.12 and 5.16].

This preliminary section is dedicated to a review of these two examples.

[1][**BND17**, Definition 2.5] uses the equivalent statement "$\mathrm{gr}(Z) = \mathrm{id}_{\mathcal{A}}$", which avoids the pro-unipotent completion on the left.

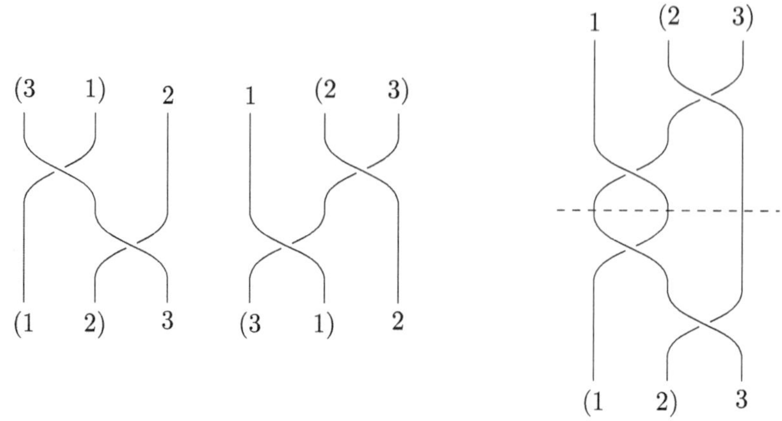

FIGURE 1. Two elements of $\mathsf{PaB}(3)$ and their stacking composition. We draw all diagrams in this paper as going from bottom to top, e.g. the braid on the far left goes from (12)3 to (31)2.

2.1. Braids. The braid group on n-strands is the fundamental group of the space of unordered configurations of n points in the complex plane. By Alexander's Theorem, every link can be represented as a braid closure, and thus braids are an important tool in the study of knots and links. The braid group has a finite presentation

$$\mathsf{B}_n = \langle \beta_1, \ldots, \beta_{n-1} \mid \beta_i\beta_j = \beta_j\beta_i \text{ when } |i-j| \geq 2;\ \beta_i\beta_{i+1}\beta_i = \beta_{i+1}\beta_i\beta_{i+1} \rangle .$$

where β_i is braid on n-strands which crosses the i-th strand over the $(i+1)$-th strand. For any $n \geq 1$, there is a homomorphism to the symmetric group

$$\mathsf{B}_n \xrightarrow{\ \pi\ } \mathcal{S}_n$$

where the map π sends the braid generator β_i to the transposition $(i, i+1)$ in the symmetric group $\mathcal{S}_n$.

The category of *parenthesised braids* with n-strands, $\mathsf{PaB}(n)$, is a category whose objects are the set of *parenthesised* permutations in n-letters and whose morphisms are braids on n-strands. Composition of morphisms in this category is given by "stacking" braids with compatible parenthesisations, as pictured in Figure 1. Every morphism β in $\mathsf{PaB}(n)$ has an underlying *skeleton*: this is the data of the source and target parenthesised permutations (which also specifies the underlying permutation $\pi(\beta)$ of β). For each skeleton s we write $\mathsf{PaB}(s)$ for the set of braids with skeleton s.

The collection of all parenthesised braids is a freely generated symmetric monoidal category, where the tensor product functor $\mathsf{PaB}(s) \otimes \mathsf{PaB}(t) \to \mathsf{PaB}(s \cup t)$ is given by the concatenation of parenthesised permutations s and t (e.g. [**JS93**, Example 1, pg 55]). On morphisms, this tensor product is the juxtaposition of braids — visually, placing the parenthesised braids next to each other. One often abuses notation and writes

$$\mathsf{PaB} = \coprod_{s \in \mathcal{S}} \mathsf{PaB}(s)$$

to describe this phenomenon.

The category of parenthesised braids is *finitely presented* in the sense that every morphism can be written as a combination of tensors, categorical compositions, strand deletions and strand doubling of two generators:

 $\in \mathrm{Hom}_{\mathsf{PaB}(2)}((12),(21))$ and  $\in \mathrm{Hom}_{\mathsf{PaB}(3)}((12)3, 1(23))$.

These generating morphisms satisfy the *pentagon* and two *hexagon* equations of braided monoidal categories. For further details on this categorical structure, we recommend [**BN98**, Section 2.1, Claim 2.6] or [**JS93**, Example 1, pg 56].

REMARK 2.3. Many readers of this article will be familiar with the fact that the collection of parenthesised braids $\{\mathsf{PaB}(n)\}$, $n \geq 1$, is a groupoid model for the operad of little 2-discs (see, for example, [**Fre17**, Chapter 6], [**Mer21**]). The tensor category described here is the prop associated to this operad. We have chosen this presentation because it allows for a more natural discussion of expansions of both braids and w-foams in the same language.

2.2. The associated graded structure: The infinitesimal braid algebra. The analogue of the group ring for parenthesised braids is the linear extension of PaB,

$$\mathbb{Q}[\mathsf{PaB}] = \bigsqcup_{s \in \mathcal{S}} \mathbb{Q}[\mathsf{PaB}](s).$$

Here $\mathbb{Q}[\mathsf{PaB}](s)$ is a vector space spanned by parenthesised braids with skeleton s, i.e. elements are linear combinations of morphisms $B = a_1\beta_1 + a_2\beta_2 + \ldots$ where β_i is a braid with skeleton s and $a_i \in \mathbb{Q}$.

The complete associated graded space of PaB is the direct product of the successive quotients of the powers of the augmentation ideal

$$\mathcal{A} := \Pi_{n=0}^{\infty} \mathcal{I}^n / \mathcal{I}^{n+1}.$$

The associated graded space of PaB has several (isomorphic) models, the most common of which is the tensor category PaCD, whose morphism spaces,

$$\mathsf{PaCD} := \bigsqcup_{s \in \mathcal{S}} \mathsf{PaCD}(s),$$

are completed graded vector spaces spanned by parenthesised *chord diagrams.*

Chord diagrams are most concisely defined as elements of the universal enveloping algebras of the *infinitesimal braid* Lie algebras. The Lie algebra of infinitesimal braids on n strands, $\mathfrak{t}_n$, $n \geq 2$, is the quotient of the free Lie algebra $\widehat{\mathfrak{lie}}_{\frac{n(n-1)}{2}}$ generated by the symbols $\{t^{i,j} = t^{j,i}\}_{1 \leq i < j \leq n}$ subject to the relations

$$[t^{i,j}, t^{k,l}] = 0 \quad \text{and} \quad [t^{i,j}, t^{i,k} + t^{k,j}] = 0$$

whenever i, j, k, l are distinct. The completed universal enveloping algebra, $\widehat{U}(\mathfrak{t}_n)$, can be viewed as a category of (non-parenthesised) *chord diagrams* on n strands, $\mathsf{CD}(n)$. This category has one object and is enriched in (completed, filtered, co-associative) co-algebras (e.g. [**Mer21**, Section 6.4], [**Fre17**, Chapter 10.2]). Morphisms of $\mathsf{CD}(n)$ are depicted as chord diagrams on n vertical strands, where $t^{i,j}$ is represented by a chord between strands i and j. Composition of morphisms is depicted by "stacking" the chord diagrams. The category of *parenthesised chord diagrams* on n-strands, $\mathsf{PaCD}(n)$, is obtained by formally replacing the single object of the category $\mathsf{CD}(n)$ with the set of parenthesised words of length n with distinct letters. A parenthesised chord diagram is shown on the right-hand side of Figure 2.

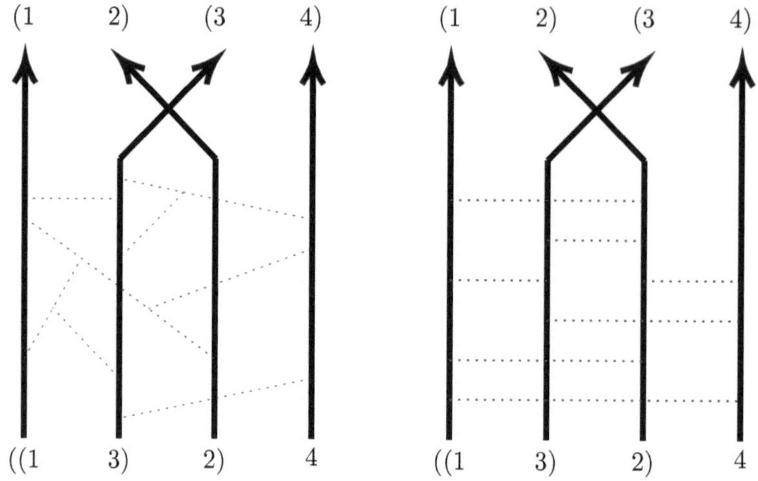

FIGURE 2. On the right is an example of an element of $\mathsf{PaCD}(4)$. The skeleton is depicted by thick black strands and the elements of $U(\mathfrak{t}_4)$ are depicted by dotted blue chords. On the left is a Jacobi diagram on the same skeleton.

An alternative combinatorial description of morphisms in PaCD, which has many computational advantages, is as a vector space of *Jacobi diagrams.* Roughly speaking, a Jacobi diagram is a graph with trivalent and univalent vertices, the latter of which are arranged along the braid skeleton, as on the left side of Figure 2.These diagrams are subject to a set of relations (STU, IHX and AS), which encode the familiar relations of Lie algebras. For a more comprehensive introduction, we recommend the survey article [**BN95**] or the book [**CDM12**] for further details.

Homomorphic expansions of braids are functors from the tensor category PaB to its associated graded $\mathcal{A} = \Pi_{n=0}^{\infty}\mathcal{I}^n/\mathcal{I}^{n+1} \cong \mathsf{PaCD}$. The filtration on PaB given by powers of the augmentation ideal turns out to agree with the Vassiliev filtration of knot theory, and thus homomorphic expansions agree with universal finite type invariants, such as the Konstevich integral ([**Kon93**], [**MO97**], [**Dan10**], etc.)

Explicitly, homomorphic expansions of parenthesised braids $Z : \widehat{\mathbb{Q}[\mathsf{PaB}]} \to \mathsf{PaCD}$ are isomorphisms of tensor categories determined by the values Z takes on the generating morphisms:

$$Z(\times) = e^{\frac{1}{2}t^{12}} \quad \text{and} \quad Z(\Phi) = \Phi(t^{12}, t^{23}),$$

where $\Phi(x, y) \in U(\widehat{\mathfrak{lie}_2})$. These values must necessarily satisfy the following equations, arising from the relations of PaB:

$$\Phi(x, y) = \Phi(y, x)^{-1}, \tag{2.1}$$

$$\Phi(t^{12}, t^{23})e^{-\frac{1}{2}t^{23}}\Phi(t^{13}, t^{23})^{-1}e^{-\frac{1}{2}t^{13}}\Phi(t^{13}, t^{12})e^{-\frac{1}{2}(t^{13}+t^{23})} = 1, \text{ in } U(\mathfrak{t}_3) \tag{2.2}$$

$$\begin{aligned} &\Phi(t^{13} + t^{23}, t^{34})\Phi(t^{12}, t^{23} + t^{24}) \\ &\qquad = \Phi(t^{12}, t^{23})\Phi(t^{12} + t^{13}, t^{24} + t^{34})\Phi(t^{23}, t^{34}), \text{ in } U(\mathfrak{t}_4). \end{aligned} \tag{2.3}$$

Drinfel'd associators are elements $\Phi(x,y) \in U(\widehat{\mathfrak{lie}_2})$ which satisfy these same equations, leading to the following theorem.

THEOREM 2.4 ([**BN98**, Proposition 3.4]). *There is a one-to-one correspondence between homomorphic expansions of parenthesised braids and the set of Drinfel'd associators.*

The (prounipotent) Grothendieck–Teichmüller groups, GT and GRT_1, defined by Drinfel'd in [**Dri90**, Section 4, 5], are known to act freely and transitively on the set of all Drinfel'd associators. It was shown by Bar-Natan [**BN98**, Propositions 4.5, 4.8] that GRT_1 is isomorphic to the (expansion-preserving) automorphism group of PaCD. It was later understood that the same result can be concisely presented using operads in [**Tam98**], [**Wil15**] and [**Fre17**, Theorem 10.3.6].

DEFINITION 2.5 ([**Dri90**]). The graded Grothendieck–Teichmüller group GRT_1 is the collection of group-like elements, $\Psi(x,y) = e^{\psi(x,y)}$ of the completed universal enveloping (Hopf) algebra $U(\widehat{\mathfrak{lie}_2})$

$$\Psi(x,y) = \Psi^{-1}(y,x), \tag{2.4}$$

$$\Psi(x,y)\Psi(y,z)\Psi(z,x) = 1 \text{ for } x+y+z=0, \tag{2.5}$$

$$\begin{aligned} \Psi(t^{13}+t^{23},t^{34})\Psi(t^{12},t^{23}+t^{24}) \\ = \Psi(t^{12},t^{23})\Psi(t^{12}+t^{13},t^{24}+t^{34})\Psi(t^{23},t^{34}) \text{ in } U(\mathfrak{t}_4). \end{aligned} \tag{2.6}$$

Automorphisms of PaCD which preserve expansions by post-composition are characterised by the fact that they fix both the skeleta of chord diagrams and the morphism $|\cdots| \in \mathrm{Hom}_{\mathsf{PaCD}(2)}((12),(12))$. We slightly abuse notation and write $\mathrm{Aut}(\mathsf{PaCD})$ for the group of *expansion preserving* automorphisms of PaCD. The following theorem can be found in [**BN98**, Proposition 4.5, Proposition 4.8], [**Tam98**] or [**Fre17**, Theorem 10.3.6]. See also [**Mer21**, Section 6.7] for a nice expository treatment.

THEOREM 2.6. *There is an isomorphism of groups* $\tau : \mathrm{Aut}(\mathsf{PaCD}) \to \mathsf{GRT}_1$.

PROOF (SKETCH). The key idea in the proof is that an automorphism F which preserves homomorphic expansions is determined by its values on the generators of PaCD. Since expansion preserving automorphisms fix the value of the morphism $|\cdots|$, an automorphism F is uniquely determined by the value $F(|\!/\!|)$, which can be written as a power series in chords t^{12} and t^{23}, namely $F(|\!/\!|) = \Psi(t^{12},t^{23})$. Understanding the equations imposed by the defining relations of PaCD leads precisely to the defining equations of GRT_1, and thus $\Psi \in \mathsf{GRT}_1$. The isomorphism τ is defined by $\tau(F) = \Psi$. □

Since homomorphic expansions of parenthesised braids are isomorphisms of tensor categories, one can conclude that for each homomorphic expansion, $Z : \widehat{\mathbb{Q}[\mathsf{PaB}]} \to \mathsf{PaCD}$, there is an isomorphism

$$\mathrm{Aut}(\widehat{\mathbb{Q}[\mathsf{PaB}]}) \overset{Z}{\cong} \mathrm{Aut}(\mathsf{PaCD}).$$

From this one can deduce the following (e.g. [**BN98**, Proposition 5.3], [**Fre17**, Proposition 11.3.5]).

COROLLARY 2.7. *There is an isomorphism of prounipotent groups:*

$$\mathsf{GT}_1 \overset{Z}{\cong} \mathrm{Aut}(\widehat{\mathbb{Q}[\mathsf{PaB}]}).$$

2.3. Welded foams. Welded braid groups [**FRR97**], also known as loop braid groups [**Dam17, BWC07**], are generalisations of the classical braid groups. They arise in various contexts in the literature, such as in the study of the configuration spaces of (Euclidean) circles in 3 space, as automorphisms of free groups, string-like defects in 4D BF theory, or as braided ribbon tubes in $\mathbb{R}^4$. A good survey of these ideas can be found in [**Dam17**] or the introduction to [**BND16**].

Welded foams, hereafter referred to as *w-foams*, are a class of four-dimensional knotted objects extending w-braids in several ways. They can be thought of as tangles of ribbon tubes, ribbon tori and (one-dimensional) strings in $\mathbb{R}^4$. As this paper deals almost entirely with the graded side, the topology of w-foams plays a limited role, primarily as motivation. We present a brief overview here for context, and refer the reader to [**BND17**, Sections 3 and 4][2] for more detail.

We follow the usual convention for presenting classes of tangles and describe w-foams using a Reidemeister, or diagrammatic, description (Definition 2.8). Specifically, as with parenthesised braids, w-foams have a presentation as a type of tensor category, called a two-coloured *wheeled prop*: this is a rigid tensor category freely generated by two objects (cf. [**DHR21**, Definition 6.1]). Morphisms in w-foams are all compositions of *crossings*, *vertices* (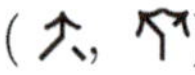) and *caps* () and subject to relations (of the Reidemeister flavour).

DEFINITION 2.8. The class of w-foams is the two-coloured *wheeled prop*:[3]

wF $= \langle$  $\mid R1^f, R2, R3, OC, CP \rangle$.

By convention and for brevity, generators are only shown with strands oriented upwards, but they are understood to come in all possible strand orientations.

An example of a w-foam depicted as tangle diagram is shown in Figure 3. In this diagram, tubes are drawn as thick black lines and strings as thin red lines. Each generator, circled with dashed lines, stands for a specific topological feature embedded in $\mathbb{R}^4$: crossings are tubes threaded through each other as in Figure 4; *vertices* represent tubes merging, , or splitting, , and the tube-string vertex  represents a string attached to the side of a tube. For more detail on vertices, see [**BND17**, Section 4.2.1]. A cap, denoted by a capped black strand , stands for a tube capped off by a disk. The crossings in Figure 3 which are *not* circled are virtual or *wiring diagram* crossings, as shown in Figure 4.

The wheeled prop of w-foams is equipped with several *auxiliary operations* in addition to circuit algebra compositions. These are *orientation switches*, *unzips* (doubling a tube along framing), *punctures* (puncture and retract a 2-dimensional tube onto a 1-dimensional string), and an *adjoint* operation. We suppress the details here (see [**BND17**, Section 4.2.3] and [**BND24b**] for more), but we present a full description of the associated graded versions of these operations in Definition 2.14.

[2]There is a recent Corrigendum [**BND24a**] to [**BND17**, Section 4]. The corrigendum's changes were incorporated into the arXiv version of [**BND17**] for readers' convenience. Hence, specific section references to [**BND17**] are pointing to the arXiv v4, rather than the published version.

[3]In [**BND24b**] and [**BNDS20**], wF is denoted $\widetilde{wTF}$.

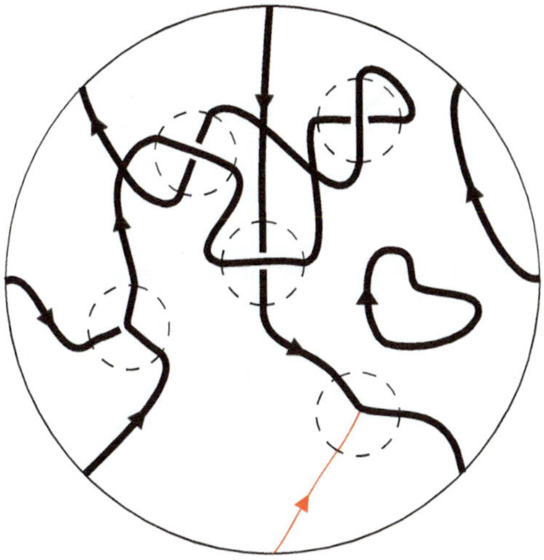

FIGURE 3. An example of a circuit algebra composition of five generators in wF.

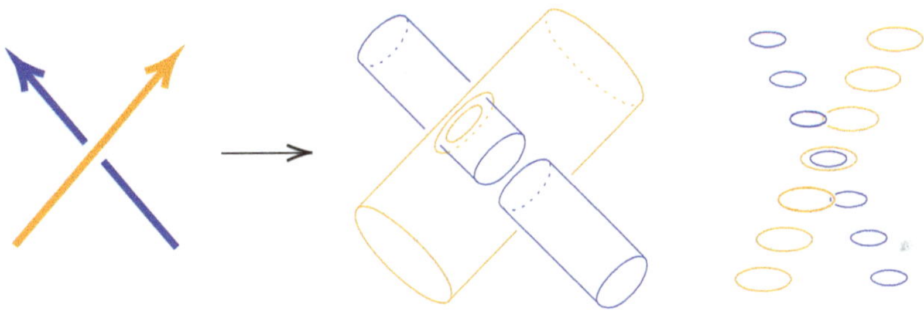

(A)

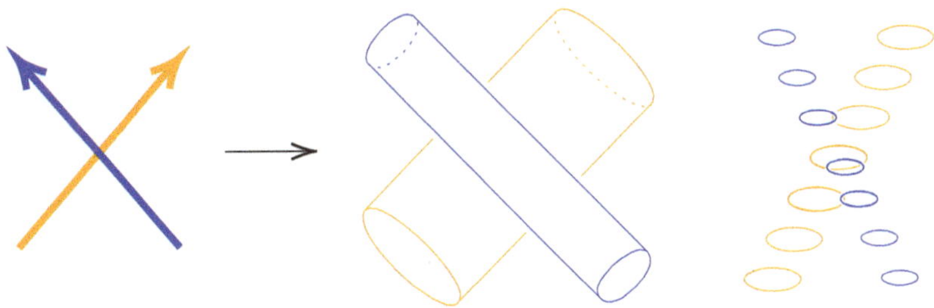

(B)

FIGURE 4. (A): A positive crossing represents a tube threaded through another tube in $\mathbb{R}^4$. In the *broken surface diagram* a surface is broken where it lies below the other in the fourth coordinate. (B): Wiring diagram crossings (also known as *virtual crossings*) represent tubes crossing in front of one another, with no interaction.

FIGURE 5. Examples of the type of vertices permitted in w-foams.

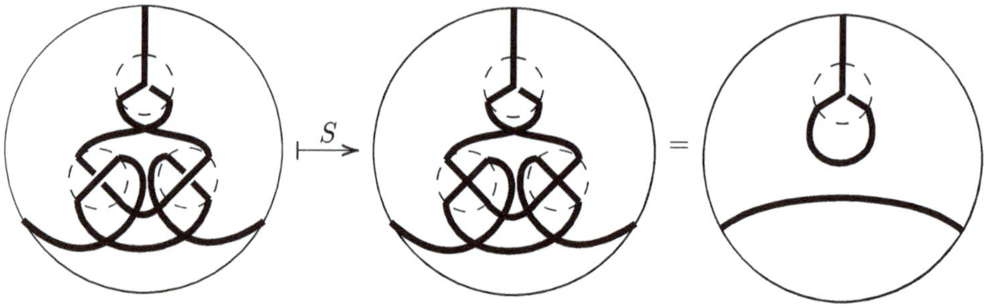

FIGURE 6. An example of the projection from a w-foam to its skeleton.

As with braids, every w-foam has an underlying w-foam *skeleton*. The skeleton of a w-foam is the combinatorial data of its features, the *vertices, caps and strands*, and how they are connected, but not the embedding (i.e. crossing) information. In other words, the skeleton of a w-foam is obtained by "flattening" all of its crossings. The collection of all w-foam skeleta is itself a tensor category, as follows.

REMARK 2.9. The *wheeled prop* of w-foam skeleta is the free wheeled prop given by:

$$\mathcal{S} = \langle\, ⅄,\ ⅄,\ ⅄,\ \downarrow \rangle$$

Here generators are understood to come in all possible orientations, and skeleta are equipped with the *orientation switch*, *unzip*, and *puncture* operations. There is a skeleton projection map $S : \mathsf{wF} \to \mathcal{S}$, given by $S(\times) = S(\times) = \times$ (see Figure 6).

2.4. The associated graded structure, tangential derivations, and cyclic words. The analogue of the group ring for w-foams is the linear extension of the circuit algebra wF, where morphisms are defined by:

$$\mathbb{Q}[\mathsf{wF}] = \bigsqcup_{s\in\mathcal{S}} \mathbb{Q}[\mathsf{wF}](s).$$

Here $\mathbb{Q}[\mathsf{wF}](s)$ is the $\mathbb{Q}$-vector space of formal linear combinations of w-foams with skeleton $s \in \mathcal{S}$. That is, $\mathbb{Q}[\mathsf{wF}](s)$ includes sums of the form $\sum_{S(T_i)=s} \alpha_i T_i$, $T_i \in \mathsf{wF}(s)$. The associated graded structure of wF is the direct product of successive quotients of powers of its augmentation ideal[4]

$$\mathcal{A} := \Pi_0^\infty \mathcal{I}^n/\mathcal{I}^{n+1}.$$

[4]Note, this equivalent to defining the pro-unipotent, or rational, completion of the wheeled prop of w-foams. For more details on this, see [**DHR23**, Section 2.3].

This associated graded $\mathcal{A}$ is a tensor category whose morphisms have a presentation in terms of a type of generalised Jacobi diagrams which we call *arrow diagrams.* In analogy with chord diagrams, arrow diagrams are closely related to the elements of universal enveloping algebras of Lie algebras of *tangential derivations*, and as we will see, homomorphic expansions of w-foams translate to solutions to the *Kashiwara-Vergne* equations.

2.4.1. *Tangential derivations and automorphisms.* Let $\widehat{\mathfrak{lie}}_n$ denote degree completion of the *free Lie algebra* on n-letters, $x_1, \ldots, x_n$. A derivation $u : \widehat{\mathfrak{lie}}_n \to \widehat{\mathfrak{lie}}_n$ is a linear map satisfying $u([a,b]) = [u(a), b] + [a, u(b)]$. A *tangential derivation* is a derivation $\mathbf{a} = (a_1, \ldots, a_n)$, $a_i \in \widehat{\mathfrak{lie}}_n$, for which $\mathbf{a}(x_i) = [x_i, a_i]$ for $i = 1, \ldots, n$ ([**AT12**, Definition 3.2, 3.5]). Tangential derivations of $\widehat{\mathfrak{lie}}_n$ form a Lie algebra, denoted $\mathfrak{tder}_n$, with bracket $[\mathbf{a}, \mathbf{b}](x_i) = \mathbf{a}(\mathbf{b}(x_i)) - \mathbf{b}(\mathbf{a}(x_i)) = \mathbf{a}([x_i, b_i]) - \mathbf{b}([x_i, a_i])$ ([**AT12**, Proposition 3.4]). Tangential derivations exponentiate to *tangential automorphisms*:

$$\mathrm{TAut}_n = \exp(\mathfrak{tder}_n).$$

These are basis conjugating, or tangential, automorphisms of $\exp(\mathfrak{lie}_n)$.

Let $\mathfrak{a}_n$ denote the n-dimensional abelian Lie algebra given by tuples of the form $(\alpha_1 x_1, \alpha_2 x_2, \ldots, \alpha_n x_n)$, for $\alpha_1, \ldots, \alpha_n \in \mathbb{Q}$. This Lie algebra arises as the kernel of the natural map $u : (\mathfrak{lie}_n)^{\oplus n} \to \mathfrak{tder}_n$, which sends a tuple $\mathbf{a} = (a_1, \ldots, a_n)$ to the derivation $x_i \mapsto [x_i, a_i]$.

EXAMPLE 2.10. An important example of a tangential derivation is the *braid derivation* $\mathbf{t}^{1,2} = (y, x) \in \mathfrak{tder}_2$. By definition, $\mathbf{t}(x) = [x, y]$ and $\mathbf{t}(y) = [y, x]$. It follows that $\mathbf{t}(x + y) = [x, y] + [y, x] = 0$. (In general, the set of derivations with this property is called *special*, and denoted $\mathfrak{sder}_n$.)

The analogously defined derivations $\mathbf{t}^{i,j}$ generate a Lie subalgebra of $\mathfrak{tder}_n$ isomorphic to the Lie algebra of infinitesimal braids $\mathfrak{t}_n$ [**AT12**, Proposition 3.11].

A solution to the *generalised Kashiwara-Vergne problem* is a tangential automorphism F of $\widehat{\mathfrak{lie}}_2$, which satisfies the two *KV equations* [**AT12**, Section 5.3]:

$$\text{(KV)}\quad \mathrm{SolKV} := \Big\{ F \in \mathrm{TAut}_2 \mid F(e^x e^y) = e^{x+y}, \text{ and}$$
$$J(F) = \mathrm{tr}\Big(r(x+y) - r(x) - r(y)\Big) \text{ for some } r \in u^2\mathbb{Q}[[u]] \Big\}.$$

To explain the meaning of the second KV equation, note first that, as a completed Hopf algebra, the universal enveloping algebra of $\widehat{\mathfrak{lie}}_n$ can be identified with the degree completed free associative algebra $\widehat{\mathfrak{ass}}_n := \mathbb{Q}\langle\langle x_1, \ldots, x_n \rangle\rangle$. The (graded) vector space of *cyclic words*[5] is the linear quotient

$$\mathrm{cyc}_n := \widehat{\mathfrak{ass}}_n / \langle ab - ba \mid \forall a, b \in \widehat{\mathfrak{ass}}_n \rangle = \widehat{\mathfrak{ass}}_n / [\widehat{\mathfrak{ass}}_n, \widehat{\mathfrak{ass}}_n].$$

Thus, there is a natural trace map $\mathrm{tr} : \widehat{\mathfrak{ass}}_n \to \mathrm{cyc}_n$.

EXAMPLE 2.11 ([**AT12**, Example 2.1]). The vector space cyc_1 is isomorphic to the space $x\mathbb{K}[[x]]$ of formal power series with no constant term. It is spanned by $\mathrm{tr}(x^k)$ for $k \geq 1$.

[5]What we call cyc_n is denoted by $\mathfrak{tr}_n$ in [**AT12**].

The natural action of $\mathfrak{tder}_n$ on $\widehat{\mathfrak{ass}}_n$ induces an action of $\mathfrak{tder}_n$ on cyc_n via the trace map. The *divergence* cocycle div : $\mathfrak{tder}_n \to \text{cyc}_n$ is defined by

$$(2.7) \qquad u = (a_1, \dots, a_n) \mapsto \sum_{k=1}^{n} \text{tr}(x_k(\partial_k a_k)),$$

where the operator ∂_k acts on $\widehat{\mathfrak{lie}}_n \subseteq \widehat{\mathfrak{ass}}_n$ by deleting words that do not end in x_k, and deleting the letter x_k from words that do. Divergence satisfies the 1-cocycle property $\text{div}([u,v]) = u \cdot \text{div}(v) - v \cdot \text{div}(u)$ and thus extends this trace action to a representation of $\mathfrak{tder}_n$ on cyc_n. This representation integrates to give a representation of tangential automorphisms on cyclic words.

DEFINITION 2.12 ([**AT12**, Section 5.1]). The non-commutative Jacobian map

$$J : \mathsf{TAut}_n \to \text{cyc}_n$$

is defined by setting

$$(2.8) \qquad J(1) = 0\,, \quad \frac{d}{dt}\Big|_{t=0} j\left(e^{tu} g\right) = \text{div}(u) + u \cdot J(g)$$

for $g \in \mathsf{TAut}_n$ and $u \in \mathfrak{tder}_n$.

In summary, solutions to the generalised Kashiwara-Vergne problem are tangential automorphisms of $\widehat{\mathfrak{lie}}_2$ which take $F(e^x e^y) = e^{x+y}$ and whose Jacobian representation satisfies a particular constraint in cyc_1. We are now ready to describe the morphisms in the tensor category of arrow diagrams which capture similar behaviour.

2.4.2. *Arrow diagrams.* The associated graded structure of w-foams is a tensor category, $\mathscr{A}$, whose morphisms are combinatorially described as *arrow diagrams.* In analogy with the relationship between chord diagrams and the Lie algebras of infinitesimal braids, certain morphism spaces in $\mathscr{A}$ are closely related to tangential automorphisms:

$$\mathscr{A}(\uparrow_n) \cong U(\mathfrak{tder}_n \oplus \mathfrak{a}_n \ltimes \text{cyc}_n).$$

Morphisms in the two-coloured wheeled prop of arrow diagrams can be formally described as linear combinations of *w-Jacobi diagrams* on w-foam skeleta.

DEFINITION 2.13. A *w-Jacobi diagram* on a w-foam skeleton $s \in \mathcal{S}$ is a (possibly infinite) linear combination of directed uni-trivalent *arrow graphs* satisfying the following conditions:

(1) trivalent arrow vertices are equipped with a cyclic orientation;
(2) each trivalent arrow vertex is incident to two incoming and one outgoing arrow edge;
(3) univalent vertices are attached to the chosen skeleton s, considered up to orientation preserving diffeomorphism of s. (That is, only the attachment order matters, not the specific locations along s.)

Such diagrams are considered up to the STU, TC, VI, CP, TF and RI relations shown in Figure 7. The *degree* of a w-Jacobi diagram is half the number of (trivalent and univalent) vertices of the arrow graph.

The STU relations for arrow diagrams are the oriented version of the classical STU relation of Jacobi diagrams, and ensures that the trivalent vertices "behave as

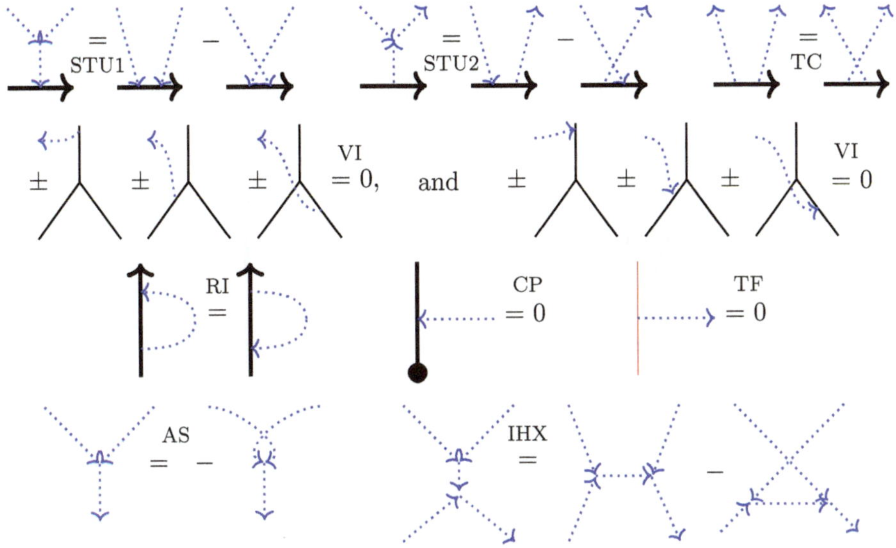

FIGURE 7. The relations for w-Jacobi diagrams.

FIGURE 8. On the left is an example of a w-Jacobi diagram. Sometimes, we denote an arbitrary w-Jacobi diagram located on a certain part of a skeleton by a box, as shown on the right.

Lie brackets."[6] In particular, the STU relation implies the AS (anti-symmetry) and IHX relations also shown in Figure 7. An example of a w-Jacobi diagram is shown in Figure 8. In figures the oriented uni-trivalent graph (arrow graph) is drawn in dashed lines, the w-foam skeleton has tubes shown as thick black lines and strings as thin red lines.

We have the following finitely presented tensor category of arrow diagrams.

DEFINITION 2.14. The wheeled prop of *arrow diagrams* is:

$$\mathscr{A} := \Big\langle$$  $$\mid 4T, TC, VI, CP, TF, RI. \Big\rangle$$

[6]The relationship between the STU relations and Lie brackets can be made precise using the notion of weight systems, which is not required for this paper. See for example [**BN95**, Section 2.4]

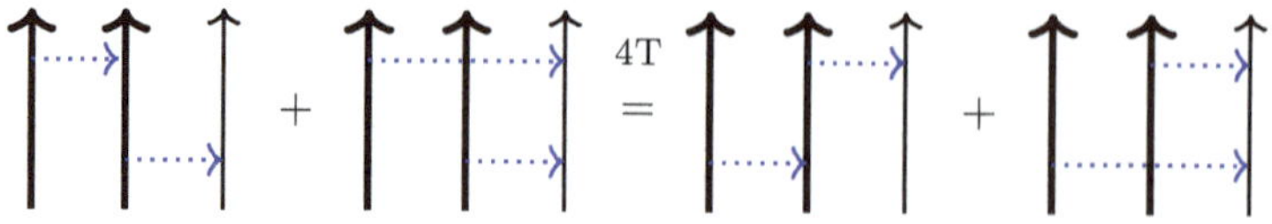

FIGURE 9. The 4T relation for arrow diagrams.

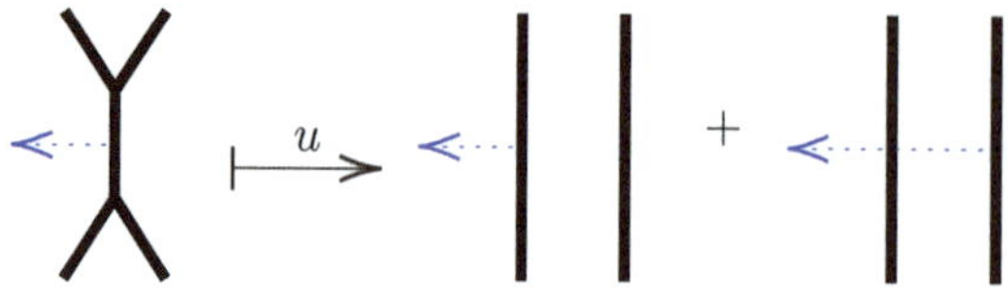

FIGURE 10. Unzip acting on an arrow tail ending on the unzipped strand. Unzip behaves similarly for arrow heads.

Here the non-skeleton generators are *arrows*, drawn in dotted lines. As before, all generators are understood to occur in all possible strand orientations. The 4T relation is shown in Figure 9. The *degree* of an arrow diagram is given by the number of arrows.

Morphisms of $\mathscr{A}$ are simply called *arrow diagrams.* There is an isomorphism between morphism spaces of $\mathscr{A}$ and w-Jacobi diagrams on the same skeleton, analogous to the classical isomorphism between chord diagrams and Jacobi diagrams [**BN95**, Theorem 6], sometimes called the "Bracket-Rise Theorem". In short, arrow diagrams are naturally also w-Jacobi diagrams, and this map turns out to descend to an isomorphism. In particular, the STU relations imply the 4T relation. For more detail see [**DHR23**, Around Theorem 3.4], [**BND17**, Theorem 3.13] and [**BND16**, Theorem 3.13].

Arrow diagrams are equipped with the following *operations* beyond the wheeled prop structure:

(1) The *orientation switch* $S_e : \mathscr{A}(s) \to \mathscr{A}(S_e(s))$ reverses the direction of a strand e of s and multiplies each arrow diagram by
$$(-1)^{\#(\text{arrow heads and tails ending on } e)}.$$
(2) The *unzip* (for a strand e ending in two vertices) and *disc unzip* (for a strand e ending in a vertex and a cap) operations are both denoted $u_e : \mathscr{A}(s) \to \mathscr{A}(u_e(s))$. The operation doubles the strand e, removes the vertex (vertices) at the end(s) of e, and maps each arrow ending on e to a sum of two new arrows, each of which ends on one of the two new strands created. This is shown in Figure 10.
(3) The *puncture* operation $p_e : \mathscr{A}(s) \to \mathscr{A}(p_e(s))$ punctures the strand e, turning it into a string (drawn as a red line). Any arrow diagram that has an arrow tail ending on e is sent to zero.
(4) The *deletion* operation $d_e : \mathscr{A}(s) \to \mathscr{A}(d_e(s))$ (for long strands e not incident to any skeleton vertex) deletes e, and an arrow diagram with any arrow ending—head or tail—on e is sent to 0.
(5) The global operation *wenjugation*, which is given by $D \mapsto (-1)^{\deg(D)} \cdot D$, for a diagram D. Note that the degree of an arrow/Jacobi diagram

coincides with the number of arrow tails in it. For an arrow diagram D, wenjugation composed with a global orientations switch (S_e applied to every strand) is called the *adjoint* of D and denoted D^*.

REMARK 2.15. These operations capture structure in $\mathscr{A}$ which is not encoded by the operad of wiring diagrams (Definition B.1). This is analogous to the type of operations that arise in the category of parenthesised braids, PaB, such as strand deletion and strand doubling.

2.4.3. *Arrow diagrams and tangential derivations.* At each skeleton $s \in \mathcal{S}$, the $\mathbb{Q}$-vector space $\mathscr{A}(s)$ can be equipped with a co-unital, co-associative co-product:

$$\Delta : \mathscr{A}(s) \longrightarrow \mathscr{A}(s) \otimes \mathscr{A}(s)$$

which sends an arrow diagram $\mathbf{D} \in \mathscr{A}(s)$ to the sum of all ways of distributing the connected components of the arrow graph of $\mathbf{D}$ between the two copies of the skeleton.

DEFINITION 2.16. An arrow diagram $\mathbf{D} \in \mathscr{A}(s)$ is said to be *primitive* if

$$\Delta(\mathbf{D}) = \mathbf{D} \otimes 1 + 1 \otimes \mathbf{D}.$$

The vector space of primitive arrow diagrams on a skeleton s is denoted by $\mathscr{P}(s)$, and $\mathscr{P} = \sqcup_{s \in \mathcal{S}} \mathscr{P}(s)$. As it turns out, a single arrow diagram is primitive if, and only if, its arrow graph is connected. Therefore, $\mathscr{P}$ is spanned by *trees* and *wheels*: a *tree* is an arrow diagram whose arrow graph is connected and has no cycles (on the right of Figure 12), and a *wheel* is an arrow diagram whose arrow graph is an n-cycle with n "spokes" all oriented towards the cycle, as shown in Figure 11 [**BND17**, Proof of Theorem 3.16].

Let $\mathscr{P}(\uparrow_n)$ denote primitive arrow diagrams on a skeleton of n black strands (that is, n tubes). As mentioned earlier, $\mathscr{P}(\uparrow_n)$ is closely related to the lie algebras $\mathfrak{tder}_n$ and cyc_n, which is made precise in the following proposition.

PROPOSITION 2.17 ([**BND17**, Proposition 3.19]). *There is a split short exact sequence of Lie algebras:*

$$0 \longrightarrow \mathrm{cyc}_n \hookrightarrow \mathscr{P}(\uparrow_n) \underset{\ell}{\overset{pr}{\rightleftarrows}} \mathfrak{tder}_n \oplus \mathfrak{a}_n \longrightarrow 0 \tag{2.9}$$

Cyclic words embed into $\mathscr{P}(\uparrow_n)$, as *wheels*, as follows. Label the skeleton strands with the letters $x_1, \ldots, x_n$. Given a cyclic word w of length k in cyc_n, associate to it a k-*wheel* as in Figure 11, oriented counterclockwise, with "spokes" (arrow tails) attached to the skeleton according to the letters of w. The order in which arrow tails are attached on each skeleton strand does not matter due to the TC relation.

The map pr sends wheels to zero. Given a single tree $\mathbf{T} \in \mathscr{P}(\uparrow_n)$, $pr(\mathbf{T}) \in \mathfrak{tder}_n \oplus \mathfrak{a}_n$ as follows.

(1) If $\mathbf{T}$ is a short arrow on strand i, then $pr(\mathbf{T}) = x_i \in \mathfrak{a}_n$.
(2) Otherwise, $pr(\mathbf{T}) \in \mathfrak{tder}_n$. Label the skeleton strands with $x_1, x_2, \ldots, x_n$. $\mathbf{T}$ determines a Lie word $w_{\mathbf{T}} \in \mathfrak{lie}_n$ by reading the generator corresponding to the skeleton strand of each leaf (tail) of the tree, and combining these with brackets corresponding to each trivalent vertex. If the *head* (root) of the tree is on strand i, then $pr(\mathbf{T})(x_i) = [x_i, w_{\mathbf{T}}]$, and $pr(T)(x_j) = 0$ for all $j \neq i$. See Figure 12 for two examples.

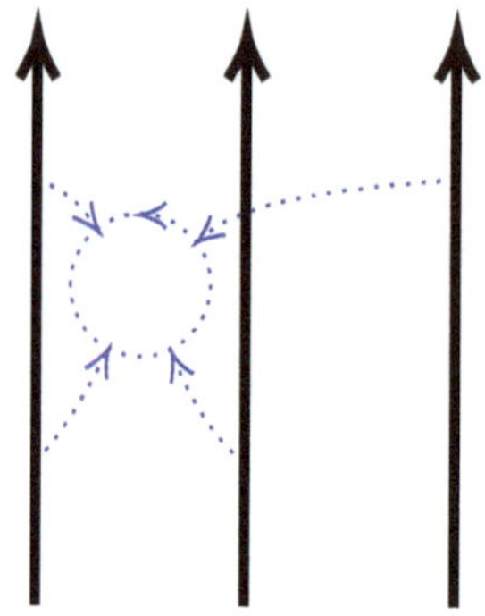

FIGURE 11. The image of $\operatorname{tr}(x^2yz)$ in $\mathcal{P}(\uparrow_3)$. We use the convention that wheels are oriented counterclockwise, unless otherwise marked.

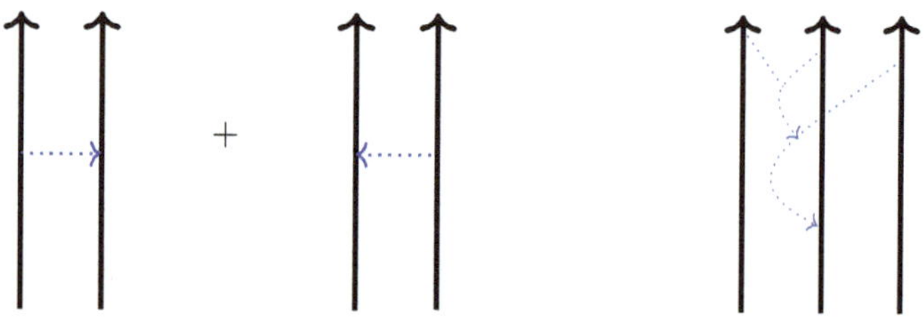

FIGURE 12. The arrow diagram on the left maps by pr to the element $(y, x) \in \mathfrak{tder}_2$ from Example 2.10. The tree on the right maps to $(0, [[x, y], z], 0) \in \mathfrak{tder}_3$.

REMARK 2.18. The map pr in (2.9) admits a section $\ell : \mathfrak{tder}_n \oplus \mathfrak{a}_n \to \mathscr{P}(\uparrow_n)$. For $x_i \in \mathfrak{a}_n$, define $\ell(x_i)$ to be a short arrow on strand i, pointing down (though this is equivalent to a short arrow pointing up via the RI relation). For $d \in \mathfrak{tder}_n$, write $d = (a_1, \ldots, a_n)$ where $d(x_i) = [x_i, a_i]$. Choose a_i so that the coefficient of the linear term x_i in a_i is 0. To each lie word in a_i, associate a binary tree oriented towards a single head, with tails labelled by the letters $x_1, \ldots, x_n$. In $\mathscr{P}(\uparrow_n)$ attach all tails to the strand labelled by the appropriate letter, and attach the head to strand i, below all the tails. See Figure 13 for an example. The element $\ell(d)$ is the sum of all the trees arising from $a_1, \ldots, a_n$ in $\mathscr{P}(\uparrow_n)$. The reader can check that ℓ is a Lie algebra map. We denote the restriction of ℓ to $\mathfrak{tder}_2$ by ℓ also.

Note that the order in which arrow tails (leaves) of each tree are attached along a strand doesn't matter, due to the TC (Tails Commute) relation. However, the attachment position of the head (root) of the tree—in relation to the tails on the same strand—does make a difference. For the map ℓ, we choose to attach tree heads below tails, but, for example, one could choose to attach the heads at the top instead. These two maps are not the same, but their difference lies in the image of cyc_n, by STU relations.

Restricting our attention to arrow diagrams on the skeleton of n vertical strands enabled us to identify $\mathfrak{tder}_n$ and cyc_n within $\mathscr{P}$. In fact, arrow diagrams on n-strands possess more structure: there is an associative, unital product on each

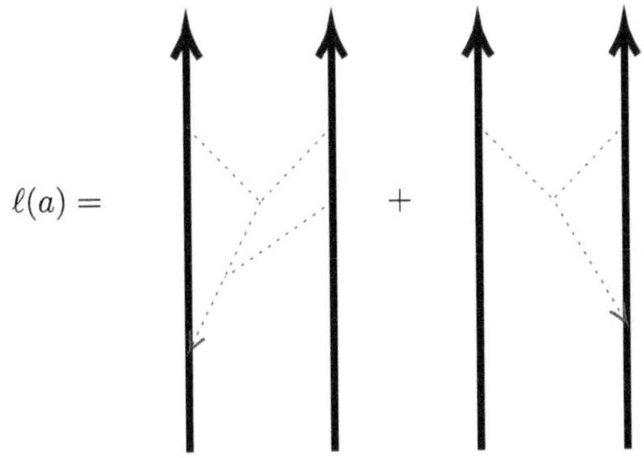

FIGURE 13. An example of $\ell(\mathbf{a})$ for $\mathbf{a} = ([[x,y],y],[x,y]) \in \mathfrak{tder}_2$.

vector space $\mathscr{A}(\uparrow_n)$ given by stacking:

$$\mathscr{A}(\uparrow_n) \otimes \mathscr{A}(\uparrow_n) \xrightarrow{stack} \mathscr{A}(\uparrow_n)$$

This product is compatible with the co-product, making $\mathscr{A}(\uparrow_n)$ into a Hopf algebra. The following fact is established in [**BND17**, Section 3.2].

PROPOSITION 2.19. *The stacking product and arrow diagram co-product make $\mathscr{A}(\uparrow_n)$ into a Hopf algebra, and, by the Milnor-Moore theorem* [**MM65**], *$\mathscr{A}(\uparrow_n$) is isomorphic to the completed universal enveloping algebra over its primitive elements. In particular,*

$$\mathscr{A}(\uparrow_n) \overset{\Upsilon}{\cong} \hat{U}(\mathfrak{tder}_n \oplus \mathfrak{a}_n \ltimes \mathrm{cyc}_n).$$

As a consequence of Proposition 2.19, it makes sense to define *group-like* arrow diagrams as exponentials of primitive elements, formally interpreted as power series: $e^{\mathbf{D}} = \sum_{k=0}^{\infty} \frac{\mathbf{D}^k}{k!}$, where $\mathbf{D} \in \mathscr{P}(\uparrow_n)$.

NOTATION 2.20. We will write $\mathscr{G}(\uparrow_n)$ for the group of group-like elements of $\mathscr{A}(\uparrow_n)$.

In contrast to chord diagrams, arrow diagrams on an arbitrary skeleton do not usually assemble into a Hopf algebra. However, the following lemma is a straightforward application of the VI relation:

LEMMA 2.21 ([**BND17**, Lemma 4.6]). *There is a canonical isomorphism of vector spaces*

$$\mathscr{A}(\text{大}) \cong \mathscr{A}(\uparrow_2) \overset{\Upsilon}{\cong} \hat{U}(\mathfrak{tder}_2 \oplus \mathfrak{a}_2 \ltimes \mathrm{cyc}_2).$$

Therefore, the Hopf algebra structure on $\mathscr{A}(\uparrow_2)$ can be pulled back along this isomorphism to give a Hopf algebra structure on $\mathscr{A}(\text{大})$.

REMARK 2.22. The projection $pr : \mathscr{P}(\uparrow_n) \to \mathfrak{tder}_n \oplus \mathfrak{a}_n$ induces a Hopf algebra map $pr : \mathscr{A}(\uparrow_n) \to \hat{U}(\mathfrak{tder}_n \oplus \mathfrak{a}_n)$. Regarding cyc_n as an abelian Lie algebra, we also have an induced inclusion $\hat{U}(\mathrm{cyc}_n) \hookrightarrow \mathscr{A}(\uparrow_n)$. Denote by $\mathscr{A}(\uparrow_n)^{tr}$ the quotient $\mathscr{A}/\operatorname{im}(\hat{U}(\mathrm{cyc}_n))$, we call this quotient the *tree-level* arrow diagrams. We write $\overrightarrow{pr}$ for the quotient map $\mathscr{A}(\uparrow_n) \to \mathscr{A}(\uparrow_n)^{tr}$, and for $\mathbf{D} \in \mathscr{A}(\uparrow_n)$ we denote $\overrightarrow{pr}(\mathbf{D})$ by

$\mathbf{D}^{tr}$. Note that the map $pr : \mathscr{A}(\uparrow_n) \to \hat{U}(\mathfrak{tder}_n \oplus \mathfrak{a}_n)$ descends to an isomorphism $\mathscr{A}(\uparrow_n)^{tr} \cong \hat{U}(\mathfrak{tder}_n \oplus \mathfrak{a}_n)$.

Next, we identify the space of arrow diagrams on one or more capped strands. Arrow diagrams with an arrow head adjacent to a cap vanish by the CP relation. When the skeleton consists only of capped strands, arrow heads can be eliminated entirely by successive applications of STU and CP relations to bring them adjacent to the cap(s). In turn, arrow diagrams with no arrow heads can be expressed as a linear combination of only *wheels*, which are encoded as cyclic words. This leads to the following lemma, which is a straightforward generalisation of [**BND17**, Lemma 4.5].

LEMMA 2.23. *There is an isomorphism of graded vector spaces* $\mathscr{A}(\downarrow_n) \cong \mathrm{cyc}_n/\mathrm{cyc}_n^1$. *Here* cyc_n^1 *denotes the degree one component of cyclic words. Cyclic words in one letter are power series, in particular,* $\mathscr{A}(\downarrow_1) \cong \mathbb{Q}[[\xi]]/\langle\xi\rangle$, *where the quotient is understood linearly.*

REMARK 2.24. In Lemma 2.23 the reason for taking the quotient by the degree one component in is the RI relation on arrow diagrams forces a degree 1 wheel (i.e. a wheel with one "spoke") to vanish.

Since $\mathrm{cyc}_n \cong \mathrm{tr}(\mathbb{Q}\langle\langle x_1, x_2, \ldots, x_n\rangle\rangle)$, we also have, for example $\mathscr{A}(\downarrow_2) \cong \mathrm{tr}(\mathbb{Q}\langle\langle\xi,\zeta\rangle\rangle)/\langle\xi,\zeta\rangle$, where the quotient is again linear.

REMARK 2.25. There is a well-defined surjective linear map $\omega : \mathscr{A}(\downarrow_n) \to \mathbb{Q}[x_1, \cdots, x_n]$ which sends an arrow diagram—expressed in a wheels form, that is, with arrow heads eliminated—to the product of "x_i to the power of the number of tails on strand i" for $i = 1, \ldots, n$.

The Jacobian map $J : \mathrm{TAut}_n \to \mathrm{cyc}_n$ also has a diagrammatic interpretation,[7] captured by the *adjoint* operation $(-)^*$. For $\mathbf{D} \in \mathscr{A}(\uparrow_n)$ recall that $\mathbf{D}^*$ denotes the *adjoint* of $\mathbf{D}$ (that is, wenjugation composed with global orientation switch).

PROPOSITION 2.26 ([**BND17**, Proposition 3.27]). *For any* $\mathbf{u} \in \mathfrak{tder}_n$,

$$e^{\ell(\mathbf{u})}\left(e^{\ell(\mathbf{u})}\right)^* = e^{-J(e^{\mathbf{u}})}.$$

2.5. Homomorphic expansions of w-foams. A map of wheeled props $Z : \mathsf{wF} \to \mathscr{A}$ is *skeleton preserving* if it restricts to a map of graded vector spaces $Z(s) : \mathsf{wF}[s] \to \mathscr{A}[s]$ for each $s \in \mathcal{S}$. Homomorphic expansions of w-foams are isomorphisms of tensor categories $Z : \mathbb{Q}[\widehat{\mathsf{wF}}] \to \mathscr{A}$, which are skeleton-preserving and compatible with all operations (including operations external to the wheeled prop structure, such as punctures and unzips) [**BND17**, Definition 2.4, 2.5]. A homomorphic expansion is *group-like* if the Z-values of generators of wF are group-like: that is, exponentials of primitive elements in $\mathscr{A}$.

By finite generation, a homomorphic expansion $Z : \mathbb{Q}[\widehat{\mathsf{wF}}] \to \mathscr{A}$ is determined by its values on the generators:

 and

[7]Note that our e^J is denoted J in [**BND17**].

The values Z takes on these generators are subject to equations given by the relations in wF, as well as equations given by the compatibility with the extra operations: orientation switch, unzip, puncture, strand deletion and the global wenjugation (or equivalently, adjoint) operation. A careful analysis of the equations leads to the following:

PROPOSITION 2.27. *A group-like homomorphic expansion $Z : \mathbb{Q}[\widehat{\mathsf{wF}}] \to \mathscr{A}$ is uniquely determined by the values*

$$Z(\nearrow\!\!\!\nwarrow) = e^{\mathrm{H}} = R \in \mathscr{A}(\uparrow_2), \quad Z(\curlywedge) = e^{\nu} = V \in \mathscr{A}(\curlywedge) \cong \mathscr{A}(\uparrow_2) \quad \textit{and}$$
$$Z(\downarrow) = e^{c} = C \in \mathscr{A}(\downarrow) \cong \mathbb{Q}[[\xi]]/\langle \xi \rangle,$$

which are subject to the following equations:

$$\text{(R4)} \qquad V^{12}R^{(12)3} = R^{23}R^{13}V^{12} \quad \textit{in} \quad \mathscr{A}(\uparrow_3)$$
$$\text{(U)} \qquad V \cdot V^{*} = 1 \quad \textit{in} \quad \mathscr{A}(\uparrow_2)$$
$$\text{(C)} \qquad C^{12}(V^{12})^{-1} = C^1C^2 \quad \textit{in} \quad \mathscr{A}(\downarrow_2).$$

This proposition is a minor generalisation of [**BND17**, Theorems 4.17, 4.23], we summarise the key points of the proof below. See also the discussion in [**DHR23**, Section 4].

PROOF (SKETCH). The main result (Theorem 4.23, building on Theorems 3.30 and 4.8) of [**BND17**] is a classification of homomorphic expansions for a class of w-foams

$$\overline{\mathsf{wF}} = \langle \nearrow\!\!\!\nwarrow, \nwarrow\!\!\!\nearrow, \curlywedge, \curlyvee, \downarrow \mid R1^f, R2, R3, OC, CP \rangle,$$

where the operations are orientation switches, unzips and wenjugation, but no *punctures*. In particular, given any group-like homomorphic expansion $\overline{Z}$ of $\overline{\mathsf{wF}}$ the $\overline{Z}$-values of the generators $\nearrow\!\!\!\nwarrow$, $\curlywedge$ and $\downarrow$ uniquely determine the values $\overline{Z}(\nwarrow\!\!\!\nearrow)$ and $\overline{Z}(\curlyvee)$. For example, the second Reidemeister relation, R2, says that $\overline{Z}(\nearrow\!\!\!\nwarrow \cdot \nwarrow\!\!\!\nearrow) = \overline{Z}(\nearrow\!\!\!\nwarrow) \cdot \overline{Z}(\nwarrow\!\!\!\nearrow) = 1$. In [**BND17**, Theorem 4.23] it is shown that $\overline{Z}$ is a homomorphic expansion of $\overline{wf}$ if and only if $\overline{Z}(\nearrow\!\!\!\nwarrow)$, $\overline{Z}(\nwarrow\!\!\!\nearrow)$ and $\overline{Z}(\curlywedge)$ satisfy the equations (R4) (U) and (C).

We deduce the Proposition by constructing a bijection between homomorphic expansions of $\overline{\mathsf{wF}}$ and homomorphic expansions of wF, via restriction and unique extension. Using the definition of circuit algebra as an algebra over the operad of wiring diagrams (Definition B.5), it is relatively straightforward to check that $\overline{\mathsf{wF}}$ is a sub-circuit algebra of wF. It follows that any homomorphic expansion $Z : \widehat{\mathsf{wF}} \to \mathscr{A}$ restricts to a homomorphic expansion of $\overline{\mathsf{wF}}$.

Conversely, any homomorphic expansion $\overline{Z}$ of $\overline{\mathsf{wF}}$ uniquely extends to a homomorphic expansion of wF, because all generators of wF with punctured strands can be obtained by applying the puncture operation to a strand of a generator of $\overline{\mathsf{wF}}$ and any homomorphic expansion Z of wF is required to commute with the puncture operation. Thus, the Z-values of the punctured generators are uniquely determined by $\overline{Z}$. □

The isomorphism $\Upsilon : \mathscr{A}(\uparrow_2) \to \widehat{U}(\mathrm{cyc}_2 \rtimes (\mathfrak{tder}_2 \oplus \mathfrak{a}_2))$ from Lemma 2.21 identifies the value $Z(\curlywedge) \in \mathscr{A}(\uparrow_2)$ with

$$\Upsilon(Z(\curlywedge)) = e^{b}e^{\nu} \in \widehat{U}(\mathrm{cyc}_2 \rtimes (\mathfrak{tder}_2 \oplus \mathfrak{a}_2)),$$

where $b \in \mathrm{cyc}_2$ and $\nu \in \mathfrak{tder}_2 \oplus \mathfrak{a}_2$. In [**BND17**, Theorem 4.23], KV solutions are shown to be in bijection with families of group-like homomorphic expansions of wF which are parameterised by the $\mathfrak{a}_2$ component of ν.

DEFINITION 2.28. A homomorphic expansion is *v-small* if the $\mathfrak{a}_2$ component of $\log \Upsilon(Z(\curlywedge))$ is zero.

THEOREM 2.29 ([**BND17**, Theorem 4.23]). *There is a one-to-one correspondence between the set of v-small, group-like homomorphic expansions $Z : \mathsf{wF} \to \mathscr{A}$, and Kashiwara-Vergne solutions.*

PROOF (SKETCH). In [**BND17**] Theorem 4.23 states this for v-small, group-like homomorphic expansions of $\overline{\mathsf{wF}}$. The main steps in this proof identify the value $Z(\curlywedge)$ with a tangential automorphism $e^{\nu} \in \hat{U}(\mathfrak{tder}_2)$ and $Z(\downarrow)$ with a power series $e^{c} \in \mathbb{Z}(\downarrow) \in u\mathbb{Q}[[u]]$. The equation R4 forces the value $Z(\curlywedge) = e^{\nu}$ to satisfy the first KV equation (KV), and the equations U and C of Proposition 2.27 force the Jacobian of e^{ν} to satisfy the second KV equation.

By unique extension and restriction (as in Proposition 2.27), homomorphic expansions for $\overline{\mathsf{wF}}$ are in bijection with homomorphic expansions for wF, and this correspondence preserves the v-small property. The statement then follows. □

2.6. Automorphisms of arrow diagrams and the KV groups. The Kashiwara-Vergne symmetry groups, KV and KRV, act freely and transitively on the set of all KV solutions. The Kashiwara-Vergne group KV ([**AET10**, after Proposition 7]) is defined as follows:

$$\mathsf{KV} := \Big\{(a,\sigma) \in \mathrm{TAut}_2(\mathbb{Q}) \times u^2\mathbb{Q}[[u]] \Big| \; a(e^x e^y) = e^x e^y \text{ and } J(a) = \mathrm{tr}\Big(\sigma(\mathfrak{bch}(x,y)) - \sigma(x) - \sigma(y)\Big)\Big\}. \tag{2.10}$$

The *graded* Kashiwara-Vergne group KRV is the group corresponding to a Lie subalgebra of the tangential derivations of $\widehat{\mathfrak{lie}}_2$ which vanish on the sum of the generators ([**AET10**, Remark 50], [**AT12**, Sec. 4.1]). Explicitly,

$$\mathsf{KRV} := \Big\{(\alpha,s) \in \mathrm{TAut}_2(\mathbb{Q}) \times u^2\mathbb{Q}[[u]] \Big| \; \alpha(e^{x+y}) = e^{x+y} \text{ and } J(\alpha) = \mathrm{tr}\Big(s(x+y) - s(x) - s(y)\Big)\Big\}. \tag{2.11}$$

The left action of KV on SolKV is given by right composition with the inverse in TAut_2, $a \cdot F = F \circ a^{-1}$, and the right action of KRV on SolKV is given by $F \cdot \alpha = \alpha^{-1} \circ F$.

The Kashiwara-Vergne symmetry groups KV and KRV can be identified with automorphisms of $\mathbb{Q}[\widehat{\mathsf{wF}}]$ and $\mathscr{A}$, respectively, which are both *skeleton* and *expansion preserving.* Explicitly, skeleton and expansion preserving automorphisms of $\mathscr{A}$ are maps $G : \mathscr{A} \to \mathscr{A}$ such that for any expansion $Z : \mathbb{Q}[\widehat{\mathsf{wF}}] \to \mathscr{A}$, the composite $G \circ Z$ is also an expansion. We denote the group of skeleton and expansion preserving automorphisms by $\mathrm{Aut}(\mathscr{A})$.

PROPOSITION 2.30 ([**DHR23**, Proposition 5.6]). *For a skeleton-preserving map $G : \mathscr{A} \to \mathscr{A}$, $G \in \mathrm{Aut}(\mathcal{A})$ if, and only if, $G(\mapsto) = R = e^{\mapsto} \in \mathscr{A}(\uparrow_2)$ and the values*

$G(\,\downarrow\,) = C = e^c$ and $G(\,\curlywedge\,) = N = e^\eta$ satisfy the following equations:

$$(\text{R4'}) \qquad (N^{12})^{-1}R^{(12)3}N^{12} = R^{(12)3} \quad in \quad \mathscr{A}(\uparrow_3)$$

$$(\text{U'}) \qquad N \cdot N^* = 1 \quad in \quad \mathscr{A}(\uparrow_2)$$

$$(\text{C'}) \qquad C^{12}(N^{12})^{-1} = C^1C^2 \quad in \quad \mathscr{A}(\,\downarrow_2).$$

REMARK 2.31. We note that [**DHR23**, Proposition 5.6] characterises automorphisms of $\mathscr{A} = \operatorname{gr}\overline{\mathsf{wF}}$, that is, arrow diagrams on skeleta without punctured strands. Nonetheless, the Proposition stated here is equivalent by the same restriction and unique extension argument that we used in the proof of Proposition 2.27.

An automorphism $G \in \operatorname{Aut}(\mathscr{A})$ is called *v-small* if the $\mathfrak{a}_2$ component of $\log \Upsilon(G(\,\curlywedge\,))$ is zero. We denote the group of v-small automorphisms by $\operatorname{Aut}_v(\mathscr{A})$. In [**DHR23**, Theorem 5.12; Proposition 5.15], two of the authors, with Halacheva, show:

THEOREM 2.32. *There are group isomorphisms*

$$\mathsf{KRV} \overset{\Theta}{\cong} \operatorname{Aut}_v(\mathscr{A}) \quad and \quad \mathsf{KV} \overset{\Theta_Z}{\cong} \operatorname{Aut}_v(\widehat{\mathsf{wF}}).$$

To define the isomorphism $\Theta : \operatorname{Aut}_v(\mathscr{A}) \longrightarrow \mathsf{KRV}$, we write, for $G \in \operatorname{Aut}_v(\mathscr{A})$,

$$pr(\log G(\,\curlywedge\,)) = \eta \in \mathfrak{tder}_2, \quad \text{and} \quad G(\,\downarrow\,) = e^c \quad \text{where} \quad c \in \mathbb{Q}[[\xi]]/\langle\xi\rangle.$$

Then

$$\Theta(G) = (e^\eta, 2c) \in \mathsf{KRV}. \tag{2.12}$$

2.6.1. *Duflo elements and the evaluation of automorphisms of vertices.* While elements of KV and KRV are defined above as pairs, in both cases the first component *uniquely* determines the second ([**AET10**, Section 6.1]). In particular, for both KV and KRV there are Duflo maps given by $a \mapsto \sigma$ and $\alpha \mapsto s$. These Duflo maps are group homomorphisms $\operatorname{TAut}_2 \to u^2\mathbb{Q}[[u]]$, where the group operation in TAut_2 is composition, and $u^2\mathbb{Q}[[u]]$ is the additive group [**AET10**, Proposition 34]. For example, if $(\alpha_1, s_1) \in \mathsf{KRV}$ and $(\alpha_2, s_2) \in \mathsf{KRV}$ then $(\alpha_2 \circ \alpha_1, s_1 + s_2) \in \mathsf{KRV}$. The corresponding fact about automorphisms of $\mathscr{A}$ is given in the following Proposition and Corollary, which are proven by standard perturbative arguments:

PROPOSITION 2.33. *Any $G \in \operatorname{Aut}_v(\mathscr{A})$ is uniquely determined by the value $N = G(\,\curlywedge\,) \in \mathscr{A}(\uparrow_2)$.*

PROOF. Given $G \in \operatorname{Aut}_v(\mathscr{A})$, we know that $G(\leftrightarrow) = \leftrightarrow$, so we only need to show that N uniquely determines the value $G(\,\downarrow\,) = C = e^c$, where $c \in \mathbb{Q}[[\xi]]/\langle\xi\rangle$. To this end, assume that there exist $G, G' \in \operatorname{Aut}_v(\mathscr{A})$ given by

$$(G(\,\curlywedge\,), G(\,\downarrow\,)) = (N, C) \quad \text{and} \quad (G'(\,\curlywedge\,), G'(\,\downarrow\,)) = (N, C')$$

where $C \neq C'$.

Let ε denote the lowest degree term of the difference $C - C'$, and assume that ε is in degree k. Note that $\varepsilon = \lambda x^k$, where $0 \neq \lambda \in \mathbb{Q}$. The Cap Equation, (C') applies to both C and C'. Writing the difference of the two cap equations in degree k we obtain

$$0 = \lambda((x+y)^r - x^r - y^r). \tag{2.13}$$

Since $\lambda \neq 0$, this implies that $r = 1$. But $C, C' \in \mathbb{Q}[[\xi]]/\langle\xi\rangle$ have no degree one terms, a contradiction. $\square$

Given a $G \in \mathrm{Aut}_v(\mathscr{A})$ with $N = G(\curlywedge)$, we write N^{tr} for $\overrightarrow{pr}(N) \in \mathscr{A}(\curlywedge)^{tr}$, as in Remark 2.22.

COROLLARY 2.34. *Any $G \in \mathrm{Aut}_v(\mathscr{A})$ is uniquely determined by N^{tr}.*

PROOF. We aim to recover $G(\curlywedge) = N$ from N^{tr}. Assume that $G, G' \in \mathrm{Aut}_v(\mathscr{A})$ with $G(\curlywedge) = N$, $G'(\curlywedge) = N'$ such that $N \neq N'$ and $N^{tr} = (N')^{tr}$. Let ε denote the lowest degree term of $N - N'$, and assume that ε is of degree k. Since ε is in the kernel of $\overrightarrow{pr}$, it is in the image of cyc_n, and thus it is a scalar multiple of a k-wheel.

By Proposition 2.27, N and N' satisfies the Unitarity equation (U'): $N^* = N^{-1}$ and $(N')^* = N'^{-1}$. Modulo degree $k+1$ we have

$$N = N' + \varepsilon \quad \text{and} \quad N^{-1} = (N')^{-1} - \varepsilon. \tag{2.14}$$

And therefore, subtracting the two (U') equations, up to degree k we have

$$\varepsilon^* = (N - N')^* = N^{-1} - (N')^{-1} = -\varepsilon \tag{2.15}$$

Now, recall that the adjoint operation $(-)^*$ reverses the direction of all strands and multiplies each arrow diagram by $(-1)^{\#\text{ heads}}$. Since ε is a scalar multiple of a k-wheel, it only has arrow tails on each strand. It follows that $\varepsilon^* = \varepsilon$ and thus $\varepsilon = -\varepsilon$ so $\varepsilon = 0$, a contradiction. $\square$

2.7. The relationship between chord diagrams and arrow diagrams. For any $n \geq 2$, there is an inclusion of Lie algebras $\mathfrak{t}_n \hookrightarrow \mathfrak{tder}_n$. Namely, $\mathfrak{t}_n$ is isomorphic to a Lie subalgebra of $\mathfrak{tder}_n$ spanned by tangential derivations of the form $t^{i,j} = (0, \ldots, x_j, \ldots, x_i, \ldots 0)$([**AT12**, Proposition 3.11] and see Example 2.10 above). This inclusion of Lie algebras extends to a Hopf algebra homomorphism

$$\mathsf{CD}(n) \cong \widehat{U}(\mathfrak{t}_n) \xrightarrow{\ \epsilon\ } \widehat{U}(\mathfrak{tder}_n \oplus \mathfrak{a}_n \ltimes \mathrm{cyc}_n) \xrightarrow[\cong]{\Upsilon^{-1}} \mathscr{A}(\uparrow_n).$$

Diagrammatically, this composition maps a chord between strands i and j to the sum of two arrows: one going from strand j to strand i (representing the x_j in the ith component) and one going from strand i to strand j to (representing x_i in the jth component). For example, the arrow diagram on the left of Figure 12 represents the image of the chord diagram $t^{12} = |\cdots|$. (See also [**DHR23**, Proof of Lemma 6.3]).

This map of Hopf algebras extends to an inclusion of tensor categories,

$$\mathsf{PaCD} \xrightarrow{\ \alpha\ } \mathscr{A}\ ,$$

defined "one skeleton at a time". Recall that $\mathsf{PaCD}(s)$ consists of the chord diagrams with underlying skeleton s. Equivalently, an element of $\mathsf{PaCD}(s)$ is a morphism in $\mathsf{PaCD}(n)$, where the data of the skeleton provides the underlying permutation and parenthesisation. The example on the right of Figure 2 showed such a morphism from $p_1 = ((13)2)4)$ to $p_2 = (12)(34)$. "Closing up" the parenthesised permutations with binary trees representing the parenthesisation produces a wF skeleton, as in Figure 14. The map α sends a chord diagram $\mathbf{D}$ to the arrow diagram $\Upsilon^{-1}\epsilon(\mathbf{D})$ on the closed skeleton, as shown in Figure 14.

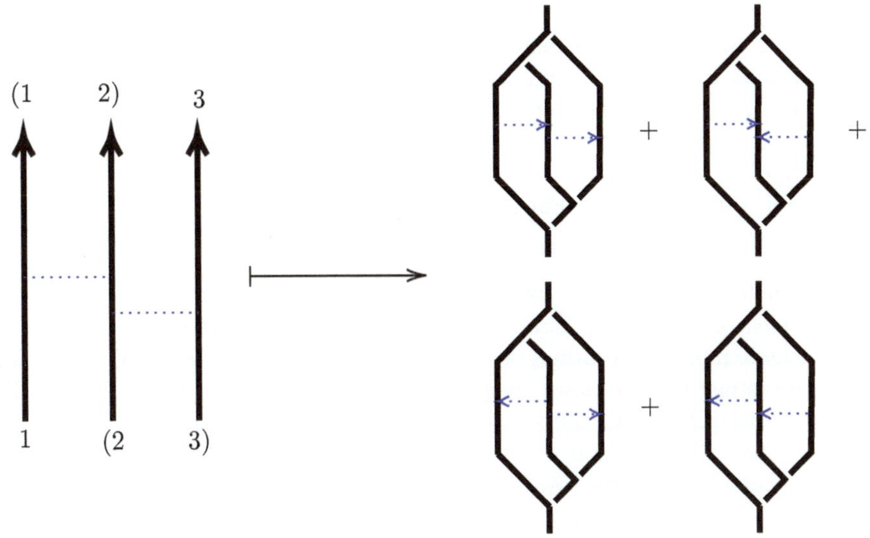

FIGURE 14. The map α applied to an element of PaCD(3).

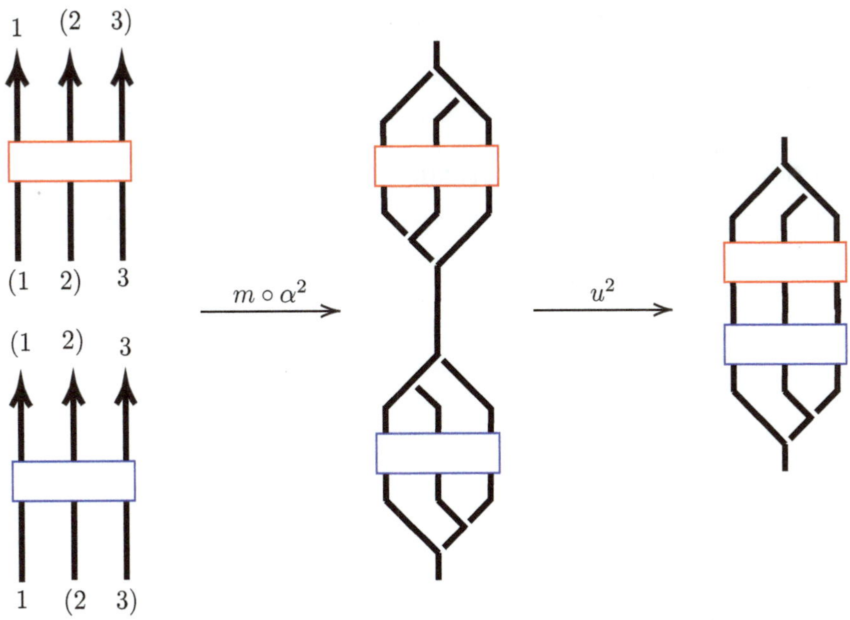

FIGURE 15. Closure and composition of chord diagrams, with m denoting the stacking map.

The unzip and stacking operations in arrow diagrams allow us to extend this inclusion of objects to a tensor functor embedding PaCD into $\mathscr{A}$ (Illustrated in Figure 15, see Theorem 6.4 [**DHR23**] for full details).

DEFINITION 2.35. The functor $\alpha : \mathsf{PaCD} \to \mathscr{A}$ is defined as follows.

- The map α sends a pair of parenthesised words (i.e. a skeleton as in Figure 2) to the w-foam skeleton that is the composition of two oriented binary, rooted trees representing the two parenthesisations. Here, each pair of parentheses corresponds to a vertex, and these are joined together to obtain each tree (see Figure 14).
- For a morphism $f \in \mathsf{PaCD}(s) \cong U(\mathfrak{t}_n)$, we define

$$\alpha(f) := \epsilon(f) \in \hat{U}(\mathfrak{tder}_n \oplus \mathfrak{a}_n \ltimes \mathrm{cyc}_n).$$

The functor $\alpha : \mathsf{PaCD} \to \mathscr{A}$ is a morphism of props (or tensor categories) in that it picks out the prop structure of PaCD inside the morphism space of $\mathscr{A}$. The symmetric monoidal product on both PaCD and $\mathscr{A}$ is given by concatenation–visually this is just placing chord (respectively, arrow) diagrams next to each other. In Theorem 6.4 [**DHR23**], the authors show that this functor actually has more structure, since the inclusion of morphisms $\mathsf{PaCD}(s) \hookrightarrow \mathscr{A}((s)^{cl})$ preserves co-products. Combining these two facts makes α a morphism of props enriched in co-associative co-algebras. However, we emphasise that PaCD is not a sub-object of $\mathscr{A}$ as the image of α is not preserved under all operations of $\mathscr{A}$.

The closure map can also be applied to a braid in PaB in the same way, and the result can then be interpreted as an element of wF: this defines a set map $a : \mathsf{PaB} \to \mathsf{wF}$, which can also be made into a tensor functor. We don't use this functor for any of the results in this paper, so we refer the reader to [**BND24b**, Section 1.2.2]) for a topological description.

Using this relationship between PaB and wF, it is possible to construct a homomorphic expansion for wF directly from a homomorphic expansion for PaB, in parallel with the [**AET10**] formula for KV solutions in terms of Drinfel'd associators. This is the main result of [**BND24b**]:

THEOREM 2.36 ([**BND24b**, Theorem 1.1]). *Every homomorphic expansion $Z^b : \mathbb{Q}[\widehat{\mathsf{PaB}}] \to \mathsf{PaCD}$ extends uniquely to a homomorphic expansion $Z : \mathbb{Q}[\widehat{\mathsf{wF}}] \to \mathscr{A}$, in the sense that the following diagram commutes:*

$$\begin{array}{ccc} \mathbb{Q}[\widehat{\mathsf{PaB}}] & \xrightarrow{\ Z^b\ } & \mathsf{PaCD} \\ \big\downarrow{\scriptstyle a} & & \big\downarrow{\scriptstyle \alpha} \\ \mathbb{Q}[\widehat{\mathsf{wF}}] & \dashrightarrow_{Z} & \mathscr{A} \end{array} \tag{2.16}$$

Theorem 1.1 of [**BND24b**] gives an explicit construction for Z from Z^b, in particular a formula for $Z(V)$ and $Z(C)$ in terms of the Drinfel'd associator $\Phi = Z^b(\not|\!\nwarrow\!|)$. We are not using this result in the current paper, however, the construction used in Section 3 to construct a diagrammatic Alekseev-Torossian map follows the techniques used in [**BND24b**] to construct Z from Z^b.

3. Deriving a map of expansion-preserving automorphisms

In [**AT12**] and [**AET10**], Alekseev, Enriquez and Torossian illustrated the deep relationship between Drinfel'd associators and solutions to the Kashiwara-Vergne conjecture. In particular, they construct an injective group homomorphism between the graded Grothendieck-Teichmüller and the graded Kashiwara-Vergne group.

THEOREM 3.1 ([**AET10**, Theorem 2.5]). *The map* $\rho : \mathsf{GRT}_1 \to \mathsf{KRV}$ *given by*

$$\Psi(x,y) \mapsto (\Psi(-x-y,x), \Psi(-x-y,y)) \tag{3.1}$$

is an injective group homomorphism.

In other words, the map ρ defines a subgroup of KRV which is *isomorphic* to GRT_1. Moreover, when restricted to the set of KV solutions constructed from Drinfel'd associators, the action of the subgroup of KRV isomorphic to GRT_1 coincides with the canonical action of GRT_1 on Drinfel'd associators ([**AT12**, Prop 9.13]).

In this section, we construct a map $\mathring{\rho} : \mathrm{Aut}(\mathsf{PaCD}) \to \mathrm{Aut}_v(\mathscr{A})$ from automorphisms of parenthesised chord diagrams to automorphisms of arrow diagrams which makes the following diagram commute:

$$\begin{array}{ccc} \mathrm{Aut}(\mathsf{PaCD}) & \overset{\mathring{\rho}}{\dashrightarrow} & \mathrm{Aut}_v(\mathscr{A}) \\ {\scriptstyle \tau}\downarrow & & \downarrow{\scriptstyle \Theta} \\ \mathsf{GRT}_1 & \xrightarrow{\quad\rho\quad} & \mathsf{KRV} \end{array} \tag{3.2}$$

In particular, we identify a subgroup of arrow diagram automorphisms isomorphic to GRT_1.

3.1. Comparing automorphisms of chord diagrams and arrow diagrams. In Section 2.7 we described a tensor functor $\alpha : \mathsf{PaCD} \to \mathscr{A}$ induced by the inclusion of Lie algebras $\mathfrak{t}_n \hookrightarrow \mathfrak{tder}_n$ (Definition 2.35). The goal for this section is to construct the map $\mathring{\rho} : \mathrm{Aut}(\mathsf{PaCD}) \to \mathrm{Aut}_v(\mathscr{A})$. The plan is to show that, given $F \in \mathrm{Aut}(\mathsf{PaCD})$, then we can construct $G \in \mathrm{Aut}_v(\mathscr{A})$, such that the following diagram commutes:

$$\begin{array}{ccc} \mathsf{PaCD} & \xrightarrow{\quad\alpha\quad} & \mathscr{A} \\ {\scriptstyle F}\downarrow & & \downarrow{\scriptstyle G} \\ \mathsf{PaCD} & \xrightarrow[\quad\alpha\quad]{} & \mathscr{A}. \end{array} \tag{3.3}$$

We will then show that setting $\mathring{\rho}(F) = G$ fits into the diagram (3.2). Before we give the construction, we need to establish some facts about arrow diagrams.

LEMMA 3.2. *There is an isomorphism*

$$\varphi : \mathscr{A}(\curlywedge_{\bullet}) \xrightarrow{\cong} \mathscr{A}(\uparrow).$$

This is a crucial fact and covered in more detail in the forthcoming [**BND24b**]. We give a proof sketch:

PROOF. The isomorphism φ is shown in Figure 16. Given any arrow diagram $\mathbf{D} \in \mathscr{A}(\curlywedge_{\bullet})$, by the TF relation, the punctured strand has only arrow heads ending on it. These arrow heads are collectively denoted by h. For any element of $\mathscr{A}(\curlywedge_{\bullet})$ there is a representative with only arrow tails on the right strand (achievable via STU and CP relations): denote these arrow endings by t. Finally, we denote arrow endings on the top strand (both heads and tails) by x. Then, by repeated applications of the VI relation, t can be "pushed" to the top strand: VI in this case does not result in a sum, as the other strand is punctured. After clearing the capped strand this way, use repeated VI relations to push h to the top strand as

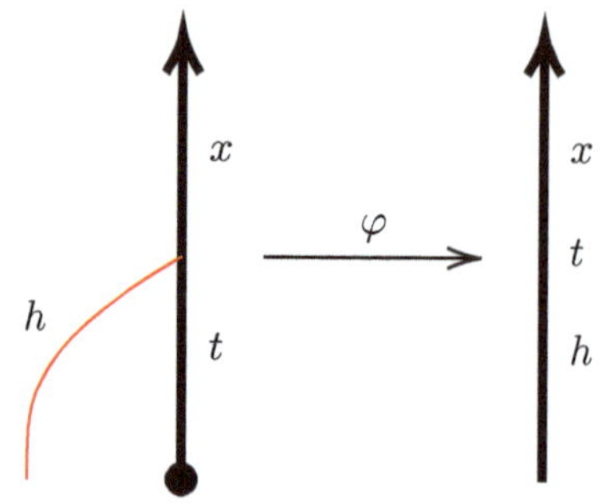

FIGURE 16. The isomorphism φ applied to the arrow diagram $\mathbf{D}$ on the left.

well. Once again, no sums appear by the CP relation, as the capped strand no longer has any arrow endings. The result of this process is $\varphi(\mathbf{D})$.

We leave it to the reader to check that φ is well-defined, and that its inverse is the natural inclusion of $\mathscr{A}(\uparrow)$ as the top strand in $\mathscr{A}(⅄)$. □

The next Lemma shows that puncturing and capping a vertex trivialises its G-value.

LEMMA 3.3. *For any* $G \in \mathrm{Aut}_v(\mathscr{A})$, $G(⅄) = 1$, *that is,* $G(⅄) = ⅄$ *with no arrows.*

PROOF. [**BND24b**, Lemma 2.9] shows that for any homomorphic expansion Z of wF, $Z(⅄) = 1$. In other words, $Z(⅄)$ is the skeleton element $⅄ \in \mathscr{A}$, with no arrows. Since automorphisms $G \in \mathrm{Aut}_v(\mathscr{A})$ are expansion-preserving, the result is immediate. □

We are now ready to construct G from F as in (3.3).

CONSTRUCTION 3.4. By Corollary 2.34, G is uniquely determined by $G(⅄)^{tr}$, and hence the idea is to compute $G(⅄)^{tr}$ from $F(\beta)$ for a specific, suitably chosen $\beta \in \mathsf{PaCD}$, depicted on the left side of Figure 17a. Given and $F \in \mathrm{Aut}(\mathsf{PaCD})$ with $F(|\!/\!|) = \Psi(t^{12}, t^{23})$, we compute $F(\beta)$:

$$F(\beta) = (\Psi^{(13)24})^{-1}\, \Psi^{132}\, \times^{23}\, (\Psi^{123})^{-1}\, \Psi^{(12)34}.$$

This is shown on the left side of Figure 17b. Using $F(\beta)$, we determine $G(⅄)^{tr}$ as follows:

(1) Applying α (Definition 2.35) to $F(\beta)$ we obtain an arrow diagram in $\mathscr{A}$ with skeleton $\beta^{cl} = \alpha(\beta)$. Any $G \in \mathrm{Aut}_v(\mathscr{A})$ fitting into the commutative diagram (3.3) satisfies $\alpha(F(\beta)) = G(\beta^{cl})$, and thus we know the value $G(\beta^{cl})$. See Figure 17b.

(2) To find the value of $G(⅄)^{tr}$, we perform a series of operations on $G(\beta^{cl})$, as shown in Figure 18 and explained below. First, we pre-compose with a capped vertex ⅄. Then we perform two unzip operations: the first on the top strand of the new vertex, and the second a disk unzip on the capped strand. Finally, we puncture the two bottom uncapped strands to obtain an arrow diagram on a skeleton that is a double capped, double punctured vertex denoted W.

(3) Since $G \in \mathrm{Aut}(\mathscr{A})$, it commutes with all the prop and auxiliary operations of $\mathscr{A}$, and hence we can compute $G(W)$ from $G(\beta^{cl})$, as shown in the bottom line of Figure 18. Due to the composition with   in the previous step, $G(W)$ contains involves copies of $N = G(\ )$ and $C = G(\ )$: this is shown in detail in Figure 18; schematically, $G(\ \ \beta^{cl}) = G(\ )G(\ )G(\beta^{cl})$ and thus $G(W) = p^2u^2\left(G(\ )G(\ )G(\beta^{cl})\right)$.

(4) By Lemma 3.3 we have $G(\ ) = 1$, and therefore the copy of $G(\ )$ in $G(W)$ from the composition with   cancels after the puncture operations. This is shown in the rightmost column of Figure 18. Thus, $G(W) = u(C)p^2G(\beta^{cl})$.

(5) On the other hand, from the fact that G is a prop automorphism, $G(W)$ is a composite of $G(\ )$, two copies of $G(\ )$ and two copies of $G(\ )$. Again by Lemma 3.3, $G(\ ) = 1$, and therefore, schematically, $G(W) = G(\ )G(\ )G(\ )$. Since $G(\ ) \in \mathscr{A}(\ )$, by Lemma 2.23 $(G(\ ))^{tr} = 1$, and therefore $(G(W))^{tr} = (G(\ ))^{tr} = N^{tr}$, which is the value we need to compute.

(6) Combining (4) and (5) we have $N^{tr} = \left(u(C)p^2G(\beta^{cl})\right)^{tr} = (p^2G(\beta^{cl}))^{tr}$.

(7) In order to express N^{tr} in its standard form as an element of $\mathscr{A}(\uparrow_2) \cong \mathscr{A}(\ )$, rather than in $\mathscr{A}(W)$, we apply the isomorphism φ of Lemma 3.2: $\mathscr{A}(W) \overset{\varphi}{\cong} \mathscr{A}(\ )$. Therefore, $N^{tr} = \varphi(p^2G(\beta^{cl}))$.

Construction 3.4 shows that given $F \in \mathrm{Aut}(\mathsf{PaCD})$, if a $G \in \mathrm{Aut}(\mathscr{A})$ fitting into the diagram (3.3) exists, then $(G(\ ))^{tr}$ is explicitly computable from F. In the following theorem, we compute the complete value of $G(\ )$:

THEOREM 3.5. *Given an automorphism* $F \in \mathrm{Aut}(\mathsf{PaCD})$ *with* $F(\ ) = \Psi = \Psi(t^{12}, t^{23})$, *then if there exists* $G \in \mathrm{Aut}_v(\mathscr{A})$ *which completes the commutative diagram* (3.3), *then* G *is uniquely determined by* F, *and* $N = G(\ ) \in \mathscr{A}(\uparrow_2)$ *and* $C = G(\ ) \in \mathbb{Q}[[\xi]]/\langle\xi\rangle$ *satisfy*

$$N = (C^1)^{-1}(C^2)^{-1}\varphi\left(\Psi^{-1}\left(a^{2(13)}, -a^{2(13)} - a^{4(13)}\right) \cdot \Psi(a^{23}, a^{43})\right) C^{12}, \tag{3.4}$$

where φ *is the isomorphism of Lemma 3.2, and* a^{ij} *denotes an arrow from strand* i *to strand* j, *and the strand numbering is as shown in Figure 17.*

To prove Theorem 3.5, we first establish a few basic facts about GRT_1.

LEMMA 3.6. *For* $\Psi = \Psi(x, y) \in \mathsf{GRT}_1$, *the following properties hold:*

(1) *If* $[x, y] = 0$ *then* $\Psi(x, y) = 1$.
(2) $\Psi(t^{ij}, t^{jk}) = \Psi(t^{ij}, -t^{ij} - t^{ik})$, *where* t^{ij} *denotes the chord between the* i*-th and* j*-th strands.*

PROOF. The first property is a direct consequence of the defining relation (2.4). The second follows from the fact that $(t^{jk} + t^{ij} + t^{ik})$ is central in $\mathfrak{t}_n$ by 4T relation of chord diagrams $[t^{jk}, t^{ij} + t^{ik}] = 0$ [**BN98**, Definition 2.7]. Because of this, for any $X \in \mathfrak{t}_n$:

$$\begin{aligned} &[t^{jk} + t^{ij} + t^{ik}, X] = 0 \\ &\implies [t^{jk}, X] = -[t^{ij} + t^{ik}, X] \\ &\implies [t^{jk}, X] = [-t^{ij} - t^{ik}, X] \end{aligned}$$

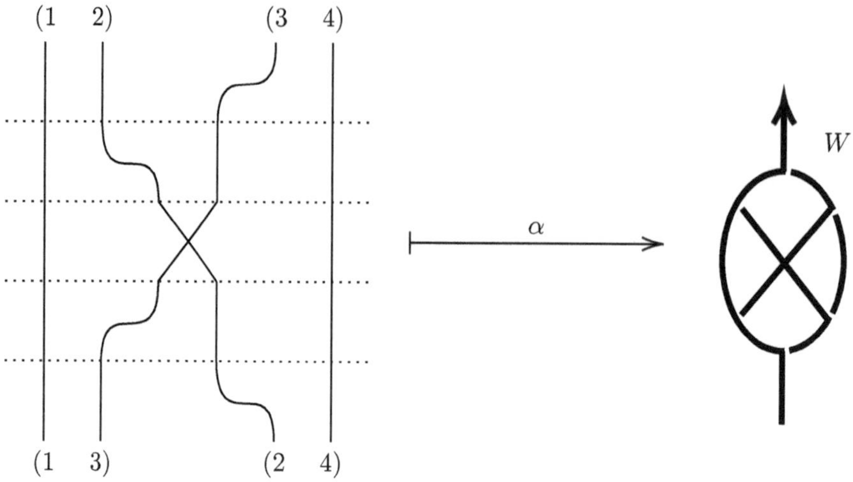

(A) The element $\beta \in \mathsf{PaCD}$ is a skeleton element of PaCD. Its image under α is its closure.

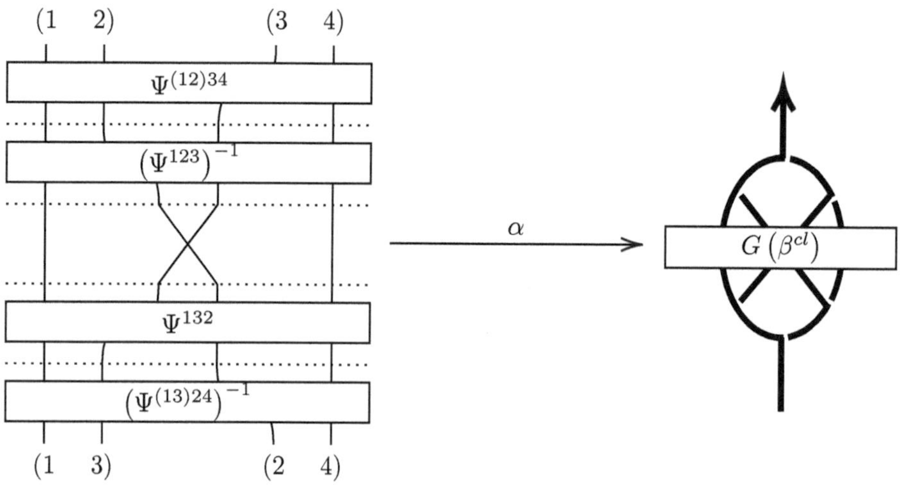

(B) $F(\beta)$, expressed in terms of Ψ, is shown on the left. By the compatibility of F and G, we know that $\alpha(F(\beta)) = G(\alpha(\beta))$.

FIGURE 17

and therefore,

$$\Psi(t^{ij}, t^{jk}) = \Psi(t^{ij}, -t^{ij} - t^{ik}) \tag{3.5}$$

as required. □

PROOF OF THEOREM 3.5. The fact that F uniquely determines G is almost immediate from Construction 3.4. By Lemma 2.34, $G \in \mathrm{Aut}_v(\mathscr{A})$ is uniquely determined by $N^{tr} = \overrightarrow{pr}(G(⅄))$. Step (6) of Construction 3.4 shows how to compute N^{tr} from $F(\beta)$. Thus, G, is uniquely determined by F.

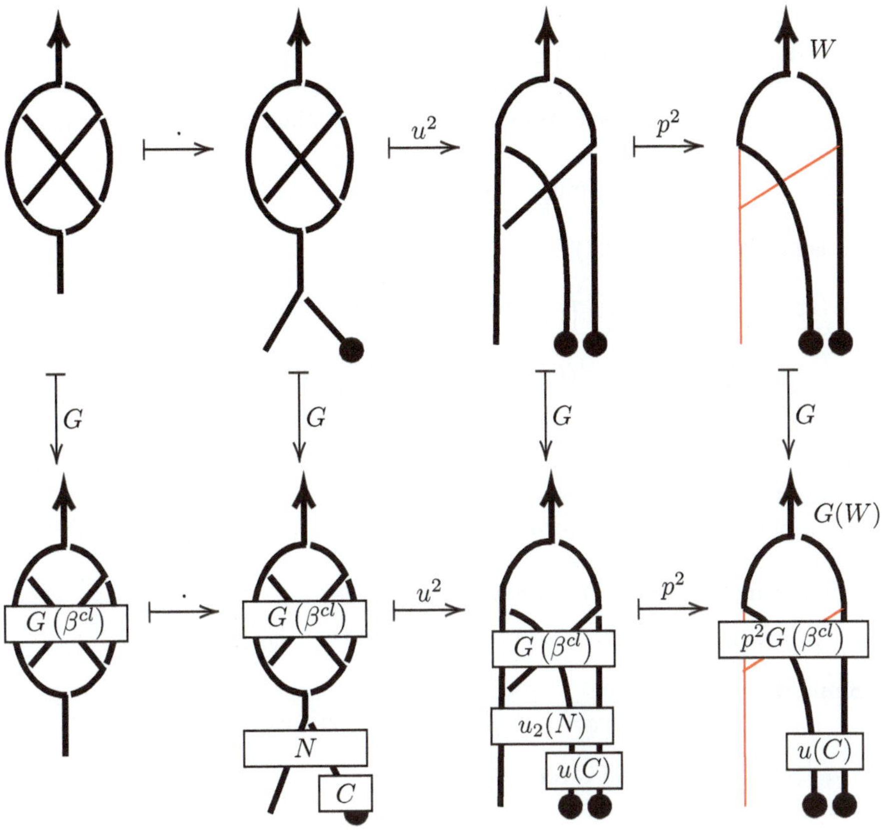

FIGURE 18. The stripping process.

Given $F(\bowtie) = \Psi(t^{12}, t^{23})$, we compute $F(\beta)$ as shown in the left-hand side of Figure 17b:

$$F(\beta) = (\Psi^{(13)24})^{-1}\,\Psi^{132} \times^{23} (\Psi^{123})^{-1}\,\Psi^{(12)34}. \tag{3.6}$$

From here on we will suppress the skeleton crossing $\times^{23}$, as it is part of the skeleton and commutes with chords. By diagram (3.3), we have

$$G(\beta^{cl}) = G(\alpha(\beta)) = \alpha(F(\beta)) = \alpha\left((\Psi^{(13)24})^{-1}\,\Psi^{132}\,(\Psi^{123})^{-1}\,\Psi^{(12)34}\right). \tag{3.7}$$

We obtain the formula (3.4) from this by the sequence of operations detailed in Construction 3.4, and shown in the first row of Figure 18 for the skeleton β^{cl}, and in the second row for $G(\beta^{cl})$. Recall that by step (6) of the construction, $\mathscr{A}(W) \cong \mathscr{A}(\curlywedge) \cong \mathscr{A}(\uparrow_2)$.

By step (5) of the construction, $G(W)$ is a circuit algebra composition of $G(\curlywedge) = N$ and two copies of $G(\downarrow) = C$. Applying $\varphi : \mathscr{A}(W) \to \mathscr{A}(\uparrow_2)$, involves sliding the C values onto the top two strands, we obtain $\varphi(G(W)) = C^1C^2N$, and therefore,

$$N = (C^1)^{-1}(C^2)^{-1}\varphi(G(W)). \tag{3.8}$$

Since G commutes with all the operations performed throughout the 'stripping' process, this allows us to compute $G(W)$, as shown in Figure 18 and in step (4) of the construction. In the strand numbering of Figure 17b, the punctured strands are 1 and 3 and the capped strands are 2 and 4, thus we make step (4) of the construction more precise:

$$G(W) = C^{(24)} p_1 p_3 G(\beta^{cl}) = C^{24} p_1 p_3 \left(\alpha(F(\beta))\right). \tag{3.9}$$

Thus, using the formula (3.7),

$$\begin{aligned} G(W) &= C^{24} p_1 p_3 \left(\alpha\left((\Psi^{(13)24})^{-1}\, \Psi^{132}\, (\Psi^{123})^{-1}\, \Psi^{(12)34}\right)\right) \\ &= C^{24} p_1 p_3 \left(\alpha\left(\Psi^{-1}(t^{(13)2}, t^{24})\, \Psi(t^{13}, t^{23})\, \Psi^{-1}(t^{12}, t^{23})\, \Psi(t^{(12)3}, t^{34})\right)\right), \end{aligned} \tag{3.10}$$

where t^{ij} is the chord from i to j and $t^{(ij)k} = t^{ik} + t^{jk}$.

By definition, $\alpha(t^{ij}) = a^{ij} + a^{ji}$, where a^{ij} denotes a horizontal arrow pointing from strand i to strand j. By the TF relation of arrow diagrams (Figure 9), arrow tails on a punctured strand are zero. In other words, $p_1 p_3(a^{1m}) = p_1 p_3(a^{3m})) = 0$ for any m.

Then, using property (1) of Lemma 3.6,

$$p_1 p_2 \alpha\left(\Psi(t^{13}, t^{23})\right) = \Psi(a^{13} + a^{31}, a^{23} + a^{32}) = \Psi(0, a^{23} + a^{32}) = 1.$$

Similarly,

$$p_1 p_3 \alpha\left(\Psi^{-1}(t^{12}, t^{23})\right) = \Psi^{-1}(a^{21}, a^{23}) = 1,$$

since $[a^{21}, a^{23}] = 0$ by the TC relation.

For $\Psi(t^{(12)3}, t^{34})$, we have

$$p_1 p_3 \alpha\left(\Psi(t^{(12)3}, t^{34})\right) = \Psi(a^{23}, a^{43}),$$

since a^{13}, a^{31}, a^{32} and a^{34} are all in the kernel of $p_1 \circ p_3$.

For $\Psi(t^{(13)2}, t^{24})^{-1}$, we first apply the symmetry property (2) from Lemma 3.6 to get $\Psi(t^{(13)2}, t^{24})^{-1} = \Psi(t^{(13)2}, -t^{(13)2} - t^{(13)4})^{-1}$. Thus,

$$p_1 p_3 \alpha\left(\Psi^{-1}(t^{(13)2}, -t^{(13)2} - t^{(13)4})\right) = \Psi^{-1}(a^{2(13)}, -a^{2(13)} - a^{4(13)}).$$

Substituting these results into equation (3.10), we obtain

$$G(W) = C^{24} \Psi^{-1}(a^{2(13)}, -a^{2(13)} - a^{4(13)}) \cdot \Psi(a^{23}, a^{43}). \tag{3.11}$$

To prove the formula (3.4), it remains to show that

$$\varphi(G(W)) = \varphi\left(\Psi^{-1}\left(a^{2(12)}, -a^{2(13)} - a^{4(13)}\right) \cdot \Psi(a^{23}, a^{43})\right) C^{12}.$$

Note that in (3.11), $G(W)$ is expressed in a form where there are only arrow tails on the capped strands 2 and 4 (and only arrow heads on the punctured strands 1 and 3). The isomorphism φ acts by "pushing" all the arrow tails from strands 2 and 4 up across the  vertices. Due to the TC relation, it is possible to start with C^{24}, followed by the arrow tails from $\Psi^{-1}\left(a^{2(13)}, -a^{2(13)} - a^{4(13)}\right) \cdot \Psi(a^{23}, a^{43})$, then—with the capped stands cleared—the arrow heads from strands 1 and 3 are pushed up as well. Thus, we have $\varphi\left(\Psi^{-1}\left(a^{2(12)}, -a^{2(13)} - a^{4(13)}\right) \cdot \Psi(a^{23}, a^{43})\right)$, as required. ∎

3.2. An embedding of automorphism groups. Our goal is to prove the following theorem, the main result of this paper:

THEOREM 3.7. *Construction* 3.4 *defines a group homomorphism*

$$\mathring{\rho} : \mathrm{Aut}(\mathsf{PaCD}) \to \mathrm{Aut}_v(\mathscr{A}),$$

given by $\mathring{\rho}(F) = G$. *This homomorphism is equivalent to the Alekseev-Torossian map* ρ *in the sense that* $\mathring{\rho}$ *fits into the commutative diagram* (3.2).

The proof of the Theorem comes down to analysing the diagram (3.12) below, where black arrows are group homomorphisms and red arrows are anti-homomorphisms. We begin with the definitions of the maps.

(3.12)

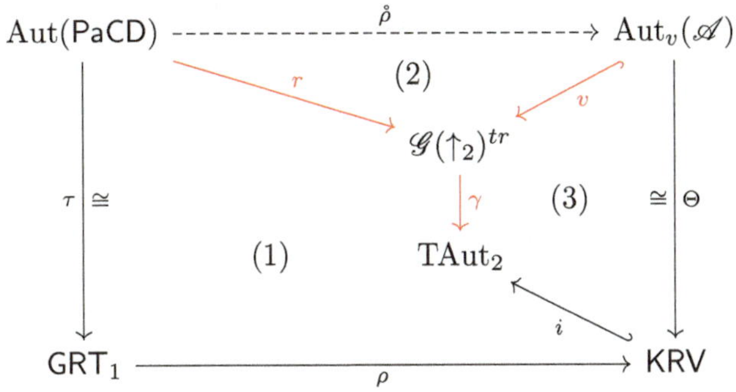

The map r. From $F \in \mathrm{Aut}(\mathsf{PaCD})$, set $r(F) = N^{tr} = (G(\curlywedge))^{tr}$. The map r is a group anti-homomorphism: see Lemma 3.11.

The map v. Given an automorphism $G \in \mathrm{Aut}_v(\mathscr{A})$, $v(G) := \overrightarrow{pr}(G(\curlywedge)) \in \mathscr{G}(\uparrow_2)^{tr}$. The map v is a group anti-homomorphism since if $v(G_1) = N_1$ and $v(G_2) = N_2$, then—using that expansion-preserving automorphisms of $\mathscr{A}$ fix the arrow—we have $v(G_1 \circ G_2) = N_2 \cdot N_1$.

The map $\mathring{\rho}$. In Theorem 3.5 we proved that if there is a $G \in \mathrm{Aut}_v(\mathscr{A})$ fitting into the diagram (3.4), then $\overrightarrow{pr}(G(\curlywedge)) = r(F)$. In turn, $\overrightarrow{pr}(G(\curlywedge))$ uniquely determines G. Thus, if there exists a $G \in \mathrm{Aut}_v(\mathscr{A})$ with $\overrightarrow{pr}(G(\curlywedge)) = r(F)$, we set $\mathring{\rho}(F) = G$. The map $\mathring{\rho}$ exists if and only if r factors through $\mathrm{Aut}_v(\mathscr{A})$, that is, if given any $F \in \mathrm{Aut}(\mathsf{PaCD})$, the value $r(F)$ is always a valid vertex value for some automorphism $G \in \mathrm{Aut}_v(\mathscr{A})$. This follows from the proof of Theorem 3.7 at the end of this section; we also give a more direct proof in Theorem A.7.

The map τ. The map τ is the isomorphism described in Theorem 2.6. It is given by $\tau(F) = \Psi$, where $F(|\!\vee\!|) = \Psi(t^{12}, t^{23})$.

The map ρ. The map ρ is the Alekseev-Torossian map defined at the beginning of this section, given by the formula $\rho(\Psi) = (\Psi(-x-y, x), \Psi(-x-y, y))$.

The map γ. There is an action[8] of $\mathscr{G}(\uparrow_2)$ on $\widehat{\mathfrak{lie}}_2$, coming from the exponentiation of the projection $\mathscr{P}(\uparrow_2) \to \mathfrak{tder}_2$ (Proposition 2.17). There is a bijection between Lie words w in x and y, and tree arrow diagrams on three strands whose tails lie on the first two strands and head lies on the third strand, as shown in Figure 19. Specifically, label the first two strands x and y, and represent the word w as a binary tree with tails on the x and y strands, vertices representing the brackets, and head on the third strand. Call this tree $\mathbf{T}_w \in \mathscr{A}(\uparrow_3)$. Given a group-like

[8]This action is described in detail in [**BND17**], see the proof of Proposition 3.19, in particular the "Conceptual argument".

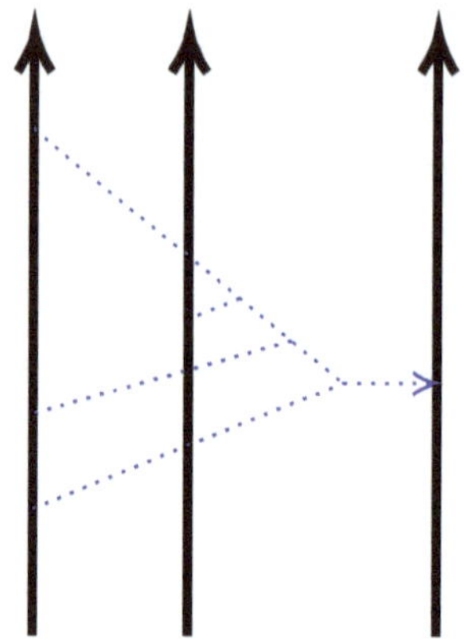

FIGURE 19. The lie word $[x,[x,[y,x]]]$ represented as an arrow diagram on 3 strands.

element $\mathbf{D} \in \mathscr{G}(\uparrow_2)$, the conjugate $(\mathbf{D}^{12})^{-1}\mathbf{T}_w\mathbf{D}^{12}$ is an (infinite) sum of trees with tails on the first two strands and heads on the third strand [**BND17**, Proposition 3.19], and thus corresponds to an element of $\widehat{\mathfrak{lie}}_2$ via the same bijection. This defines the action of $\mathbf{D}$ on w. Since wheels in $\mathscr{G}(\uparrow_2)$ act trivially by the TC relation, the action descends to an action of $\mathscr{G}(\uparrow_2)^{tr}$ on $\widehat{\mathfrak{lie}}_2$.

Furthermore, it turns out that $\mathscr{G}(\uparrow_2)^{tr}$ acts on $\widehat{\mathfrak{lie}}_2$ by *tangential* automorphisms. Namely, if $\mathbf{D} \in \mathscr{G}(\uparrow_2)^{tr}$, then modulo wheels $\mathbf{D} = e^{\delta}$, where δ is a linear combination of trees. Furthermore, $\delta = \delta_1 + \delta_2$, where δ_1 is a sum of the terms with heads ending on strand 1, and δ_2 is a sum of terms with heads on the second strand. Each term of δ_1 corresponds to a Lie word in x and y, similarly to Figure 19 but now with head on strand 1, thus δ_1 corresponds to a sum $u_1 \in \widehat{\mathfrak{lie}}_2$. Similarly, δ_2 corresponds to $u_2 \in \widehat{\mathfrak{lie}}_2$. Then $(u_1, u_2) \in \mathfrak{tder}_2$, and $\mathbf{D}$ acts as $\exp((u_1,u_2)) \in \mathrm{TAut}_2$. Note that where the head of each tree in $\mathbf{D}$ is attached to the strand 1 or 2 in relation to its own tails does not make a difference in $\gamma(\mathbf{D})$.

The map γ is a group anti-homomorphism since if $\gamma(N_1) = a$ and $\gamma(N_2) = b$, then $\gamma(N_1 \cdot N_2) = b \circ a = \gamma(N_2) \circ \gamma(N_1)$ in TAut_2.

The map i. By definition, KRV is a subgroup of TAut_2, this is the inclusion i.

The map Θ. The map Θ is the isomorphism of [**DHR23**, Theorem 5.12], recalled as Theorem 2.32. Namely, for $G \in \mathrm{Aut}_v(\mathscr{A})$, let $G(\,⅄\,) = N \in \mathscr{G}(\uparrow_2)$. Then, modulo wheels, we have $\overrightarrow{pr}(N) = N^{tr} \in \mathscr{G}(\uparrow_2)^{tr}$. By [**DHR23**, Theorem 5.12], $\gamma(N^{tr})$ then satisfies the KRV equations and $\Theta(G) := \gamma(N^{tr}) \in \mathsf{KRV}$.

To prove Theorem 3.7, we need to show that the map $\mathring{\rho}$ exists, and that the diagram (3.12) commutes. A significant step is to describe the composite $\gamma \circ r$, and prove that r is an anti-homomorphism.

LEMMA 3.8. *Given $F \in \mathrm{Aut}(\mathsf{PaCD})$ with $F(|\!\!\nearrow\!\!|) = \Psi(t^{12}, t^{13})$, we have*

$$(\gamma \circ r(F))\,(x) = \Psi(-x-y,x)^{-1}\,x\,\Psi(-x-y,x), \tag{3.13}$$

$$(\gamma \circ r(F))\,(y) = \Psi(-x-y,y)^{-1}\,y\,\Psi(-x-y,y). \tag{3.14}$$

PROOF. Recall that, by definition of r and γ,

$$(\gamma \circ r(F))\,(x) = (N^{tr})^{-1}\,x\,N^{tr}, \quad \text{and} \quad (\gamma \circ r(F))\,(y) = (N^{tr})^{-1}\,y\,N^{tr},$$

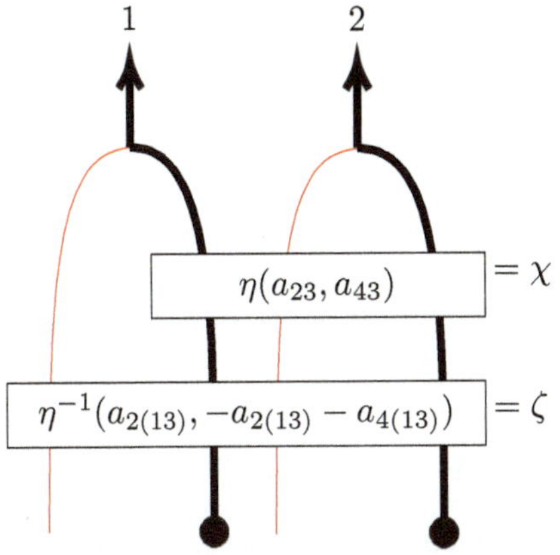

(A) N^{tr} shown in the usual box notation, with names χ and ζ assigned to each arrow diagram component.

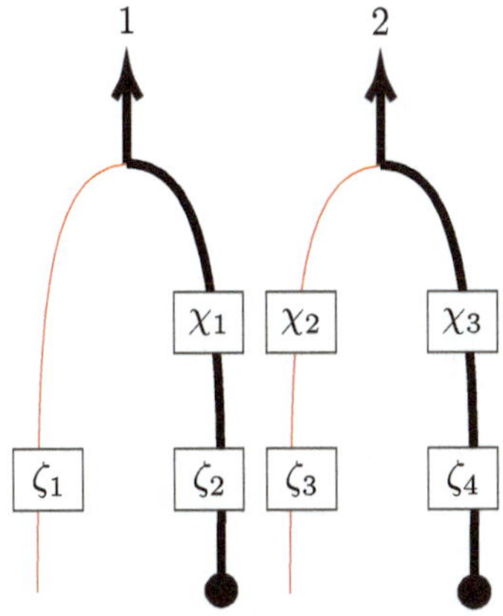

(B) N^{tr} shown in placement notation, with arrow endings numbered according to the strand numbering.

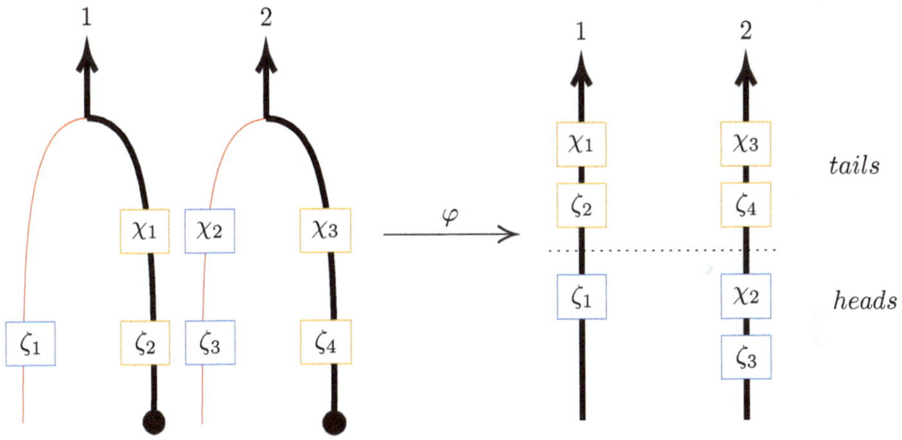

(C) Applying the isomorphism φ using the placement notation. Arrow tails are shown in blue, and heads in orange.

FIGURE 20

where $N = G(⋏)$ is the value obtained by the "stripping" process and N^{tr} is the tree projection $\overrightarrow{pr}(N)$ of N. Specifically, by Theorem 3.5,

$$N^{tr} = \varphi\left(\Psi^{-1}\left(a^{2(12)}, -a^{2(13)} - a^{4(13)}\right) \cdot \Psi(a^{23}, a^{43})\right). \tag{3.15}$$

The main step still missing is to compute the value of the isomorphism φ of Lemma 3.2. To do this, we introduce a diagrammatic *placement notation*, as shown in Figures 20a and 20b. In this notation, the two Ψ-factors are named ζ and χ for short, and their arrow endings on each strand are numbered by the strand numbers: ζ_1, ζ_2, etc, for arrow endings of ζ on the first, second, etc strand. The value of $\varphi\left(\Psi^{-1}\left(a^{2(12)}, -a^{2(13)} - a^{4(13)}\right) \cdot \Psi(a^{23}, a^{43})\right)$ in $\mathscr{A}(\uparrow_2)$ is shown in Figure 20c.

Due to the TC relation on arrow diagrams, only the tree components of N^{tr} which have heads on the first strand (after applying φ) act non-trivially on x. The only such component – written in the placement notation – is ζ_1, coming from

$$\zeta = \Psi^{-1}(a^{2(13)}, -a^{2(13)} - a^{4(13)}) = \Psi^{-1}(a^{21} + a^{23}, -a^{21} - a^{23} - a^{41} - a^{43}).$$

The arrows $\varphi(a^{23})$ and $\varphi(a^{43})$ have heads lying on strand 2: this is where ζ_3 arises from in Figure 20c. Therefore, these arrows do not act on x, so x is only conjugated by $\varphi(\Psi^{-1}(a^{21}, -a^{21} - a^{41}))$, which, by the anti-symmetry relation of GRT_1 equals $\varphi(\Psi(-a^{21} - a^{41}, a^{21}))$.

In summary

$$(3.16)\qquad (\gamma \circ r(F))\,(x) = \varphi(\Psi(-a^{21} - a^{41}, a^{21}))^{-1}\, x\, \varphi(\Psi(-a^{21} - a^{41}, a^{21})).$$

Now recall that the value of γ—that is, the conjugation action—only depends on the Lie words represented by the tree components of $\varphi(\Psi(-a^{21} - a^{41}, a^{21}))$, not the specific placement of arrow heads and tails. Let $\Psi = e^{\psi}$ for some Lie element $\psi \in \mathfrak{grt}_1$, and let $w = w(-a^{21} - a^{41}, a^{21})$ be a Lie word component of ψ. Since $\varphi(a^{21}) = a^{11}$ and $\varphi(a^{41}) = a^{21}$, the Lie word represented by $\varphi(w)$ is[9] $w(-x - y, y)$. Exponentiating, we obtain

$$(3.17)\qquad (\gamma \circ r(F))\,(x) = \Psi(-x - y, x)^{-1}\, x\, \Psi(-x - y, x).$$

We proceed similarly to compute the action on y. Due to the TC relation, only components of N^{tr} with heads on the second strand act non-trivially on y. There are two such components: χ_3 and ζ_3 in placement notation, where χ_3 comes from the term $\Psi(a^{23}, a^{43})$, and ζ_3 comes from $\Psi^{-1}(a^{2(13)}, -a^{2(13)} - a^{4(13)})$. Cancelling arrows which act trivially on y (i.e. those with heads on strand 1), we get that y is conjugated by

$$(3.18)\qquad \varphi\left(\Psi^{-1}(a^{23}, -a^{23} - a^{43}) \cdot \Psi(a^{23}, a^{43})\right).$$

To simplify, we apply the hexagon relation (2.5) of GRT_1 with $x = -a^{23} - a^{43}, y = a^{23}, z = a^{43}$ and obtain that y is conjugated by

$$(3.19)\qquad \varphi(\Psi(-a^{23} - a^{43}, a^{43})).$$

And therefore we have

$$(3.20)\qquad (N^{12})^{-1}\, y\, N^{12} = \varphi(\Psi(-a^{23} - a^{43}, a^{43}))^{-1}\, y\, \varphi(\Psi(-a^{23} - a^{43}, a^{43})).$$

Then, computing the value of γ as before, we have

$$(3.21)\qquad (\gamma \circ r(F))\,(y) = \Psi(-x - y, y)^{-1}\, y\, \Psi(-x - y, y),$$

as required. □

LEMMA 3.9. *The map γ is injective*[10] *on the image of r.*

[9] Note that $\varphi(w(-a^{21} - a^{41}, a^{21})) \neq w(-a^{11} - a^{21}, a^{11})$: at the level of diagrams the tail/head placements matter, and in fact a short arrow a^{11} (with no other arrow endings between its tail and head) is central in $\mathscr{A}(\uparrow_2)$. However, the Lie word represented by the tree $\varphi(w(-a^{21} - a^{41}, a^{21}))$ does have an x for each a^{21} and a y for each a^{41}.

[10] Equivalently, $r(F)$ is free of short arrows for any $F \in \mathrm{Aut}(\mathsf{PaCD})$. That is, $r(F) = e^{\zeta}$ and ζ is an (infinite) sum of tree arrow diagrams with no short arrow summands.

PROOF. To see that γ is injective on the image of r, we recall that the kernel of γ is $\exp(\mathfrak{a}_2)$, where $\mathfrak{a}_2$ is the abelian lie algebra spanned by single *short arrows* a^{11} and a^{22} [**BND17**, Proposition 3.19]. On the other hand, $F \in \mathrm{Aut}(\mathsf{PaCD})$ is given by $F(|\!\nearrow\!|) = \Psi = e^{\psi} \in \mathsf{GRT}_1$, and ψ vanishes in degree one [**Dri90**, Remark after Proposition 6.2]. Hence, $r(F)$ does not include short arrows. ☐

The following lemma, due to Drinfel'd, will be useful several times through the rest of the argument:

LEMMA 3.10 ([**Dri90**, Equations 5.14 and 5.19]). *For* $\Psi = e^{\psi} \in \mathsf{GRT}_1$, *if* $X + Y + Z = 0$ *then*

$$[X, \psi(Z, X)] + [Y, \psi(Z, Y)] = 0 \tag{3.22}$$

and

$$X + \Psi(X, Y)^{-1}\, Y\, \Psi(X, Y) + \Psi(X, Z)^{-1}\, Z\, \Psi(X, Z) = 0. \tag{3.23}$$

LEMMA 3.11. *The map* $r : \mathrm{Aut}(\mathsf{PaCD}) \to \mathscr{G}(\uparrow_2)^{tr}$ *is a group anti-homomorphism.*

PROOF. Since γ is a group anti-homomorphism and injective on the image of r, we can prove that r is an anti-homomorphism by showing that $\gamma \circ r$, as computed in Lemma 3.8, is a group homomorphism.

First, recall that composition in $\mathrm{Aut}(\mathsf{PaCD})$ does not translate to multiplication in PaCD, but is given by the following formula [**BN98**, Proposition 5.1]: if $F_1(|\!\nearrow\!|) = \Psi_1(t^{12}, t^{23})$ and $F_2(|\!\nearrow\!|) = \Psi_2(t^{12}, t^{23})$, then

$$(F_1 \cdot F_2)(|\!\nearrow\!|) = \Psi_1(t^{12}, t^{23})\Psi_2\left(t^{12}\,,\ \Psi_1^{-1}(t^{12}, t^{23})\, t^{23}\, \Psi_1(t^{12}, t^{23})\right).$$

By Lemma 3.8, $\gamma \circ r(F_1)$ conjugates x by $\Psi_1(-x-y, x) =: \Psi_{1,x}$ and y by $\Psi_1(-x-y, y) =: \Psi_{1,y}$. Similarly, $\gamma \circ r(F_2)$ conjugates x by $\Psi_2(-x-y, x) =: \Psi_{2,x}$ and y by $\Psi_2(-x-y, y) =: \Psi_{2,y}$.

When acting with $(\gamma \circ r(F_1)) \circ (\gamma \circ r(F_2))$ on x, this is first a conjugation by $\Psi_2(-x-y, x)$, followed by the action of $\gamma \circ r(F_1)$, which acts on each letter of the result. Thus, the composite action on x is conjugation by

$$\Psi_{1,x} \cdot \Psi_2(-\Psi_{1,x}^{-1}\, x\, \Psi_{1,x} - \Psi_{1,y}^{-1}\, y\, \Psi_{1,y}\,,\ \Psi_{1,x}^{-1}\, x\, \Psi_{1,x}). \tag{3.24}$$

Applying Lemma 3.10 with the substitution $X = -x-y, Y = x, Z = y$, we obtain

$$-\Psi_{1,x}^{-1} x \Psi_{1,x} - \Psi_{1,y}^{-1} y \Psi_{1,y} = -x - y$$

Thus, $(\gamma \circ r(F_1)) \circ (\gamma \circ r(F_2))$ conjugates x by

$$\Psi_1(-x-y, x) \cdot \Psi_2(-x-y, \Psi_1^{-1}(-x-y, x) x \Psi_1(-x-y, x)). \tag{3.25}$$

On the other hand, by Lemma 3.8 we also have that $\gamma \circ r(F_1 \cdot F_2)$ conjugates x by

$$\Psi_1(-x-y, x) \cdot \Psi_2(-x-y, \Psi_1^{-1}(-x-y, x) x \Psi_1(-x-y, x)), \tag{3.26}$$

therefore

$$(\gamma \circ r(F_1)) \circ (\gamma \circ r(F_2))\,(x) = (\gamma \circ r(F_1 \cdot F_2))\,(x).$$

After an analogous computation on y, we have that $(\gamma \circ r(F_1 \cdot F_2)) = (\gamma \circ r(F_1)) \cdot (\gamma \circ r(F_2))$. Hence, $\gamma \circ r$ is a group homomorphism, and it follows from Lemma 3.9 that r is a group anti-homomorphism as required. ☐

Now, given that we know the value of $\gamma \circ r(F)$ and that r is an anti-homomorphism, we can prove the following:

LEMMA 3.12. *The pentagon (1) in diagram* (3.12) *commutes:*

$$\begin{array}{ccc} \mathsf{Aut}(\mathsf{PaCD}) & \xrightarrow{\quad r \quad} & \mathscr{G}(\uparrow_2)^{tr} \\ & & \big\downarrow{\scriptstyle \gamma} \\ \big\downarrow{\scriptstyle \tau} & & \mathrm{TAut}_2 \\ & & \big\uparrow \\ \mathsf{GRT}_1 & \xrightarrow[\quad \rho \quad]{} & \mathsf{KRV}_2 \end{array} \tag{3.27}$$

PROOF. Let F be an element of $\mathsf{Aut}(\mathsf{PaCD})$ such that $F(\lvert\!/\!\rvert) = \Psi(t^{12}, t^{23})$. Then $\tau(F) = \Psi(x, y)$, and $\rho(\tau(F)) = (\Psi(-x-y, x), \Psi(-x-y, y))$.

In Lemma 3.8, we showed that $\gamma(r(F))$ acts on x and y by conjugation with $(\Psi(-x-y, x), \Psi(-x-y, y))$, which indeed agrees with $\rho(\Psi)$. $\square$

LEMMA 3.13. *The square (3) in the diagram* (3.12) *commutes:*

$$\begin{array}{ccc} \mathscr{G}(\uparrow_2)^{tr} & \xleftarrow{\quad v \quad} & \mathrm{Aut}_v(\mathscr{A}) \\ \big\downarrow{\scriptstyle \gamma} & & \big\downarrow{\scriptstyle \Theta} \\ \mathrm{TAut}_2 & \longleftarrow & \mathsf{KRV}_2 \end{array} \tag{3.28}$$

PROOF. For $G \in \mathrm{Aut}_v(\mathscr{A})$ given by $G(\curlywedge) = N$, $\Theta(G)$ is the conjugation action by N^{tr} on $\widehat{\mathfrak{lie}}_2$, which is a priori an element of TAut_2, but proven in [**DHR23**, Theorem 5.12] to lie in KRV. This agrees with $\gamma(v(G))$ by definition. $\square$

This implies the Main Theorem:

PROOF OF THEOREM 3.7. Given Lemmas 3.12 and 3.13, the only statement we need to show is that r factors through $\mathrm{Aut}_v(\mathscr{A})$. Indeed, by the commutativity of the pentagon (1) and the square (3), $r = v \circ (\Theta)^{-1} \circ \rho \circ \tau$. Thus, for any $F \in \mathsf{Aut}(\mathsf{PaCD})$, the value $r(F)$ is the tree projection of a valid vertex value[11] for an expansion preserving, v-small automorphism of arrow diagrams, namely $G = \Theta^{-1}\rho\tau(F)$. $\square$

In Appendix A we present an alternative, more direct proof that r factors through $\mathrm{Aut}_v(\mathscr{A})$, which relies less on the Alekseev-Torossian map, and in Remark A.8 we explain an avenue for making the proof entirely self-contained.

COROLLARY 3.14. *$\mathring{\rho} : \mathsf{Aut}(\mathsf{PaCD}) \to \mathrm{Aut}_v(\mathscr{A})$ is an injective group homomorphism.*

PROOF. It's enough to show that both v and r are injective.

The map v is injective since by Corollary 2.34 an automorphism is uniquely determined by its vertex value modulo wheels.

[11] Readers of [**DHR23**] may note that N^{tr} in this paper does not look the same as a vertex value of an automorphism in the image of Θ^{-1}. However, by Lemma A.5, N^{tr} is equal to a vertex value of the correct form.

The injectivity of r follows from the in injectivity of $\gamma \circ r$. If we have $\gamma \circ r(F_1) = \gamma \circ r(F_2)$ with $F_1(\text{⊬}) = \Psi_1(t^{12}, t^{23})$ and $F_2(\text{⊬}) = \Psi_2(t^{12}, t^{23})$ then we must have,

$$(\Psi_1(-x-y, x), \Psi_1(-x-y, y)) = (\Psi_2(-x-y, x), \Psi_2(-x-y, y)). \tag{3.29}$$

By Lemma 3.12, $\gamma \circ r$ is in the image of ρ. Then, by [**AT12**, Theorem 4.6], ρ is injective, and therefore Equation (3.29) is only possible if $\Psi_1 = \Psi_2$ which gives $F_1 = F_2$. By Lemma 3.9 it follows that r is injective, and therefore so is $\mathring{\rho}$. □

Corollary 3.14 states that there exists a subgroup of arrow diagram automorphisms isomorphic to GRT_1. As each homomorphic expansion of w-foams $Z : \widehat{\mathbb{Q}[\mathsf{wF}]} \to \mathscr{A}$ is an isomorphism of (completed) wheeled props, each one induces an isomorphism

$$\mathrm{Aut}_v(\widehat{\mathbb{Q}[\mathsf{wF}]}) \overset{Z}{\cong} \mathrm{Aut}_v(\mathscr{A}).$$

COROLLARY 3.15. *There exists a subgroup of* $\mathrm{Aut}_v(\widehat{\mathbb{Q}[\mathsf{wF}]})$ *which is (non-canonically) isomorphic to* GT_1.

Ideally, one would construct a topological map $\tilde{\rho} : \mathrm{Aut}(\widehat{\mathbb{Q}[\mathsf{PaB}]}) \hookrightarrow \mathrm{Aut}_v(\widehat{\mathbb{Q}[\mathsf{wF}]})$ which agrees with the embedding of $\mathsf{GT}_1 \hookrightarrow \mathsf{KV}$ from [**AET10**]. This construction is more subtle due to the completions, and is the topic of ongoing work of the second author.

Appendix A. Alternative proof of the Main Theorem

To prove directly that r factors through $\mathrm{Aut}_v(\mathscr{A})$, we need to show that given $F \in \mathrm{Aut}(\mathsf{PaCD})$, the value $N^{tr} = r(F)$ lifts to a pair $(N, C) \in \mathscr{A}(\uparrow_2) \times \mathscr{A}(\downarrow)$ satisfying the equations $R4', C'$ and U' of Proposition 2.30. The value $N = G(\text{⅄})$ is computed in Theorem 3.5, and given in terms of N^{tr} and a yet-to-be-determined cap value $C = G(\downarrow)$ in equation (3.4), which we recall here for convenience:

$$N = (C^1)^{-1}(C^2)^{-1}N^{tr}C^{12}.$$

LEMMA A.1. *The value N of Theorem 3.5 satisfies the R4' equation:*

$$N^{12}R^{(12)3}(N^{12})^{-1} = R^{(12)3} \quad \in \mathscr{A}(\uparrow_3) \tag{R4'}$$

where $R^{(12)3} = e^{a^{13}+a^{23}}$.

PROOF. Note that for the conjugation action of N on $R^{(12)3}$, only the tree-projection of N matters. That is, N satisfies $R4'$ if and only if N^{tr} does. Note that $(N^{tr})^{-1}e^{a^{13}+a^{23}}N^{tr}$ maps to $(N^{tr}) \cdot e^{x+y} \in \widehat{\mathfrak{lie}}_2$ via the map γ, where $\cdot$ denotes the conjugation action. Thus, the proof that N satisfies the R4' equation reduces to the following computation, where we denote $\Psi_x = \Psi(-x-y, x)$ and $\Psi_y = \Psi(-x-y, y)$, and explain the steps below.

$$\begin{aligned}(N^{12})^{-1}\exp(a^{13}+a^{23})N^{12} &= \exp\left((N^{12})^{-1}a^{13}N^{12} + (N^{12})^{-1}a^{23}N^{12}\right)\\ &= \exp\left(\Psi_x^{-1}\, x\, \Psi_x + \Psi_y^{-1}\, y\, \Psi_y\right)\\ &= \exp(x+y)\\ &= e^{a^{13}+a^{23}}\end{aligned}$$

Here the first equality is by the definition of the exponential, the second is by Lemma 3.8, and the third is by Lemma 3.10 with the substitution $X = -x-y, Y = x, Z = y$, and the fourth is by definition of the map γ. □

LEMMA A.2. *For any choice of $C \in \mathscr{A}(\downarrow)$ the value N of (3.4) satisfies the cap equation,*

$$C^{12}(N^{12})^{-1} = C^1C^2 \quad \text{in} \quad \mathscr{A}(\downarrow_2). \tag{C'}$$

PROOF. Substituting the value of N (3.4) into the Cap equation (C') we obtain:

$$C^{12}(C^{12})^{-1}(N^{tr})^{-1}C^2C^1 = C^1C^2. \tag{A.1}$$

We cancel $C^{12}(C^{12})^{-1}$ on the left. Since $\mathscr{A}(\downarrow_2)$ is a right $\mathscr{A}(\uparrow_2)$-module by stacking, we multiply on the right by $(C^1)^{-1}(C^2)^{-1}$. Note that C^1 and C^2 commute.

Thus, we have reduced the equation to

$$N^{tr} = 1 \quad \text{in} \quad \mathscr{A}(\downarrow_2).$$

Then, observe that N^{tr} is in the image of φ, and as such, all arrow heads in every summand in N^{tr} are below all arrow tails. Thus, all positive degree components of N^{tr} vanish in $\mathscr{A}(\downarrow)$ by the CP relation, and this completes the proof. □

In the proof of the unitarity equation U', the following three Lemmas are useful.

LEMMA A.3. $N = W \cdot N^{tr}$, *where* $W = C^{12}(C^1)^{-1}(C^2)^{-1}$.

PROOF. By repeated applications of the STU relation, the C^{12} factor at the right-hand side of N can be commuted to the left-hand side, at the possible "cost" of additional wheels. Thus, N can be written in the form $W \cdot N^{tr}$ for some product of wheels W. To calculate W, we apply that the pair (N, C) satisfies the cap equation by Lemma A.2 and hence in $\mathscr{A}(\downarrow_2)$ we can deduce (explained below):

$$\begin{aligned} C^{12}N^{-1} &= C^1C^2 \\ C^{12}(N^{tr})^{-1}W^{-1} &= C^1C^2 \\ C^{12}(N^{tr})^{-1} &= C^1C^2W \\ N^{tr}C^{12}(N^{tr})^{-1} &= C^1C^2W. \end{aligned}$$

Here the first step is substituting $N = WN^{tr}$; the second step is right multiplication by W using the right $\mathscr{A}(\uparrow_2)$-module structure, and the last step inserts N^{tr} at the bottom caps, using that $N^{tr} = 1 \in \mathscr{A}(\downarrow_2)$ by the CP relation.

Now recall that $C = e^{c(x)}$, where $c \in \mathbb{Q}[[\xi]]/\langle\xi\rangle$ (Lemma 2.23), and thus $C^{12} = e^{c(x+y)}$. By the (R4') equation, $N^{tr}(x+y)(N^{tr})^{-1} = x + y$. It is tempting to use this to deduce that $N^{tr}C^{12}(N^{tr})^{-1} = C^{12}$, but unfortunately in this case $e^{c(x+y)}$ denotes a power series in wheels, rather than a tree element corresponding to $\widehat{\mathfrak{lie}}_2$ as in the (R4') equation.

To overcome this issue, we use a linear trace map $\sigma : \mathscr{A}(\uparrow_3) \to \mathscr{A}(\downarrow_2)$ defined in [**BND17**, Section 4.5] and illustrated in Figure 21. The trace σ "closes up" the third strand and converts it from a skeletal to an internal strand, then caps the first two strands at the bottom, as shown. This map is well-defined (it kills almost all relations, and converts one STU relation into an IHX relation). If $f(x, y)$ is a word in $\hat{U}(\mathfrak{lie}_2)$ represented by an arrow diagram where $x = a^{13}$ and $y = a^{23}$, then $\sigma(f(x, y))$ is the wheel arrow diagram representing the cyclic word $\operatorname{tr}(f(x, y))$. In particular, for a power series $c \in \mathbb{Q}[[\xi]]/\langle\xi\rangle$, we have that $\sigma(c(x + y)) = c(x + y)$.

Thus, $\sigma(R4')$ states that $N^{12}C^{12}(N^{12})^{-1} = C^{12}$, and thus we have $C^{12} = C^1C^2W$. Since all factors on both sides of this equation are now wheel arrow

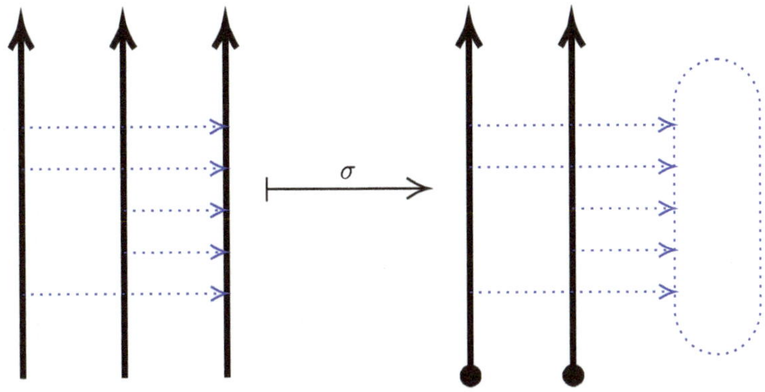

FIGURE 21. The map σ applied to the of $\mathscr{A}(\uparrow_3)$ representing xy^2x^2.

diagrams, they all commute, and thus we can use the right $\mathscr{A}(\uparrow_2)$-module structure to obtain

$$W = C^{12}(C^1)^{-1}(C^2)^{-1} = e^{c^{12}-c^1-c^2}. \tag{A.2}$$

We have identified the value of W in $\mathscr{A}(\downarrow_2)$, rather than in $\mathscr{A}(\uparrow_2)$. Although $\mathscr{A}(\downarrow_2)$ is a factor of $\mathscr{A}(\uparrow_2)$, wheels embed in $\mathscr{A}(\downarrow_2)$, and therefore if $W = C^{12}(C^1)^{-1}(C^2)^{-1}$ in $\mathscr{A}(\downarrow_2)$, then the same is true in $\mathscr{A}(\uparrow_2)$. This completes the proof. □

LEMMA A.4. *Let $\psi \in \mathfrak{grt}_1$ and $\eta = (\psi(-x-y,x), \psi(-x-y,y)) \in \mathfrak{tder}_2$. Then, the action of η (as a tangential derivation) on itself (as an element of $\widehat{\mathfrak{lie}}_2^{\oplus 2}$) is trivial.*

PROOF. Let $\psi_x = \psi(-x-y,x)$ and $\psi_y = \psi(-x-y,y)$. The action of η on the Lie series ψ_x and ψ_y is – by definition – given by:

(A.3)

$$\eta(\psi_x) = \psi([\psi_x,-x]+[\psi_y,-y],[\psi_x,x]) \quad \text{and} \quad \eta(\psi_y) = \psi([\psi_x,-x]+[\psi_y,-y],[\psi_y,y])$$

Now applying Equation (3.22) of Lemma 3.10 we have,

$$[\psi(-x-y,x),x] + [\psi(-x-y,y),y] = 0,$$

and thus the first component of both $\eta(\psi_x)$ and $\eta(\psi_y)$ given in (A.3) is zero. Thus, $\eta(\psi_x) = \eta(\psi_y) = 0$, as stated. □

LEMMA A.5. *Given $F \in \mathrm{Aut}(\mathsf{PaCD})$ with $F(\text{⩘}) = \Psi(t^{12},t^{13}) = e^{\psi(t^{12},t^{23})}$, we have*

$$G(\text{人})^{tr} = N^{tr} = e^{\ell(\eta)} \in \mathscr{A}(\uparrow_2)$$

where $\eta = (\psi(-x-y,x), \psi(-x-y,y)) \in \mathfrak{tder}_2$, and $\ell : \mathfrak{tder}_n \to \mathscr{P}(\uparrow_n)$ is the map described in Remark 2.18.

PROOF. Even though N^{tr} and $e^{\ell(\eta)}$ look different as arrow diagrams, $\gamma(N^{tr}) = \gamma(e^{\ell(\eta)})$ (Lemma 3.8). Therefore, the difference $N^{tr} - e^{\ell(\eta)}$ is in the kernel of γ, and is hence given by some number of short arrows. The point of this proof is to show that these short arrows cancel.

The distinction between N^{tr} and $e^{\ell(\eta)}$ is that terms of N^{tr} include several trees with *all* heads attached below *all* tails, as on the left side of Figure 22. On the other hand, terms of $e^{\ell(\eta)}$ are products of trees, where the head of each individual

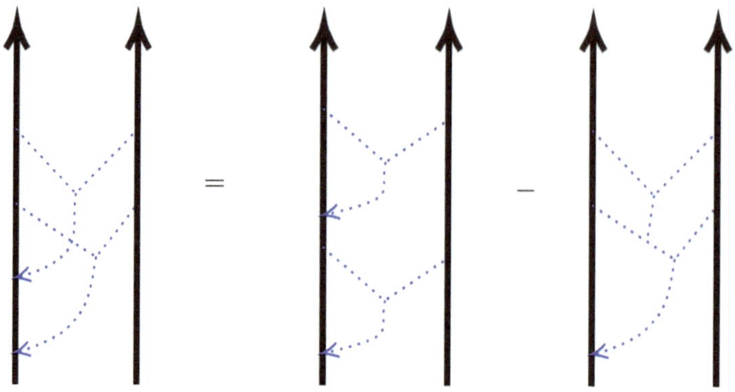

FIGURE 22. A typical term of N^{tr} is shown on the left. We perform an STU relation to separate the trees, producing a term of $e^{\ell(\eta)}$, and a term which cancels in the sum.

tree is attached below its tails, but tree factors are separated, as the first term on the right-hand side of Figure 22. Converting from the N^{tr} form to the $e^{\ell(\eta)}$ form requires separating the trees using STU relations. This results in extra tree summands, as shown in Figure 22. In fact, performing each STU relation amounts to a higher tree acting on a leaf of a lower one by the adjoint action (equivalently, as a tangential derivation).

Consider the sum of terms of N^{tr} with 2 trees. Separating the upper tree from the lower in each term is equivalent to η acting on itself. By Lemma A.4, the action of η on itself is trivial. Therefore, the sum of the resulting additional tree summands vanish.

Iterating this process for the sums of terms with n trees thus completes the proof that $N^{tr} = e^{\ell(\eta)}$. □

We are now ready to prove the unitarity equation (U'):

LEMMA A.6. *There is an appropriate choice of $C \in \mathbb{Q}[[\xi]]/\langle \xi \rangle$, so that $N = C^{12}C^1C^2N^{tr}$ satisfies the unitarity equation* (U')*:*

$$N \cdot N^* = 1 \quad \textit{in} \quad \mathscr{A}(\uparrow_2) \tag{A.4}$$

PROOF. First, we prove that unitarity is satisfied modulo wheels, that is,

$$N^{tr} \cdot (N^{tr})^* = 1 \quad \text{in} \quad \overrightarrow{pr}(\mathscr{A}(\uparrow_2)). \tag{A.5}$$

Since by Lemma A.5, $N^{tr} = e^{\ell(\eta)}$, the left-hand side of the equation reads as

$$e^{\ell(\eta)} \cdot (e^{\ell(\eta)})^* = e^{\ell(\eta)} \cdot e^{(\ell(\eta))^*}. \tag{A.6}$$

By definition, $*$ acts on a single tree by reversing the order of the arrow endings along the strands, and multiplying with (-1), as each tree has a single arrow head. Reversing the order of leaves means that the head of the tree is now attached above its tails. Using STU relations, the head can be commuted back to its original position below its tails, at the "cost" of wheels. Thus, $(\ell(\eta))^* = -\ell(\eta)$ in $\mathscr{A}(\uparrow_2)^{tr}$, and therefore

$$e^{\ell(\eta)} \cdot e^{(\ell(\eta))^*} = e^{\ell(\eta)} \cdot e^{-\ell(\eta)} = 1 \quad \text{in} \quad \mathscr{A}(\uparrow_2)^{tr} \tag{A.7}$$

Therefore, N satisfies (U') modulo wheels. Showing that N satisfies (U') in general is slightly more complex.

By applying Lemma A.3 and Lemma A.5, the left-hand side of the unitarity equation reads

$$W \cdot e^{\ell(\eta)} \cdot (W \cdot e^{\ell(\eta)})^* = W \cdot e^{\ell(\eta)} \cdot (e^{\ell(\eta)})^* \cdot W, \tag{A.8}$$

using that $*$ reverses multiplication order, and acts trivially on wheels.

By Proposition 2.26,

$$e^{\ell(\eta)} \cdot (e^{\ell(\eta)})^* = e^{-J(e^{\eta})}.$$

Thus, in order to satisfy the (U') equation, we must be able to choose a cap value $C = e^c$ such that

$$e^{J(e^{\eta})} = W^2 = e^{2(c^{12}-c^1-c^2)}. \tag{A.9}$$

e^{η} is equal to $\rho(\Psi)$, and hence $e^{\eta} \in \mathsf{KRV}$ [**AT12**, Proposition 4.8]. Thus, by definition (see (2.11)), $J(e^{\eta})$ is of the form $\mathrm{tr}(s(x) + s(y) - s(x+y))$ for some $s \in x^2\mathbb{Q}[[x]]$. Choose $C = e^c$ with $c = \frac{1}{2}s(x)$, and with this choice N satisfies the (U') equation. □

The following corollary concludes the alternative proof on the Main Theorem 3.7:

COROLLARY A.7. *The map* $r : \mathrm{Aut}(\mathsf{PaCD}) \to \mathscr{G}(\uparrow_2)^{tr}$ *factors through* $\mathrm{Aut}_v(\mathscr{A})$. *That is, there exists a map* $\mathring{\rho} : \mathrm{Aut}(\mathsf{PaCD}) \to \mathrm{Aut}_v(\mathscr{A})$ *such that the triangle (2) in diagram* (3.12) *commutes:*

(A.10)

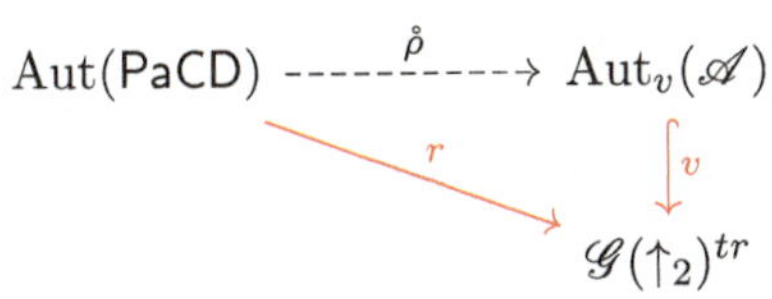

PROOF. Define $\mathring{\rho}(F)$ by sending F to the automorphism G defined by the pair (N, C) where N is as in Theorem 3.5 and C is as defined in the proof of Lemma A.6.

By Lemmas A.1, A.2, and A.6 the pair (N, C) satisfies the equations of Proposition 2.30, and by Lemma 3.9 N is free of short arrows. Therefore, $G \in \mathrm{Aut}_v(\mathscr{A})$. It is immediate that $r(F) = v(\mathring{\rho}(F))$ since $v(G) = N^{tr} = r(F)$. □

REMARK A.8. The fact that the image of the Alekseev-Torossian map ρ lies in KRV was used in order to prove that the (U') equation is satisfied in general (with wheels taken into account). It is possible to show this directly and give a self-contained topological proof: this is more complex and done in [**BND24b**].

Appendix B. Circuit algebras

An essential ingredient in the topological interpretation of the Kashiwara-Vergne problem is the construction of homomorphic expansions for a class of w-foams ([**BND17**], [**DHR23**]). This construction relies on the fact that w-foams—and, in general, classes of knotted objects—can be described as finitely presented algebraic structures called (oriented) *circuit algebras*. Circuit algebras are a generalisation of Jones' planar algebras, where one drops the condition of "planarity". In this appendix, we give a brief overview of circuit algebras and recommend [**DHR21**, Section 2] for full details.

In a circuit algebra, operations are parameterised by *wiring diagrams.*

DEFINITION B.1. An *oriented wiring diagram* is a triple $D = (\mathcal{A}, M, f)$ consisting of:

(1) A set $\mathcal{A} = \{A_0^{\text{out}}, A_0^{\text{in}}, A_1^{\text{out}}, A_1^{\text{in}}, \ldots, A_r^{\text{out}}, A_r^{\text{in}}\}$ of sets of labels, for some non-negative integer r. The elements of the sets A_i^{out} are referred to as *outgoing labels* and the elements of A_i^{in} are *incoming labels.* The sets A_0^{out} and A_0^{in} play a distinguished role: their elements are called the *output labels* of the diagram, while the sets A_1^{out}, A_1^{in}, $\ldots$, $A_r^{\text{out}}, A_r^{\text{in}}$ contain *input labels* of the diagram.
(2) An oriented compact 1-manifold M, with boundary ∂M, regarded up to orientation-preserving homeomorphism. The connected components of M are homeomorphic to either an oriented circle (with no boundary) or an oriented interval with one beginning and one ending point. We write ∂M^{out} for the set of *beginning* boundary points of M, and ∂M^{in} for the set of *ending* boundary points, so $\partial M = \partial M^{\text{out}} \sqcup \partial M^{\text{in}}$.
(3) Bijections[12]

$$\partial M^{\text{out}} \xrightarrow{f} \cup_{i=0}^{r} A_i^{\text{out}} \quad \text{and} \quad \partial M^{\text{in}} \xrightarrow{f} \cup_{i=0}^{r} A_i^{\text{in}}.$$

Wiring diagrams can be pictorially represented as "disks with r holes," as in Figure 23. A disc with r holes is a closed disc with the interiors of r disjoint small discs removed $D_0 \setminus (\mathring{D}_1 \cup \ldots \cup \mathring{D}_r)$. A set of marked points on the disc boundaries $\mathcal{L} = \{\mathcal{L}_0, \mathcal{L}_1, \ldots, \mathcal{L}_r\}$ are arranged so that $\mathcal{L}_i$ lies on ∂D_i, and $\mathcal{L}_i$ is in bijection with $A_i^{\text{out}} \cup A_i^{\text{in}}$. The manifold is represented as an immersion $M \subset D$, where ∂M is mapped to the set of marked points on the boundary of D, so that interval components of M connect the points paired by the bijection f. The set ∂M is partitioned to incoming and outgoing boundary points: incoming points are mapped to marked points corresponding to $\bigcup_i A_i^{\text{in}}$ and outgoing points to $\bigcup_i A_i^{\text{out}}$.

Note that the choice of a particular immersion of M is only for notational convenience, it is not part of the data. Different immersions of M represent the same wiring diagram as long as they induce the same bijection between incoming and outgoing marked points. See [**DHR21**, Lemma 5.3] for further details.

EXAMPLE B.2. Figure 23 depicts the wiring diagram represented by the triple

$$\mathfrak{D} = \{\{\{l_1, l_3\}, \{l_2, l_4, l_5\}, \{a_1, a_3, a_4\}, \{a_2\}, \{b_4, b_5\}, \{b_1, b_2, b_3\}\}, \sqcup_{i=1}^{7} I,$$
$$\{(l_1, l_2), (l_3, b_2), (a_1, l_4), (a_3, b_1), (a_4, l_5), (b_4, b_3), (b_5, a_2)\}\}$$

DEFINITION B.3. Given a set of colours $\mathfrak{C}$, a $\mathfrak{C}$*-coloured operad* $\mathsf{P} = \{\mathsf{P}(c_1, \ldots, c_r; c_0)\}$ consists of a collection of sets (or vector spaces) $\mathsf{P}(c_1, \ldots, c_r; c_0)$, one for each sequence $(c_1, \ldots, c_r; c_0) \in \mathfrak{C}$, which are equipped with a right action of the symmetric group $\mathcal{S}_r$ which permutes the "inputs" $c_1, \ldots, c_r$. Moreover, P is equipped with a family of equivariant, associative and unital partial compositions:

$$\circ_i : \mathsf{P}(c_1, \ldots, c_r; c_0) \times \mathsf{P}(d_1, \ldots, d_s; d_0)$$
$$\longrightarrow \mathsf{P}(c_1, \ldots, c_{i-1}, d_1, \ldots, d_s, c_{i+1}, \ldots, c_r; c_0),$$

[12]If the sets $\{A_i^{\text{out/in}}\}$ are not pairwise disjoint, replace the unions $(\cup_{i=0}^{r} A_i^{\text{out}})$ and $(\cup_{i=0}^{r} A_i^{\text{in}})$ by the set of triples $\{(a, i, \text{out/in}) \mid a \in A_i^{\text{out/in}}, 0 \leq i \leq r\}$.

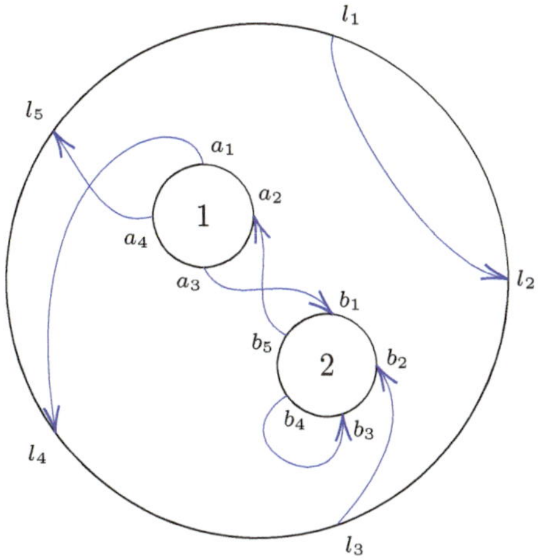

FIGURE 23. An example wiring diagram

whenever $d_0 = c_i$.

For more details on coloured operads see, for example, [**BM07**, Definition 1.1]. An *algebra* over an $\mathfrak{C}$-coloured operad P is a $\mathfrak{C}$-indexed family of vector spaces $\mathsf{A} = \{\mathsf{A}(c)\}_{c\in\mathfrak{C}}$ together with an action by P ([**BM07**, Definition 1.2]).

DEFINITION B.4. The collection of all wiring diagrams

$$\mathsf{WD} = \{\mathsf{WD}(A_0^{\text{in/out}}; A_1^{\text{in/out}}, \ldots, A_r^{\text{in/out}})\}$$

forms a discrete operad called the *operad of oriented wiring diagrams*. For $\mathfrak{D} = (\mathcal{A}, M, f)$ and $\mathfrak{D}' = (\mathcal{B}, N, g)$, if $A_i^{\text{in}} = B_0^{\text{out}}$ and $A_i^{\text{out}} = B_0^{\text{in}}$ then the partial composition $\mathfrak{D} \circ_i \mathfrak{D}'$ is defined by the label set

$$\{A_1^{\text{in/out}}, \ldots, A_{i-1}^{\text{in/out}}, B_1^{\text{in/out}}, \ldots, B_s^{\text{in/out}}, A_{i+1}^{\text{in/out}} \ldots A_r^{\text{in/out}}\};$$

the manifold $M \sqcup N/\sim$ obtained from M and N by gluing along the boundary identification $A_i^{\text{in}} = B_0^{\text{out}}$ and $A_i^{\text{out}} = B_0^{\text{in}}$; and the set bijection $f \sqcup g/\sim$ induced by f and g in the natural way.

Pictorially, wiring diagram composition shrinks the wiring diagram $\mathfrak{D}'$ and glues it into the ith input circle of $\mathfrak{D}$ in such a way that the labels match and the boundary points of the 1-manifolds are identified: see Figure 24. Note that composition in the operad WD is similar to the operad of planar tangles in [**Jon99**] and [**HPT23**, Definition 2.1].

DEFINITION B.5. An oriented circuit algebra is an algebra over the operad of wiring diagrams WD.

We briefly unravel the definition of an algebra over an operad (see also [**DHR21**, Definition 2.4]). An oriented circuit algebra is a collection of sets $\mathsf{V} = \{\mathsf{V}[A^{\text{in}}; A^{\text{out}}]\}$, where the pairs $(A^{\text{in}}, A^{\text{out}})$ run over all pairs of label sets, together with a family of multiplication operations parameterised by oriented wiring diagrams. This means that, for each wiring diagram $\mathfrak{D} = (\mathcal{A}, M, f)$, there is a

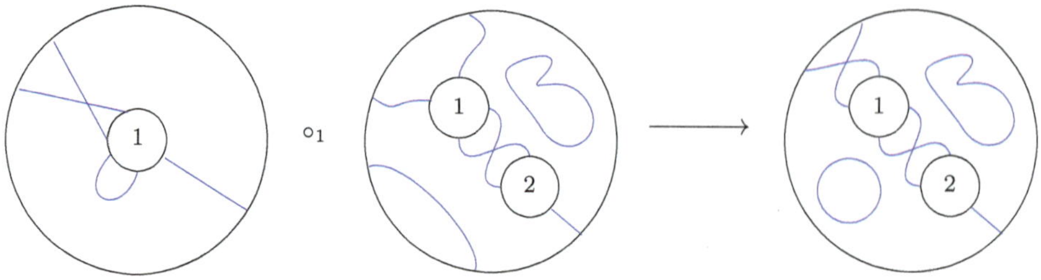

FIGURE 24. The operadic composition of two wiring diagrams.

corresponding function

$$F_{\mathfrak{D}} : \mathsf{V}[A_1^{\text{in}}; A_1^{\text{out}}] \times \ldots \times \mathsf{V}[A_r^{\text{in}}; A_r^{\text{out}}] \to \mathsf{V}[A_0^{\text{out}}; A_0^{\text{in}}].$$

Moreover, the assignment $\mathfrak{D} \mapsto F_{\mathfrak{D}}$ must be compatible with the operadic composition of wiring diagrams.

DEFINITION B.6. A *homomorphism* of circuit algebras is a map of algebras over the operad of wiring diagrams. In other words, a homomorphism $\Phi : \mathsf{V} \to \mathsf{W}$ is a family of maps $\{\Phi_{A^{\text{in}};A^{\text{out}}} : \mathsf{V}[A^{\text{in}}; A^{\text{out}}] \to \mathsf{W}[A^{\text{in}}; A^{\text{out}}]\}_{(A^{\text{in}},A^{\text{out}})}$, which commutes with the action of wiring diagrams. For details, see [**DHR21**, Definition 2.5].

Wiring diagrams and circuit algebras can be naturally upgraded to a *coloured* context:

DEFINITION B.7. Given a finite set $\mathcal{C}$ of *colours*, an oriented $\mathcal{C}$*-coloured* wiring diagram is a wiring diagram $\mathfrak{D} = (\mathcal{A}, M, f)$, along with a map $c : \pi_0(M) \to \mathcal{C}$. Here $\pi_0(M)$ is the set of connected components of M. Composition of $\mathcal{C}$-coloured wiring diagrams

$$(\mathcal{A}, M, f, c) \circ_i (\mathcal{B}, N, g, d)$$

is only defined where the gluing of M and N respects colours. In other words, only connected components of the same colour may be glued. An *oriented coloured circuit algebra* is an algebra over the coloured operad of oriented coloured wiring diagrams.

Elements of a circuit algebra can be viewed as decorations of wiring diagrams by elements of V, where the operations are "wiring elements together": this is illustrated in the important example of the circuit algebra of w-foams described in Section 2.3 (e.g. Figure 3).

Circuit algebras can be given by a presentation $\mathsf{V} = \langle g_1, \ldots, g_k \mid r_1, \ldots, r_l \rangle$ where the free circuit algebra on the generators $\{g_1, \ldots, g_l\}$ is factored by the ideal generated by the relations $\{r_1, \ldots, r_1\}$. The notions of *free* and *ideal* are the standard definitions for algebras over and operad (e.g.: [**LV12**]). Examples of circuit algebras in this paper will also be equipped with *auxiliary operations.* These are additional operations which are not parameterised by wiring diagrams.

B.1. Circuit algebras as wheeled props. Circuit algebras also possess an equivalent description as wheeled props [**DHR21**]. A (coloured) *prop*[13] is a

[13]Some authors prefer to capitalise the word PROP to emphasise that prop refers to "PROduct and Permutation category."

strict symmetric tensor category in which the *monoid of objects* is freely generated ([**Mac65**, Chapter V], [**HR15**] or [**YJ15**]). This means that a one-coloured prop is a symmetric tensor category equipped with a distinguished object x, such that every object is a tensor power $x^{\otimes n}$, for some $n \geq 0$. Hence, morphisms in a prop are of the form $f : x^{\otimes n} \to x^{\otimes m}$. We picture these morphisms as graphs with n-inputs and m-outputs. Composition of morphisms, also called *vertical composition*, is modelled diagrammatically by stacking the outputs of one graph to the inputs of another. The tensor product of the prop is realised by taking disjoint unions of graphs (stacking them next to each other), and is called *horizontal composition.*

EXAMPLE B.8. The collection of all braid groups $\mathsf{B} = \coprod \mathsf{B}_n$ forms a one-coloured prop (e.g. [**JS93**, Example 1, page 55]). Similarly, parenthesised braids $\mathsf{PaB} = \coprod_n \mathsf{PaB}(n)$ form a prop ([**BN98**]).

A *wheeled prop* is a prop in which every object has a dual. This gives rise to a family of linear "trace" or "contraction" maps

$$\mathrm{tr}_i^j : x^{*\otimes m} \otimes x^{\otimes n} \longrightarrow x^{*\otimes m-1} \otimes x^{\otimes n-1}.$$

They arise, for example, in the Batalin-Vilkovisky quantization formalism, where formal germs of SP-manifolds are identified as representations of a certain wheeled prop [**MMS09**, Theorem 3.4.3]. We note that in the literature, a strict symmetric tensor category with duals is also called a *rigid* symmetric[14] tensor category.

DEFINITION B.9. Let $\mathcal{I}$ denote a fixed countable alphabet. A *wheeled prop* $\mathsf{E} := \left(\mathsf{E}, *, \mathrm{tr}_i^j\right)$ consists of:

(1) a collection of vector spaces $\mathsf{E} = \{\mathsf{E}[I;J]\}$, $I, J \subset \mathcal{I}$;
(2) a *horizontal composition*

$$* : \mathsf{E}[I;J] \otimes \mathsf{E}[K;L] \longrightarrow \mathsf{E}[I \cup K; J \cup L],$$

where $I \cap K = \emptyset$ and $J \cap L = \emptyset$;
(3) a linear map $1_\emptyset : \Bbbk \to \mathsf{E}[\emptyset;\emptyset]$ called the *empty unit*;
(4) a *contraction* operation

$$\mathsf{E}[I;J] \xrightarrow{\mathrm{tr}_i^j} \mathsf{E}[I \setminus \{i\}; J \setminus \{j\}],$$

for every pair $i \in I$ and $j \in J$;
(5) a linear map

$$1_i : \Bbbk \to \mathsf{E}[\{i\};\{i\}],$$

for every $i \in \mathcal{I}$, called the *unit.*

Moreover, the horizontal composition and contractions commute with each other and are associative, equivariant and unital (see [**DHR21**, Definition 6.1] for a full list of axioms).

EXAMPLE B.10. The tensor category of tangles described by Turaev [**Tur89**] (see also the expository description in [**JS93**, Section 11]) is a wheeled prop. Similarly, the categories of virtual and welded tangles, as described in [**DHR21**, Section 3.4].

[14]Rigid symmetric tensor categories are also called compact closed categories.

THEOREM B.11 ([**DHR21**, Theorem 5.5]). *There exists an equivalence of categories between the category of ($\mathfrak{C}$-coloured) circuit algebras and the category of ($\mathfrak{C}$-coloured) wheeled props.*

Acknowledgments

This paper was prepared in honour of Ezra Getzler on the occasion of his 60th birthday. Much of this work follows his influential work using operads and props to study geometric problems and quantum algebra.

References

[AET10] A. Alekseev, B. Enriquez, and C. Torossian, *Drinfeld associators, braid groups and explicit solutions of the Kashiwara-Vergne equations*, Publ. Math. Inst. Hautes Études Sci. **112** (2010), 143–189, DOI 10.1007/s10240-010-0029-4. MR2737979

[AT12] Anton Alekseev and Charles Torossian, *The Kashiwara-Vergne conjecture and Drinfeld's associators*, Ann. of Math. (2) **175** (2012), no. 2, 415–463, DOI 10.4007/annals.2012.175.2.1. MR2877064

[BM07] Clemens Berger and Ieke Moerdijk, *Resolution of coloured operads and rectification of homotopy algebras*, Categories in algebra, geometry and mathematical physics, Contemp. Math., vol. 431, Amer. Math. Soc., Providence, RI, 2007, pp. 31–58, DOI 10.1090/conm/431/08265. MR2342815

[BN95] Dror Bar-Natan, *On the Vassiliev knot invariants*, Topology **34** (1995), no. 2, 423–472, DOI 10.1016/0040-9383(95)93237-2. MR1318886

[BN97] Dror Bar-Natan, *Non-associative tangles*, Geometric topology (Athens, GA, 1993), AMS/IP Stud. Adv. Math., vol. 2, Amer. Math. Soc., Providence, RI, 1997, pp. 139–183, DOI 10.1090/amsip/002.1/09. MR1470726

[BN98] Dror Bar-Natan, *On associators and the Grothendieck-Teichmuller group. I*, Selecta Math. (N.S.) **4** (1998), no. 2, 183–212, DOI 10.1007/s000290050029. MR1669949

[BND16] Dror Bar-Natan and Zsuzsanna Dancso, *Finite-type invariants of w-knotted objects, I: w-knots and the Alexander polynomial*, Algebr. Geom. Topol. **16** (2016), no. 2, 1063–1133, DOI 10.2140/agt.2016.16.1063. MR3493416

[BND17] Dror Bar-Natan and Zsuzsanna Dancso, *Finite type invariants of w-knotted objects II: Tangles, foams and the Kashiwara-Vergne problem*, Math. Ann. **367** (2017), no. 3-4, 1517–1586, DOI 10.1007/s00208-016-1388-z. MR3623232

[BND24a] Dror Bar-Natan and Zsuzsanna Dancso, *Correction to: Finite type invariants of w-knotted objects II: Tangles, foams and the Kashiwara-Vergne problem*, Math. Ann. **389** (2024), no. 3, 3257–3259, DOI 10.1007/s00208-024-02811-1. MR4753087

[BND24b] Dror Bar-Natan and Zsuzsanna Dancso, *Finite type invariants of w-knotted objects iii: the double tree construction*, `http://drorbn.net/AcademicPensieve/Projects/WKO3/index.html`, 2024.

[BNDS20] Dror Bar-Natan, Zsuzsanna Dancso, and Nancy Scherich, *Ribbon 2-knots, $1+1=2$ and Duflo's theorem for arbitrary Lie algebras*, Algebr. Geom. Topol. **20** (2020), no. 7, 3733–3760, DOI 10.2140/agt.2020.20.3733. MR4194292

[Bro19] Adrien Brochier, *Virtual tangles and fiber functors*, J. Knot Theory Ramifications **28** (2019), no. 7, 1950044, 17, DOI 10.1142/s0218216519500445. MR3975573

[BWC07] John C. Baez, Derek K. Wise, and Alissa S. Crans, *Exotic statistics for strings in 4D BF theory*, Adv. Theor. Math. Phys. **11** (2007), no. 5, 707–749, DOI 10.4310/atmp.2007.v11.n5.a1. MR2362007

[CDM12] S. Chmutov, S. Duzhin, and J. Mostovoy, *Introduction to Vassiliev knot invariants*, Cambridge University Press, Cambridge, 2012, DOI 10.1017/CBO9781139107846. MR2962302

[Dam17] Celeste Damiani, *A journey through loop braid groups*, Expo. Math. **35** (2017), no. 3, 252–285, DOI 10.1016/j.exmath.2016.12.003. MR3689901

[Dan10] Zsuzsanna Dancso, *On the Kontsevich integral for knotted trivalent graphs*, Algebr. Geom. Topol. **10** (2010), no. 3, 1317–1365, DOI 10.2140/agt.2010.10.1317. MR2661529

[DHR21] Zsuzsanna Dancso, Iva Halacheva, and Marcy Robertson, *Circuit algebras are wheeled props*, J. Pure Appl. Algebra **225** (2021), no. 12, Paper No. 106767, 33, DOI 10.1016/j.jpaa.2021.106767. MR4265709

[DHR23] Zsuzsanna Dancso, Iva Halacheva, and Marcy Robertson, *A topological characterisation of the Kashiwara-Vergne groups*, Trans. Amer. Math. Soc. **376** (2023), no. 5, 3265–3317, DOI 10.1090/tran/8761. MR4577332

[Dri90] V. G. Drinfel′d, *On quasitriangular quasi-Hopf algebras and on a group that is closely connected with* $\mathrm{Gal}(\overline{\mathbf{Q}}/\mathbf{Q})$ (Russian), Algebra i Analiz **2** (1990), no. 4, 149–181; English transl., Leningrad Math. J. **2** (1991), no. 4, 829–860. MR1080203

[Fre17] Benoit Fresse, *Homotopy of operads and Grothendieck-Teichmüller groups. Part 1*, Mathematical Surveys and Monographs, vol. 217, American Mathematical Society, Providence, RI, 2017. The algebraic theory and its topological background, DOI 10.1090/surv/217.1. MR3643404

[FRR97] Roger Fenn, Richárd Rimányi, and Colin Rourke, *The braid-permutation group*, Topology **36** (1997), no. 1, 123–135, DOI 10.1016/0040-9383(95)00072-0. MR1410467

[FY89] Peter J. Freyd and David N. Yetter, *Braided compact closed categories with applications to low-dimensional topology*, Adv. Math. **77** (1989), no. 2, 156–182, DOI 10.1016/0001-8708(89)90018-2. MR1020583

[Hal15] Brian Hall, *Lie groups, Lie algebras, and representations*, 2nd ed., Graduate Texts in Mathematics, vol. 222, Springer, Cham, 2015. An elementary introduction, DOI 10.1007/978-3-319-13467-3. MR3331229

[HPT23] André Gil Henriques, David Penneys, and James Tener, *Planar algebras in braided tensor categories*, Mem. Amer. Math. Soc. **282** (2023), no. 1392, vi+106, DOI 10.1090/memo/1392. MR4528312

[HR15] Philip Hackney and Marcy Robertson, *On the category of props*, Appl. Categ. Structures **23** (2015), no. 4, 543–573, DOI 10.1007/s10485-014-9369-4. MR3367131

[Jon99] V. F. R. Jones, *Planar algebras, I*, New Zealand J. Math. **52** (2021 [2021), 1–107, DOI 10.53733/172. MR4374438

[JS93] André Joyal and Ross Street, *Braided tensor categories*, Adv. Math. **102** (1993), no. 1, 20–78, DOI 10.1006/aima.1993.1055. MR1250465

[Kon93] Maxim Kontsevich, *Vassiliev's knot invariants*, I. M. Gel′fand Seminar, Adv. Soviet Math., vol. 16, Amer. Math. Soc., Providence, RI, 1993, pp. 137–150. MR1237836

[KV78] Masaki Kashiwara and Michèle Vergne, *The Campbell-Hausdorff formula and invariant hyperfunctions*, Invent. Math. **47** (1978), no. 3, 249–272, DOI 10.1007/BF01579213. MR492078

[LV12] Jean-Louis Loday and Bruno Vallette, *Algebraic operads*, Grundlehren der mathematischen Wissenschaften [Fundamental Principles of Mathematical Sciences], vol. 346, Springer, Heidelberg, 2012, DOI 10.1007/978-3-642-30362-3. MR2954392

[Mac65] Saunders Mac Lane, *Categorical algebra*, Bull. Amer. Math. Soc. **71** (1965), 40–106, DOI 10.1090/S0002-9904-1965-11234-4. MR171826

[Mer21] Sergei Merkulov, *Grothendieck-Teichmüller group, operads and graph complexes: A survey*, Integrability, quantization, and geometry II. Quantum theories and algebraic geometry, Proc. Sympos. Pure Math., vol. 103, Amer. Math. Soc., Providence, RI, [2021] ©2021, pp. 383–445, DOI 10.1090/pspum/103.2/01863. MR4285704

[MM65] John W. Milnor and John C. Moore, *On the structure of Hopf algebras*, Ann. of Math. (2) **81** (1965), 211–264, DOI 10.2307/1970615. MR174052

[MMS09] M. Markl, S. Merkulov, and S. Shadrin, *Wheeled PROPs, graph complexes and the master equation*, J. Pure Appl. Algebra **213** (2009), no. 4, 496–535, DOI 10.1016/j.jpaa.2008.08.007. MR2483835

[MO97] J. Murakami and T. Ohtsuki, *Topological quantum field theory for the universal quantum invariant*, Comm. Math. Phys. **188** (1997), no. 3, 501–520, DOI 10.1007/s002200050176. MR1473309

[Tam98] Dmitry E. Tamarkin, *Operadic proof of M. Kontsevich's formality theorem*, ProQuest LLC, Ann Arbor, MI, 1999. Thesis (Ph.D.)–The Pennsylvania State University. MR2699544

[Tur88] V. G. Turaev, *The Yang-Baxter equation and invariants of links*, Invent. Math. **92** (1988), no. 3, 527–553, DOI 10.1007/BF01393746. MR939474

[Tur89] V. G. Turaev, *The category of oriented tangles and its representations* (Russian), Funktsional. Anal. i Prilozhen. **23** (1989), no. 3, 93–94, DOI 10.1007/BF01079545; English transl., Funct. Anal. Appl. **23** (1989), no. 3, 254–255 (1990). MR1027006

[Wil15] Thomas Willwacher, *M. Kontsevich's graph complex and the Grothendieck-Teichmüller Lie algebra*, Invent. Math. **200** (2015), no. 3, 671–760, DOI 10.1007/s00222-014-0528-x. MR3348138

[YJ15] Donald Yau and Mark W. Johnson, *A foundation for PROPs, algebras, and modules*, Mathematical Surveys and Monographs, vol. 203, American Mathematical Society, Providence, RI, 2015, DOI 10.1090/surv/203. MR3329226

School of Mathematics and Statistics, The University of Sydney, Sydney, NSW, Australia

Email address: zsuzsanna.dancso@sydney.edu.au

School of Mathematics and Statistics, The University of Melbourne, Melbourne, Victoria, Australia

Email address: hogant@student.unimelb.edu.au

School of Mathematics and Statistics, The University of Melbourne, Melbourne, Victoria, Australia

Email address: marcy.robertson@unimelb.edu.au

Contemporary Mathematics
Volume **841**, 2026
https://doi.org/10.1090/conm/841/16836

Affine BPS algebras, W algebras, and the cohomological Hall algebra of $\mathbb{A}^2$

Ben Davison

ABSTRACT. We introduce affinizations and deformations of the BPS Lie algebra associated to a tripled quiver with its canonical cubic potential, and use them to precisely determine the T-equivariant cohomological Hall algebra $\mathcal{A}^T_{\mathbb{A}^2}$ of compactly supported coherent sheaves on $\mathbb{A}^2_{\mathbb{C}}$, acted on by a torus T. In particular we show that this algebra is spherically generated for all T.

1. Introduction

1.1. Hall algebras. Let $\mathfrak{M}_d(\mathbb{C}[x,y])$ denote the moduli stack of coherent sheaves of length d on $\mathbb{A}^2 := \mathbb{A}^2_{\mathbb{C}}$. We also consider the quotient $\mathfrak{M}^T_d(\mathbb{C}[x,y])$, by the induced action of various tori T acting linearly on the two coordinates of $\mathbb{A}^2$. We consider the Borel–Moore homology of the stack

$$\mathcal{A}^T_{\mathbb{A}^2} := \bigoplus_{d\in\mathbb{Z}_{\geq 0}} \mathbf{H}^{\mathrm{BM}}(\mathfrak{M}^T_d(\mathbb{C}[x,y]),\mathbb{Q}),$$

which carries the cohomological Hall algebra (CoHA) product defined in [**SV13**]. This algebra plays a key intermediate role in the geometric representation theory of $\mathrm{Hilb}_d(\mathbb{A}^2)$ [**Nak97, Gro96**], W-algebras [**SV13, Mik07, AS13, Neg16**], Yangians, and Hall algebras [**RSYZ20, RSYZ23, LY20, GY22**]. Setting $T = \{1\}$, the resulting algebra is important in the study of the cohomological Hall algebra of coherent sheaves with zero-dimensional support for an arbitrary smooth surface [**KV23**], as well as partially categorifying the degree zero motivic Donaldson–Thomas theory of $\mathbb{A}^3$, as studied in [**BBS13**]. In particular, via dimensional reduction [**Dav17**] this algebra is isomorphic to a particular case of the critical cohomological Hall algebra introduced by Kontsevich and Soibelman [**KS11**].

In this paper we fully describe the algebra $\mathcal{A}^T_{\mathbb{A}^2}$, for various T, via the introduction and study of deformed and affinized BPS Lie algebras $\hat{\mathfrak{g}}^T_{\tilde{Q},\tilde{W}}$ of tripled quivers with their canonical cubic potentials, extending the original BPS Lie algebra $\mathfrak{g}_{\tilde{Q},\tilde{W}}$ of [**DM20**] for this class of quivers with potential.

2020 *Mathematics Subject Classification.* Primary 17B37, 16G20.
The author was supported by the European Research Council starting grant "Categorified Donaldson–Thomas theory" No. 759967 and a Royal Society University Research Fellowship during the writing of this paper.

©2026 American Mathematical Society

1.2. Donaldson–Thomas theory. Donaldson–Thomas theory [**Tho00**], is a theory that assigns to a projective Calabi–Yau threefold X, a Chern class $\alpha \in \mathbf{H}(X,\mathbb{Z})$, and a stability condition ζ, a number ω_α^ζ "counting" ζ-semistable sheaves on X of Chern class α. The inverted commas are due to the fact that the moduli space $\mathcal{M}_\alpha^\zeta(X)$ of such sheaves will generally have strictly positive dimension, i.e. there are infinitely many such sheaves, and so ω_α^ζ is defined by first constructing a virtual fundamental class of the right expected dimension (zero) and then taking its degree.

Subsequently it was realised by Kai Behrend [**Beh09**] that every quasi-projective complex scheme Y carries a canonical $\mathbb{Z}$-valued constructible function ν_Y, and the above virtual count can be computed as the Euler characteristic of $Y = \mathcal{M}_\alpha^\zeta(X)$, weighted by ν_Y. Furthermore, in the event that a scheme Y can be written[1] as the critical locus of a function f on a smooth d-dimensional ambient variety U, its ν_Y-weighted Euler characteristic is equal to the Euler characteristic of the vanishing cycle cohomology $\mathbf{H}(U, {}^{\mathfrak{p}}\phi_f\mathbb{Q}^{\mathrm{vir}})$, where we define $\mathbb{Q}^{\mathrm{vir}} = \mathbb{Q}[d]$, the cohomological shift of the locally constant sheaf on U. Firstly, these facts give us a way to define DT invariants for noncompact moduli spaces, and secondly, they suggest a natural (partial) "categorification" of the DT invariant: instead of considering the *number* $\chi(\mathbf{H}(U, {}^{\mathfrak{p}}\phi_f\mathbb{Q}^{\mathrm{vir}})) = (-1)^d\sum_{i\in\mathbb{Z}}(-1)^i \dim(\mathbf{H}^i(U, {}^{\mathfrak{p}}\phi_f\mathbb{Q}))$, we study the cohomology $\mathbf{H}(U, {}^{\mathfrak{p}}\phi_f\mathbb{Q}^{\mathrm{vir}})$ itself. Being a vector space, and not a mere number, this cohomology is expected to have (and indeed does have, in cases of interest in this paper) a rich algebraic structure.

1.3. Noncommutative Donaldson–Thomas theory. Let Q be a quiver, by which we mean a directed graph, which we always assume to be finite, and which we allow to have multiple edges and loops. Let $W \in \mathbb{C}Q/[\mathbb{C}Q,\mathbb{C}Q]_{\mathrm{vect}}$ be a *potential*: an element of the quotient of the free path algebra of Q by the vector space spanned by commutators. Associated to the quiver with potential (Q,W), we may define the Jacobi algebra $\mathrm{Jac}(Q,W)$ as in §3.2 below. This is a kind of noncommutative threefold, in the sense that the category of finite-dimensional modules over $\mathrm{Jac}(Q,W)$ behaves in many ways like the category of compactly supported coherent sheaves on a 3CY variety.[2] We refer to [**Gin06, VdB15**] for background and details on this point of view. The moduli stack of finite-dimensional $\mathrm{Jac}(Q,W)$-modules is realised as the critical locus of the function $\mathrm{Tr}(W)$ on the smooth stack of $\mathbb{C}Q$-modules. From the discussion in §1.2, it is thus natural to study the vanishing cycle cohomology

$$\mathcal{A}_{Q,W} := \bigoplus_{\mathbf{d}\in\mathbb{N}^{Q_0}} \mathbf{H}(\mathfrak{M}_{\mathbf{d}}(Q), {}^{\mathfrak{p}}\phi_{\mathrm{Tr}(W)}\mathbb{Q}^{\mathrm{vir}}).$$

This is a categorification of the noncommutative DT theory associated to the pair (Q,W), as initiated in [**Sze08**] in the case of the noncommutative conifold. We continue to use the vir superscript to denote a shift in cohomological degree (see §3.1). The cohomology $\mathcal{A}_{Q,W}$ carries an action of $\mathbf{H}_{\mathbb{C}_u^*} := \mathbf{H}(\mathrm{B}\mathbb{C}^*,\mathbb{Q}) \cong \mathbb{Q}[u]$ by

[1]For arbitrary moduli spaces of stable coherent sheaves on Calabi–Yau threefolds, this global critical locus description is too much to ask for, but can still be achieved locally: see [**BBJ19, BBBBJ15**].

[2]By which we mean a smooth three-dimensional variety X satisfying $\mathcal{O}_X \cong \omega_X$.

multiplying by tautological classes, where $u \bullet -$ is a derivation: see (19). Furthermore, the object $\mathcal{A}_{Q,W}$ carries the cohomological Hall algebra structure introduced by Kontsevich and Soibelman [**KS11**].

The connection with the algebra $\mathcal{A}^T_{\mathbb{A}^2}$ arises by considering the quiver $Q^{(3)}$ with one vertex and three loops a, a^*, ω, and considering the potential

$$\tilde{W} = \omega[a, a^*].$$

Then there is a natural isomorphism of Hall algebras (see §3.9)

$$\mathcal{A}_{Q^{(3)},\tilde{W}} \cong \mathcal{A}_{\mathbb{A}^2}.$$

If T scales the x, y directions of $\mathbb{A}^2$ with weights w_1, w_2 respectively, we may consider the T-action on $Q^{(3)}$ that scales the arrows a, a^*, ω with weights $w_1, w_2, -w_1 - w_2$ respectively, and there is a similar isomorphism of *deformed* CoHAs $\mathcal{A}^T_{Q^{(3)},\tilde{W}} \cong \mathcal{A}^T_{\mathbb{A}^2}$.

Via a purity result of [**Dav23b**], the graded dimensions of $\mathcal{A}_{Q^{(3)},\tilde{W}}$ encode the refined DT invariants of the Jacobi algebra $\mathrm{Jac}(Q^{(3)}, \tilde{W}) \cong \mathbb{C}[x, y, z]$ (equivalently, the DT theory of compactly supported coherent sheaves on $\mathbb{A}^3$) as defined in [**KS08**, **KS11**], and calculated in [**BBS13**]. This means that the DT theory of $\mathrm{Jac}(Q^{(3)}, \tilde{W})$ enables us to calculate the characteristic function encoding the graded dimensions of $\mathcal{A}_{\mathbb{A}^2}$ (and from there, using purity again, the characteristic function of $\mathcal{A}^T_{\mathbb{A}^2}$ — see Proposition 3.6). More than this, the cohomological DT theory of the category of $\mathrm{Jac}(Q^{(3)}, \tilde{W})$-modules, and a thorough study of the deformed affine BPS Lie algebra defined in this paper, enable us to precisely determine the algebra structure on $\mathcal{A}^T_{\mathbb{A}^2}$.

1.4. Results. The algebra $W_{1+\infty}$ is defined to be the universal central extension of the Lie algebra of algebraic differential operators on the complex torus $\mathbb{C}^*$. A basis for $W_{1+\infty}$ is provided by the set

$$\{z^m D^n \mid m \in \mathbb{Z}, n \in \mathbb{Z}_{\geq 0}\} \cup \{c\}$$

where we set $D = z(d/dz)$, and c accounts for the central extension, which will not feature in this paper. We use this basis to consider $W_{1+\infty}$ as a Lie algebra over $\mathbb{Q}$. We define

$$W^+_{1+\infty} = \mathrm{Span}_{\mathbb{Q}}(z^m D^n \mid m \in \mathbb{Z}_{\geq 1}, n \in \mathbb{Z}_{\geq 0}), \tag{1}$$

which is a Lie subalgebra of $W_{1+\infty}$. The Lie bracket is given by

$$[z^m D^a, z^n D^b] = z^{m+n}((D+n)^a D^b - D^a(D+m)^b). \tag{2}$$

The Lie algebra $W^+_{1+\infty}$ is $\mathbb{Z}_{\geq 0}$-graded, with nth graded piece spanned by the operators $z^n D^a$ for $a \geq 0$. The Lie algebra $W^+_{1+\infty}$ has a $\mathbb{Z}$-filtration given by setting[3]

$$F_i W^+_{1+\infty} = \mathrm{Span}(z^m D^a \mid 0 \leq a \leq (i+2)/2, m \in \mathbb{Z}_{\geq 1}).$$

We denote by $\mathbf{Gr}^F_\bullet W^+_{1+\infty}$ the associated graded Lie algebra. From (2), the induced Lie bracket on $\mathbf{Gr}^F_\bullet W^+_{1+\infty}$ is nontrivial; it is given by

$$[z^m D^a, z^n D^b] = (an - bm) z^{m+n} D^{a+b-1}. \tag{3}$$

So the associated graded Lie algebra $\mathbf{Gr}^F_\bullet W^+_{1+\infty}$ is isomorphic to (part of) the *quantum torus* associated to the lattice $\mathbb{Z}^2$ with skew-symmetric form $\omega((m, a), (n, b)) = an - bm$.

---

[3]There is a doubling of degree in our definition, so that the associated graded object lives only in even degrees, in order to match a later, cohomological grading of Hall algebras.

THEOREM A (Theorem 5.4 and Corollary 5.5). *There is an isomorphism of Lie algebras*

$$\hat{\mathfrak{g}}_{Q^{(3)},\tilde{W}} \cong \mathbf{Gr}^F_\bullet \mathrm{W}^+_{1+\infty}$$

between the affine BPS Lie algebra[4] *for the pair* $(Q^{(3)}, \tilde{W})$ *and the associated graded Lie algebra of* $\mathrm{W}^+_{1+\infty}$ *with respect to the filtration* $F_\bullet$. *As a result, there is an isomorphism of algebras*

$$\mathcal{A}_{\mathbb{A}^2} \cong \mathbf{U}(\mathbf{Gr}^F_\bullet \mathrm{W}^+_{1+\infty}).$$

The algebras $\mathcal{A}^T_{\mathbb{A}^2}$ (for various T) have appeared in a great deal of work in geometric representation theory in the last decade. One often restricts to the *spherical subalgebra* of $\mathcal{A}^T_{\mathbb{A}^2}$, i.e., the algebra generated by the subspace $\mathcal{A}^T_{\mathbb{A}^2,1}$, corresponding to coherent sheaves on $\mathbb{A}^2$ of length one. One of the main consequences of Theorem A is the following.

THEOREM B (Theorem 5.6). *Let* T *act on the quiver* $Q^{(3)}$, *leaving the potential* $\tilde{W}$ *invariant. Then the CoHA* $\mathcal{A}^T_{\mathbb{A}^2}$ *is spherically generated, i.e. it is generated by the piece* $\mathcal{A}^T_{\mathbb{A}^2,1}$ *corresponding to coherent sheaves of length one. In particular,* $\mathcal{A}_{\mathbb{A}^2,1}$ *is spherically generated.*

It may come as a temporary disappointment to the reader that the associated graded algebra of the filtered algebra $W^+_{1+\infty}$ appears in Theorem A, as opposed to $W^+_{1+\infty}$ itself. This is, however, somewhat inevitable: the algebra $\mathcal{A}_{Q^{(3)},\tilde{W}}$ is *bi*graded, with one grading coming from the decomposition of the stack of finite-dimensional $\mathbb{C}[x,y]$-modules according to their dimension, and one grading coming from cohomological degree. The Lie algebra $W^+_{1+\infty}$, by contrast, is graded according to the exponent of z, but only *filtered* according to the exponent of D.

To improve upon this situation, we let $\mathbb{C}^*$ act on $\mathbb{A}^2$ by scaling the first coordinate, and leaving the second invariant. This induces a $\mathbb{C}^*$-action on $\mathfrak{M}_d(\mathbb{C}[x,y])$ for all d, and we take the equivariant Borel–Moore homology of this stack with respect to this action, and define the cohomological Hall algebra

$$\mathcal{A}^{\mathbb{C}^*}_{\mathbb{A}^2} := \bigoplus_{d\in\mathbb{N}} \mathbf{H}^{\mathrm{BM}}(\mathfrak{M}^{\mathbb{C}^*}_d(\mathbb{C}[x,y]),\mathbb{Q})$$

to be this equivariant Borel–Moore homology, equipped with the usual multiplication via correspondences. The algebra $\mathbb{Q}[t] \cong \mathbf{H}_{\mathbb{C}^*} := \mathbf{H}(\mathrm{B}\mathbb{C}^*,\mathbb{Q})$ acts on $\mathcal{A}^{\mathbb{C}^*}_{\mathbb{A}^2}$, with the action induced by the $\mathbb{C}^*$-action on $\mathbb{A}^2$, and the Hall algebra multiplication is $\mathbf{H}_{\mathbb{C}^*}$-linear.

We denote by

$$\mathbf{R}_F[W^+_{1+\infty}] := \sum_{i\geq -1} F_{2i}W^+_{1+\infty}t^i\mathbb{Q}[t] \subset W^+_{1+\infty}\otimes_{\mathbb{Q}} t^{-1}\mathbb{Q}[t] \tag{4}$$

the Rees Lie algebra. This is a $\mathbb{Q}[t]$-linear Lie algebra, and so may be thought of as a family of Lie algebras over the affine line, deforming the algebra $\mathbf{Gr}^F_\bullet W^+_{1+\infty}$ to $W^+_{1+\infty}$: for $\epsilon \neq 0$ the specialisation at $t=\epsilon$ is isomorphic to $W^+_{1+\infty}$, and the specialisation at $t=0$ is isomorphic to $\mathbf{Gr}^F_\bullet W^+_{1+\infty}$.

[4] As defined in §3.10.

Let A be a commutative ring. For $\mathfrak{g}$ an A-linear Lie algebra, we define the tensor algebra

$$T_A(\mathfrak{g}) := \bigoplus_{m\geq 0} \underbrace{\mathfrak{g}\otimes_A \dots \otimes_A \mathfrak{g}}_{m \text{ times}}$$

and define

$$\mathbf{U}_A(\mathfrak{g}) = T_A(\mathfrak{g})/\langle \alpha\cdot\beta - \beta\cdot\alpha - [\alpha,\beta]_{\mathfrak{g}} \mid \alpha,\beta\in\mathfrak{g}\rangle.$$

THEOREM C (Corollary 5.9). *Let $\mathbb{C}^*$ act on $Q^{(3)}$ with weights $1, 0, -1$ on the arrows a, a^*, ω respectively. There is an isomorphism of $\mathbf{H}_{\mathbb{C}^*} = \mathbb{Q}[t]$-linear Lie algebras*

$$\hat{\mathfrak{g}}^{\mathbb{C}^*}_{Q^{(3)},\tilde{W}} \cong \mathbf{R}_F[W^+_{1+\infty}]$$

between the deformed affine BPS Lie algebra for $(Q^{(3)},\tilde{W})$, introduced in §3.10, and the Rees Lie algebra of $W^+_{1+\infty}$. As a result, there is an isomorphism of $\mathbb{Q}[t]$-linear algebras:

$$\mathcal{A}^{\mathbb{C}^*}_{\mathbb{A}^2} \cong \mathbf{U}_{\mathbb{Q}[t]}(\mathbf{R}_F[W^+_{1+\infty}]).$$

Let $T = (\mathbb{C}^*)^2$ act on $\mathbb{A}^2$, by scaling the two coordinates independently. From the above results, it is very natural to seek an algebraic description of the fully deformed CoHA $\mathcal{A}^T_{\mathbb{A}^2}$. We denote by $\mathcal{Y}_{t_1,t_2,t_3}(\widehat{\mathfrak{gl}(1)})^+$ the strictly positive part of the affine Yangian; a presentation is recalled in §3.13. Our final theorem is a strengthening of a result of Rapčák, Soibelman, Yang and Zhao [**RSYZ20**]:

THEOREM D (Theorem 5.11). *For $T = (\mathbb{C}^*)^2$ as above, there is an isomorphism of $\mathbf{H}_T$-linear algebras*

$$\mathcal{A}^T_{\mathbb{A}^2} \cong \mathcal{Y}_{t_1,t_2,t_3}(\widehat{\mathfrak{gl}(1)})^+.$$

1.5. Notation and conventions.

- Every cohomological Hall algebra that we consider throughout the paper will carry a $\mathbf{H}(\mathrm{B}\mathbb{C}^*,\mathbb{Q})$-action arising from the morphism from the stack of finite-dimensional modules over an algebra to $\mathrm{B}\mathbb{C}^*$ given by taking a module ρ to $\wedge^{\mathrm{top}}\rho$. We denote this copy of $\mathrm{B}\mathbb{C}^*$ by $\mathrm{B}\mathbb{C}^*_u$, to distinguish it from other instances of $\mathrm{B}\mathbb{C}^*$ arising from e.g. torus actions on quivers. We continue to write, as in the introduction, $\mathbf{H}_{\mathbb{C}^*_u} := \mathbf{H}(\mathrm{B}\mathbb{C}^*_u,\mathbb{Q}) \cong \mathbb{Q}[u]$, with u in cohomological degree 2.
- More generally, if $T = (\mathbb{C}^*)^l$ is a torus, we define $\mathbf{H}_T := \mathbf{H}(\mathrm{B}T,\mathbb{Q})$.
- If a torus T acts on a stack $\mathfrak{X}$ with quotient $\mathfrak{Y}$, we define $({}^{\mathfrak{p}'}\tau^{\leq i}, {}^{\mathfrak{p}'}\tau^{\geq i})$ to be the shift of the usual perverse t structure on the derived category of constructible complexes on $\mathfrak{Y}$, so that $\mathcal{F}$ is in the heart if and only if $(\mathfrak{X}\to\mathfrak{Y})^*\mathcal{F}$ is perverse (see §3.1).
- We define $\mathbb{N} := \mathbb{Z}_{\geq 0}$.

2. Algebraic background

2.1. Heisenberg algebras. We denote by **Heis** the Lie algebra over $\mathbb{Q}$ having basis $\{p,q,c\}$, subject to the relations

$$[q,p] = c$$
$$[c,p] = [c,q] = 0.$$

We denote by **Heis** −mod the category of modules over **Heis**. We give **Heis** a (cohomological) $\mathbb{Z}$-grading, putting p, c, q in degrees $2, 0, -2$ respectively. A cohomologically graded **Heis**-module is a **Heis**-module ρ, along with a direct sum decomposition according to cohomological degree of the underlying vector space

$$\rho = \bigoplus_{n\in\mathbb{Z}} \rho^n$$

such that p maps ρ^n to ρ^{n+2}, q maps ρ^n to ρ^{n-2}, and c preserves the decomposition. We denote by $\mathbf{Heis}-\mathrm{mod}_{\mathbb{Z}}$ the category of cohomologically graded **Heis**-modules. A morphism of cohomologically graded **Heis**-modules is a morphism of the underlying **Heis**-modules that preserves the grading. If ρ is a cohomologically graded **Heis** module, and $n \in \mathbb{Z}$, we denote by $\rho[n]$ the cohomologically graded **Heis**-module obtained by setting $\rho[n]^i = \rho^{n+i}$.

For a quiver Q with vertex set Q_0, we define the Lie algebra

$$\mathbf{Heis}'_Q := \bigoplus_{i\in Q_0} \mathbf{Heis}\,.$$

For $i \in Q_0$ we denote by p_i, q_i, c_i the basis elements in $\mathbf{Heis}'_Q$ corresponding to the ith summand. We define the Lie subalgebra $\mathbf{Heis}_Q \subset \mathbf{Heis}'_Q$ to be the span of $\{p_i \mid i \in Q_0\} \cup \{c_i \mid i \in Q_0\} \cup \{q_\Sigma := \sum_{i\in Q_0} q_i\}$.

For $\mathbf{c} \in \mathbb{Z}$ we say that a **Heis**-representation ρ has central charge $\mathbf{c}$ if c acts on ρ via scalar multiplication by $\mathbf{c}$. For $\mathbf{c} \in \mathbb{Z}^{Q_0}$ we say that a $\mathbf{Heis}_Q$-representation ρ has central charge $\mathbf{c}$ if c_i acts by multiplication by $\mathbf{c}_i$. There is a canonical inclusion

$$\mathbf{Heis} \hookrightarrow \mathbf{Heis}_Q \tag{5}$$

sending $p \mapsto \sum_{i\in Q_0} p_i$, $q \mapsto q_\Sigma$ and $c \mapsto \sum_{i\in Q_0} c_i$. We can consider a $\mathbf{Heis}_Q$-module as a **Heis**-module via this inclusion. Placing all p_i, c_i, q_i in degrees $2, 0, -2$ respectively, we give $\mathbf{Heis}_Q$ a $\mathbb{Z}$-grading, and define the category $\mathbf{Heis}_Q-\mathrm{mod}_{\mathbb{Z}}$ of graded modules as before.

The vector space $\mathbb{Q}_0 = \mathbb{Q}$ forms a $\mathbf{Heis}_Q$-module of central charge 0 if we let all p_i, q_i, c_i act via the zero matrix. The categories $\mathbf{Heis}_Q-\mathrm{mod}$ and $\mathbf{Heis}_Q-\mathrm{mod}_{\mathbb{Z}}$ are symmetric tensor categories with monoidal unit $\mathbb{Q}_0$. If $\mathscr{C}$ is a tensor category, a *$\mathscr{C}$-algebra object* is an object $\mathcal{F}$ in $\mathscr{C}$, along with morphisms $\mathbf{1}_{\mathscr{C}} \to \mathcal{F}$ from the monoidal unit, and a multiplication morphism $\mathcal{F} \otimes \mathcal{F} \to \mathcal{F}$, satisfying the standard properties. If $\mathscr{C}$ is a symmetric tensor category, a *$\mathscr{C}$-Lie algebra object* is an object $\mathcal{F}$ in $\mathscr{C}$ along with an antisymmetric morphism $\mathcal{F} \otimes \mathcal{F} \to \mathcal{F}$ satisfying the Jacobi identity. If ρ is a Lie algebra object in $\mathbf{Heis}_Q$, the universal enveloping algebra $\mathbf{U}(\rho)$ is an algebra object in $\mathbf{Heis}_Q$, with underlying algebra the universal enveloping algebra of the underlying Lie algebra of ρ.

We say that a $\mathbf{Heis}_Q$-module ρ is integrable if there is a decomposition

$$\rho = \bigoplus_{\mathbf{d}\in\mathbb{N}^{Q_0}} \rho_{[\mathbf{d}]} \tag{6}$$

with each $\rho_{[\mathbf{d}]}$ a $\mathbf{Heis}_Q$-module of central charge $\mathbf{d}$. Similarly, we say that a **Heis**-module ρ is integrable if it admits a decomposition $\rho = \bigoplus_{d\in\mathbb{N}} \rho_{[d]}$ with $\rho_{[d]}$ of central charge d.

Let $\mathbf{Heis}_Q$ act on the $\mathbb{N}^{Q_0}$-graded algebra $\mathcal{A}$, where the $\mathbb{N}^{Q_0}$-grading is induced by a decomposition of $\mathcal{A}$ into sub-modules of fixed central charge, as in (6). Then $\mathcal{A}$ is an algebra object in $\mathbf{Heis}_Q$ if each p_i for $i \in Q_0$, and q_Σ, act by derivations on $\mathcal{A}$.

Similarly, if $\mathfrak{n}$ is a $\mathbb{N}^{Q_0}$-graded Lie algebra, and also an integrable $\mathbf{Heis}_Q$-module for which the $\mathbb{N}^{Q_0}$-graded decomposition is provided by the decomposition according to central charge, then $\mathfrak{n}$ is a Lie algebra object in $\mathbf{Heis}_Q$ if and only if for every $i \in Q_0$, p_i and q_Σ act by Lie algebra derivations.

2.2. The Lie algebra $W^+_{1+\infty}$. Recall from (1) the definition of $W^+_{1+\infty} \subset W_{1+\infty}$. The Lie bracket on $W^+_{1+\infty}$ is given by

$$[z^m f(D), z^n g(D)] = z^{m+n}(f(D+n)g(D) - f(D)g(D+m)). \tag{7}$$

We set

$$\tilde{D}^{(a)} := D^a t^{a-1} \in \mathbf{R}_F[W^+_{1+\infty}].$$

The Lie algebra $\mathbf{R}_F[W^+_{1+\infty}]$ (defined as in (4)) is spanned by elements $z^m\tilde{D}^{(a)}t^r$ with $m \geq 1, a \geq 0, r \geq 0$. We give $z^m\tilde{D}^{(a)}t^r = z^m D^a t^{r+a-1}$ cohomological degree $2r + 2a - 2$, making $\mathbf{R}_F[W^+_{1+\infty}]$ into a cohomologically graded Lie algebra.

PROPOSITION 2.1.

- *There is an action of* **Heis** *on $W^+_{1+\infty}$ by derivations, given by sending p to $[D^2, -]$ and sending q to the endomorphism*

$$\partial_D \colon W^+_{1+\infty} \to W^+_{1+\infty}$$
$$z^m f(D) \mapsto z^m f'(D).$$

  For $m \geq 1$, the subspace $\mathrm{Span}(z^m D^n \mid n \in \mathbb{N})$ *is preserved by this action, on which it has central charge $2m$.*
- *There is an action of* **Heis** *on $\mathbf{R}_F[W^+_{1+\infty}]$ by derivations, sending p to $t[D^2, -]$ and q to the endomorphism $t^{-1}\partial_D$, making $\mathbf{R}_F[W^+_{1+\infty}]$ into a cohomologically graded* **Heis**-*module. For $m \geq 1$, the subspace* $\mathrm{Span}(z^m\tilde{D}^n t^r \mid n, r \in \mathbb{N})$ *is preserved by this action, on which it has central charge $2m$.*

PROOF. We prove the first part, the second is similar. The fact that $[D^2, -]$ is a derivation follows from the Jacobi identity, while the fact that ∂_D is a derivation follows from the formula (7). Now we calculate

$$\begin{aligned}
[\partial_D, [D^2, -]](z^m D^n) &= \partial_D[D^2, z^m D^n] - [D^2, nz^m D^{n-1}] \\
&= [2D, z^m D^n] + [D^2, nz^m D^{n-1}] - [D^2, nz^m D^{n-1}] \\
&= 2z^m((D+m)D^n - D^{1+n}) \\
&= 2mz^m D^n
\end{aligned}$$

as required. □

For $m \in 2\cdot\mathbb{Z}_{\geq 1}$ we denote by $V_{[m]} \subset W^+_{1+\infty}$ the **Heis**-submodule with central charge m. By the above proposition, it has a basis given by elements $z^{m/2}D^n$. If we let $\mathbb{Q}[u]$ act on $V_{[m]}$ by letting u act via the raising operator $p = [D^2, -]$, then all of the $V_{[m]}$ are isomorphic to $\mathbb{Q}[u]$, considered as $\mathbb{Q}[u]$-modules.

We endow $V_{[2m]}$ with a filtration by setting

$$F_i V_{[2m]} = \begin{cases} \mathrm{Span}_{\mathbb{Q}}(z^m, z^m D, \ldots, z^m D^{i/2+1}) & \text{if } i \text{ is even} \\ F_{i-1}V_{[2m]} & \text{if } i \text{ is odd.} \end{cases}$$

So if i is odd, the ith piece of the associated graded object $\mathbf{Gr}^F_i V_{[2m]}$ is zero, and if $i \in \mathbb{Z}_{\geq -1}$ then a basis of $\mathbf{Gr}^F_{2i} V_{[2m]}$ is given by $\{z^m D^{i+1}\}$. We have $pF_i V_{[2m]} \subset F_{i+2} V_{[2m]}$ and $qF_i V_{[2m]} \subset F_{i-2} V_{[2m]}$, so that $\mathbf{Gr}^F_\bullet V_{[2m]}$ becomes a cohomologically graded **Heis**-module, where the cohomological degree of $z^m D^n$ is $2n-2$. Taking the direct sum of the filtrations above, we obtain the *order* filtration $F_\bullet$ on $W^+_{1+\infty}$, considered in the introduction. The Lie bracket on $W^+_{1+\infty}$ preserves the order filtration, and there is an isomorphism of Lie algebra objects in the category $\mathbf{Heis}-\mathrm{mod}_{\mathbb{Z}}$:

$$\begin{aligned}\Psi\colon \mathbf{Gr}^F_\bullet W^+_{1+\infty} &\to \mathbf{R}_F[W^+_{1+\infty}] \otimes_{\mathbb{Q}[t]} (\mathbb{Q}[t]/(t)) \\ z^m D^n &\mapsto z^m \tilde{D}^{(n)} \otimes 1.\end{aligned}$$

LEMMA 2.2. *The Lie algebras* $\mathbf{Gr}^F_\bullet W^+_{1+\infty}$ *and* $\mathbf{R}_F[W^+_{1+\infty}]$ *are spherically generated: they are generated by their respective* **Heis**-*submodules of central charge* 2.

PROOF. Let $\mathfrak{g} \subset \mathbf{R}_F[W^+_{1+\infty}]$ be the Lie subalgebra generated by $\mathbf{R}_F[W^+_{1+\infty}]_{[2]} = \mathrm{Span}(z\tilde{D}^{(a)}t^r \mid a \geq 0, r \geq 0)$. For $a, m \geq 1$, the identity (2) yields

$$\begin{aligned}[z\tilde{D}^{(a)}, z^m \tilde{D}^{(0)}] =& t^{a-2} z^{m+1}((D+m)^a - D^a) \\ =& z^{m+1}(ma\tilde{D}^{(a-1)} + \text{lower order terms in } \tilde{D}).\end{aligned} \tag{8}$$

We induct on m, and assume that $z^{m'}\tilde{D}^{(a')} \in \mathbf{R}_F[W^+_{1+\infty}]$ for every $1 \leq m' < m$ and $a' \geq 0$. Substituting $m \mapsto m-1$ and $a \mapsto 1$ in (8) we deduce that $z^m \tilde{D}^{(0)} \in \mathfrak{g}$. Then we induct on a, and assume that $z^m \tilde{D}^{(a')} \in \mathbf{R}_F[W^+_{1+\infty}]$ for every $0 \leq a' < a$. Substituting $a \mapsto a+1$ and $m \mapsto m-1$ in (8) we deduce that $z^m \tilde{D}^{(a)} \in \mathfrak{g}$ as required for the inductive step.

Spherical generation of $\mathbf{Gr}^F_\bullet W^+_{1+\infty}$ follows directly from (3), or from the spherical generation of $\mathbf{R}_F[W^+_{1+\infty}]$ and the isomorphism Ψ. □

We record a variant of the result, for later. It follows from (3).

LEMMA 2.3. *The elements, defined for all* $n, m \geq 0$ *by*

$$\begin{cases} (\mathbf{ad}_{zD})^m(zD^n) & \text{if } m < n \text{ or } n = 0 \\ ((\mathbf{ad}_{zD})^{m-n} \circ \mathbf{ad}_D \circ (\mathbf{ad}_{zD})^{n-1})(zD^{n+1}) & \text{otherwise} \end{cases}$$

provide a basis for $W^+_{1+\infty}$.

The Lie algebra $\mathbf{Gr}^F_\bullet W^+_{1+\infty}$ carries two gradings, one by central charge, and one given by the cohomological grading, and $\mathbf{Gr}^F_\bullet W^+_{1+\infty}$ is a Lie algebra object in $\mathbf{Heis}-\mathrm{mod}_{\mathbb{Z}}$. It turns out that this object satisfies a simple universal property: the Lie algebra $\mathbf{Gr}^F_\bullet W^+_{1+\infty}$ is the universal Lie algebra object in $\mathbf{Heis}-\mathrm{mod}_{\mathbb{Z}}$ containing $\mathbf{Gr}^F_\bullet V_{[2]}$, and containing only representations $\mathbf{Gr}^F_\bullet V_{[m]}$ for various m as subrepresentations (i.e. no copies of $\mathbf{Gr}^F_\bullet V_{[m]}[j]$ with $j \neq 0$)). More precisely,

LEMMA 2.4. *If* $\mathcal{H}$ *is a Lie algebra object in* $\mathbf{Heis}-\mathrm{mod}_{\mathbb{Z}}$ *containing only unshifted representations* $\mathbf{Gr}^F_\bullet V_{[m]}$ *as cohomologically graded* **Heis**-*subrepresentations, and* $\iota\colon \mathbf{Gr}^F_\bullet V_{[2]} \to \mathcal{H}$ *is an inclusion in* $\mathbf{Heis}-\mathrm{mod}_{\mathbb{Z}}$, *then* ι *extends uniquely to a morphism of* **Heis**-*Lie algebra objects* $\mathbf{Gr}^F_\bullet W^+_{1+\infty} \to \mathcal{H}$.

PROOF. The lemma is equivalent to the claim that $\mathbf{Gr}^F_\bullet W^+_{1+\infty}$ is isomorphic to $\mathcal{L}' = \mathcal{L}/\mathcal{I}$, where $\mathcal{L}$ is the free $\mathbb{Z}$-graded Lie algebra $\mathcal{L}$ generated by $\mathbf{Gr}^F_\bullet V_{[2]}$, and $\mathcal{I}$ is the Lie ideal containing all the $\mathbb{Z}$-graded **Heis**-submodules of $\mathcal{L}$ that are not isomorphic to unshifted copies of $\mathbf{Gr}^F_\bullet V_{[m]}$ for various m.

The Lie algebra $\mathcal{L}'$ carries a grading by tensor degree, and we claim that $\mathcal{L}'_m \cong \mathbf{Gr}^F_\bullet V_{[2m]}$, where on the left hand side of this isomorphism the subscript denotes the degree with respect to the tensor degree grading. This obviously implies the lemma. We prove the claim by induction. The base case $m = 1$ is trivial, and so we assume the claim is proved for $i < m$, with $m \geq 2$. By the inductive hypothesis, the mth summand $\mathcal{L}'_m$ is either a quotient of

$$\mathbf{Gr}^F_\bullet V_{[2]} \wedge \mathbf{Gr}^F_\bullet V_{[2]} \cong \bigoplus_{i\geq 0} \mathbf{Gr}^F_\bullet V_{[4]}[-4i]$$

if $m = 2$, or

$$\mathbf{Gr}^F_\bullet V_{[2]} \otimes \mathbf{Gr}^F_\bullet V_{[2m-2]} \cong \bigoplus_{i\geq 0} \mathbf{Gr}^F_\bullet V_{[2m]}[2-2i]$$

if $m \geq 3$. In either case, at most a single unshifted copy of $\mathbf{Gr}^F_\bullet V_{[2m]}$ survives after we quotient out by all of the shifted copies of $\mathbf{Gr}^F_\bullet V_{[2m]}$. On the other hand, $\mathcal{L}'_m$ contains at least one copy of $\mathbf{Gr}^F_\bullet V_{[2m]}$, since by its universal property, and spherical generation of $\mathbf{Gr}^F_\bullet W^+_{1+\infty}$, there is a surjective morphism $\mathcal{L}' \to \mathbf{Gr}^F_\bullet W^+_{1+\infty}$. The inductive claim follows. □

Given an integrable cohomologically-graded **Heis**-module ρ, we define the characteristic function

$$\chi(\rho) := \sum_{i,j\in\mathbb{Z}} \dim(\rho^j_{[i]})v^i q^j,$$

where $\rho^j_{[i]}$ denotes the jth cohomologically graded piece of the summand of ρ with central charge i.

COROLLARY 2.5. *Let $\mathcal{H}$ be an integrable $\mathbb{Z}$-graded* **Heis**-*module generated as a Lie algebra by $\mathbf{Gr}^F_\bullet V_{[2]} \subset \mathcal{H}$ and satisfying*

$$\begin{aligned}\chi(\mathcal{H}) =&\chi(\mathbf{Gr}^F_\bullet W^+_{1+\infty})\\ =&\sum_{i\geq 1, j\geq -1} v^{2i}q^{2j}\\ =&v^2q^{-2}(1-v^2)^{-1}(1-q^2)^{-1}.\end{aligned}$$

Then $\mathcal{H} \cong \mathbf{Gr}^F_\bullet W^+_{1+\infty}$.

PROOF. By the universal property of $\mathbf{Gr}^F_\bullet W^+_{1+\infty}$ (Lemma 2.4) there is a morphism of Lie algebra objects in **Heis** $-\mathrm{mod}_{\mathbb{Z}}$

$$f\colon \mathbf{Gr}^F_\bullet W^+_{1+\infty} \to \mathcal{H}$$

sending the copy of $\mathbf{Gr}^F_\bullet V_{[2]}$ inside $\mathbf{Gr}^F_\bullet W^+_{1+\infty}$ isomorphically to the given copy inside $\mathcal{H}$. By supposition, f is surjective. On the other hand, the source and the target have the same characteristic function, so f is an isomorphism. □

3. DT theory background

3.1. Vanishing cycle toolkit. If $\mathfrak{X}$ is a smooth stack, we denote by $\mathbb{Q}_{\mathfrak{X}}^{\mathrm{vir}} = \mathbb{Q}_{\mathfrak{X}}[\dim(\mathfrak{X})]$ the constant perverse sheaf on $\mathfrak{X}$. More generally, if an extra torus T acts on a smooth stack $\mathfrak{X}$ with T-quotient $\mathfrak{Y}$, we define[5] $\mathbb{Q}_{\mathfrak{Y}}^{\mathrm{vir}} = \mathbb{Q}_{\mathfrak{Y}}[\dim(\mathfrak{X})]$. Under the same conditions, we modify the usual perverse truncation functors for constructible complexes of sheaves on $\mathfrak{Y}$ by setting ${}^{\mathfrak{p}'}\tau^{\leq i} = {}^{\mathfrak{p}}\tau^{\leq i-\dim(T)}$.

Given a function f on a smooth stack X, we define the vanishing cycle functor ${}^{\mathfrak{p}}\phi_f$ as in [**KS90**]. It is an exact functor with respect to the perverse t structure, and in particular ${}^{\mathfrak{p}}\phi_f\mathbb{Q}_X^{\mathrm{vir}}$ is a perverse sheaf on X, which we may refer to just as the sheaf of vanishing cycles on X. We will often abuse notation by just writing $\mathbf{H}(X, {}^{\mathfrak{p}}\phi_f)$ instead of $\mathbf{H}(X, {}^{\mathfrak{p}}\phi_f\mathbb{Q}_X^{\mathrm{vir}})$.

Let f, g be functions on smooth stacks X, Y respectively. The Thom–Sebastiani theorem [**Mas01**] states that there is a natural bifunctorial equivalence of constructible complexes of sheaves on $X \times Y$

$$ {}^{\mathfrak{p}}\phi_{f\boxplus g}(\mathcal{F} \boxtimes \mathcal{G})|_{f^{-1}(0)\times g^{-1}(0)} \cong {}^{\mathfrak{p}}\phi_f\mathcal{F} \boxtimes {}^{\mathfrak{p}}\phi_g\mathcal{G}. $$

Assuming that $\mathrm{supp}({}^{\mathfrak{p}}\phi_{f\boxplus g}(\mathcal{F}\boxtimes\mathcal{G})) \subset f^{-1}(0) \times g^{-1}(0)$, applying the derived global sections functor to the above isomorphism yields the isomorphism

$$ \mathtt{TS}\colon \mathbf{H}(X, {}^{\mathfrak{p}}\phi_f\mathcal{F}) \otimes \mathbf{H}(Y, {}^{\mathfrak{p}}\phi_g\mathcal{G}) \cong \mathbf{H}(X \times Y, {}^{\mathfrak{p}}\phi_{f\boxplus g}(\mathcal{F} \boxtimes \mathcal{G})), $$

the analogue in vanishing cycle cohomology of the Künneth isomorphism in ordinary cohomology.

Given a morphism $q\colon X \to Y$ and a function f on Y there is a natural transformation $\theta\colon {}^{\mathfrak{p}}\phi_f q_* \to q_* {}^{\mathfrak{p}}\phi_{fq}$, which is an isomorphism if q is projective. The natural transformation θ induces the natural transformation $\eta\colon {}^{\mathfrak{p}}\phi_f \to q_*{}^{\mathfrak{p}}\phi_{fq}q^*$ by adjunction. Applying the derived global sections functor, η induces the pullback morphism

$$ q^{\star}\colon \mathbf{H}(Y, {}^{\mathfrak{p}}\phi_f\mathcal{F}) \to \mathbf{H}(X, {}^{\mathfrak{p}}\phi_{fq}q^*\mathcal{F}). \tag{9} $$

Given a proper morphism $p\colon X \to Y$ between smooth stacks and a function f on Y, we obtain the pushforward morphism

$$ p_{\star}\colon \mathbf{H}(X, {}^{\mathfrak{p}}\phi_{pf}\mathbb{Q}_X)[2\dim(X)] \to \mathbf{H}(Y, {}^{\mathfrak{p}}\phi_f\mathbb{Q})[2\dim(Y)] \tag{10} $$

by applying ${}^{\mathfrak{p}}\phi_f$ to $p_*\mathbb{Q}_X[2\dim(X)] \to \mathbb{Q}_Y[2\dim(Y)]$ (the Verdier dual of $\mathbb{Q}_Y \to p_*\mathbb{Q}_X$), and composing with θ^{-1}.

Let $\mathfrak{X}$ be a smooth stack, f be a function on $\mathfrak{X}$, and $\mathcal{F} \in \mathcal{D}_c^b(\mathfrak{X})$ be a constructible complex. Let $\Delta\colon \mathfrak{X} \to \mathfrak{X} \times \mathfrak{X}$ be the diagonal embedding. Then $f = (f \boxplus 0) \circ \Delta$ and there is a natural isomorphism $\mathcal{F} \cong \Delta^*(\mathcal{F} \boxtimes \mathbb{Q}_{\mathfrak{X}})$. The pullback morphism $\Delta^{\star}$, composed with the Thom-Sebastiani isomorphism, yields

$$ \mathbf{H}(\mathfrak{X}, {}^{\mathfrak{p}}\phi_f\mathcal{F}) \otimes \mathbf{H}(\mathfrak{X}, \mathbb{Q}) \to \mathbf{H}(\mathfrak{X}, {}^{\mathfrak{p}}\phi_f\mathcal{F}), $$

and an action of the cohomology of $\mathfrak{X}$ on $\mathbf{H}(\mathfrak{X}, {}^{\mathfrak{p}}\phi_f\mathcal{F})$. If $f = 0$ this is just the usual action of $\mathbf{H}(\mathfrak{X}, \mathbb{Q})$ on $\mathbf{H}(\mathfrak{X}, \mathcal{F})$, and if in addition $\mathcal{F} = \mathbb{Q}_{\mathfrak{X}}$ this is the usual

---

[5] The motivation for this slightly mixed convention is that we would like $\mathbb{Q}_{\mathfrak{Y}}^{\mathrm{vir}}$ to correspond to a T-equivariant perverse sheaf on $\mathfrak{Y}$, i.e. we would like the pullback $(\mathfrak{X} \to \mathfrak{Y})^*\mathbb{Q}_{\mathfrak{Y}}^{\mathrm{vir}}$ to be perverse.

associative algebra structure on $\mathbf{H}(\mathfrak{X},\mathbb{Q})$. If $q\colon \mathfrak{X}\to\mathfrak{Y}$ is a morphism of smooth stacks, from the commutativity of

$$\begin{array}{ccc} \mathfrak{X} & \xrightarrow{\quad q \quad} & \mathfrak{Y} \\ {\scriptstyle (\mathrm{id}\times q)\circ\Delta}\big\downarrow & & \big\downarrow{\scriptstyle \Delta} \\ \mathfrak{X}\times\mathfrak{Y} & \xrightarrow{q\times \mathrm{id}} & \mathfrak{Y}\times\mathfrak{Y} \end{array}$$

we deduce that (9) respects the $\mathbf{H}(\mathfrak{Y},\mathbb{Q})$-actions on the domain and target. If q is projective, one may show as in [**Dav17**, Prop.2.14] that (10) respects the $\mathbf{H}(\mathfrak{Y},\mathbb{Q})$ actions on the domain and target.

3.2. Quivers. Throughout the paper, Q will denote a finite quiver, i.e. a pair of finite sets Q_1 and Q_0 (the arrows and vertices, respectively) along with a pair of morphisms $s,t\colon Q_1\to Q_0$. The free path algebra kQ of Q over the field k is the algebra having as k-basis the paths in Q, including "lazy" paths e_i of length zero at each of the vertices $i\in Q_0$.

We define the Euler form

$$\begin{aligned} \chi_Q\colon \mathbb{N}^{Q_0}\times\mathbb{N}^{Q_0} &\to \mathbb{Z} \\ (\mathbf{d},\mathbf{e}) &\mapsto (\mathbf{d},\mathbf{e})_Q := \sum_{i\in Q_0}\mathbf{d}_i\mathbf{e}_i-\sum_{a\in Q_1}\mathbf{d}_{s(a)}\mathbf{e}_{t(a)}. \end{aligned}$$

The *double* of Q is obtained by setting $\overline{Q}_0=Q_0$ and $\overline{Q}_1=Q_1\coprod Q_1^*$, where $Q_1^*:=\{a^*\mid a\in Q_1\}$ and the orientation of a^* is the opposite of the orientation of a. The preprojective algebra is defined by

$$\Pi_Q:=\mathbb{C}\overline{Q}/\langle\sum_{a\in Q_1}[a,a^*]\rangle.$$

The quiver $\tilde{Q}$ is defined by adding a loop ω_i to each vertex $i\in\overline{Q}_0$.

EXAMPLE 3.1. If $Q=Q^{(1)}$ is the Jordan quiver with one vertex and one loop a, then $\tilde{Q}=Q^{(3)}$ is the three loop quiver, with one vertex and loops labelled a,a^*,ω.

A *potential* $W\in\mathbb{C}Q/[\mathbb{C}Q,\mathbb{C}Q]_{\mathrm{vect}}$ is a linear combination of cyclic words in Q. If $W=a_1\dots a_l$ is a single cyclic word then

$$\partial W/\partial a:=\sum_{a_i=a}a_{i+1}\dots a_la_1\dots a_{i-1}.$$

We extend this definition to general potentials by linearity. The *Jacobi algebra* associated to the pair (Q,W) is defined by

$$\mathrm{Jac}(Q,W):=\mathbb{C}Q/\langle\partial W/\partial a\mid a\in Q_1\rangle.$$

If Q is an arbitrary quiver, the tripled quiver $\tilde{Q}$ carries the *canonical cubic potential*

$$\tilde{W}:=\left(\sum_{i\in Q_0}\omega_i\right)\left(\sum_{a\in Q_1}[a,a^*]\right).$$

Let $\Pi_Q[\omega]$ denote the polynomial algebra in one variable ω with coefficients in Π_Q. There is an isomorphism

$$j\colon\Pi_Q[\omega]\to\mathrm{Jac}(\tilde{Q},\tilde{W}) \tag{11}$$

defined by $j(a)=a$, $j(a^*)=a^*$, $j(\omega)=\sum_{i\in Q_0}\omega_i$.

3.3. Moduli spaces. Given a quiver Q and dimension vector $\mathbf{d} \in \mathbb{N}^{Q_0}$ we define

$$\mathbb{A}_{Q,\mathbf{d}} := \prod_{a\in Q_1} \operatorname{Hom}(\mathbb{C}^{\mathbf{d}(s(a))}, \mathbb{C}^{\mathbf{d}(t(a))}); \qquad \operatorname{GL}_{\mathbf{d}} := \prod_{i\in Q_0} \operatorname{GL}_{\mathbf{d}(i)}.$$

The group $\operatorname{GL}_{\mathbf{d}}$ acts on $\mathbb{A}_{Q,\mathbf{d}}$ by conjugation. The moduli stack $\mathfrak{M}_{\mathbf{d}}(Q)$, when evaluated on a scheme X, yields the groupoid of coherent sheaves on X, flat over X, equipped with an action of $\mathbb{C}Q$, such that at every geometric K-point of X, with $K \supset \mathbb{C}$, the induced KQ-module is $\mathbf{d}$-dimensional. This stack is naturally isomorphic to the stack quotient $\mathbb{A}_{Q,\mathbf{d}}/\operatorname{GL}_{\mathbf{d}}$.

Let $N = \mathbb{Z}^r$ be a lattice, and fix a function $\tau\colon Q_1 \to N$. We fix the torus $T = \operatorname{Hom}_{\mathrm{Grp}}(N, \mathbb{C}^*)$. The torus T acts on $\mathbb{A}_{Q,\mathbf{d}}$ by sending

$$(\mathbf{t}, (\rho_a)_{a\in Q_1}) \mapsto (\mathbf{t}(\tau(a)) \cdot \rho_a)_{a\in Q_1},$$

and this action commutes with the action of $\operatorname{GL}_{\mathbf{d}}$ by conjugation. We denote by

$$\mathfrak{M}^T_{\mathbf{d}}(Q) \cong \mathbb{A}_{Q,\mathbf{d}}/(\operatorname{GL}_{\mathbf{d}} \times T)$$

the quotient of $\mathfrak{M}_{\mathbf{d}}(Q)$ by the T-action. Precisely, a homomorphism $X \to \mathfrak{M}^T_{\mathbf{d}}(Q)$ is provided by a T-torsor $E \to X$, along with a flat family $\mathcal{F}$ of $\mathbf{d}$-dimensional $\mathbb{C}Q$-modules on E, and a T-equivariant structure on $\mathcal{F}$, such that for every $a \in Q_1$ the following diagram of coherent sheaves on $T \times E$ commutes:

$$\begin{array}{ccc} l^*\mathcal{F} & \xrightarrow{\quad l^*\rho(a) \quad} & l^*\mathcal{F} \\ \downarrow \cong & & \downarrow \cong \\ p^*\mathcal{F} & \xrightarrow[\prod_{i\leq r} z_i^{\tau(a)_i}\cdot p^*\rho(a)]{} & p^*\mathcal{F}. \end{array}$$

Here, $l\colon T\times E \to E$ is the action, $p\colon T\times E \to E$ is the projection, $\rho(a)$ is the action of a on $\mathcal{F}$, z_i are coordinates on T, and the vertical isomorphisms are defined by the T-equivariant structure on $\mathcal{F}$. We will use the morphism

$$(12) \qquad \jmath\colon \mathfrak{M}^T_{\mathbf{d}'}(Q) \times_{\mathrm{B}T} \mathfrak{M}^T_{\mathbf{d}''}(Q) \to \mathfrak{M}^T_{\mathbf{d}'}(Q) \times \mathfrak{M}^T_{\mathbf{d}''}(Q)$$

induced by the structure map $\mathrm{B}T \to \mathrm{pt}$.

We denote by

$$\mathcal{M}_{\mathbf{d}}(Q) = \operatorname{Spec}(\Gamma(\mathbb{A}_{Q,\mathbf{d}})^{\operatorname{GL}_{\mathbf{d}}})$$

the coarse moduli space. For $K \supset \mathbb{C}$ a field, K-points of $\mathcal{M}_{\mathbf{d}}(Q)$ correspond to isomorphism classes of semisimple $\mathbf{d}$-dimensional KQ-modules. Since the T-action commutes with the $\operatorname{GL}_{\mathbf{d}}$-action, $\mathcal{M}_{\mathbf{d}}(Q)$ carries a natural T-action. We define the stack-theoretic quotient

$$\mathcal{M}^T_{\mathbf{d}}(Q) := \mathcal{M}_{\mathbf{d}}(Q)/T.$$

In the case of trivial T, there is a natural morphism

$$\mathtt{JH}\colon \mathfrak{M}_{\mathbf{d}}(Q) \to \mathcal{M}_{\mathbf{d}}(Q)$$

which, at the level of points, takes modules to their semisimplifications, and at the level of algebraic geometry is just the affinization morphism. This extends naturally to a morphism

$$(13) \qquad \mathtt{JH}^T : \mathfrak{M}^T_{\mathbf{d}}(Q) \to \mathcal{M}^T_{\mathbf{d}}(Q).$$

Precisely, if the morphism $f\colon X \to \mathfrak{M}^T_{\mathbf{d}}(Q)$ is represented by the T-torsor $E \to X$ and the morphism $\tilde{f}\colon E \to \mathfrak{M}_{\mathbf{d}}(Q)$, the morphism $\mathtt{JH}^T \circ f$ is represented by the same torsor E, along with the morphism $\mathtt{JH} \circ \tilde{f}$.

Given two dimension vectors $\mathbf{d}', \mathbf{d}'' \in \mathbb{N}^{Q_0}$ with $\mathbf{d} = \mathbf{d}' + \mathbf{d}''$ we define $\mathbb{A}_{Q,\mathbf{d}',\mathbf{d}''} \subset \mathbb{A}_{Q,\mathbf{d}}$ and $\mathrm{P}_{\mathbf{d}',\mathbf{d}''} \subset \mathrm{GL}_{\mathbf{d}}$ to be the space of representations, and changes of basis, respectively, preserving the flags $0 \subset \mathbb{C}^{\mathbf{d}'(i)} \subset \mathbb{C}^{\mathbf{d}(i)}$ for each $i \in Q_0$. Denoting by $\mathfrak{M}_{\mathbf{d}',\mathbf{d}''}(Q)$ the stack of short exact sequences $0 \to \rho' \to \rho \to \rho'' \to 0$ of $\mathbb{C}Q$-modules, for which the dimension vectors of ρ' and ρ'' are $\mathbf{d}', \mathbf{d}''$ respectively, there is a natural isomorphism

$$\mathfrak{M}_{\mathbf{d}',\mathbf{d}''}(Q) \cong \mathbb{A}_{Q,\mathbf{d}',\mathbf{d}''} / \mathrm{GL}_{\mathbf{d}',\mathbf{d}''}.$$

We define $\mathfrak{M}^T_{\mathbf{d}',\mathbf{d}''}(Q) := \mathbb{A}_{Q',\mathbf{d}',\mathbf{d}''}/(\mathrm{GL}_{\mathbf{d}',\mathbf{d}''} \times T)$.

3.4. Cohomological Hall algebras. Let Q be a quiver. Let

$$W \in \mathbb{C}Q/[\mathbb{C}Q, \mathbb{C}Q]_{\mathrm{vect}}$$

be a potential, which we assume to be T-invariant in the following sense: for every cyclic word $a_1 \ldots a_c$ appearing in W

$$\sum_{i=1}^{c} \tau(a_i) = 0. \tag{14}$$

The potential W induces a function $\mathrm{Tr}(W)$ on $\mathbb{A}_{Q,\mathbf{d}}$ which is $\mathrm{GL}_{\mathbf{d}}$-invariant by cyclic invariance of the trace map, and T-invariant by (14). We define

$$\mathcal{A}^T_{Q,W} := \bigoplus_{\mathbf{d} \in \mathbb{N}^{Q_0}} \mathbf{H}(\mathfrak{M}^T_{\mathbf{d}}(Q), {}^{\mathfrak{p}}\phi_{\mathrm{Tr}(W)}) \tag{15}$$

the underlying $\mathbb{N}^{Q_0}$-graded vector space of the critical cohomological Hall algebra, defined by Kontsevich and Soibelman in [**KS11**]. Consider the convolution diagram

$$\mathfrak{M}^T_{\mathbf{d}'}(Q) \times_{\mathrm{B}T} \mathfrak{M}^T_{\mathbf{d}''}(Q) \xleftarrow{\pi_1 \times \pi_3} \mathfrak{M}^T_{\mathbf{d}',\mathbf{d}''}(Q) \xrightarrow{\pi_2} \mathfrak{M}^T_{\mathbf{d}}(Q)$$

where π_1, π_2, π_3 take a flag $0 \subset \mathcal{F}' \subset \mathcal{F}$ of flat families of $\mathbb{C}Q$-modules to the families $\mathcal{F}'$, $\mathcal{F}$, $\mathcal{F}/\mathcal{F}'$ respectively. We define

$$\begin{aligned} m'_{\mathbf{d}',\mathbf{d}''}\colon \mathbf{H}\left(\mathfrak{M}^T_{\mathbf{d}'}(Q) \times_{\mathrm{B}T} \mathfrak{M}^T_{\mathbf{d}''}(Q), {}^{\mathfrak{p}}\phi_{\mathrm{Tr}(W) \boxplus \mathrm{Tr}(W)}\right) &\to \mathbf{H}\left(\mathfrak{M}^T_{\mathbf{d}}(Q), {}^{\mathfrak{p}}\phi_{\mathrm{Tr}(W)}\right) \\ &= \pi_{2,\star} \circ (\pi_1 \times \pi_3)^{\star} \end{aligned} \tag{16}$$

and $m_{\mathbf{d}',\mathbf{d}''} = m'_{\mathbf{d}',\mathbf{d}''} \circ \jmath^{\star} \circ \mathtt{TS}$ where $\jmath$ is as in (12) and $\mathtt{TS}$ is the Thom–Sebastiani isomorphism. We then define the multiplication

$$m\colon \mathcal{A}^T_{Q,W} \otimes \mathcal{A}^T_{Q,W} \to \mathcal{A}^T_{Q,W}$$

by summing $m_{\mathbf{d}',\mathbf{d}''}$ over all pairs of dimension vectors.

The multiplication m makes $\mathcal{A}^T_{Q,W}$ into a $\mathbb{N}^{Q_0}$-graded associative algebra. If we assume that Q is *symmetric*, meaning that for every pair of vertices $i, j \in Q_0$ there are as many arrows from i to j as there are from j to i, then the morphism m preserves the cohomological degree, and $\mathcal{A}^T_{Q,W}$ is an associative algebra in the category of $\mathbb{N}^{Q_0} \oplus \mathbb{Z}$-graded vector spaces.

Let $\mathcal{F}$ be a flat family of $\mathbf{d}$-dimensional $\mathbb{C}Q$-modules on a scheme X. For each $i \in Q_0$ we obtain a vector bundle $\mathcal{F}_i = e_i \cdot \mathcal{F}$ of rank $\mathbf{d}_i$. There is a determinant line bundle $\mathrm{Det}(\mathcal{F}_i)$ on X, which is defined to be the top exterior power of the underlying bundle of $\mathcal{F}_i$. This defines a morphism of stacks for each $i \in Q_0$:

$$\mathrm{Det}_{(i)}\colon \mathfrak{M}^T_{\mathbf{d}}(Q) \to \mathrm{B}\mathbb{C}^*_u.$$

Alternatively, we can construct this morphism as the composition

$$\begin{aligned}\mathrm{Det}_{(i)} = &(\mathrm{Det}\colon \mathrm{B}\,\mathrm{GL}_{\mathbf{d}_i} \to \mathrm{B}\mathbb{C}^*_u) \circ \left(\pi_{\mathrm{pt}\,/\,\mathrm{GL}_{\mathbf{d}_i}}\colon \mathrm{pt}\,/(\mathrm{GL}_{\mathbf{d}_i} \times T) \to \mathrm{B}\,\mathrm{GL}_{\mathbf{d}_i}\right)\\ &\circ (\mathbb{A}_{Q,\mathbf{d}}/(\mathrm{GL}_{\mathbf{d}} \times T) \to \mathrm{pt}\,/(\mathrm{GL}_{\mathbf{d}} \times T)).\end{aligned}$$

We denote by $u_{(i)} \in \mathbf{H}(\mathfrak{M}^T_{\mathbf{d}}(Q), \mathbb{Q})$ the image of $u \in \mathbf{H}_{\mathbb{C}^*_u}$ under $\mathrm{Det}^\star_{(i)}$. Consider the morphism

$$l_i = (\mathrm{Det}_{(i)}, \mathrm{id})\colon \mathfrak{M}^T_{\mathbf{d}}(Q) \to \mathrm{B}\mathbb{C}^*_u \times \mathfrak{M}^T_{\mathbf{d}}(Q).$$

Since the function $\mathrm{Tr}(W)$ is pulled back from the function $0 \boxplus \mathrm{Tr}(W)$ on the target of l_i we obtain the morphism in vanishing cycle cohomology

$$\mathbf{H}(\mathrm{B}\mathbb{C}^*_u, \mathbb{Q}) \otimes \mathbf{H}(\mathfrak{M}^T_{\mathbf{d}}(Q), {}^{\mathfrak{p}}\phi_{\mathrm{Tr}(W)}) \xrightarrow{l_i^\star} \mathbf{H}(\mathfrak{M}^T_{\mathbf{d}}(Q), {}^{\mathfrak{p}}\phi_{\mathrm{Tr}(W)})$$

which defines an action of the algebra $\mathbb{Q}[u] = \mathbf{H}_{\mathbb{C}^*_u}$ on $\mathcal{A}^T_{Q,W}$, for each $i \in Q_0$:

$$p(u) \bullet_i \alpha := l_i^\star(p(u) \otimes \alpha). \tag{17}$$

Similarly, we define the morphism

$$l = (\bigotimes_{i \in Q_0} \mathrm{Det}_{(i)}, \mathrm{id})\colon \mathfrak{M}^T_{\mathbf{d}}(Q) \to \mathrm{B}\mathbb{C}^*_u \times \mathfrak{M}^T_{\mathbf{d}}(Q) \tag{18}$$

and an action of $\mathbf{H}_{\mathbb{C}^*_u}$ on $\mathcal{A}^T_{Q,W}$ by

$$p(u) \bullet \alpha = l^\star(p(u) \otimes \alpha). \tag{19}$$

3.5. PBW isomorphism. In this section we assume $T = \{1\}$ for simplicity, and also assume that Q is symmetric. It follows that there exists a (non-unique) bilinear form

$$\psi\colon \mathbb{Z}^{Q_0} \times \mathbb{Z}^{Q_0} \to \mathbb{Z}^{Q_0}$$

satisfying

$$\psi(\mathbf{d}', \mathbf{d}'') + \psi(\mathbf{d}'', \mathbf{d}') = \chi_Q(\mathbf{d}', \mathbf{d}')\chi_Q(\mathbf{d}'', \mathbf{d}'') + \chi_Q(\mathbf{d}', \mathbf{d}'') \bmod 2$$

for all $\mathbf{d}', \mathbf{d}''$. We fix a choice of such ψ in what follows, and denote by $\mathcal{A}^\psi_{Q,W}$ the cohomological Hall algebra $\mathcal{A}_{Q,W}$ with the multiplication twisted by setting

$$m^{\mathtt{tw}}_{\mathbf{d}',\mathbf{d}''} = (-1)^{\psi(\mathbf{d}',\mathbf{d}'')} m_{\mathbf{d}',\mathbf{d}''}$$

where $m_{\mathbf{d}',\mathbf{d}''}$ and $m^{\mathtt{tw}}_{\mathbf{d}',\mathbf{d}''}$ are the restrictions of the products in $\mathcal{A}_{Q,W}$ (resp. $\mathcal{A}^\psi_{Q,W}$) to the pieces of degree $\mathbf{d}', \mathbf{d}''$. In our main application, we will have that $Q = Q^{(3)}$, so that χ_Q is even, and we can (and will) set $\psi = 0$.

REMARK 3.2. In what follows, if we write that something is true of $\mathcal{A}^{(\psi)}_{Q,W}$, we mean that it is true of both $\mathcal{A}^\psi_{Q,W}$ and $\mathcal{A}_{Q,W}$. Note that the underlying graded vector spaces of these algebras are identical.

By [**DM20**] the direct image $\mathrm{JH}_* {}^{\mathfrak{p}}\phi_{\mathrm{Tr}(W)}\mathbb{Q}^{\mathrm{vir}}_{\mathfrak{M}_{\mathbf{d}}(Q)}$ splits: there is an isomorphism in $\mathcal{D}^+(\mathbf{Perv}(\mathcal{M}_{\mathbf{d}}(Q)))$:

$$\mathrm{JH}_* {}^{\mathfrak{p}}\phi_{\mathrm{Tr}(W)}\mathbb{Q}^{\mathrm{vir}}_{\mathfrak{M}_{\mathbf{d}}(Q)} \cong \bigoplus_{i\geq 1} {}^{\mathfrak{p}}\mathcal{H}^i(\mathrm{JH}_* {}^{\mathfrak{p}}\phi_{\mathrm{Tr}(W)}\mathbb{Q}^{\mathrm{vir}}_{\mathfrak{M}_{\mathbf{d}}(Q)})[-i]. \tag{20}$$

There is a natural morphism

$$\mathfrak{P}_i \mathcal{A}^{(\psi)}_{Q,W,\mathbf{d}} := \mathbf{H}(\mathcal{M}_{\mathbf{d}}(Q), {}^{\mathfrak{p}}\tau^{\leq i}\mathrm{JH}_* {}^{\mathfrak{p}}\phi_{\mathrm{Tr}(W)}\mathbb{Q}^{\mathrm{vir}}_{\mathfrak{M}_{\mathbf{d}}(Q)}) \to \mathcal{A}^{(\psi)}_{Q,W,\mathbf{d}} \tag{21}$$

induced by the natural transformation ${}^{\mathfrak{p}}\tau^{\leq i} \to \mathrm{id}$ and it follows from the splitting (20) that (21) is injective, and the subspaces $\mathfrak{P}_i \mathcal{A}^{(\psi)}_{Q,W,\mathbf{d}}$ provide a filtration of $\mathcal{A}^{(\psi)}_{Q,W,\mathbf{d}}$, called the *perverse filtration.*

By [**DM20**, Sec.5.3] the multiplication on $\mathcal{A}^{(\psi)}_{Q,W}$ respects the perverse filtration. There is vanishing

$$\mathfrak{P}_0\mathcal{A}^{(\psi)}_{Q,W} = 0, \tag{22}$$

and furthermore the subspace

$$\mathfrak{g}_{Q,W} := \mathfrak{P}_1\mathcal{A}^{\psi}_{Q,W} \subset \mathcal{A}^{\psi}_{Q,W}$$

is preserved by the commutator Lie bracket $[-,-]$ by [**DM20**, Cor.6.11] (here is where it starts to become necessary to choose the ψ-twisted multiplication). There are isomorphisms

$$\mathfrak{g}_{Q,W,\mathbf{d}} = \mathbf{H}(\mathcal{M}_{\mathbf{d}}(Q), \mathcal{BPS}_{Q,W,\mathbf{d}})[-1]$$

where $\mathcal{BPS}_{Q,W,\mathbf{d}} = ({}^{\mathfrak{p}}\tau^{\leq 1}\mathrm{JH}_* {}^{\mathfrak{p}}\phi_{\mathrm{Tr}(W)}\mathbb{Q}^{\mathrm{vir}}_{\mathfrak{M}_{\mathbf{d}}(Q)})[1]$ is the *BPS sheaf* defined in [**DM20**, Theorem A]. This is a perverse sheaf, with a natural lift to the category of mixed Hodge modules on $\mathcal{M}_{\mathbf{d}}(Q)$, so that $\mathfrak{g}_{Q,W}$ carries a natural mixed Hodge structure, which is respected by the Lie algebra structure on it; we refer to [**DM20**, Section 2] for details. We call $\mathfrak{g}_{Q,W}$ the *BPS Lie algebra* associated to (Q, W). It inherits a cohomological grading from $\mathcal{A}^{\psi}_{Q,W}$, and for $i \in \mathbb{Z}$ we denote by $\mathfrak{g}^i_{Q,W}$ the summand of cohomological degree i.

THEOREM 3.3 ([**DM20**, Theorem C]). *The PBW morphism*

$$\Psi\colon \mathbf{Sym}(\mathfrak{g}_{Q,W} \otimes \mathbf{H}_{\mathbb{C}^*_u}) \to \mathcal{A}^{\psi}_{Q,W} \tag{23}$$

is an isomorphism of cohomologically graded $\mathbb{N}^{Q_0}$-graded vector spaces.

The morphism Ψ is constructed as follows. Firstly, we consider the composition

$$\Psi'\colon \mathbf{H}(\mathrm{B}\mathbb{C}^*_u, \mathbb{Q}) \otimes \mathfrak{g}_{Q,W} \hookrightarrow \mathbf{H}(\mathrm{B}\mathbb{C}^*_u, \mathbb{Q}) \otimes \mathcal{A}^{\psi}_{Q,W} \xrightarrow{l^*} \mathcal{A}^{\psi}_{Q,W}$$

with l as defined in (18). Since the target carries an algebra structure, Ψ' extends uniquely to a morphism of algebras

$$\Phi\colon \mathrm{T}\left(\mathbf{H}(\mathrm{B}\mathbb{C}^*_u, \mathbb{Q}) \otimes \mathfrak{g}_{Q,W}\right) \to \mathcal{A}^{\psi}_{Q,W}$$

from the free associative algebra generated by the domain of Ψ'. Then Ψ is defined to be the restriction of Φ to the subspace of (supersymmetric) tensors. Note that the morphism Ψ will typically not be a morphism of algebras; the CoHA $\mathcal{A}^{\psi}_{Q,W}$ will in general have an interesting algebra structure, and not be a free (super)commutative algebra.

3.6. Deformed BPS Lie algebras. Let Q be a quiver, which we continue to assume is symmetric. Let T be a torus acting on the arrows of Q, and let $W \in \mathbb{C}Q/[\mathbb{C}Q, \mathbb{C}Q]_{\mathrm{vect}}$ be a T-invariant potential. Recall the map $\mathtt{JH}^T$ from (13). We define[6] the $\mathbf{H}_T$-module

$$\mathfrak{g}^T_{Q,W,\mathbf{d}} := \mathbf{H}(\mathcal{M}^T_{\mathbf{d}}(Q), {}^{\mathfrak{p}'}\tau^{\leq 1}\mathtt{JH}^T_* {}^{\mathfrak{p}}\phi_{\mathrm{Tr}(W)}\mathbb{Q}^{\mathrm{vir}}_{\mathfrak{M}^T_{\mathbf{d}}(Q)}).$$

Again, this cohomologically graded vector space admits a natural lift to the category of cohomologically graded mixed Hodge structures, for which we refer to [**DM20**, Sec.2] for details. Applying the derived global sections functor to the natural morphism of complexes ${}^{\mathfrak{p}'}\tau^{\leq 1}\mathtt{JH}^T_* {}^{\mathfrak{p}}\phi_{\mathrm{Tr}(W)}\mathbb{Q}^{\mathrm{vir}}_{\mathfrak{M}^T_{\mathbf{d}}(Q)} \to \mathtt{JH}^T_* {}^{\mathfrak{p}}\phi_{\mathrm{Tr}(W)}\mathbb{Q}^{\mathrm{vir}}_{\mathfrak{M}^T_{\mathbf{d}}(Q)}$ we obtain the morphism

$$\iota\colon \mathfrak{g}^T_{Q,W,\mathbf{d}} \to \mathcal{A}^{T,\psi}_{Q,W,\mathbf{d}}\,.$$

THEOREM 3.4. *The morphism ι is injective, so that there is a natural inclusion of $\mathbb{N}^{Q_0}$-graded subspaces $\mathfrak{g}^T_{Q,W} \subset \mathcal{A}^{T,\psi}_{Q,W}$. The subspace $\mathfrak{g}^T_{Q,W}$ is closed under the commutator Lie bracket on $\mathcal{A}^{T,\psi}_{Q,W}$. Assume, furthermore, that $\mathfrak{g}^T_{Q,W}$ is pure, as a mixed Hodge structure. Then there is an isomorphism of vector spaces*

$$\mathfrak{g}^T_{Q,W} \cong \mathfrak{g}_{Q,W} \otimes \mathbf{H}_T, \tag{24}$$

and the PBW morphism

$$\mathbf{Sym}_{\mathbf{H}_T}(\mathfrak{g}^T_{Q,W} \otimes \mathbf{H}_{\mathbb{C}^*_u}) \to \mathcal{A}^{T,\psi}_{Q,W}$$

constructed from the Hall algebra product on the target is an isomorphism.

PROOF. The proof is a minor modification of the analogous statements from [**DM20**], to which we refer for more details. Firstly, we claim that $\mathtt{JH}^T$ is approximated by projective morphisms in the sense of [**DM20**, Sec.4.1]. For this the proof is as in [**DM20**], using the quotient of moduli spaces of framed $\mathbb{C}Q$-modules by the T-action as auxiliary spaces. From this it follows as in [**DM20**, Prop.4.4] that there is a splitting

$$\mathbf{H}(\mathcal{M}^T_{\mathbf{d}}(Q), \mathtt{JH}^T_* {}^{\mathfrak{p}}\phi_{\mathrm{Tr}(W)}\mathbb{Q}^{\mathrm{vir}}_{\mathfrak{M}^T_{\mathbf{d}}(Q)}) \cong \bigoplus_{i\in\mathbb{Z}} \mathbf{H}(\mathcal{M}^T_{\mathbf{d}}(Q), {}^{\mathfrak{p}}\mathcal{H}^i\left(\mathtt{JH}^T_* {}^{\mathfrak{p}}\phi_{\mathrm{Tr}(W)}\mathbb{Q}^{\mathrm{vir}}_{\mathfrak{M}^T_{\mathbf{d}}(Q)}\right)[-i])$$

and the morphism

$$\mathbf{H}(\mathcal{M}^T_{\mathbf{d}}(Q), {}^{\mathfrak{p}'}\tau^{\leq 1}\mathtt{JH}^T_* {}^{\mathfrak{p}}\phi_{\mathrm{Tr}(W)}\mathbb{Q}^{\mathrm{vir}}_{\mathfrak{M}^T_{\mathbf{d}}(Q)}) \to \mathbf{H}(\mathcal{M}^T_{\mathbf{d}}(Q), \mathtt{JH}^T_* {}^{\mathfrak{p}}\phi_{\mathrm{Tr}(W)}\mathbb{Q}^{\mathrm{vir}}_{\mathfrak{M}^T_{\mathbf{d}}(Q)})$$

has a left inverse, and the injectivity part of the theorem follows.

The monoid morphism $\mathcal{M}(Q) \times \mathcal{M}(Q) \to \mathcal{M}(Q)$ taking a pair of semisimple $\mathbb{C}Q$-modules to their direct sum is T-equivariant, so that we can define the monoid structure

$$\oplus\colon \mathcal{M}^T(Q) \times_{\mathrm{B}T} \mathcal{M}^T(Q) \to \mathcal{M}^T(Q) \tag{25}$$

by passing to the stack-theoretic quotient by the T-action. Given a pair of complexes of sheaves $\mathcal{F}$ and $\mathcal{G}$ on two stacks $\mathfrak{X}$ and $\mathfrak{Y}$ over a stack $\mathfrak{Z}$ we define the external product, a complex of sheaves on $\mathfrak{X} \times_{\mathfrak{Z}} \mathfrak{Y}$ in the usual way:

$$\mathcal{F} \boxtimes_{\mathfrak{Z}} \mathcal{G} := \pi^*_{\mathfrak{X}}\mathcal{F} \otimes \pi^*_{\mathfrak{Y}}\mathcal{G}.$$

[6]The perverse truncation functors are modified according to the conventions introduced in §3.1.

This defines a symmetric tensor product on complexes of sheaves on $\mathcal{M}^T(Q)$: we define

$$\mathcal{F} \boxtimes_{\oplus} \mathcal{G} = \oplus_*(\mathcal{F} \boxtimes_{\mathrm{B}T} \mathcal{G}).$$

Let $p\colon \mathcal{M}^T(Q) \to \mathrm{B}T$ be the canonical map. This is a morphism of monoid objects over $\mathrm{B}T$, where $\mathrm{B}T$ is considered as a monoid over itself via the identity morphism. For arbitrary complexes of sheaves on $\mathcal{M}^T(Q)$ it is too much to hope that there is an isomorphism

$$\mathbf{H}(\mathcal{M}^T(Q), \mathcal{F} \boxtimes_{\oplus} \mathcal{G}) \cong \mathbf{H}(\mathcal{M}^T(Q), \mathcal{F}) \otimes_{\mathbf{H}_T} \mathbf{H}(\mathcal{M}^T(Q), \mathcal{G}). \tag{26}$$

However, let us assume that the complexes $p_*\mathcal{F}$ and $p_*\mathcal{G}$ lift to pure complexes of mixed Hodge modules. Then via the usual spectral sequence argument (see [**Dav23b**, Sec.9]), there are isomorphisms

$$\begin{aligned}
\mathbf{H}(\mathcal{M}^T(Q), \mathcal{F}) &\cong \mathbf{H}(\mathcal{M}(Q), \overline{\mathcal{F}}) \otimes \mathbf{H}_T \\
\mathbf{H}(\mathcal{M}^T(Q), \mathcal{G}) &\cong \mathbf{H}(\mathcal{M}(Q), \overline{\mathcal{G}}) \otimes \mathbf{H}_T \\
\mathbf{H}(\mathcal{M}^T(Q), \mathcal{F} \boxtimes_{\oplus} \mathcal{G}) &\cong (\mathbf{H}(\mathcal{M}(Q), \overline{\mathcal{F}}) \otimes \mathbf{H}(\mathcal{M}(Q), \overline{\mathcal{G}})) \otimes \mathbf{H}_T
\end{aligned}$$

where $\overline{\mathcal{F}}$ and $\overline{\mathcal{G}}$ denote the inverse images of $\mathcal{F}$ and $\mathcal{G}$ along the canonical morphism $\mathcal{M}(Q) \to \mathcal{M}^T(Q)$, and a *natural* isomorphism (26).

Via the same morphisms as in [**DM20**, Sec.5.1] (see also §3.4), the complex $\mathrm{JH}^T_* {}^{\mathfrak{p}}\phi_{\mathrm{Tr}(W)} \mathbb{Q}^{\mathrm{vir}}_{\mathfrak{M}^T(Q)}$ carries a relative Hall algebra structure, i.e. for $\mathbf{d} = \mathbf{d}' + \mathbf{d}''$ there is a morphism

$$\mathrm{JH}^T_* {}^{\mathfrak{p}}\phi_{\mathrm{Tr}(W)} \mathbb{Q}^{\mathrm{vir}}_{\mathfrak{M}^T_{\mathbf{d}'}(Q)} \boxtimes_{\oplus} \mathrm{JH}^T_* {}^{\mathfrak{p}}\phi_{\mathrm{Tr}(W)} \mathbb{Q}^{\mathrm{vir}}_{\mathfrak{M}^T_{\mathbf{d}''}(Q)} \to \mathrm{JH}^T_* {}^{\mathfrak{p}}\phi_{\mathrm{Tr}(W)} \mathbb{Q}^{\mathrm{vir}}_{\mathfrak{M}^T_{\mathbf{d}}(Q)}$$

satisfying the obvious associativity condition. We claim that

- The perverse truncation ${}^{\mathfrak{p}'}\tau^{\leq 0} \mathrm{JH}^T_* {}^{\mathfrak{p}}\phi_{\mathrm{Tr}(W)} \mathbb{Q}^{\mathrm{vir}}_{\mathfrak{M}^T_{\mathbf{d}}(Q)}$ vanishes.
- The morphism

$$\begin{aligned}
&{}^{\mathfrak{p}'}\tau^{\leq 1} \mathrm{JH}^T_* {}^{\mathfrak{p}}\phi_{\mathrm{Tr}(W)} \mathbb{Q}^{\mathrm{vir}}_{\mathfrak{M}^T_{\mathbf{d}}(Q)}[1] \boxtimes_{\oplus} {}^{\mathfrak{p}'}\tau^{\leq 1} \mathrm{JH}^T_* {}^{\mathfrak{p}}\phi_{\mathrm{Tr}(W)} \mathbb{Q}^{\mathrm{vir}}_{\mathfrak{M}^T_{\mathbf{d}}(Q)}[1] \\
&\qquad \to {}^{\mathfrak{p}'}\mathcal{H}^2\left(\mathrm{JH}^T_* {}^{\mathfrak{p}}\phi_{\mathrm{Tr}(W)} \mathbb{Q}^{\mathrm{vir}}_{\mathfrak{M}^T_{\mathbf{d}}(Q)}\right)
\end{aligned}$$

  given by restricting the commutator bracket on the relative CoHA $\mathrm{JH}^T_* {}^{\mathfrak{p}}\phi_{\mathrm{Tr}(W)} \mathbb{Q}^{\mathrm{vir}}_{\mathfrak{M}^T_{\mathbf{d}}(Q)}$ (with sign twisted by ψ) vanishes.

Taking derived global sections, these two claims prove that $\mathfrak{g}^T_{Q,W}$ is closed under the commutator Lie bracket. The first claim concerns the vanishing of certain perverse sheaves on $\mathcal{M}^T(Q)$, and the second concerns the vanishing of certain morphisms of perverse sheaves on $\mathcal{M}^T(Q)$. It follows from [**BBDG18**, Sec.4] that the forgetful functor $\mathbf{Perv}(\mathcal{M}^T(Q)) \to \mathbf{Perv}(\mathcal{M}(Q))$ is faithful, so both these claims reduce to the analogous claims on $\mathcal{M}(Q)$, which are proved as [**DM20**, Thm.A], [**DM20**, Cor.6.11] respectively.

The isomorphism (24) follows from the usual spectral sequence argument (as in [**Dav23b**, Sec.9.2]). Since p is a morphism of monoids over $\mathrm{B}T$, the complex $p_*\mathrm{JH}^T_* {}^{\mathfrak{p}}\phi_{\mathrm{Tr}(W)} \mathbb{Q}^{\mathrm{vir}}_{\mathfrak{M}^T(Q)}$ inherits a (ψ-twisted) Hall algebra structure, along with an action of $\mathbf{H}_{\mathbb{C}^*_u}$. We claim that the morphism of complexes on $\mathrm{B}T$

$$\mathbf{Sym}_{\boxtimes_{\mathrm{B}T}}\left(p_* {}^{\mathfrak{p}'}\tau^{\leq 1} \mathrm{JH}^T_* {}^{\mathfrak{p}}\phi_{\mathrm{Tr}(W)} \mathbb{Q}^{\mathrm{vir}}_{\mathfrak{M}^T(Q)} \otimes \mathbf{H}_{\mathbb{C}^*_u}\right) \to p_* \mathrm{JH}^T {}^{\mathfrak{p}}\phi_{\mathrm{Tr}(W)} \mathbb{Q}^{\mathrm{vir}}_{\mathfrak{M}^T(Q)}$$

is an isomorphism. This may be checked one perverse degree at a time; then using the fact that the forgetful morphism $\mathbf{Perv}^T(\mathrm{pt}) \to \mathbf{Perv}(\mathrm{pt})$ is faithful, the statement reduces to the non-equivariant PBW theorem (Theorem 3.3). The equivariant PBW theorem then follows. □

We refer to $\mathfrak{g}^T_{Q,W}$ as the *deformed* BPS Lie algebra. By the above theorem, as long as $\mathfrak{g}_{Q,W}$ is pure, this deformed Lie algebra provides a flat family of Lie algebras over $\mathfrak{t} \cong \mathrm{Spec}(\mathbf{H}_T)$ specialising to the original BPS Lie algebra $\mathfrak{g}_{Q,W}$ at the origin.

3.7. Kac polynomials and Kac–Moody Lie algebras. Given a quiver Q without loops we denote by $\mathfrak{n}^-_Q$ the negative half of the associated Kac–Moody Lie algebra. This is the free $\mathbb{N}^{Q_0}$-graded Lie algebra with generators F_i in $\mathbb{N}^{Q_0}$-degree 1_i, subject to the Serre relations

$$[F_i, -]^{a_{ij}+1}(F_j) = 0$$

for $i \neq j$. Here a_{ij} denotes the number of edges joining i and j in the graph obtained from Q by forgetting orientations of edges. If the underlying graph of Q is a type ADE Dynkin diagram, $\mathfrak{n}^-_Q$ is (one half of) the associated simple Lie algebra.

Given a quiver Q and a field K, a KQ-module ρ is called indecomposable if it cannot be written as $\rho = \rho' \oplus \rho''$ with $\rho', \rho'' \neq 0$. It is called absolutely indecomposable if $\rho \otimes_K \overline{K}$ is an indecomposable $\overline{K}Q$-module. For $q = p^n$ a prime power, let $\mathtt{a}_{Q,\mathbf{d}}(q)$ denote the number of isomorphism classes of absolutely indecomposable $\mathbf{d}$-dimensional $\mathbb{F}_qQ$-modules. Victor Kac showed [**Kac83**] that $\mathtt{a}_{Q,\mathbf{d}}(q)$ is a polynomial in q, and moreover an element of $\mathbb{Z}[q]$.

If Q is an orientation of a finite ADE graph, then

$$\mathtt{a}_{Q,\mathbf{d}}(q) = \dim(\mathfrak{n}^-_{Q,\mathbf{d}}). \tag{27}$$

In particular, it is constant. Conversely, if Q is any other type of quiver, $\mathtt{a}_{Q,\mathbf{d}}(q)$ is nonconstant for at least some values of $\mathbf{d}$.

3.8. Preprojective CoHAs. We consider the class of special cases of quiver with potential $(Q, W) = (\tilde{Q}', \tilde{W})$ for a finite quiver Q'. Although we do not assume that Q' is symmetric, Q obviously is. We call these CoHAs *preprojective CoHAs*, for reasons that will become evident in §3.9. By [**Dav23b**, Thm.4.7] there is an equality of generating series

$$\sum_{i\in\mathbb{Z}} \dim(\mathfrak{g}^i_{Q,\tilde{W},\mathbf{d}})q^{i/2} = \mathtt{a}_{Q',\mathbf{d}}(q^{-1}). \tag{28}$$

where $\mathtt{a}_{Q',\mathbf{d}}(q)$ is the Kac polynomial of Q'. So in particular, $\mathfrak{g}_{Q,\tilde{W}}$ is concentrated in even, negative cohomological degrees.

PROPOSITION 3.5 ([**Dav25**, Thm.6.6]). *Let Q' be a finite quiver, and denote by Q^{re} the quiver obtained from Q' by removing all vertices that support a loop, along with any arrows beginning or ending at such a vertex. Set $Q = \tilde{Q}'$. There is an isomorphism of $\mathbb{N}^{Q_0}$-graded Lie algebras*

$$\mathfrak{g}^0_{Q,\tilde{W}} \cong \mathfrak{n}^-_{Q^{\mathrm{re}}}. \tag{29}$$

The Lie algebra on the right hand side a priori carries a $\mathbb{N}^{Q_0^{\mathrm{re}}}$-grading. We promote this to a $\mathbb{N}^{Q_0}$-grading via extension by zero along the embedding $\mathbb{N}^{Q_0^{\mathrm{re}}} \subset \mathbb{N}^{Q_0}$ induced by the inclusion $Q_0^{\mathrm{re}} \subset Q_0$.

Choose a torus $T = (\mathbb{C}^*)^l$, and a weighting function $\tau\colon Q_1 \to \mathbb{Z}^l$ defining an action of T on the category of $\mathbb{C}Q$-modules, for which $\tilde{W}$ is invariant. The following theorem is a consequence of the *purity* of the natural mixed Hodge structure on $\mathcal{A}_{Q,\tilde{W}}$, see [**Dav23b**] for details.

PROPOSITION 3.6 ([**Dav23b**, Theorem 9.6]). *Let $l\colon \mathbb{Z}^r \to \mathbb{Z}^{r'}$ be a split surjection of lattices, with complement $\mathbb{Z}^{r''}$, inducing the split inclusion of tori*

$$T' = \mathrm{Hom}_{\mathrm{Grp}}(\mathbb{Z}^{r'}, \mathbb{C}^*) \hookrightarrow T = \mathrm{Hom}_{\mathrm{Grp}}(\mathbb{Z}^r, \mathbb{C}^*)$$

with complement $T'' = \mathrm{Hom}_{\mathrm{Grp}}(\mathbb{Z}^{r''}, \mathbb{C}^*)$. There is an isomorphism of $\mathbb{N}^{Q_0}$-graded, cohomologically graded $\mathbf{H}_T$-modules[7]

$$\mathcal{A}^T_{Q,\tilde{W}} \cong \mathcal{A}^{T'}_{Q,\tilde{W}} \otimes \mathbf{H}_{T''} \tag{30}$$

and an isomorphism of algebras

$$\mathcal{A}^{T',(\psi)}_{Q,\tilde{W}} \cong \mathcal{A}^{T,(\psi)}_{Q,\tilde{W}} \otimes_{\mathbf{H}_T} \mathbf{H}_{T'}. \tag{31}$$

For the special case $r' = 0$, the above proposition yields an isomorphism of $\mathbf{H}_T$-modules

$$\mathcal{A}^T_{Q,\tilde{W}} \cong \mathcal{A}_{Q,\tilde{W}} \otimes \mathbf{H}_T \tag{32}$$

and an isomorphism of algebras

$$\mathcal{A}^{(\psi)}_{Q,\tilde{W}} \cong \mathcal{A}^{T,(\psi)}_{Q,\tilde{W}} / \mathfrak{m}_T \cdot \mathcal{A}^{T,(\psi)}_{Q,\tilde{W}},$$

where we denote by $\mathfrak{m}_T$ the maximal homogeneous ideal in $\mathbf{H}_T$. We will be particularly interested in the action of the torus $T = (\mathbb{C}^*)^2 = \mathrm{Hom}_{\mathrm{Grp}}(\mathbb{Z}^2, \mathbb{C}^*)$ acting via the weighting function

$$\begin{aligned} \tilde{\tau}\colon Q_1 &\to \mathbb{Z}^2 \\ a \mapsto (1,0); \qquad a^* &\mapsto (0,1); \qquad \omega_i \mapsto (-1,-1). \end{aligned} \tag{33}$$

3.9. Dimensional reduction. We recall from [**Dav17**, Appendix.A] the following result.

THEOREM 3.7 ([**Dav17**, Thmeorem A.1]). *Let $X = Y \times \mathbb{A}^n$ be a smooth variety, let f be a function on X of weight one for the scaling action on $\mathbb{A}^n$, so that we can write*

$$f = \sum_{i=1}^n x_i f_i$$

for $x_1, \ldots, x_n$ a system of coordinates on $\mathbb{A}^n$ and $f_1, \ldots, f_n$ functions on Y. Set $Z = Z(f_1, \ldots, f_n)$, let $i\colon Z \hookrightarrow Y$ be the inclusion, and let $\pi\colon X \to Y$ be the projection. Then there is a natural isomorphism

$$i_* i^! \mathcal{F} \to \pi_* {}^{\mathfrak{p}}\phi_f \pi^* \mathcal{F}.$$

Applying the result to $\mathbb{D}\mathbb{Q}_Y$ and passing to global sections yields the isomorphism

$$\mathbf{H}^{\mathrm{BM}}(Z, \mathbb{Q}) \to \mathbf{H}(X, {}^{\mathfrak{p}}\phi_f \mathbb{Q})[2\dim(Y)].$$

The theorem extends in the obvious way to affine fibrations over stacks [**Dav17**, Corollary A.9]. In particular, for a quiver Q (not assumed symmetric) and the affine fibration

$$\mathfrak{M}_{\mathbf{d}}(\tilde{Q}) \to \mathfrak{M}_{\mathbf{d}}(\overline{Q})$$

[7] We omit the ψ superscript because the ψ-twist does not feature in the $\mathbf{H}_T$-module structure.

the function $\mathrm{Tr}(\tilde{W})$ has weight one with respect to the action that scales the fibres (i.e. the space of choices of matrices for each of the loops ω_i). As such, we obtain an isomorphism

$$\Psi_{\mathbf{d}}\colon \mathbf{H}^{\mathrm{BM}}(\mathfrak{M}^T_{\mathbf{d}}(\Pi_Q),\mathbb{Q})[-2\chi_Q(\mathbf{d},\mathbf{d})] \cong \mathbf{H}(\mathfrak{M}^T_{\mathbf{d}}(\tilde{Q}),{}^{\mathfrak{p}}\phi_{\mathrm{Tr}(\tilde{W})}).$$

We slightly twist this isomorphism[8] by setting $\Psi'_{\mathbf{d}} = (-1)^{\binom{|\mathbf{d}|}{2}}\Psi_{\mathbf{d}}$. We use these sign-twisted isomorphisms, and the algebra structure on $\mathcal{A}^{T,\chi}_{\tilde{Q},\tilde{W}}$, to induce an algebra structure on

$$\mathcal{A}^T_{\Pi_Q} := \bigoplus_{\mathbf{d}\in\mathbb{N}^{Q_0}} \mathbf{H}^{\mathrm{BM}}(\mathfrak{M}^T_{\mathbf{d}}(\Pi_Q),\mathbb{Q}^{\mathrm{vir}}).$$

Here, in the definition of the sign twist in the definition of $\mathcal{A}^{T,\chi}_{\tilde{Q},\tilde{W}}$, we have chosen $\psi = \chi = \chi_Q$, noting that

$$\chi_Q(\mathbf{d}',\mathbf{d}'') + \chi_Q(\mathbf{d}'',\mathbf{d}') = \chi_{\tilde{Q}}(\mathbf{d}',\mathbf{d}')\chi_{\tilde{Q}}(\mathbf{d}'',\mathbf{d}'') + \chi_{\tilde{Q}}(\mathbf{d}',\mathbf{d}'') \bmod 2.$$

Recall that we have defined the complex $\mathbb{Q}^{\mathrm{vir}}$ on the quotient of a smooth stack by a T-action to be the constant sheaf, shifted on each component of the stack by the dimension of the stack. Since $\mathfrak{M}_{\mathbf{d}}(\Pi_Q)$ is not smooth, we need a new convention to make sense of the above definition of $\mathcal{A}^T_{\Pi_Q}$: on $\mathfrak{M}^T_{\mathbf{d}}(\Pi_Q)$ we define $\mathbb{Q}^{\mathrm{vir}} := \mathbb{Q}[-2\chi(\mathbf{d},\mathbf{d})]$.

The algebra structure on $\mathcal{A}^T_{\Pi_Q}$ induced by the algebra structure on $\mathcal{A}^{T,\chi}_{\tilde{Q},\tilde{W}}$ and the isomorphism Ψ' is the same as the one defined by Schiffmann and Vasserot in [**SV13**], see [**RS17**, Appendix], [**YZ20**] for proofs. Our algebra structure is thus isomorphic to theirs, with the added sign twist provided by χ_Q.

3.10. Affinizing BPS Lie algebras. For pairs $(\tilde{Q},\tilde{W})$ of a tripled quiver with potential there is extra structure on the CoHA $\mathcal{A}^{\chi}_{\tilde{Q},\tilde{W}}$, ultimately derived from a factorization structure on the stack of representations of the Jacobi algebra. We briefly describe this here; full details for more general 3-Calabi–Yau completions of 2-Calabi–Yau categories will appear in joint work with Hennecart, Kinjo, Schiffmann and Vasserot. In the special case in which Q is the one-loop quiver, i.e. the case we need for the main results of this paper, this structure can be extracted from [**KV23**] via dimensional reduction.

Under the isomorphism j from (11) the element $j(\omega)$ is central in $\mathrm{Jac}(\tilde{Q},\tilde{W})$, so that every $\mathrm{Jac}(\tilde{Q},\tilde{W})$-module ρ admits a canonical decomposition as a $\mathrm{Jac}(\tilde{Q},\tilde{W})$-module

$$\rho \cong \bigoplus_{\lambda\in\mathbb{C}} \rho_\lambda$$

where $j(\omega)$ acts on ρ_λ with unique generalised eigenvalue λ. From this one deduces (see [**Dav23b**, Lemma 4.1]) that for $U\subset\mathbb{C}$ an open ball, if one denotes by $\mathfrak{M}^U_{\mathbf{d}}(\tilde{Q})$ the open substack satisfying the condition that each ω_i acts with generalised eigenvalues inside U, the restriction

$$\mathbf{H}(\mathfrak{M}_{\mathbf{d}}(\tilde{Q}),{}^{\mathfrak{p}}\phi_{\mathrm{Tr}(\tilde{W})}) \to \mathbf{H}(\mathfrak{M}^U_{\mathbf{d}}(\tilde{Q}),{}^{\mathfrak{p}}\phi_{\mathrm{Tr}(\tilde{W})})$$

[8] See [**RS17**, Lem.4.1] for the origin of this sign.

is an isomorphism. Let U' and U'' be disjoint open balls in $\mathbb{C}$, then via the Thom–Sebastiani isomorphism one constructs the composition of morphisms

$$\begin{aligned}\Delta\colon \mathbf{H}(\mathfrak{M}(\tilde{Q}), {}^{\mathfrak{p}}\phi_{\mathrm{Tr}(\tilde{W})}\mathbb{Q}^{\mathrm{vir}}) \to & \mathbf{H}(\mathfrak{M}^{U'\coprod U''}(\tilde{Q}), {}^{\mathfrak{p}}\phi_{\mathrm{Tr}(\tilde{W})}\mathbb{Q}^{\mathrm{vir}}) \\ &\cong \mathbf{H}(\mathfrak{M}^{U'}(\tilde{Q}), {}^{\mathfrak{p}}\phi_{\mathrm{Tr}(\tilde{W})}\mathbb{Q}^{\mathrm{vir}}) \\ &\quad \otimes \mathbf{H}(\mathfrak{M}^{U''}(\tilde{Q}), {}^{\mathfrak{p}}\phi_{\mathrm{Tr}(\tilde{W})}\mathbb{Q}^{\mathrm{vir}}) \\ &\cong \mathcal{A}^{\chi}_{\tilde{Q},\tilde{W}} \otimes \mathcal{A}^{\chi}_{\tilde{Q},\tilde{W}}\end{aligned}$$

defining a coproduct on $\mathcal{A}^{\chi}_{\tilde{Q},\tilde{W}}$. The second isomorphism is not quite canonical: there is a choice of sign, with the geometric origin explained in [**RS17**, Lem.4.1]. It is almost cocommutative since the space of choices of U' and U'' is connected, so that we can continuously swap them. Because of the signs appearing in dimensional reduction, in order for this coproduct to be actually cocommutative, we again multiply the component $\Delta_{\mathbf{d}',\mathbf{d}''}$ by $(-1)^{\chi_Q(\mathbf{d}',\mathbf{d}'')}$. The resulting coproduct is compatible with m^{tw}, since m^{tw} lifts to an algebra structure on the object

$$(\mathfrak{M}(\tilde{Q}) \xrightarrow{\lambda} \mathbf{Sym}(\mathbb{A}^1))_* {}^{\mathfrak{p}}\phi_{\mathrm{Tr}(\tilde{W})}\mathbb{Q}^{\mathrm{vir}}$$

where λ is the morphism recording the generalised eigenvalues of $j(\omega)$. It follows by the Milnor–Moore theorem that $\mathcal{A}^{\chi}_{\tilde{Q},\tilde{W}}$ is a universal enveloping algebra. Furthermore, by the support lemma [**Dav23b**, Lem.4.1] $\lambda_*\mathcal{BPS}_{\tilde{Q},\tilde{W}} \otimes \mathbf{H}_{\mathbb{C}^*_u}$ is supported on the subspace of $\mathcal{M}(\tilde{Q})$ for which all of the generalised eigenvalues of $j(\omega)$ are identical, and so

$$\hat{\mathfrak{g}}_{\tilde{Q},\tilde{W}} := \mathfrak{g}_{\tilde{Q},\tilde{W}} \otimes \mathbf{H}_{\mathbb{C}^*_u} \subset \mathcal{A}^{\chi}_{\tilde{Q},\tilde{W}}$$

is primitive for the coproduct Δ (by its construction), i.e. for $\alpha \in \hat{\mathfrak{g}}_{\tilde{Q},\tilde{W}}$, $\Delta(\alpha) = 1\otimes\alpha+\alpha\otimes 1$. By Theorem 3.3 and the PBW theorem for $\mathcal{A}^{\chi}_{\tilde{Q},\tilde{W}}$ the subspace $\hat{\mathfrak{g}}_{\tilde{Q},\tilde{W}}$ has the same graded dimensions as the space of primitive elements for $\mathcal{A}^{\chi}_{\tilde{Q},\tilde{W}}$. Since all elements of $\hat{\mathfrak{g}}_{\tilde{Q},\tilde{W}}$ are primitive, we find

PROPOSITION 3.8. *The subspace $\hat{\mathfrak{g}}_{\tilde{Q},\tilde{W}} \subset \mathcal{A}^{\chi}_{\tilde{Q},\tilde{W}}$ is closed under the commutator Lie bracket, and $\mathcal{A}^{\chi}_{\tilde{Q},\tilde{W}} = \mathbf{U}(\hat{\mathfrak{g}}_{\tilde{Q},\tilde{W}})$.*

DEFINITION 3.9. We call $\hat{\mathfrak{g}}_{\tilde{Q},\tilde{W}}$ the *affine BPS Lie algebra* for $(\tilde{Q},\tilde{W})$.

REMARK 3.10. So far there are no examples in the literature of quivers with potential $\mathcal{A}^{\psi}_{Q',W}$ for which $\mathfrak{g}_{Q',W} \otimes \mathbf{H}_{\mathbb{C}^*}$ has been shown *not* to be closed under the commutator Lie bracket. It would be interesting to know what class of BPS Lie algebras can be affinized this way, and in particular whether the "partially fermionized" cousins [**Dav23a**] of the BPS Lie algebras $\mathfrak{g}_{\tilde{Q},\tilde{W}}$ belong to this class.

Let $\mathbb{C}^*$ act on $\mathbb{C}\tilde{Q}$-modules via the weighting function $\tau\colon \tilde{Q}_1 \to \mathbb{Z}$ which sends a to 1 and a^* to -1 for all $a \in Q_1$, and ω_i to 0 for all $i \in Q_0$. Then the morphism λ is $\mathbb{C}^*$-invariant, so that we can repeat the construction of the commutative coproduct above for the CoHA

$$\mathcal{A}^{\mathbb{C}^*,\chi}_{\tilde{Q},\tilde{W}},$$

considered as an algebra object in the category of $\mathbf{H}_{\mathbb{C}^*}$-modules.

REMARK 3.11. Via the symmetry that permutes the arrows a, a^*, ω, the algebra $\mathcal{A}^{\mathbb{C}^*,\chi}_{Q^{(3)},\tilde{W}}$ carries a cocommutative coproduct in the same category, i.e. a morphism

$$\mathcal{A}^{\mathbb{C}^*}_{Q^{(3)},\tilde{W}} \to \mathcal{A}^{\mathbb{C}^*}_{Q^{(3)},\tilde{W}} \otimes_{\mathbf{H}_{\mathbb{C}^*}} \mathcal{A}^{\mathbb{C}^*,\chi}_{Q^{(3)},\tilde{W}},$$

and is a universal enveloping algebra, as long as $\mathbb{C}^*$ acts on one of a, a^*, ω with weight zero.

Returning to the case of general tripled quivers, since the coproduct on $\mathcal{A}^{\mathbb{C}^*,\chi}_{\tilde{Q},\tilde{W}}$ is $\mathbf{H}_{\mathbb{C}^*}$-linear by construction, we obtain a $\mathbf{H}_{\mathbb{C}^*}$-linear Lie algebra of primitive elements $\hat{\mathfrak{g}}^{\mathbb{C}^*}_{\tilde{Q},\tilde{W}}$.

DEFINITION 3.12. We denote by $\hat{\mathfrak{g}}^{\mathbb{C}^*}_{\tilde{Q},\tilde{W}} \subset \mathcal{A}^{\mathbb{C}^*,\chi}_{\tilde{Q},\tilde{W}}$ the Lie algebra of primitives, which we call the *deformed affinized BPS Lie algebra* for $(\tilde{Q}, \tilde{W})$.

The following is a basic application of the Milnor–Moore theorem:

PROPOSITION 3.13. *There is an isomorphism of* $\mathbf{H}_{\mathbb{C}^*}$*-algebra objects:*

$$\mathbf{U}_{\mathbf{H}_{\mathbb{C}^*}}(\hat{\mathfrak{g}}^{\mathbb{C}^*}_{\tilde{Q},\tilde{W}}) \to \mathcal{A}^{\mathbb{C}^*,\chi}_{\tilde{Q},\tilde{W}}. \tag{34}$$

The "deformed" in the definition is justified by the following

PROPOSITION 3.14. *There is an isomorphism of* $\mathbb{N}^{Q_0}$*-graded, cohomologically graded* $\mathbf{H}_{\mathbb{C}^*}$*-modules.*

$$\hat{\mathfrak{g}}^{\mathbb{C}^*}_{\tilde{Q},\tilde{W}} \cong \hat{\mathfrak{g}}_{\tilde{Q},\tilde{W}} \otimes \mathbf{H}_{\mathbb{C}^*} \tag{35}$$

as well as an isomorphism of Lie algebras

$$\hat{\mathfrak{g}}_{\tilde{Q},\tilde{W}} \cong \hat{\mathfrak{g}}^{\mathbb{C}^*}_{\tilde{Q},\tilde{W}} / \mathfrak{m}_{\mathbb{C}^*} \cdot \hat{\mathfrak{g}}^{\mathbb{C}^*}_{\tilde{Q},\tilde{W}}.$$

PROOF. Before we start, we remind the reader that $\mathbf{H}_{\mathbb{C}^*} \neq \mathbf{H}_{\mathbb{C}^*_u}$, although these algebras are isomorphic: the first algebra has to do with the extra $\mathbb{C}^*$-action on $\mathbb{C}\tilde{Q}$, the second has to do with the action studied at the end of §3.4.

Since the coproduct is $\mathbf{H}_{\mathbb{C}^*}$-linear, and $\mathcal{A}^{\mathbb{C}^*,\chi}_{\tilde{Q},\tilde{W}}$ is free as a $\mathbf{H}_{\mathbb{C}^*}$-module by Proposition 3.6, it follows that α is primitive if and only if $t\cdot\alpha$ is primitive, for t a generator of $\mathfrak{m}_{\mathbb{C}^*}$, the maximal homogeneous ideal in $\mathbf{H}_{\mathbb{C}^*}$. Therefore, the space of primitives for the coproduct (i.e. $\hat{\mathfrak{g}}^{\mathbb{C}^*}_{\tilde{Q},\tilde{W}}$) is free as a $\mathbf{H}_{\mathbb{C}^*}$-module. The graded dimensions of this module are determined by the existence of the isomorphisms (34) and (30), yielding (35).

Again by $\mathbf{H}_{\mathbb{C}^*}$-linearity of the coproduct, the composition

$$\hat{\mathfrak{g}}^{\mathbb{C}^*}_{\tilde{Q},\tilde{W}} / \mathfrak{m}_{\mathbb{C}^*} \cdot \hat{\mathfrak{g}}^{\mathbb{C}^*}_{\tilde{Q},\tilde{W}} \hookrightarrow \mathcal{A}^{\mathbb{C}^*,\chi}_{\tilde{Q},\tilde{W}} / \mathfrak{m}_{\mathbb{C}^*} \cdot \mathcal{A}^{\mathbb{C}^*,\chi}_{\tilde{Q},\tilde{W}} \xrightarrow{\cong} \mathcal{A}^{\chi}_{\tilde{Q},\tilde{W}}$$

realises $\hat{\mathfrak{g}}^{\mathbb{C}^*}_{\tilde{Q},\tilde{W}} / \mathfrak{m}_{\mathbb{C}^*} \cdot \hat{\mathfrak{g}}^{\mathbb{C}^*}_{\tilde{Q},\tilde{W}}$ as a subspace of primitive elements in $\mathcal{A}^{\chi}_{\tilde{Q},\tilde{W}}$. But the graded dimensions of $\hat{\mathfrak{g}}^{\mathbb{C}^*}_{\tilde{Q},\tilde{W}} / \mathfrak{m}_{\mathbb{C}^*} \cdot \hat{\mathfrak{g}}^{\mathbb{C}^*}_{\tilde{Q},\tilde{W}}$ match those of $\hat{\mathfrak{g}}_{\tilde{Q},\tilde{W}}$, by the first statement, and the second statement follows. □

3.11. Affinized finite type BPS algebras. Let Q be an orientation of a finite type, ADE Dynkin diagram. We calculate the (undeformed) affine BPS Lie algebra $\hat{\mathfrak{g}}_{\tilde{Q},\tilde{W}}$. We start with an easy lemma:

LEMMA 3.15. *Let $\mathcal{H}$ be an algebra object in* **Heis** $-\mathrm{mod}_{\mathbb{Z}}$*, such that there is an isomorphism of underlying* $\mathbf{Heis}_{\mathbb{Z}}$*-modules*

$$\mathcal{H} \cong \bigoplus_{m \in 2\cdot\mathbb{Z}_{\geq 1}} \mathbf{Gr}^F_{\bullet}(V_{[m]})[-2]^{\oplus a_m}$$

for integers a_m. Then up to isomorphism, $\mathcal{H}$ is determined by $\mathcal{H}^0$, the summand of $\mathcal{H}$ of cohomological degree zero, considered as an ordinary associative algebra. Similarly, if $\mathfrak{g}$ is a Lie algebra object in **Heis** $-\mathrm{mod}_{\mathbb{Z}}$ *such that there is an isomorphism of underlying* $\mathbf{Heis}_{\mathbb{Z}}$*-modules*

$$\mathfrak{g} \cong \bigoplus_{m \in 2\cdot\mathbb{Z}_{\geq 1}} \mathbf{Gr}^F_{\bullet}(V_{[m]})[-2]^{\oplus b_m}$$

for integers b_m, then $\mathfrak{g}$ is determined up to isomorphism by $\mathfrak{g}^0$, considered as an ordinary Lie algebra.

PROOF. Under the conditions of the lemma, $\mathcal{H}$ is concentrated in cohomological degrees indexed by $2{\cdot}\mathbb{N}$. Moreover, the lowering operator $q\colon \mathcal{H}^i \to \mathcal{H}^{i-2}$ is injective for $i > 0$, since $\mathcal{H}^{\geq 2}$ is spanned by elements $z^m D^n$, with $m, n \geq 1$, on which q acts by sending $z^m D^n \mapsto n z^m D^{n-1}$. As such, if $\alpha, \beta \in \mathcal{H}$ are not both of cohomological degree zero, $\alpha \star \beta$ is determined by $q(\alpha \star \beta) = q(\alpha) \star \beta + \alpha \star q(\beta)$, and the result follows by induction. The proof for $\mathfrak{g}$ is identical. □

Given a $\mathbb{N}^{Q_0}$-graded Lie algebra $\mathfrak{g}$ we form the $\mathbb{N}^{Q_0}$-graded Lie algebra $\mathfrak{g} \otimes_{\mathbb{Q}} \mathbb{Q}[D]$ by $\mathbb{Q}[D]$-linear extension. We make this into a $\mathbf{Heis}_Q$-module by setting $q_\Sigma(g \otimes D^m) = mg \otimes D^{m-1}$ and $p_i(g \otimes D^m) = \mathbf{d}_i g \otimes D^{m+1}$, for g of homogeneous $\mathbb{N}^{Q_0}$-degree $\mathbf{d}$. The $\mathbf{Heis}_Q$-module $\mathfrak{g} \otimes_{\mathbb{Q}} \mathbb{Q}[D]$ inherits a **Heis**-module structure via (5), and is $\mathbb{Z}$-graded by (doubling the) D-degree. For Q a finite type ADE Dynkin diagram with associated Lie algebra $\mathfrak{g}$ and negative half $\mathfrak{n}^-_Q \subset \mathfrak{g}$, the Lie algebra object $\mathfrak{n}^-_Q \otimes_{\mathbb{Q}} \mathbb{Q}[D]$ may thus be considered as a Lie algebra object in **Heis** $-\mathrm{mod}_{\mathbb{Z}}$ satisfying the assumptions of Lemma 3.15.

PROPOSITION 3.16. *Let Q be a quiver obtained by orienting a finite type Dynkin diagram. There is an isomorphism of Lie algebras*

$$\hat{\mathfrak{g}}_{\tilde{Q},\tilde{W}} \cong \mathfrak{n}^-_Q \otimes_{\mathbb{Q}} \mathbb{Q}[D]$$

where $\mathfrak{n}^-_Q$ is half of the simple Lie algebra associated to the underlying Dynkin diagram of Q.

PROOF. It follows from (27) and (28) that $\mathfrak{g}^i_{\tilde{Q},\tilde{W}} = 0$ for $i \neq 0$, and so $\hat{\mathfrak{g}}^i_{\tilde{Q},\tilde{W}} = 0$ for $i < 0$. By Proposition 3.5 there is an isomorphism $\hat{\mathfrak{g}}^0_{\tilde{Q},\tilde{W}} \cong \mathfrak{n}^-_Q$, and then the result follows from Lemma 3.15. □

We leave it to the enthusiastic reader to calculate the deformed affine BPS Lie algebra $\mathfrak{g}^{\mathbb{C}^*}_{\tilde{Q},\tilde{W}}$.

3.12. Nakajima quiver varieties. Fix a finite quiver Q, and a framing vector $\mathbf{f} \in \mathbb{N}^{Q_0}$. We define the quiver $Q_{\mathbf{f}}$ to be the quiver obtained by adding a single vertex ∞ to the set Q_0, and for each $i \in Q_0$, adding $\mathbf{f}_i$ arrows from ∞ to i, which we label $a_{i,m}$ for $m = 1, \ldots, \mathbf{f}_i$. We define $\widetilde{Q_{\mathbf{f}}}$ to be the quiver obtained by tripling

$Q_{\mathbf{f}}$, and set $Q_{\mathbf{f}}^+$ to be the quiver obtained by removing the loop ω_∞ from $\widetilde{Q_{\mathbf{f}}}$. We define

$$W^+ = \left(\sum_{a\in Q_1}[a,a^*] + \sum_{i\in Q_0}\sum_{m=1}^{\mathbf{f}_i} a_{i,m}a_{i,m}^*\right)\left(\sum_{i\in Q_0}\omega_i\right).$$

Given a dimension vector $\mathbf{d}\in\mathbb{N}^{Q_0}$ we extend $\mathbf{d}$ to a dimension vector $\mathbf{d}^+$ for $Q_{\mathbf{f}}^+$ by setting $\mathbf{d}_\infty^+ = 1$. We define $\mathbb{A}_{Q_{\mathbf{f}}^+,\mathbf{d}^+}^{\zeta-\mathrm{ss}}\subset\mathbb{A}_{Q_{\mathbf{f}}^+,\mathbf{d}^+}$ to be the subspace of representations $\mathcal{F}$ such that the one-dimensional vector space $e_\infty\cdot\mathcal{F}$ generates $\mathcal{F}$ under the action of $\mathbb{C}Q_{\mathbf{f}}^+$. We call such modules ζ-stable. Then we define the scheme

$$\mathcal{M}_{\mathbf{f},\mathbf{d}}(Q) = \mathbb{A}_{Q_{\mathbf{f}}^+,\mathbf{d}^+}^{\zeta-\mathrm{ss}}/\operatorname{GL}_{\mathbf{d}}.$$

This is the moduli scheme parameterising pairs of a $\mathbf{d}^+$-dimensional ζ-stable $\mathbb{C}Q_{\mathbf{f}}^+$-module $\mathcal{F}$, along with a trivialisation $e_\infty\cdot\mathcal{F}\cong\mathbb{C}$.

We identify $\mathbb{A}_{\overline{Q_{\mathbf{f}}},\mathbf{d}^+}$ with the space of tuples $((A_a,A_a^*)_{a\in Q_1},(I_{i,m},J_{i,m})_{i\in Q_0,m\leq\mathbf{f}_i})$ where $A_a\in\operatorname{Hom}(\mathbb{C}^{\mathbf{d}_{s(a)}},\mathbb{C}^{\mathbf{d}_{t(a)}})$, $A_a^*\in\operatorname{Hom}(\mathbb{C}^{\mathbf{d}_{t(a)}},\mathbb{C}^{\mathbf{d}_{s(a)}})$, $I_{i,m}\in(\mathbb{C}^{\mathbf{d}_i})^*$ and $J_{i,m}\in\mathbb{C}^{\mathbf{d}_i}$. We define $\mathbb{A}_{\overline{Q_{\mathbf{f}}},\mathbf{d}^+}^{\zeta-\mathrm{ss}}\subset\mathbb{A}_{\overline{Q_{\mathbf{f}}},\mathbf{d}^+}$ to be the subspace of representations $\mathcal{F}$ of the doubled framed quiver such that the one-dimensional space $e_\infty\cdot\mathcal{F}$ generates $\mathcal{F}$ under the action of $\mathbb{C}\overline{Q_{\mathbf{f}}}$. We consider the function

$$\begin{aligned}\mu_{\mathbf{d}}\colon \mathbb{A}_{\overline{Q_{\mathbf{f}}},\mathbf{d}^+}^{\zeta-\mathrm{ss}} &\to \mathfrak{gl}_{\mathbf{d}}\\ ((A_a,A_a^*)_{a\in Q_1},(I_{i,m},J_{i,m})_{i\in Q_0,m\leq\mathbf{f}_i}) &\mapsto \sum_{a\in Q_1}[A_a,A_a^*] + \sum_{i\in Q_0,m\leq\mathbf{f}_i} I_{i,m}(J_{i,m}).\end{aligned}$$

The group $\operatorname{GL}_{\mathbf{d}}$ acts freely on $\mathbb{A}_{\overline{Q_{\mathbf{f}}},\mathbf{d}^+}^{\zeta-\mathrm{ss}}$, and thus on $\mu_{\mathbf{d}}^{-1}(0)$, and we define the (smooth) *Nakajima quiver variety* $\mathcal{N}_{\mathbf{f},\mathbf{d}}(Q)=\mu_{\mathbf{d}}^{-1}(0)/\operatorname{GL}_{\mathbf{d}}$ [**Nak94**]. Composing the morphisms $\mu_{\mathbf{d}}^{-1}(0)/\operatorname{GL}_{\mathbf{d}}\to\operatorname{B}\operatorname{GL}_{\mathbf{d}}\xrightarrow{\operatorname{Det}}\operatorname{B}\mathbb{C}_u^*$ defines an action of $\mathbf{H}_{\mathbb{C}_u^*}$ on $\mathbf{H}(\mathcal{N}_{\mathbf{f},\mathbf{d}}(Q),\mathbb{Q}^{\mathrm{vir}})$ as in (19); by abuse of notation, we denote this action also by $\bullet$. Via dimensional reduction, there is a natural isomorphism (see [**Dav25**, Prop.6.3])

$$\mathbf{H}(\mathcal{N}_{\mathbf{f},\mathbf{d}}(Q),\mathbb{Q}^{\mathrm{vir}})\cong\mathbf{H}(\mathcal{M}_{\mathbf{f},\mathbf{d}}(Q),{}^{\mathfrak{p}}\phi_{\operatorname{Tr}(W^+)}). \tag{36}$$

Let $\mathbf{d}',\mathbf{d}''\in\mathbb{N}^{Q_0}$ be a pair of dimension vectors satisfying $\mathbf{d}'+\mathbf{d}''=\mathbf{d}$. We define $\mathcal{M}_{\mathbf{f},\mathbf{d}',\mathbf{d}''}(Q)$ to be the moduli scheme of pairs of a short exact sequence

$$0\to\mathcal{F}'\to\mathcal{F}\to\mathcal{F}''\to 0$$

along with a trivialisation $e_\infty\cdot\mathcal{F}\cong\mathbb{C}$, where $\mathcal{F}$ is a $\mathbf{d}^+$-dimensional ζ-stable $\mathbb{C}Q_{\mathbf{f}}^+$-module, and $\mathcal{F}'$ is a $\mathbf{d}'$-dimensional $\mathbb{C}\tilde{Q}$-module, considered as a $\mathbb{C}Q_{\mathbf{f}}^+$-module with dimension vector zero at the vertex ∞. The module $\mathcal{F}''$ appearing in such a short exact sequence is a $\mathbf{d}''^+$-dimensional ζ-stable $\mathbb{C}Q_{\mathbf{f}}^+$-module. We consider the standard correspondence diagram

$$\mathfrak{M}_{\mathbf{d}'}(\tilde{Q})\times\mathcal{M}_{\mathbf{f},\mathbf{d}''}(Q)\xleftarrow{\pi_1\times\pi_3}\mathcal{M}_{\mathbf{f},\mathbf{d}',\mathbf{d}''}(Q)\xrightarrow{\pi_2}\mathcal{M}_{\mathbf{f},\mathbf{d}}(Q).$$

Composing the Thom–Sebastiani isomorphism with the pushforward $\pi_{2,\star}$ and pullback $(\pi_1\times\pi_3)^\star$ we define the operation

$$\star\colon\mathbf{H}\left(\mathfrak{M}_{\mathbf{d}'}(\tilde{Q}),{}^{\mathfrak{p}}\phi_{\operatorname{Tr}(\tilde{W})}\right)\otimes\mathbf{H}\left(\mathcal{M}_{\mathbf{f},\mathbf{d}''}(Q),{}^{\mathfrak{p}}\phi_{\operatorname{Tr}(W^+)}\right)\to\mathbf{H}\left(\mathcal{M}_{\mathbf{f},\mathbf{d}}(Q),{}^{\mathfrak{p}}\phi_{\operatorname{Tr}(W^+)}\right).$$

PROPOSITION 3.17. *Let Q be a finite quiver.*

- *Via the isomorphism* (36), *the operation* $\star$ *defines an action of the CoHA* $\mathcal{A}^{\chi}_{\tilde{Q},\tilde{W}}$ *on the vector space*

$$M_{\mathbf{f}}(Q) := \bigoplus_{\mathbf{d}\in\mathbb{N}^{Q_0}} \mathbf{H}(\mathcal{N}_{\mathbf{f},\mathbf{d}}(Q), \mathbb{Q}^{\mathrm{vir}})$$

- *The operation is* $\mathbf{H}_{\mathbb{C}^*_u}$*-linear: there are equalities* $u\bullet(\alpha\star\beta) = (u\bullet\alpha)\star\beta + \alpha\star(u\bullet\beta)$.

PROOF. For the first part, see e.g. [**Soi16, YZ18**]. The second part is proved as in Proposition 4.1 below. □

Now fix $Q = Q^{(1)}$ the one loop quiver, and fix the framing vector $\mathbf{f} = 1$. Then we have $\mathcal{A}_{\tilde{Q},\tilde{W},1} \cong \mathbf{H}_{\mathbb{C}^*} \cong \mathbb{Q}[x]$. One shows as in [**Dav25**] that for 1_1 the copy of $1 \in \mathbb{Q}[x] \cong \mathcal{A}_{\tilde{Q},\tilde{W},1}$ we have

$$1_1 \star - = \mathfrak{p}_1 \tag{37}$$

where $\mathfrak{p}_1$ is the Nakajima raising operator [**Nak97**] on

$$M_1(Q) \cong \bigoplus_{n\in\mathbb{N}} \mathbf{H}(\mathrm{Hilb}_n(\mathbb{A}^2), \mathbb{Q}^{\mathrm{vir}}).$$

From the second part of Proposition 3.17 it follows that

$$(u\bullet\alpha)\star(-) = [u\bullet, \alpha\star](-) \tag{38}$$

as operators on $M_1(Q)$. We recall the following theorem of Manfred Lehn:

THEOREM 3.18 ([**Leh99**, Theorem 3.10]). *Write* $\mathfrak{p}_n$ *for Nakajima's* n*th raising operator on* $\mathrm{Hilb}(\mathbb{A}^2)$. *Write* $\mathfrak{p}'_n = [u\bullet, \mathfrak{p}_n]$. *Then*

$$[\mathfrak{p}'_1, \mathfrak{p}_n] = -n\mathfrak{p}_{n+1}.$$

We refer to [**Nak97, Leh99**] for the definition of the operators $\mathfrak{p}_n$. Alternatively, via (38), (37) and Theorem 3.18, these operators may be defined in terms of the action of $x^0 \in \mathcal{A}_{Q^{(3)},\tilde{W},1}$ and $u\bullet$ on $M_1(Q)$.

3.13. Affine Yangian and shuffle algebra embedding. Let T be a torus, acting on the quiver Q. The superpotential $W = 0$ is obviously T-invariant, and the vanishing cycle sheaf ${}^{\mathfrak{p}}\phi_{\mathrm{Tr}(W)}\mathbb{Q}^{\mathrm{vir}}_{\mathfrak{M}^T_{\mathbf{d}}(Q)}$ on $\mathfrak{M}^T_{\mathbf{d}}(Q)$ is isomorphic to the constant (shifted) perverse sheaf $\mathbb{Q}^{\mathrm{vir}}_{\mathfrak{M}^T_{\mathbf{d}}(Q)}$. It is thus straightforward to describe the vanishing cycle cohomology (leaving out the overall cohomological shifts for now):

$$\begin{aligned}\mathbf{H}(\mathfrak{M}^T_{\mathbf{d}}(Q), {}^{\mathfrak{p}}\phi_{\mathrm{Tr}(W)}) \cong& \mathbf{H}(\mathrm{pt}/(\mathrm{GL}_{\mathbf{d}}\times T), \mathbb{Q})\\ \cong& \mathbb{Q}[x_{1,1},\dots x_{1,\mathbf{d}_1}, x_{2,1},\dots,x_{r,\mathbf{d}_r}]^{\mathfrak{S}_{\mathbf{d}}}\otimes\mathbb{Q}[t_1,\dots t_l]\end{aligned}$$

where $Q_0 = \{1,\dots,r\}$ and $\mathfrak{S}_{\mathbf{d}} = \prod_{i\in Q_0}\mathfrak{S}_{\mathbf{d}_i}$ is the product of symmetric groups generated by permutations of the variables preserving their first subscript.

PROPOSITION 3.19. *Let the torus* $T = \mathrm{Hom}_{\mathrm{Grp}}(\mathbb{Z}, (\mathbb{C}^*)^2)$ *act on* $\tilde{Q}$ *via the function* $\tilde{\tau}$ *from* (33). *The natural morphism* $\Phi\colon \mathcal{A}^T_{\tilde{Q},\tilde{W}} \to \mathcal{A}^T_{\tilde{Q}}$ *of algebras is an embedding of algebras.*

The definition of ϕ and proof that Φ preserves the algebra product is given in [**SV13**]. The fact that it is an embedding of algebras is proved in [**Dav23b**, Theorem 10.2], slightly modifying and fixing a gap in the proof in the published version of [**SV20**, Proposition 4.6].

By (32) there is a (non-canonical) isomorphism of $\mathbf{H}_{T'}$-modules

$$\mathcal{A}_{\mathbb{A}^2}^{T'} \cong \mathcal{A}_{\mathbb{A}^2} \otimes \mathbf{H}_{T'} \,. \tag{39}$$

We denote by $\alpha_1^{(0)} \in \hat{\mathfrak{g}}_{\mathbb{A}^2,1}^{T} = \mathcal{A}_{\mathbb{A}^2,1}^{T'}$ the unique (up to scaling) non-zero element of cohomological degree 2, and set $\alpha_1^{(m)} := u^m \bullet \alpha_1^{(0)}$.

For what follows, we introduce (the strictly positive part of) the affine Yangian $\mathcal{Y}_{t_1,t_2,t_3}(\widehat{\mathfrak{gl}(1)})$, for which our reference is [**Tsy17**]. Let t_1, t_2, t_3 be formal parameters satisfying $t_1 + t_2 + t_3 = 0$. Then $\mathcal{Y}_{t_1,t_2,t_3}(\widehat{\mathfrak{gl}(1)})^+$ is the algebra over $\mathbb{Q}[t_1,t_2,t_3]/(t_1+t_2+t_3)$ generated by symbols e_i with $i \in \mathbb{N}$ subject to the two relations

$$\begin{aligned} &[e_{i+3}, e_j] - 3[e_{i+2}, e_{j+1}] + 3[e_{i+1}, e_{j+2}] - [e_i, e_{j+3}] \\ &\quad + \sigma_2([e_{i+1}, e_j] - [e_i, e_{j+1}]) = -\sigma_3(e_ie_j + e_je_i) \end{aligned} \tag{40}$$

$$\mathrm{Sym}_{\mathfrak{S}_3}[e_{i_1}, [e_{i_2}, e_{i_3+1}]] = 0 \tag{41}$$

where $\sigma_2 = t_1t_2 + t_1t_3 + t_2t_3$ and $\sigma_3 = t_1t_2t_3$.

LEMMA 3.20. *As a $\mathbf{H}_T$-module, $\mathcal{Y}_{t_1,t_2,t_3}(\widehat{\mathfrak{gl}(1)})$ is free.*

PROOF. We define a filtration G on $\mathcal{Y}_{t_1,t_2,t_3}(\widehat{\mathfrak{gl}(1)})$ by putting e_i in degree $-2+2i$. So $G^r\mathcal{Y}_{t_1,t_2,t_3}(\widehat{\mathfrak{gl}(1)})$ is spanned by Lie words $[e_{i_1}, [e_{i_2}, [\ldots, [e_{i_{l-1}}, e_{i_l}]\ldots]$ with $\sum_{j=1}^{l}(2i_j - 2) \le r$. Set $\mathcal{A} = \mathbf{Gr}^G\,\mathcal{Y}_{t_1,t_2,t_3}(\widehat{\mathfrak{gl}(1)})$. Then $\mathcal{A} = \mathbf{U}(\mathcal{L}) \otimes \mathbf{H}_T$ where $\mathcal{L}$ is the Lie algebra over $\mathbb{Q}$ generated by symbols $\overline{e}_i$ with $i \in \mathbb{N}$ subject to the relations

$$[\overline{e}_{i+3}, \overline{e}_j] - 3[\overline{e}_{i+2}, \overline{e}_{j+1}] + 3[\overline{e}_{i+1}, \overline{e}_{j+2}] - [\overline{e}_i, \overline{e}_{j+3}] \tag{42}$$

$$\mathrm{Sym}_{\mathfrak{S}_3}[\overline{e}_{i_1}, [\overline{e}_{i_2}, \overline{e}_{i_3+1}]] = 0. \tag{43}$$

In particular, $\mathcal{A}$ is free as a $\mathbf{H}_T$-module, from which it follows that $\mathcal{Y}_{t_1,t_2,t_3}(\widehat{\mathfrak{gl}(1)})$ is too. □

PROPOSITION 3.21 ([**RSYZ20**, Theorem 7.1.1]). *There is an algebra morphism*

$$\iota\colon \mathcal{Y}_{t_1,t_2,t_3}(\widehat{\mathfrak{gl}(1)})^+ \to \mathcal{A}_{\mathbb{A}^2}^{T}$$

sending e_i to $\alpha_1^{(i)}$.

Note that [**RSYZ20**, Theorem 7.1.1] uses in an essential way the injectivity statement of Proposition 3.19, meaning that it is sufficient to check that the defining relations of $\mathcal{Y}_{t_1,t_2,t_3}(\widehat{\mathfrak{gl}(1)})^+$ hold after applying Φ. Since we work with the full rank 2-torus T we may indeed apply [**Dav23b**, Theorem 10.2] for this injectivity. It is furthermore claimed in [**RSYZ20**] that ι is injective, although the proof seems to be missing;[9] injectivity of ι follows from the following result of Tsymbaluk, as we will see in §5.

[9] Thanks go to the anonymous referee for spotting this gap in the literature.

PROPOSITION 3.22 ([**Tsy17**, Appendix C]). *The morphism*

$$\iota' \colon (\mathcal{Y}_{t_1,t_2,t_3}(\widehat{\mathfrak{gl}(1)})^+)_{t_2=0} \to \mathbf{U}_{\mathbb{Q}[t]}(\mathbf{R}_F[W^+_{1+\infty}])$$

sending e_i to $z\tilde{D}^{(i)}$ is an injection.

We include here the explicit description of the shuffle algebra product on $\mathcal{A}^T_{Q^{(3)}}$. Since there is only one vertex, we can simplify the subscripts and write

$$\mathbf{H}(\mathfrak{M}^T_d(Q^{(3)}), \mathbb{Q}) \cong \mathbb{Q}[t_1, t_2] \otimes \mathbb{Q}[x_1, \dots, x_d]^{\mathfrak{S}_d}.$$

We define $\zeta(z) := (z+t_1)(z+t_2)(z-t_1-t_2)z^{-1}$. Given two functions $f(x_1, \dots, x_d) \in \mathbf{H}(\mathfrak{M}^T_d(Q^{(3)}), \mathbb{Q})$ and $g(x_1, \dots, x_e) \in \mathbf{H}(\mathfrak{M}^T_e(Q^{(3)}), \mathbb{Q})$ we have

$$(44)\quad f \star g(x_1, \dots, x_{d+e}) = \sum_{\pi \in \mathrm{Sh}(d,e)} \pi \left(f(x_1, \dots, x_d) g(x_{d+1}, \dots, g_{d+e}) \prod_{\substack{1 \leq i \leq d \\ d+1 \leq j \leq d+e}} \zeta(x_i - x_j) \right)$$

where $\mathrm{Sh}(d,e) \subset \mathfrak{S}_{d+e}$ is the subset of the permutation group containing those π such that $\pi(i) \leq \pi(i+1)$ for $i \neq d$.

In [**Neg16**] a related shuffle algebra is studied, in reference to the AGT relations.[10] Firstly, define $\zeta'(z) := (z+t_1)(z+t_2)(z+t_1+t_2)^{-1}z^{-1}$. Then we fix the coefficient field $\mathbb{F} = \mathbb{Q}(t_1, t_2)$ and for each d define

$$\mathcal{A}'^T_{\mathbb{A}^2,d,\mathrm{loc}} := \mathbb{F}(x_1, \dots x_d)^{\mathfrak{S}_d}$$

the field of symmetric rational functions. Then $\mathcal{A}'^T_{\mathbb{A}^2,\mathrm{loc}} := \bigoplus_{d \in \mathbb{Z}_{\geq 0}} \mathcal{A}'^T_{\mathbb{A}^2,d,\mathrm{loc}}$ carries a shuffle algebra structure defined as in (44) but with $\zeta'(z)$ replacing $\zeta(z)$. We define the algebra isomorphism

$$\begin{aligned} J \colon \mathcal{A}^T_{\mathbb{A}^2,\mathrm{loc}} &\to \mathcal{A}'^T_{\mathbb{A}^2,\mathrm{loc}} \\ f(x_1, \dots, x_d) &\mapsto \prod_{1 \leq i < j \leq d} (z^2 - (t_1+t_2)^2)^{-1} f(x_1, \dots, x_d)(t_1 t_2)^{-\deg(f)}. \end{aligned}$$

Composing with the injections

$$\mathcal{A}^T_{\mathbb{A}^2} \xrightarrow{i_1} \mathbb{Q}[t_1, t_2] \otimes \mathbb{Q}[x_1, \dots, x_d]^{\mathfrak{S}_d} \xrightarrow{i_2} \mathcal{A}^T_{\mathbb{A}^2,\mathrm{loc}}$$

we obtain the injection $\Gamma = J \circ i_2 \circ i_1$. In [**Neg16**], explicit elements $\tilde{B}_d$ and $\tilde{L}_d$ of a Heisenberg–Virasoro algebra are given: it is easy to see that $\tilde{B}_1 = \Gamma(1)$ and $\tilde{L}_1 = \Gamma(x_1)$. Then $\tilde{B}_d$ is equal to a generator of the 1-dimensional vector space $\Gamma(\mathfrak{g}_{\mathbb{A}^2,d})$ by Equation (2.43) of [**Neg16**] and Proposition 5.2 below.

4. Heisenberg actions on $\mathcal{A}^{(\psi)}_{Q,W}$

It turns out that $\mathbf{Heis}_Q$-actions on cohomological Hall algebras exist at a very high level of generality[11] We explain this first, before zooming in on the CoHAs that this paper is most concerned with, i.e. $\mathcal{A}^T_{\mathbb{A}^2}$ for various choices of T.

[10] Many thanks to Andrei Neguţ for patiently explaining the following to me.

[11] Note that the Heisenberg algebra appearing in this section has a different origin to the half Heisenberg algebra occurring inside $\hat{\mathfrak{g}}_{\mathbb{A}^2}$, studied in §5.

4.1. Raising operators.

PROPOSITION 4.1. *Let Q be an arbitrary quiver, with a T-action for some torus $T = \mathrm{Hom}_{\mathrm{Grp}}(N, \mathbb{C}^*)$, and with T-invariant potential $W \in \mathbb{C}Q/[\mathbb{C}Q, \mathbb{C}Q]_{\mathrm{vect}}$. For each $i \in Q_0$ the morphism $\alpha \mapsto u \bullet_i \alpha$, defined in (17) defines a derivation of $\mathcal{A}^{T,(\psi)}_{Q,W}$.*

PROOF. Recall that the bracketed superscript ψ in the statement of the proposition means that we claim the statement is true for both the algebras $\mathcal{A}^{T}_{Q,W}$ and $\mathcal{A}^{T,\psi}_{Q,W}$. We first prove the version without the sign-twist, i.e. for the multiplication in $\mathcal{A}^{T}_{Q,W}$.

Fix $\mathbf{d}', \mathbf{d}'' \in \mathbb{N}^{Q_0}$ and set $\mathbf{d} = \mathbf{d}' + \mathbf{d}''$. Consider the following commutative diagram

$$\begin{array}{ccccc} \mathfrak{M}^T_{\mathbf{d}'}(Q) \times_{\mathrm{B}T} \mathfrak{M}^T_{\mathbf{d}''}(Q) & \xleftarrow{\pi_1 \times \pi_3} & \mathfrak{M}^T_{\mathbf{d}',\mathbf{d}''}(Q) & \xrightarrow{\pi_2} & \mathfrak{M}^T_{\mathbf{d}}(Q) \\ \Big\downarrow{\scriptstyle \mathrm{Det}_{(i)} \times \mathrm{Det}_{(i)}} & & & & \Big\downarrow{\scriptstyle \mathrm{Det}_{(i)}} \\ \mathrm{B}\mathbb{C}^*_u \times \mathrm{B}\mathbb{C}^*_u & & \xrightarrow{\quad\otimes\quad} & & \mathrm{B}\mathbb{C}^*_u. \end{array}$$

The morphisms $(\pi_1 \times \pi_3)^\star$ and $\pi_{2,\star}$ respect the $\mathbf{H}_{\mathbb{C}^*_u}$-actions induced by the morphism to $\mathrm{B}\mathbb{C}^*_u$ in the above diagram. On the other hand, under the morphism

$$\mathbb{Q}[u] = \mathbf{H}(\mathrm{B}\mathbb{C}^*_u, \mathbb{Q}) \xrightarrow{\otimes^\star} \mathbb{Q}[u_1, u_2] = \mathbf{H}(\mathrm{B}\mathbb{C}^*_u \times \mathrm{B}\mathbb{C}^*_u, \mathbb{Q})$$

we have

$$\otimes^\star(u) = u_1 + u_2. \tag{45}$$

So

$$\begin{aligned} u \bullet_i (m(\alpha \otimes \beta)) =& m\left((\mathrm{Det}_{(i)} \otimes \mathrm{Det}_{(i)})^\star((u_1 \otimes 1) + (1 \otimes u_2)) \cdot (\alpha \otimes \beta)\right) \\ =& m(((u \bullet_i \alpha) \otimes \beta)) + m((\alpha \otimes (u \bullet_i \beta))). \end{aligned}$$

If the multiplication is instead calculated in $\mathcal{A}^{T,\psi}_{Q,W}$, all of the terms in the above expression are multiplied by $(-1)^{\psi(\mathbf{d}',\mathbf{d}'')}$, and so equality also holds in $\mathcal{A}^{T,\psi}_{Q,W}$. □

4.2. Lowering operators. The morphism $\otimes^\star$ in the proof of Proposition 4.1 makes $\mathbf{H}_{\mathbb{C}^*_u}$ into a cocommutative coalgebra. Let $\mathcal{L}$ be a line bundle on a scheme X, and let $\mathcal{F}$ be a flat family of $\mathbf{d}$-dimensional $\mathbb{C}Q$-modules on X. Then $\mathcal{L} \otimes \mathcal{F}$ is a flat family of $\mathbf{d}$-dimensional $\mathbb{C}Q$-modules, where the $\mathbb{C}Q$-action is induced by the given action on $\mathcal{F}$ and the natural isomorphism $\mathcal{E}nd_X(\mathcal{L} \otimes \mathcal{F}, \mathcal{L} \otimes \mathcal{F}) \cong \mathcal{E}nd_X(\mathcal{F}, \mathcal{F})$. Recall that an X-point of $\mathfrak{M}^T_{\mathbf{d}}(Q)$ is defined by a T-torsor $p \colon E \to X$ along with a flat family $\mathcal{F}$ of $\mathbf{d}$-dimensional $\mathbb{C}Q$-modules on E compatible with the T-action. We define a morphism

$$\begin{aligned} A \colon \mathrm{B}\mathbb{C}^*_u \times \mathfrak{M}^T_{\mathbf{d}}(Q') &\to \mathfrak{M}^T_{\mathbf{d}}(Q) \\ (\mathcal{L}, (E \xrightarrow{p} X, \mathcal{F})) &\mapsto (E \xrightarrow{p} X, p^*\mathcal{L} \otimes \mathcal{F}) \end{aligned}$$

which makes $\mathfrak{M}^T_{\mathbf{d}}(Q)$ into a module over the algebra object $\mathrm{B}\mathbb{C}^*_u$ in stacks. We obtain an induced morphism in vanishing cycle cohomology

$$A^\star \colon \mathbf{H}(\mathfrak{M}^T_{\mathbf{d}}(Q), {}^{\mathfrak{p}}\phi_{\mathrm{Tr}(W)}) \to \mathbf{H}_{\mathbb{C}^*_u} \otimes \mathbf{H}(\mathfrak{M}^T_{\mathbf{d}}(Q), {}^{\mathfrak{p}}\phi_{\mathrm{Tr}(W)})$$

making $\mathcal{A}^T_{Q,W,\mathbf{d}}$ into a comodule for the coalgebra $\mathbf{H}_{\mathbb{C}^*_u}$. Via our fixed identification $\mathbf{H}_{\mathbb{C}^*_u} = \mathbb{Q}[u]$, for $\alpha \in \mathcal{A}_{Q,W}$, we define coefficients α_n via the expansion

$$A^\star(\alpha) = \sum_{n\geq 0} u^n \otimes \alpha_n. \tag{46}$$

PROPOSITION 4.2. *The morphism*

$$\begin{aligned}\partial_u \colon \mathcal{A}^{T,(\psi)}_{Q,W} &\to \mathcal{A}^{T,(\psi)}_{Q,W}\\ \alpha &\mapsto \alpha_1\end{aligned}$$

defines a derivation of $\mathcal{A}^{T,(\psi)}_{Q,W}$, *where* α_1 *is as defined in* (46).

PROOF. We abbreviate $\mathfrak{M}^T_{\mathbf{d}} = \mathfrak{M}^T_{\mathbf{d}}(Q)$ etc. to reduce clutter. As in the proof of Proposition 4.1, it is sufficient to prove the statement for $\mathcal{A}^T_{Q,W}$. Consider the commutative diagram

$$\begin{array}{ccccc}
\mathrm{B}\mathbb{C}^*_u \times \mathfrak{M}^T_{\mathbf{d}'} \times_{\mathrm{B}T} \mathfrak{M}^T_{\mathbf{d}''} & \longleftarrow & \mathrm{B}\mathbb{C}^*_u \times \mathfrak{M}^T_{\mathbf{d}',\mathbf{d}''} & \longrightarrow & \mathrm{B}\mathbb{C}^*_u \times \mathfrak{M}^T_{\mathbf{d}}\\
\downarrow{\scriptstyle a} & & \downarrow{\scriptstyle b} & & \downarrow{\scriptstyle A}\\
\mathfrak{M}^T_{\mathbf{d}'} \times_{\mathrm{B}T} \mathfrak{M}^T_{\mathbf{d}''} & \longleftarrow & \mathfrak{M}^T_{\mathbf{d}',\mathbf{d}''} & \longrightarrow & \mathfrak{M}^T_{\mathbf{d}}
\end{array}$$

with morphisms of points over a scheme X defined by

$a\colon (\mathcal{L}, (E \xrightarrow{p} X, \mathcal{F}', \mathcal{F}'')) \mapsto (E \xrightarrow{p} X, p^*\mathcal{L} \otimes \mathcal{F}', p^*\mathcal{L} \otimes \mathcal{F}'')$

$b\colon (\mathcal{L}, (E \xrightarrow{p} X, \mathcal{F}' \to \mathcal{F} \to \mathcal{F}'')) \mapsto (E \xrightarrow{p} X, (p^*\mathcal{L} \otimes \mathcal{F}' \to p^*\mathcal{L} \otimes \mathcal{F} \to p^*\mathcal{L} \otimes \mathcal{F}''))$.

This induces the commutative diagram

$$\begin{array}{ccc}
\mathbf{H}_{\mathbb{C}^*_u} \otimes \mathbf{H}(\mathfrak{M}^T_{\mathbf{d}'} \times_{\mathrm{B}T} \mathfrak{M}^T_{\mathbf{d}''}, {}^{\mathfrak{p}}\phi_{\mathrm{Tr}(W)}) & \xrightarrow{\mathrm{id} \otimes m'_{\mathbf{d}',\mathbf{d}''}} & \mathbf{H}_{\mathbb{C}^*_u} \otimes \mathbf{H}(\mathfrak{M}^T_{\mathbf{d}}, {}^{\mathfrak{p}}\phi_{\mathrm{Tr}(W)})\\
\uparrow{\scriptstyle a^\star} & & \uparrow{\scriptstyle A^\star}\\
\mathbf{H}(\mathfrak{M}^T_{\mathbf{d}'} \times_{\mathrm{B}T} \mathfrak{M}^T_{\mathbf{d}''}, {}^{\mathfrak{p}}\phi_{\mathrm{Tr}(W)}) & \xrightarrow{m'_{\mathbf{d}',\mathbf{d}''}} & \mathbf{H}(\mathfrak{M}^T_{\mathbf{d}}, {}^{\mathfrak{p}}\phi_{\mathrm{Tr}(W)})
\end{array}$$

with $m'_{\mathbf{d}',\mathbf{d}''}$ as in (16), from which we deduce that

$$(m'_{\mathbf{d}',\mathbf{d}''}(\alpha \otimes \beta))_1 = m'_{\mathbf{d}',\mathbf{d}''}((\alpha \otimes \beta)_1)$$

with $(\alpha \otimes \beta)_1$ defined via $a^*(\alpha \otimes \beta) = \sum_{n\geq 0} u^n \otimes (\alpha \otimes \beta)_n$. Here we have also used that the pullback $\jmath^\star$ commutes with the $\mathbf{H}_{\mathbb{C}^*_u}$-action. On the other hand a factorises as

$$\mathrm{B}\mathbb{C}^*_u \times \mathfrak{M}^T_{\mathbf{d}'} \times_{\mathrm{B}T} \mathfrak{M}^T_{\mathbf{d}''} \xrightarrow{\Delta \times \mathrm{id} \times \mathrm{id}} \mathrm{B}\mathbb{C}^*_u \times \mathrm{B}\mathbb{C}^*_u \times \mathfrak{M}^T_{\mathbf{d}'} \times_{\mathrm{B}T} \mathfrak{M}^T_{\mathbf{d}''} \xrightarrow{A \times A} \mathfrak{M}^T_{\mathbf{d}'} \times_{\mathrm{B}T} \mathfrak{M}^T_{\mathbf{d}''}$$

and so since $\Delta^\star(p(u_1, u_2)) = p(u, u)$ the u coefficient of $a^*(\alpha \otimes \beta)$ is equal to $\alpha_1 \otimes \beta + \alpha \otimes \beta_1$. □

PROPOSITION 4.3. *For each* $\mathbf{d} \in \mathbb{N}^{Q_0}$ *there is an action of* $\mathbf{Heis}_Q$ *on* $\mathcal{A}^T_{Q,W,\mathbf{d}}$ *defined by letting* p_i *act by* $u\bullet_i$, *and letting* q_Σ *act by* ∂_u. *This action has central charge* $2\mathbf{d}$.

PROOF. Consider the composition

$$\mathrm{B}\mathbb{C}^*_u \times \mathfrak{M}^T_{\mathbf{d}}(Q) \xrightarrow{A} \mathfrak{M}^T_{\mathbf{d}}(Q) \xrightarrow{l_i} \mathrm{B}\mathbb{C}^*_u \times \mathfrak{M}^T_{\mathbf{d}}(Q),$$

which on X-points is given by

$$(\mathcal{L}, (E \xrightarrow{p} X, \mathcal{F})) \mapsto (\mathcal{L}^{\otimes \mathbf{d}_i} \otimes \mathrm{Det}(\mathcal{F}_i), (E \xrightarrow{p} X, p^*\mathcal{L} \otimes \mathcal{F})).$$

The induced endomorphism $F = (l_i \circ A)^\star$ of $\mathbf{H}_{\mathbb{C}_u^*} \otimes \mathbf{H}(\mathfrak{M}_{\mathbf{d}}(Q), {}^{\mathfrak{p}}\phi_{\mathrm{Tr}(W)})$ satisfies

$$F(u \otimes \alpha) = (\mathbf{d}_i(u \otimes 1) + (1 \otimes u)) \bullet_i A^\star(\alpha)$$

from which we deduce that

$$\begin{aligned}\partial_u(u \bullet_i \alpha) =& (u \bullet_i \alpha)_1\\ =& \mathbf{d}_i \alpha_0 + u \bullet_i \alpha_1\\ =& \mathbf{d}_i \alpha + u \bullet_i (\partial_u \alpha)\end{aligned}$$

as required. □

Combining Propositions 4.1, 4.2 and 4.3 yields

THEOREM 4.4. *Let Q be symmetric, so that $\mathcal{A}_{Q,W}^{T,(\psi)}$ is a cohomologically graded algebra. Then $\mathcal{A}_{Q,W}^{T,(\psi)}$ is an algebra object in the category $\mathbf{Heis}_Q - \mathrm{mod}_{\mathbb{Z}}$, for which the underlying $\mathbf{Heis}_Q$-module is integrable. The decomposition*

$$\mathcal{A}_{Q,W}^{T,(\psi)} \cong \bigoplus_{\mathbf{d} \in \mathbb{N}^{Q_0}} (\mathcal{A}_{Q,W}^{T,(\psi)})_{[2\mathbf{d}]}$$

according to central charge of the $\mathbf{Heis}_Q$-action is given by the standard decomposition (15) *according to dimension vector.*

PROPOSITION 4.5. *Let Q be a symmetric quiver with potential*

$$W \in \mathbb{C}Q/[\mathbb{C}Q, \mathbb{C}Q]_{\mathrm{vect}}.$$

The inclusion $\mathfrak{g}_{Q,W}^T \otimes \mathbf{H}_{\mathbb{C}_u^*} \hookrightarrow \mathcal{A}_{Q,W}^{T,\psi}$ is an inclusion of $\mathbb{N}^{Q_0} \oplus \mathbb{Z}$-graded $\mathbf{Heis}_Q$-modules.

PROOF. Since the image of $\mathfrak{g}_{Q,W}^T \otimes \mathbf{H}_{\mathbb{C}_u^*}$ is just the closure of the image of $\mathfrak{g}_{Q,W}^T \subset \mathcal{A}_{Q,W}^{T,\psi}$ under the raising operator $u\bullet$, the proposition follows from the claim that the lowering operator ∂_u annihilates $\mathfrak{g}_{Q,W}^T$. Consider the following diagram

$$\begin{array}{ccc} \mathrm{B}\mathbb{C}_u^* \times \mathfrak{M}_{\mathbf{d}}^T(Q) & \xrightarrow{A} & \mathfrak{M}_{\mathbf{d}}^T(Q) \\ & \searrow^{p := \mathtt{JH}^T \circ \pi_2} & \downarrow \mathtt{JH}^T \\ & & \mathcal{M}_{\mathbf{d}}^T(Q) \end{array}$$

where π_2 is projection onto $\mathfrak{M}_{\mathbf{d}}^T(Q)$. We claim that the diagram commutes: this is easy to see at the level of geometric points, from which the claim follows. Thus there is a natural isomorphism

$$\mathtt{JH}_*^T A_*(\mathbb{Q} \boxtimes {}^{\mathfrak{p}}\phi_{\mathrm{Tr}(W)}\mathbb{Q}^{\mathrm{vir}}) \cong \mathbf{H}(\mathrm{B}\mathbb{C}_u^*, \mathbb{Q}) \otimes \mathtt{JH}_*^T {}^{\mathfrak{p}}\phi_{\mathrm{Tr}(W)}\mathbb{Q}^{\mathrm{vir}}$$

and so passing to the first perverse truncation and using the perverse vanishing result (22) the adjunction morphism

$${}^{\mathfrak{p}'}\tau^{\leq 1} \mathtt{JH}_*^T {}^{\mathfrak{p}}\phi_{\mathrm{Tr}(W)}\mathbb{Q}^{\mathrm{vir}} \to {}^{\mathfrak{p}'}\tau^{\leq 1} \mathtt{JH}_*^T A_*(\mathbb{Q} \boxtimes {}^{\mathfrak{p}}\phi_{\mathrm{Tr}(W)}\mathbb{Q}^{\mathrm{vir}})$$

is an isomorphism. It follows that the morphism

$${}^{\mathfrak{p}'}\tau^{\leq 1} \mathtt{JH}_*^T \left({}^{\mathfrak{p}}\phi_{\mathrm{Tr}(W)}\mathbb{Q}^{\mathrm{vir}} \to A_*(\mathbb{Q}_{\mathrm{B}\mathbb{C}_u^*} \boxtimes {}^{\mathfrak{p}}\phi_{\mathrm{Tr}(W)}\mathbb{Q}^{\mathrm{vir}})\right)$$

factors through the morphism

$$\mathbf{H}^0(\mathrm{B}\mathbb{C}^*_u, \mathbb{Q}) \otimes {}^{\mathfrak{p}}\tau'^{\leq 1}\mathrm{JH}^T_*{}^{\mathfrak{p}}\phi_{\mathrm{Tr}(W)}\mathbb{Q}^{\mathrm{vir}} \to \mathbf{H}(\mathrm{B}\mathbb{C}^*_u, \mathbb{Q}) \otimes {}^{\mathfrak{p}}\tau'^{\leq 1}\mathrm{JH}^T_*{}^{\mathfrak{p}}\phi_{\mathrm{Tr}(W)}\mathbb{Q}^{\mathrm{vir}}$$

and $A^\star(\alpha) = 1 \otimes \alpha$ for $\alpha \in \mathfrak{P}_1\mathcal{A}^{T,\psi}_{Q,W}$. As such, $\alpha_1 = 0$, as required. □

5. Proofs of main results

Let $Q = Q^{(1)}$ be the Jordan quiver, with one vertex and one loop, so $\tilde{Q} = Q^{(3)}$ is the quiver with three loops. Then for every $q = p^n$ and every $d \geq 1$ there are q absolutely indecomposable d-dimensional $\mathbb{F}_qQ$-modules, so $\mathtt{a}_{Q,d}(q) = q$, and

$$\mathfrak{g}_{Q^{(3)},\tilde{W},d} \cong \mathbb{Q}[2] \tag{47}$$

by (28). We define

$$\mathfrak{g}^T_{\mathbb{A}^2} := \mathfrak{g}^T_{Q^{(3)},\tilde{W}}; \qquad \hat{\mathfrak{g}}^T_{\mathbb{A}^2} := \hat{\mathfrak{g}}^T_{Q^{(3)},\tilde{W}}$$

the deformed and the deformed affinized BPS Lie algebras for the pair $(Q^{(3)}, \tilde{W})$, respectively. For $d \in \mathbb{N}$, let $\alpha_d \in \mathfrak{g}^T_{\mathbb{A}^2,d'}$ and $\alpha^{(0)}_d \in \hat{\mathfrak{g}}^T_{\mathbb{A}^2}$ be the unique (up to scalar) nonzero elements of cohomological degree -2.

PROPOSITION 5.1. *Let the torus $T = \mathrm{Hom}_{\mathrm{Grp}}(\mathbb{Z}, (\mathbb{C}^*)^r)$ act on the quiver $Q^{(3)}$, with $\tilde{W}$ T-invariant. Then the Lie bracket on $\mathfrak{g}^T_{\mathbb{A}^2}$ vanishes.*

PROOF. By (47) and Theorem 3.4 there is an isomorphism of cohomologically graded vector spaces

$$\mathfrak{g}^T_{\mathbb{A}^2,d} \cong \mathbf{H}_T[2]$$

for every $d \geq 1$. Then

$$[\alpha_{d'}, \alpha_{d''}] \in \mathfrak{g}^T_{\mathbb{A}^2,d'+d''}$$

has cohomological degree -4, and so is zero. For arbitrary $p(\mathbf{t}), q(\mathbf{t}) \in \mathbf{H}_T$ it follows that

$$[p(\mathbf{t})\cdot\alpha_{d'}, q(\mathbf{t})\cdot\alpha_{d''}] = p(\mathbf{t})q(\mathbf{t})\cdot[\alpha_{d'}, \alpha_{d''}] = 0.$$ □

By (47) we have the equality of generating series

$$\begin{aligned} \chi(\hat{\mathfrak{g}}_{\mathbb{A}^2}) :=& \sum_{i,j\in\mathbb{Z}} \dim(\hat{\mathfrak{g}}^j_{\mathbb{A}^2,i})v^iq^j \\ =& q^{-2}v(1-v)^{-1}(1-q^2)^{-1}. \end{aligned} \tag{48}$$

From the isomorphism $\hat{\mathfrak{g}}^{\mathbb{C}^*}_{\mathbb{A}^2} \cong \hat{\mathfrak{g}}_{\mathbb{A}^2} \otimes \mathbf{H}_{\mathbb{C}^*}$ (Proposition 3.14) we deduce the equality

$$\chi(\hat{\mathfrak{g}}^{\mathbb{C}^*}_{\mathbb{A}^2}) = q^{-2}v(1-v)^{-1}(1-q^2)^{-2}. \tag{49}$$

We continue to denote by $\alpha^{(0)}_d$ a basis element of $\hat{\mathfrak{g}}^{-2}_{\mathbb{A}^2,d'}$, and set $\alpha^{(m)}_d := u^m \bullet \alpha^{(0)}_d$. Then the elements $\alpha^{(m)}_d$ with $d \geq 1$ and $m \geq 0$ form a basis for $\hat{\mathfrak{g}}_{\mathbb{A}^2}$.

PROPOSITION 5.2. *For every $d \geq 1$ we have an inequality*

$$[\alpha^{(1)}_1, \alpha^{(0)}_d] = \lambda_d\alpha^{(0)}_{d+1} \tag{50}$$

for some nonzero $\lambda_d \in \mathbb{Q}$.

PROOF. We consider the $\mathcal{A}_{\mathbb{A}^2}$-action on $M_1(Q) \cong \bigoplus_{d\in\mathbb{N}} \mathbf{H}(\mathrm{Hilb}_d(\mathbb{C}^2), \mathbb{Q}^{\mathrm{vir}})$ recalled in §3.12. We denote by $\rho\colon \hat{\mathfrak{g}}_{\mathbb{A}^2} \to \mathrm{End}(M_1(Q))$ the action restricted to $\hat{\mathfrak{g}}_{\mathbb{A}^2}$. Then $\rho(\alpha_0^{(0)}) = \mathfrak{p}_1$ is the action of the Nakajima raising operator and from

$$u \bullet (\alpha \star m) = (u \bullet \alpha) \star m + \alpha \star (u \bullet m)$$

we deduce that

$$\rho(\alpha_1^{(1)}) = [u\bullet, \rho(\alpha_1^{(0)})] = \mathfrak{p}_1'.$$

From Theorem 3.18 we deduce inductively that

$$\rho\left(\mathbf{ad}^n_{\alpha_1^{(1)}}(\alpha_1^{(0)})\right) \neq 0. \tag{51}$$

Set $\alpha = \mathbf{ad}^d_{\alpha_1^{(1)}}(\alpha_1^{(0)}) \in \widehat{\mathfrak{g}}_{\mathbb{A}^2,d+1}$. By (51), $\alpha \neq 0$. On the other hand, $\alpha_{d+1}^{(0)}$ is the unique nonzero element of $\widehat{\mathfrak{g}}_{\mathbb{A}^2,d+1}$ of cohomological degree -2, which is the cohomological degree of α, and the result follows by induction. □

COROLLARY 5.3. *The affinized BPS Lie algebra $\hat{\mathfrak{g}}_{\mathbb{A}^2}$ is generated by $\alpha_1^{(m)}$ for $m \geq 0$.*

PROOF. We assume, for an inductive argument, that the Lie subalgebra $\mathfrak{g}' \subset \hat{\mathfrak{g}}_{\mathbb{A}^2}$ generated by $\alpha_1^{(m)}$ for $m \geq 0$ contains all of the elements $\alpha_{d'}^{(m)}$ for $d' < d+1$ and $m \in \mathbb{N}$. From Proposition 5.2 we deduce that, possibly after replacing $\alpha_{d+1}^{(0)}$ by itself multiplied by some $\lambda \in \mathbb{Q} \setminus \{0\}$, we have the equality $[\alpha_1^{(1)}, \alpha_d^{(0)}] = \alpha_{d+1}^{(0)}$. Acting by $u^m\bullet$, we find

$$\sum_{0\leq r\leq m} \binom{m}{r} [\alpha_1^{(1+r)}, \alpha_d^{(m-r)}] = \alpha_{d+1}^{(m)}.$$ □

The point of the next theorem and corollary is that the structure of $\mathcal{A}_{\mathbb{A}^2}$ follows from dimension counting, spherical generation, its upgrade to an algebra object in $\mathbf{Heis}-\mathrm{mod}_{\mathbb{Z}}$ (§4) and from the construction of the affinized BPS Lie algebra (§3.10).

THEOREM 5.4. *There is an isomorphism $F\colon \hat{\mathfrak{g}}_{\mathbb{A}^2} \cong \mathbf{Gr}^F_\bullet W^+_{1+\infty}$.*

PROOF. This is a corollary of (48), Theorem 4.4, Corollary 5.3, which states that $\hat{\mathfrak{g}}_{\mathbb{A}^2}$ is generated by $\hat{\mathfrak{g}}_{\mathbb{A}^2,1} = \mathbf{Gr}^F_\bullet V_{[2]}$, and Corollary 2.5. □

The next corollary follows from Proposition 3.8. The second statement in the corollary follows from Lemma 2.2.

COROLLARY 5.5. *There is an isomorphism of algebras*

$$\mathcal{A}_{\mathbb{A}^2} \cong \mathbf{U}(\mathbf{Gr}^F_\bullet W^+_{1+\infty}).$$

In particular, $\mathcal{A}_{\mathbb{A}^2}$ is spherically generated (i.e. it is generated by $\mathcal{A}_{\mathbb{A}^2,1}$).

Consider the torus $T = (\mathbb{C}^*)^2$, acting by independently rescaling the two coordinates of $\mathbb{A}^2$ (via the weighting $\tilde{\tau}$ of (33)), and the fully equivariant CoHA $\mathcal{A}^T_{\mathbb{A}^2}$.

THEOREM 5.6. *If T' is a torus acting on $Q^{(3)}$, leaving $\tilde{W}$ invariant, then $\mathcal{A}^{T'}_{Q^{(3)},\tilde{W}}$ is spherically generated. In particular, the algebra $\mathcal{A}^T_{\mathbb{A}^2}$ is spherically generated.*

PROOF. For $d \geq 1$ we define a morphism of $\mathbf{H}_{T'}$-modules

$$\Psi_d \colon \mathbb{Q}[u] \otimes \mathbf{H}_{T'} \to \mathcal{A}^{T'}_{\mathbb{A}^2}$$

$$u^m \otimes t^l \mapsto \begin{cases} t^l\Big([\alpha_1^{(1)}, -]^{d-1}(\alpha_1^{(m)})\Big) & \text{if } d-1 < m \text{ or } m = 0 \\ t^l\Big(([\alpha_1^{(1)}, -]^{d-1-m} \circ [\alpha_1^{(0)}, -] \circ [\alpha_1^{(1)}, -]^{m-1})(\alpha_1^{(m+1)})\Big) & \text{otherwise} \end{cases}$$

and define

$$\Psi = \oplus_{d \geq 1} \Psi_d \colon \bigoplus_{d \geq 1} \mathbb{Q}[u] \otimes \mathbf{H}_{T'} \to \mathcal{A}^{T'}_{\mathbb{A}^2}.$$

Finally we define

$$\Phi \colon \mathbf{Sym}_{\mathbf{H}_{T'}}\left(\bigoplus_{d \geq 1} \mathbb{Q}[u] \otimes \mathbf{H}_{T'}\right) \to \mathcal{A}^{T'}_{\mathbb{A}^2}$$

via Ψ and the CoHA product on the target. This is a morphism of free $\mathbf{H}_{T'}$-modules that (via Lemma 2.3, Corollary 5.5 and Proposition 3.6 (31)) becomes an isomorphism after applying $\otimes_{\mathbf{H}_{T'}} (\mathbf{H}_{T'} / \mathfrak{m}_{T'})$, where $\mathfrak{m}_{T'} \subset \mathbf{H}_{T'}$ is the maximal homogeneous ideal. It follows that Φ is an isomorphism, and the algebra $\mathcal{A}^{T'}_{\mathbb{A}^2}$ is spherically generated. □

There is a partial converse to Theorem 5.6, provided by the next two propositions:

PROPOSITION 5.7. *Let Q be a quiver without loops that is not an orientation of a finite type ADE Dynkin diagram, and let T act on $\tilde{Q}$, leaving $\tilde{W}$ invariant. Then $\mathcal{A}^T_{\tilde{Q},\tilde{W}}$ is not spherically generated, i.e. it is not generated by the subspaces $\mathcal{A}^T_{\tilde{Q},\tilde{W},1_i}$ for $i \in Q_0$.*

PROOF. This follows for cohomological degree reasons. The conditions on Q imply that $\mathtt{a}_{Q,\mathbf{d}}(q)$ is not a constant for some $\mathbf{d} \in \mathbb{N}^{Q_0}$. From (32) and (28) we deduce that $\mathcal{A}^T_{\tilde{Q},\tilde{W}}$ has a summand in strictly negative cohomological degree. For each $i \in Q_0$, since there are no loops supported at i, there is an isomorphism

$$\mathcal{A}^T_{\tilde{Q},\tilde{W},1_i} \cong \mathbf{H}_T$$

as a cohomologically graded vector space. So the sum of these spaces for $i \in Q_0$ generates an algebra lying in positive cohomological degrees. □

The next proposition is in the same vein. Before we state it, we introduce some notation. We label the three loops of $Q^{(3)}$ by the symbols x, y, z. We let $\mathfrak{M}^{\mathcal{SN}}(Q^{(3)}) \subset \mathfrak{M}(Q^{(3)})$ be the reduced closed substack, the points of which correspond to representations for which z acts via a nilpotent operator, and let $\mathfrak{M}^{\mathcal{N}}(Q^{(3)}) \subset \mathfrak{M}(Q^{(3)})$ be the reduced closed substack, the points of which correspond to representations for which both y and z act via nilpotent operators. Then

$$\mathcal{A}^{\heartsuit}_{Q^{(3)},\tilde{W}} := \bigoplus_d \mathbf{H}\Big(\mathfrak{M}^{\heartsuit}_d(Q^{(3)}), ({}^{\mathfrak{p}}\phi_{\mathrm{Tr}(\tilde{W})} \mathbb{Q}^{\mathrm{vir}}_{\mathfrak{M}_d(Q^{(3)})})|_{\mathfrak{M}^{\heartsuit}(Q^{(3)})}\Big)$$

carry Hall algebra structures defined as in §3.4, for $\heartsuit = \mathcal{SN}, \mathcal{N}$. If the torus $T' \cong (\mathbb{C}^*)^n$ acts on $\mathbb{A}^3$, preserving the 3-form $dx \wedge dy \wedge dz$ then there is again an induced action on $\mathcal{A}^{T,\heartsuit}_{\mathbb{A}^2} := \mathcal{A}^{T,\heartsuit}_{Q^{(3)},\tilde{W}}$. By the PBW theorem [**DM20**, Thm.C], there exist Lie subalgebras $\mathfrak{g}^{T,\heartsuit}_{\mathbb{A}^2} \subset \mathcal{A}^{T,\heartsuit}_{\mathbb{A}^2}$ such that the induced morphism

$$\mathbf{Sym}_{\mathbf{H}_T}\left(\mathfrak{g}^{T,\heartsuit}_{\mathbb{A}^2} \otimes \mathbf{H}_{\mathbb{C}^*_u}\right) \to \mathcal{A}^{T,\heartsuit}_{\mathbb{A}^2} \tag{52}$$

is an isomorphism. The generating functions of these Lie algebras are given by

$$\chi(\mathfrak{g}^{T,\heartsuit}_{\mathbb{A}^2}) = vq^{f(\heartsuit)}(1-v)^{-1}(1-q^2)^{-n} \tag{53}$$

where $f(\mathcal{SN}) = 0$ and $f(\mathcal{N}) = 2$; see [**BSV20, Dav23b**].

PROPOSITION 5.8. *The algebras* $\mathcal{A}^{T,\heartsuit}_{\mathbb{A}^2}$ *for* $\heartsuit = \mathcal{SN}, \mathcal{N}$ *are not spherically generated.*

PROOF. Both results follow by dimension counting. Firstly, both algebras are concentrated entirely in non-negative cohomological degrees. Then (52) and (53) imply that the subspace $V \subset \mathcal{A}^{T,\mathcal{SN}}_{\mathbb{A}^2,2}$ in cohomological degree zero is 2 dimensional, while the cohomological degree zero piece of $U \subset \mathcal{A}^{T,\mathcal{SN}}_{\mathbb{A}^2,1}$ is 1-dimensional, so cannot generate V. The argument for $\mathcal{A}^{T,\mathcal{N}}_{\mathbb{A}^2}$ is similar. □

It would be interesting to have some neat description of the images of any one of the injections $\mathcal{A}^{T,\heartsuit}_{\mathbb{A}^2} \hookrightarrow \bigoplus_{d\geq 1} \mathbb{Q}[t_1,t_2] \otimes \mathbb{Q}[x_1,\ldots,x_d]^{\mathfrak{S}_d}$, for $\heartsuit = \emptyset, \mathcal{SN}, \mathcal{N}$ to compare with the main result of [**Neg24**] in K-theory for the case $\heartsuit = \mathcal{SN}$ — we leave this to future work.

THEOREM 5.9. *Let* $\mathbb{C}^*$ *act with weights* $(1,0)$ *on the two coordinates of* $\mathbb{A}^2$. *Then there is a* $\mathbf{H}_{\mathbb{C}^*}$*-linear isomorphism*

$$F\colon \hat{\mathfrak{g}}^{\mathbb{C}^*}_{\mathbb{A}^2} \cong \mathbf{R}_F[W^+_{1+\infty}]$$

between the deformed affinized BPS Lie algebra for the pair $(Q^{(3)}, \tilde{W})$ *and the Rees Lie algebra of* $W^+_{1+\infty}$, *uniquely determined by setting* $F(\alpha^{(m)}_1) = z\tilde{D}^{(m)}$. *Thus, there is an isomorphism of algebras*

$$\mathcal{A}^{\mathbb{C}^*}_{\mathbb{A}^2} \cong \mathbf{U}_{\mathbb{Q}[t]}(\mathbf{R}_F[W^+_{1+\infty}]). \tag{54}$$

PROOF. Let $T = \mathrm{Hom}_{\mathrm{Grp}}(\mathbb{Z}, (\mathbb{C}^*)^2)$ act on $\tilde{Q}$ via the weighting $\tilde{\tau}\colon \tilde{Q}_1 \to \mathbb{Z}^2$ from (33). By Proposition 3.6 there is an isomorphism of algebras

$$\mathcal{A}^{\mathbb{C}^*}_{\mathbb{A}^2} \cong \mathcal{A}^T_{\mathbb{A}^2} \otimes_{\mathbf{H}_T} \mathbb{Q}[t_1,t_2]/(t_2). \tag{55}$$

Then by Proposition 3.21 there is a morphism of algebras

$$G\colon (\mathcal{Y}_{t_1,t_2,t_3}(\widehat{\mathfrak{gl}(1)})^+)_{t_2=0} \to \mathcal{A}^{\mathbb{C}^*}_{\mathbb{A}^2} \tag{56}$$

setting $t_1 = t$ and sending e_i to $\alpha^{(i)}_1$. This morphism is *surjective* by Theorem 5.6. By Proposition 3.22 there is an *injection* $\iota'\colon (\mathcal{Y}_{t_1,t_2,t_3}(\widehat{\mathfrak{gl}(1)})^+)_{t_2=0} \hookrightarrow \mathbf{U}_{\mathbb{Q}[t]}(\mathbf{R}_F[W^+_{1+\infty}])$. Now by (49) we have that

$$\chi(\hat{\mathfrak{g}}^{\mathbb{C}^*}_{\mathbb{A}^2}) = q^{-2}v(1-v)^{-1}(1-q^2)^{-2} = \chi(\mathbf{R}_F[W^+_{1+\infty}])_{v^2\mapsto v}$$

with the doubling of exponents as in Theorem 4.4. By Proposition 3.13 we deduce

$$\chi(\mathcal{A}^{\mathbb{C}^*}_{\mathbb{A}^2}) = \chi(\mathbf{U}_{\mathbb{Q}[t]}(\hat{\mathfrak{g}}^{\mathbb{C}^*}_{\mathbb{A}^2})) = \chi(\mathbf{U}_{\mathbb{Q}[t]}(\mathbf{R}_F[W^+_{1+\infty}]))_{v^2\mapsto v}$$

and both G and ι' are isomorphisms. The isomorphism (54) is provided by $\iota' \circ G^{-1}$. Composing $\iota' \circ G^{-1}$ with the PBW isomorphism (Proposition 3.13 again) yields the isomorphism

$$\mathbf{U}_{\mathbb{Q}[t]}(\hat{\mathfrak{g}}^{\mathbb{C}^*}_{\mathbb{A}^2}) \xrightarrow{\cong} \mathbf{U}_{\mathbb{Q}[t]}(\mathbf{R}_F[W^+_{1+\infty}])$$

restricting to the claimed isomorphism F. □

PROPOSITION 5.10. *The morphism ι from Proposition 3.21 is injective, so that $\mathcal{Y}_{t_1,t_2,t_3}(\widehat{\mathfrak{gl}(1)})^+$ may be identified with the spherical subalgebra $\mathcal{SA}^T_{\mathbb{A}^2} \subset \mathcal{A}^T_{\mathbb{A}^2}$ under ι, i.e. the subalgebra generated by $\mathcal{A}^T_{\mathbb{A}^2,1}$.*

PROOF. By construction, ι induces a surjection $\mathcal{Y}_{t_1,t_2,t_3}(\widehat{\mathfrak{gl}(1)})^+ \to \mathcal{SA}^T_{\mathbb{A}^2}$. The morphism ι is injective, since it is a morphism of free $\mathbf{H}_T$-modules (by Proposition 3.6 and Lemma 3.20) and it is injective after setting $t_2 = 0$ by Proposition 3.22 and Theorem 5.9. □

Finally we can completely describe the fully equivariant CoHA:

THEOREM 5.11. *There is an isomorphism of algebras $\iota\colon \mathcal{Y}_{t_1,t_2,t_3}(\widehat{\mathfrak{gl}(1)})^+ \to \mathcal{A}^T_{\mathbb{A}^2}$ sending e_i to $\alpha_1^{(i)}$.*

PROOF. This is an immediate corollary of Proposition 5.10 and Theorem 5.6. □

Acknowledgments

Special thanks go to Ezra Getzler, who encouraged me in this endeavour along with many others over many years. Thanks go to Kevin Costello, Brian Williams, Sasha Minets and Shivang Jindal for helpful conversations about the results of this paper, and encouragement to write them down correctly. Thanks also to Andrei Neguţ for numerous helpful comments regarding an earlier draft, and Lucien Hennecart for spotting a typo in an advanced draft. Thanks are also due to KIAS, and Minhyong Kim, for organizing an extremely stimulating stay in Seoul in Summer 2022, where this paper was finally written. Finally, many thanks are due to the anonymous referee for their many useful corrections and suggestions.

References

[AS13] Noah Arbesfeld and Olivier Schiffmann, *A presentation of the deformed $W_{1+\infty}$ algebra*, Symmetries, integrable systems and representations, Springer Proc. Math. Stat., vol. 40, Springer, Heidelberg, 2013, pp. 1–13, DOI 10.1007/978-1-4471-4863-0_1. MR3077678

[BBBBJ15] Oren Ben-Bassat, Christopher Brav, Vittoria Bussi, and Dominic Joyce, *A 'Darboux theorem' for shifted symplectic structures on derived Artin stacks, with applications*, Geom. Topol. **19** (2015), no. 3, 1287–1359, DOI 10.2140/gt.2015.19.1287. MR3352237

[BBDG18] Alexander Beilinson, Joseph Bernstein, Pierre Deligne, and Ofer Gabber, *Faisceaux pervers*, Astérisque **100** (2018), vi+180. MR4870047

[BBJ19] Christopher Brav, Vittoria Bussi, and Dominic Joyce, *A Darboux theorem for derived schemes with shifted symplectic structure*, J. Amer. Math. Soc. **32** (2019), no. 2, 399–443, DOI 10.1090/jams/910. MR3904157

[BBS13] Kai Behrend, Jim Bryan, and Balázs Szendrői, *Motivic degree zero Donaldson-Thomas invariants*, Invent. Math. **192** (2013), no. 1, 111–160, DOI 10.1007/s00222-012-0408-1. MR3032328

[Beh09] Kai Behrend, *Donaldson-Thomas type invariants via microlocal geometry*, Ann. of Math. (2) **170** (2009), no. 3, 1307–1338, DOI 10.4007/annals.2009.170.1307. MR2600874

[BSV20] Tristan Bozec, Olivier Schiffmann, and Eric Vasserot, *On the number of points of nilpotent quiver varieties over finite fields* (English, with English and French summaries), Ann. Sci. Éc. Norm. Supér. (4) **53** (2020), no. 6, 1501–1544, DOI 10.24033/asens.2452. MR4203037

[Dav17] Ben Davison, *The critical CoHA of a quiver with potential*, Q. J. Math. **68** (2017), no. 2, 635–703, DOI 10.1093/qmath/haw053. MR3667216

[Dav23a] Ben Davison, *A boson-fermion correspondence in cohomological Donaldson-Thomas theory*, Glasg. Math. J. **65** (2023), no. S1, S28–S52, DOI 10.1017/S001708952200009X. MR4594274

[Dav23b] Ben Davison, *The integrality conjecture and the cohomology of preprojective stacks*, J. Reine Angew. Math. **804** (2023), 105–154, DOI 10.1515/crelle-2023-0065. MR4661532

[Dav25] Ben Davison, *BPS Lie algebras and the less perverse filtration on the preprojective CoHA*, Adv. Math. **463** (2025), Paper No. 110114, 75, DOI 10.1016/j.aim.2025.110114. MR4853317

[DM20] Ben Davison and Sven Meinhardt, *Cohomological Donaldson-Thomas theory of a quiver with potential and quantum enveloping algebras*, Invent. Math. **221** (2020), no. 3, 777–871, DOI 10.1007/s00222-020-00961-y. MR4132957

[Gin06] Victor Ginzburg, *Calabi–Yau algebras*, `arXiv:math/0612139`, 2006.

[Gro96] Ian Grojnowski, *Instantons and affine algebras. I. The Hilbert scheme and vertex operators*, Math. Res. Lett. **3** (1996), no. 2, 275–291, DOI 10.4310/MRL.1996.v3.n2.a12. MR1386846

[GY22] Dmitry Galakhov and Masahito Yamazaki, *Quiver Yangian and supersymmetric quantum mechanics*, Comm. Math. Phys. **396** (2022), no. 2, 713–785, DOI 10.1007/s00220-022-04490-y. MR4504934

[Kac83] Victor G. Kac, *Root systems, representations of quivers and invariant theory*, Invariant theory (Montecatini, 1982), Lecture Notes in Math., vol. 996, Springer, Berlin, 1983, pp. 74–108, DOI 10.1007/BFb0063236. MR718127

[KS90] Masaki Kashiwara and Pierre Schapira, *Sheaves on manifolds*, Grundlehren der mathematischen Wissenschaften [Fundamental Principles of Mathematical Sciences], vol. 292, Springer-Verlag, Berlin, 1990. With a chapter in French by Christian Houzel, DOI 10.1007/978-3-662-02661-8. MR1074006

[KS08] Maxim Kontsevich and Yan Soibelman, *Stability structures, motivic Donaldson–Thomas invariants and cluster transformations*, `arXiv:math.AG/08112435`, 2008.

[KS11] Maxim Kontsevich and Yan Soibelman, *Cohomological Hall algebra, exponential Hodge structures and motivic Donaldson-Thomas invariants*, Commun. Number Theory Phys. **5** (2011), no. 2, 231–352, DOI 10.4310/CNTP.2011.v5.n2.a1. MR2851153

[KV23] Mikhail Kapranov and Eric Vasserot, *The cohomological Hall algebra of a surface and factorization cohomology*, J. Eur. Math. Soc. (JEMS) **25** (2023), no. 11, 4221–4289, DOI 10.4171/jems/1264. MR4662292

[Leh99] Manfred Lehn, *Chern classes of tautological sheaves on Hilbert schemes of points on surfaces*, Invent. Math. **136** (1999), no. 1, 157–207, DOI 10.1007/s002220050307. MR1681097

[LY20] Wei Li and Masahito Yamazaki, *Quiver Yangian from crystal melting*, J. High Energy Phys. **11** (2020), 035, 124, DOI 10.1007/jhep11(2020)035. MR4204250

[Mas01] David B. Massey, *The Sebastiani-Thom isomorphism in the derived category*, Compositio Math. **125** (2001), no. 3, 353–362, DOI 10.1023/A:1002608716514. MR1818986

[Mik07] Kei Miki, *A (q,γ) analog of the $W_{1+\infty}$ algebra*, J. Math. Phys. **48** (2007), no. 12, 123520, 35, DOI 10.1063/1.2823979. MR2377852

[Nak94] Hiraku Nakajima, *Instantons on ALE spaces, quiver varieties, and Kac-Moody algebras*, Duke Math. J. **76** (1994), no. 2, 365–416, DOI 10.1215/S0012-7094-94-07613-8. MR1302318

[Nak97] Hiraku Nakajima, *Heisenberg algebra and Hilbert schemes of points on projective surfaces*, Ann. of Math. (2) **145** (1997), no. 2, 379–388, DOI 10.2307/2951818. MR1441880

[Neg16] Andrei Neguţ, *Exts and the AGT relations*, Lett. Math. Phys. **106** (2016), no. 9, 1265–1316, DOI 10.1007/s11005-016-0865-3. MR3533570

[Neg24] Andrei Neguţ, *An integral form of quantum toroidal* $\mathfrak{gl}_1$, Math. Rep. (Bucur.) **26(76)** (2024), no. 3-4, 183–205. MR4836214

[RS17] Jie Ren and Yan Soibelman, *Cohomological Hall algebras, semicanonical bases and Donaldson-Thomas invariants for 2-dimensional Calabi-Yau categories (with an appendix by Ben Davison)*, Algebra, geometry, and physics in the 21st century, Progr. Math., vol. 324, Birkhäuser/Springer, Cham, 2017, pp. 261–293, DOI 10.1007/978-3-319-59939-7_7. MR3727563

[RSYZ20] Miroslav Rapčák, Yan Soibelman, Yaping Yang, and Gufang Zhao, *Cohomological Hall algebras, vertex algebras and instantons*, Comm. Math. Phys. **376** (2020), no. 3, 1803–1873, DOI 10.1007/s00220-019-03575-5. MR4104538

[RSYZ23] Miroslav Rapčák, Yan Soibelman, Yaping Yang, and Gufang Zhao, *Cohomological Hall algebras and perverse coherent sheaves on toric Calabi-Yau 3-folds*, Commun. Number Theory Phys. **17** (2023), no. 4, 847–939, DOI 10.4310/cntp.2023.v17.n4.a2. MR4704941

[Soi16] Yan Soibelman, *Remarks on cohomological Hall algebras and their representations*, Arbeitstagung Bonn 2013, Progr. Math., vol. 319, Birkhäuser/Springer, Cham, 2016, pp. 355–385, DOI 10.1007/978-3-319-43648-7_13. MR3618057

[SV13] Olivier Schiffmann and Eric Vasserot, *Cherednik algebras, W-algebras and the equivariant cohomology of the moduli space of instantons on* $\mathbf{A}^2$, Publ. Math. Inst. Hautes Études Sci. **118** (2013), 213–342, DOI 10.1007/s10240-013-0052-3. MR3150250

[SV20] Olivier Schiffmann and Eric Vasserot, *On cohomological Hall algebras of quivers: Generators*, J. Reine Angew. Math. **760** (2020), 59–132, DOI 10.1515/crelle-2018-0004. MR4069884

[Sze08] Balázs Szendrői, *Non-commutative Donaldson-Thomas invariants and the conifold*, Geom. Topol. **12** (2008), no. 2, 1171–1202, DOI 10.2140/gt.2008.12.1171. MR2403807

[Tho00] Richard P. Thomas, *A holomorphic Casson invariant for Calabi-Yau 3-folds, and bundles on* $K3$ *fibrations*, J. Differential Geom. **54** (2000), no. 2, 367–438. MR1818182

[Tsy17] Alexander Tsymbaliuk, *The affine Yangian of* $\mathfrak{gl}_1$ *revisited*, Adv. Math. **304** (2017), 583–645, DOI 10.1016/j.aim.2016.08.041. MR3558218

[VdB15] Michel Van den Bergh, *Calabi-Yau algebras and superpotentials*, Selecta Math. (N.S.) **21** (2015), no. 2, 555–603, DOI 10.1007/s00029-014-0166-6. MR3338683

[YZ18] Yaping Yang and Gufang Zhao, *The cohomological Hall algebra of a preprojective algebra*, Proc. Lond. Math. Soc. (3) **116** (2018), no. 5, 1029–1074, DOI 10.1112/plms.12111. MR3805051

[YZ20] Yaping Yang and Gufang Zhao, *On two cohomological Hall algebras*, Proc. Roy. Soc. Edinburgh Sect. A **150** (2020), no. 3, 1581–1607, DOI 10.1017/prm.2018.162. MR4091073

School of Mathematics and Maxwell Institute for Mathematical Sciences, University of Edinburgh, Edinburgh EH9 3FD, United Kingdom

Email address: Ben.Davison@ed.ac.uk

Contemporary Mathematics
Volume **841**, 2026
https://doi.org/10.1090/conm/841/16837

PROBs and perverse sheaves II.
Ran spaces and 0-cycles with coefficients

Mikhail Kapranov and Vadim Schechtman

To Ezra Getzler for his sixtieth birthday

ABSTRACT. We consider the space $Z(\mathbb{C}, L)$ of 0-cycles on the complex line $\mathbb{C}$ with coefficients in a commutative monoid L subject to certain conditions. Such spaces include the symmetric products (for $L = \mathbb{Z}_+$) and the Ran space (for $L = \mathbb{T} = \{\texttt{True}, \texttt{False}\}$ being the Boolean algebra of truth values). We describe the appropriately defined category of perverse sheaves on $Z(\mathbb{C}, L)$ in terms of the braided category (PROB) generated by the components of the universal L-graded bialgebra. We give another description in terms of so-called Janus sheaves which are objects of mixed functoriality (data covariant in one direction and contravariant in the other) on a category formed by certain matrices with entries in L. The matrices in question are analogs of contingency tables familiar in statistics.

1. Introduction

1.1. Description of the main result. This is the second part of the series, see [**23**], where we develop a new type of "quiver description" of the categories of perverse sheaves on certain configuration spaces. The "quiver" in our description is a block (direct summand) of the braided monoidal category (colored PROB) governing graded bialgebras of appropriate type. We refer to [**23**] and references therein for discussion of the terms PROP and PROB.

More precisely, we start with a commutative monoid $(L, +, 0)$ with operation written additively and unit element 0, satisfying natural monotonicity conditions (Assumptions 2.2) and consider the space $Z(\mathbb{C}, L)$ formed by 0-cycles (or divisors) $\mathbf{z} = \sum_{z \in \mathbb{C}} n_z \cdot z$ on the complex line $\mathbb{C}$ with coefficients $n_z \in L$. This space splits into connected components $Z(\mathbb{C}, L)_n$ labelled by $n \in L$ (the degree of the 0-cycle), and we consider each $Z(\mathbb{C}, L)_n$ separately. Among the $Z(\mathbb{C}, L)_n$ we find:

- The symmetric products $\mathrm{Sym}^n(\mathbb{C}) = Z(\mathbb{C}, \mathbb{Z}_+)_n$, corresponding to $L = \mathbb{Z}_+$. This is the case considered in [**23**].
- The Ran space $\mathrm{Ran}(\mathbb{C}) = Z(\mathbb{C}, \mathbb{T})_{\overline{1}}$. It corresponds to $L = \mathbb{T} = \{\overline{0}, \overline{1}\}$ being the Boolean algebra of truth values $\overline{0} = \texttt{False}$, $\overline{1} = \texttt{True}$, with the operation $+ = \vee = \texttt{or}$ being the Boolean addition.

2020 *Mathematics Subject Classification.* Primary 14F08.

This work was supported by the World Premier International Research Center Initiative (WPI Initiative), MEXT, Japan. The work of M.K. was also supported by the JSPS KAKENHI grant 20H01794.

©2026 American Mathematical Society

- Various flag Ran spaces and spaces of colored divisors, see §2.1

The space $Z(\mathbb{C}, L)_n$ may be infinite-dimensional but it has a natural structure of an inductive limit of stratified spaces with strata being complex manifolds. So fixing a base field $\mathbf{k}$, we can consider perverse sheaves (of $\mathbf{k}$-vector spaces, with middle perversity) on $Z(\mathbb{C}, L)_n$ supported on a finite-dimensional skeleton, similarly to the approach common, e.g., in the theory of the Geometric Satake Correspondence [**28**]. We denote this category $\mathrm{Perv}^{<\infty}(Z(\mathbb{C}, L)_n)$.

On the other hand, for any $\mathbf{k}$-linear braided monoidal category $(\mathcal{V}, \otimes, \mathbf{1}, R)$ we can consider L-graded bialgebras $A = \bigoplus_{n \in L} A_n$ in $\mathcal{V}$, with $A_0 = \mathbf{1}$ and all (co)multiplications involving A_0 being the identities. Note that we can avoid speaking of direct sums or assuming their existence in $\mathcal{V}$, formulating everything in terms of data involving the individual A_i. With this understanding, there is a *universal L-graded bialgebra* $\mathfrak{a}$ belonging to a similar universal braided category $\mathfrak{B}^L$ which is generated (as a braided category) by formal generating objects $\mathbf{a}_n$, $n \in L$ and by generating morphisms which are the components of the multiplication and comultiplication, subject only to the axioms of an L-graded bialgebra, see §6.2 We refer to $\mathfrak{B}^L$ as the *(colored) PROB of L-graded bialgebras.* Our main result, Theorem 7.14, says that $\mathrm{Perv}^{<\infty}(Z(\mathbb{C}, L)_n)$ is identified with the category of functors $\mathfrak{B}^L_n \to \mathrm{Vect}_{\mathbf{k}}$ which vanish on all but finitely many objects. In fact, we consider a nominally more general case where $\mathrm{Vect}_{\mathbf{k}}$ is replaced by an arbitrary abelian category.

1.2. The main tool: Contingency matrices. Our main tool relating the topology of $Z(\mathbb{C}, L)$ and L-graded bialgebras is the concept of a contingency matrix with values in L. By this we mean, generalizing [**21**] , a matrix $M = \|m_{ij}\|$ over L of arbitrary size $p \times q$ without zero rows or columns. Remarkably, they appear in two contexts:

- They parametrize cells U_M of a natural *contingency cell decomposition* $\mathcal{S}^{(2)}$ of $Z(\mathbb{C}, L)$. More precisely, we we record the real and imaginary parts of every complex number separately, then every finite subset of $\mathbb{C}$ can be seen as lying naturally on a finite grid. The multiplicities $n_z \in L$ of any 0-cycle $\mathbf{z} = \sum_{z \in \mathbb{C}} n_z \cdot z$ arrange therefore into a contingency matrix $M(\mathbf{z})$, see Figure 1 and §4.1 The cell U_M consists of $\mathbf{z}$ with $M(\mathbf{z}) = M$.
- They appear naturally (and classically for $L = \mathbb{Z}_+$) in formulas expressing the compatibility between multiplication and comultiplication in graded bialgebras as the level of graded components, see §6.1.

In general, the cell decomposition $\mathcal{S}^{(2)}$ is not quasi-regular, i.e., informally, a bigger cell can approach a smaller cell from several different directions. This type of behavior is formalized by Mac Pherson's concept of the exit path category [**8**, **9**, **31**]. We describe such category for $\mathcal{S}^{(2)}$ in terms of the two augmented semi-simplicial structures on the set of all contingency matrices, given by adding rows and by adding columns. See Proposition 4.5 (where we consider 0-cycles in arbitrary $\mathbb{R}^p$, not just in $\mathbb{C} = \mathbb{R}^2$). This allows us to describe the category of $\mathcal{S}^{(2)}$-constructible sheaves as representations of the exit path category.

Note that perverse sheaves on $Z(\mathbb{C}, L)_n$ are, in particular, $\mathcal{S}^{(2)}$-constructible complexes since the natural "complex stratification by multiplicities" $\mathcal{S}^{(0)}$ is refined by $\mathcal{S}^{(2)}$.

1.3. Second main result: Janus sheaves. Our proof of Theorem 7.14 is based on another result, Theorem 8.4, which is interesting in its own right and analogous to the main result of [**22**]. It describes the category $\mathrm{Perv}^{<\infty}(Z(\mathbb{C}, L)_n, \mathcal{V})$ in terms of combinatorial diagrams of bivariant nature (covariant in one direction, contravariant in the other) which we call *Janus sheaves.* Such diagrams are analogous to mixed Bruhat sheaves introduced in [**22**] in the context of Coxeter groups. A nontrivial intersection between the two concepts appears for $L = \mathbb{Z}_+$, when Janus sheaves are the same as mixed Bruhat sheaves for the symmetric group S_n acting on $\mathbb{R}^n$. In this case we have a *third appearance* of contingency matrices: as parametrizing the parabolic Bruhat decompositions for the space of partial classical (GL_n-) flags, see [**6**, §4.3].

If we view a perverse sheaf $\mathcal{F} \in \mathrm{Perv}^{<\infty}(Z(\mathbb{C}, L)_n)$ as a complex of $\mathcal{S}^{(2)}$-constructible sheaves so as a complex $F^\bullet$ of functors on the exit path category, then the Janus sheaf corresponding to $\mathcal{F}$ can be obtained from $F^\bullet$ by an appropriate formal version of *relative Verdier duality.* By "relative" we mean duality taken along the fibers of the imaginary part map $\mathrm{Im} : Z(\mathbb{C}, L) \to \mathbb{Z}(\mathbb{R}, L)$ to the space of 0-cycles on $\mathbb{R}$. A remarkable fact is that the perverse t-structure on constructible complexes is taken by such duality into the standard t-structure on the derived category, so the Janus sheaf associated to a perverse sheaf $\mathcal{F}$ is a non-derived object, i.e., consists of vector spaces and not complexes.

More generally, it seems to us that the two main points of view on perverse sheaves used in algebraic geometry—as complexes of sheaves and as "linear algebra data" with values in an abelian category—are almost invariably connected by an appropriate relative version of Verdier duality. We plan to develop this point of view in a future paper.

1.4. Ran spaces and logic? Specializing to Ran spaces, we have a cell decomposition $\mathcal{S}^{(2)}$ of $\mathrm{Ran}(\mathbb{C})$ labelled by *Boolean contingency matrices* $M = \|m_{ij}\|$, $m_{ij} \in \mathbb{T} = \{\overline{0}, \overline{1}\}$. The cell attachment structure (i.e., the category of exit paths) of this cell decomposition is described in terms of Boolean addition of rows and columns of such matrices. This adds an extra touch to the "logical flavor" of Ran spaces in general. Thus, for any finite connected CW-complex X the space $\mathrm{Ran}(X)$ formed by finite nonempty subsets $I \subset X$, possesses an idempotent semigroup structure $(I, J) \mapsto I \cup J$. This implies that $\mathrm{Ran}(X)$ is weakly contractible [**4**, §3.4.1], a fact whose algebro-geometric analog (for X an algebraic curve) plays an important role in the study of the moduli stack of G-bundles on X; see [**12**].

It is therefore tempting to "marry" Ran spaces with more general algebraic structures appearing in mathematical logic. For example, considering 0-cycles with coefficients in the "algebra of truth values of the N-valued logic" (see Example 2.1(f)) leads to the flag Ran space formed by flags of finite subsets $X \supset I^1 \supset \cdots \supset I^{N-1}$. Further, one can likely weaken Assumption 2.2(b) so as to allow more general distributive lattices of interest in logic such as the lattice of vector subspaces in $\mathbb{F}_q^N$ with addition being the sum of subspaces ("quantum logic"). The corresponding versions of Ran spaces seem worth studying.

1.5. Further directions: Non-finitistic perverse sheaves and factorization. It is desirable (and we believe possible) to develop a more general concept of perverse sheaves on ind-stratified spaces such as $Z(\mathbb{C}, L)$ (in particular, on ind-schemes such as the affine Grassmannian). We mean "perverse sheaves" not necessarily supported on a finite skeleton and not represented as limits of perverse subsheaves thus supported. One reason for attempting this is to incorporate factorization structures, i.e., to speak about *factorizing systems* $(\mathcal{F}_n)_{n\in L}$ of perverse sheaves on all the $Z(\mathbb{C}, L)_n$ at once. Such systems should correspond to actual L-graded bialgebras A, i.e., to braided monoidal functors from the PROB $\mathfrak{B}^L = \bigoplus_{n\in L} \mathfrak{B}^L_n$. Note that for a general A (with all $A_n \neq 0$) each $\mathcal{F}_n$ must be supported on the entire $Z(\mathbb{C}, L)_n$ and so cannot be supported on a finite skeleton unless $Z(\mathbb{C}, L)_n$ is finite-dimensional (as in the case $L = \mathbb{Z}_+$, $Z(\mathbb{C}, \mathbb{Z}_+)_n = \mathrm{Sym}^n(\mathbb{C})$ treated in [**20, 23**]).

Note that the naive approach of considering all constructible complexes on $Z(\mathbb{C}, L)_n$ and imposing the standard geometric perversity conditions does not seem to work as it is not clear whether this gives a t-structure. Still, we believe that there is a natural and satisfactory way to address this issue and hope to take it up in the future. This may simplify the approach to Etingof-Kazhdan quantization of bialgebras developed by Kalugin [**18**], where a different, more sophisticated version of the Ran space is used.

The main phenomenon underlying this paper and [**23**], is the Koszul duality between bialgebras and E_2-algebras. That is, applying the derived Koszul duality (cobar-construction taking dg-algebras to dg-coalgebras) to the multiplication in a bialgebra, one gets a dg-vector space with two homotopy commuting comultiplications, i.e., an E_2-coalgebra. This fundamental observation is due to Lurie and Gaitsgory; see [**11**]. In their turn, E_2-coalgebras can be described as factorizing complexes of sheaves on $\mathrm{Ran}(\mathbb{C})$; see [**25**]. From this point of view, the additional feature studied here and in [**23**] (and also implicit in [**11**]) is that the perversity condition on complexes of sheaves corresponds to the condition on an E_2-coalgebra that the partial Koszul dual is a single vector space, i.e., a (non-dg) bialgebra. In the present paper and [**23**] we do not assume factorizability, so E_2-coalgebras do not appear as such. Nevertheless, one can say that partial Koszul duality as above corresponds, in our approach, to the partial Verdier duality (along the "horizontal" real direction) for complexes of sheaves.

From this point of view one can look at any E_n-coalgebra and "Koszul dualize" any p-element subset of the n commuting comultiplications, resulting in an algebraic structure with p multiplications and $n-p$ comultiplications, all compatible in a natural sense. Such a structure is a priori a complex, and for $n > 2$ there do not seem to be interesting new examples where this complex reduces to a single vector space.

1.6. Organization of the paper. In §2 we recall the basic properties of the space $Z(X, L)$ of 0-cycles in a space X with coefficients in an abelian monoid L. Taken by itself, this generality is not new: much more general "configuration spaces with summable labels" were studied by P. Salvatore [**30**] and we refer to his work for foundational material. We focus, however, on monoids satisfying certain monotonicity properties (Assumptions 2.2) which allow us to treat $Z(X, L)$ for X a manifold as an (ind-)stratified space and to speak about perverse sheaves on it later on. These assumotions also allow us to identify, for connected X, the connected components $Z(X, L)_n$ with elements $n \in L$ (the degree of 0-cycles).

§3 is devoted to the concept of contingency matrices with coefficients in a monoid L. We consider "p-dimensional contingency matrices" (i.e., arrays) $\|m_{i_1,\cdots,i_p}\|$ for any p. The set of such matrices with *content* (sum of all the elements) equal to a given $n \in L$, is organized into a p-fold augmented semi-simplicial set $\underline{\mathrm{CM}}^p_n(L)$ by adding together adjacent slices in each of the p directions.

In §4 we construct a cell decomposition $\mathcal{S}^{(2)}$ of $Z(\mathbb{R}^p, L)_n$ labelled by p-dimensional contingency matrices of content n and describe the exit path category of this cell decomposition as the Grothendieck construction of $\underline{\mathrm{CM}}_n^p(L)$ (Proposition 4.5).

In §5 we specialize to the case $X = \mathbb{C} = \mathbb{R}^2$ and consider some related stratifications of $Z(\mathbb{C}, L)$: the Fox-Neuwirth-Fuchs cell decomposition $\mathcal{S}^{(1)}$ and the decomposition $\mathcal{I}$ into imaginary strata $Z_{\mathbf{n}}^{\mathrm{Im}}$.

In §6 we define the concept of L-graded bialgebras in braided monoidal categories and consider the PROB $\mathfrak{B}^L$ generated by the components of the universal such bialgebra. In the case $L = \mathbb{T}$, relevant for Ran spaces, we relate $\mathbb{T}$-graded bialgebras with more familiar ungraded bialgebras with unit and counit.

In §7 we define the category $\mathrm{Perv}^{<\infty}(Z(\mathbb{C}, L)_n)$ and formulate the main result (Theorem 7.14) relating it with the block $\mathfrak{B}_n^L$ of $\mathfrak{B}^L$. We also prove some technical results describing various features of and operations on constructible complexes in terms of their combinatorial description via the category of exit paths.

In §8 we introduce the concept of Janus sheaves and give an example of a Janus sheaf associated to an L-graded bialgebra (Proposition 8.3). We also formulate our second main result, Theorem 8.4 relating Janus sheaves with perverse sheaves.

§9 is devoted to the proof of Theorem 8.4. Our main tool, similar to [**22**], is the concept of a Cousin-type complexes which on one hand, are algebraically in a one-to-one correspondence with Janus (pre)sheaves and on the other, appear as Cousin resolutions of perverse sheaves.

Finally, in §10 we deduce Theorem 7.14 from Theorem 8.4. This is done by considering the *universal Janus sheaf* taking values in a braided category $\mathfrak{CM}(L)$ specially constructed for this purpose. We then identify $\mathfrak{B}^L$ and $\mathfrak{CM}(L)$ as braided categories (Theorem 10.7) by constructing a natural L-graded bialgebra in $\mathfrak{CM}(L)$.

2. 0-cycles with coefficients in a monoid

2.1. Spaces of 0-cycles. Let $(L, +)$ be a commutative monoid (semigroup with unit element, denoted 0). Let X be a topological space which we assume homeomorphic to an open subset of a finite CW-complex. By a *0-cycle* on X with coefficients in L we will mean a formal linear combination $\mathbf{z} = \sum_{x \in X} n_x \cdot x$ of points $x \in X$ with multiplicities $n_x \in L$ such that $n_x = 0$ for almost all x. Denote by $Z(X, L)$ the set of all 0-cycles on X with coefficients in L. This set has a natural topology induced by the topology of X and the additive structure on L ("when points merge together, their multiplicities are added"); see [**30**]. With respect to this topology and the obvious addition of 0-cycles, $Z(X, L)$ is a commutative topological monoid with unit element 0, the 0-cycle with all $n_x = 0$.

EXAMPLES 2.1.

(a) Suppose that L is an abelian group. In this case the classical Dold-Thom theorem says that $\pi_i Z(X, L) \simeq H_i(X, L)$ for all $i \geqslant 0$. In particular, taking $X = S^n$ to be the n-dimensional sphere, we get $Z(S^n, L) \simeq L \times K(L, n)$, where $K(L, n)$ is the Eilenberg-MacLane space.

(b) If X is connected, then for any monoid L as above we have $\pi_0 Z(X, L) \simeq L$. The identification is given by associng to a 0-cycle $\mathbf{z} = \sum n_x \cdot x$ its *degree* $\deg(\mathbf{z}) = \sum n_x \in L$. That is, the component $Z(X, L)_n$ corresponding to $n \in L$ consists of cycles of degree n.

(c) Let $L = \mathbb{Z}_+$. Then $Z(X, \mathbb{Z}_+) = \bigsqcup_{n \geqslant 0} \mathrm{Sym}^n(X)$ is the disjoint union of the symmetric powers of X. Thus $Z(X, \mathbb{Z}_+)_n = \mathrm{Sym}^n(X)$.

(d) More generally, let $L = \Lambda_+$ be the dominant (co)weight lattice of a reductive group. Then $L \simeq \mathbb{Z}_+^I$ where I is the set of fundamental (co)weights. If X is an algebraic curve over $\mathbb{C}$, then $Z(X, \Lambda_+)$ is the space of I-colored divisors on X considered in [**10**] and [**11**].

(e) Let $L = \mathbb{T} = \{\overline{0} = \texttt{False}, \overline{1} = \texttt{True}\}$ be the Boolean algebra of truth values, with $+ = \vee$ being the Boolean addition: $\overline{0} + x = x$, $\overline{1} + x = \overline{1}$. Suppose X connected. Then $Z(X, \mathbb{T})$ consists of two components: $Z(X, \mathbb{T})_{\overline{0}} = \{0\}$ and $Z(X, \mathbb{T})_{\overline{1}} = \mathrm{Ran}(X)$ being the *Ran space* of X, i.e., the space of all nonempty finite subsets $I \subset X$ equipped with natural (Vietoris) topology; see [**4**, §3.4.1].

(f) More generally, let $N \geqslant 2$ be an integer and

$$L = \mathbb{T}_N = \{0, 1/(N-1), 2/(N-1), \cdots, 1\}$$

be the "algebra of truth values of N-valued logic", with the operation $x + y = \max(x, y)$ (maximum with respect to the order induced from $\mathbb{R}$). Then $Z(X, \mathbb{T}_N)$ is the *flag Ran space* of X, i.e., the space formed by flags of finite subsets $X \supset I^1 \supset \cdots \supset I^{N-1}$ with the topology induced from $(\{0\} \sqcup \mathrm{Ran}(X))^{N-1}$, in which it is embedded as a closed subset given by the incidence conditions. Explicitly, let us put $I^0 = X$, $I^N = \varnothing$. Then a flag $I^1 \supset \cdots \supset I^{N-1}$ corresponds to the 0-cycle $\sum n_x \cdot x$ where $n_x = k/(N-1)$, if $x \in I^k \setminus I^{k+1}$.

2.2. Skeleta of $Z(X, L)$. From now on we impose the following.

ASSUMPTIONS 2.2.

(a) L is finitely generated as a monoid.

(b) L admits a total order $\leqslant$ compatible with the addition such that 0 is the minimal element and each lower interval $L_{\leqslant n} = \{m \in L \mid m \leqslant n\}$ is finite.

EXAMPLES 2.3. Assumptions 2.2 exclude part (a) of Examples 2.1 but allow parts (c)–(f). Indeed:

(a) An abelian group other than $\{0\}$ does not admit a compatible order with 0 being minimal.

(c)(d) The free monoid $\mathbb{Z}_+^d$ admits a compatible order as in Assumption 2.2 (b). Indeed, let us choose positive real numbers $\alpha_1, \cdots, \alpha_d$ which are linearly independent over $\mathbb{Q}$. Then the map

$$\mathbb{Z}_+^d \longrightarrow \mathbb{R}, \quad (a_1, \cdots, a_d) \mapsto \sum a_i \alpha_i$$

is an embedding. Pulling back the order on $\mathbb{R}$ under this embedding, we get an order on $\mathbb{Z}_+^d$ satisfying the assumption.

(e)(f) The monoid $\mathbb{T}_N$ (of which $\mathbb{T} = \mathbb{T}_2$ is a particular case) is finite and comes equipped with the order induced from $\mathbb{R}$ which satisfies the assumption.

PROPOSITION 2.4. *For any $d \geqslant 1$ and $n \in L$ the set*

$$\{(n_1, \cdots, n_d) \in L^d \mid n_1 + \cdots + n_d = n\}$$

is finite.

PROOF. The equality $n_1 + \cdots + n_d = n$ implies that each $n_i \leqslant n$, so our set is embedded into the Cartesian power $L_{\leqslant n}^d$ which is finite. □

We now study the space $Z(X, L)$ in more detail. First, note that our assumptions imply that 0 is an isolated point of $Z(X, L)$.

Second, let us choose once and for all a system of generators $(\xi_k)_{k\in K}$ of L so we have a surjective morphism of monoids $\mathbb{Z}_+^K \to L$. Let $X_K = K \times X$ be the disjoint union of K copies of X; its points are pairs (k, x), $k \in K$, $x \in X$. For each $r \geqslant 0$ we have the continuous map

$$p_l : X_K^l \to Z(X,L), \quad ((k_1, x_1), \cdots, (k_l, x_l)) \mapsto \sum \xi_{k_i} \cdot x_i. \tag{2.5}$$

PROPOSITION 2.6. *Each p_l is a proper map with finite fibers.*

PROOF. The fact that p_r has finite fibers follows from Proposition 2.4. The fact that p_r is proper follows from this and from the definition of topology on $Z(X, L)$. □

Proposition 2.6 implies that the image of p_l is a closed subset of $Z(X, L)$. We denote by $Z_l(X, L)$ the union of the images of the $p_i, i \leqslant l$. Thus we have an exhaustion

$$Z_0(X,L) \subset Z_1(X,L) \subset \cdots \subset Z(X,L) = \varinjlim_l{}^{\mathcal{T}op} Z_l(X,L) \tag{2.7}$$

by closed subsets so that the topology on $Z(X, L)$ is that of their inductive limit. We will refer to $Z_l(X, L)$ as the *lth skeleton* of $Z(X, L)$.

2.3. Strata of $Z(X, L)$ labelled by partitions. From now on we further assume that X is a C^∞-manifold of dimension p (without boundary).

Let $n \in L$ be a nonzero element. A *partition* of n is, by definition, a sequence $\alpha = (\alpha_1, \cdots, \alpha_d)$ of nonzero elements such that $\sum n_i = n$, considered up to permutations. We denote $\mathcal{P}_n^L$ the set of all partitions of n. We put $\mathcal{P}_0^L = \{\varnothing\}$ (the set consisting of the empty partition) and $\mathcal{P}^L = \bigsqcup_{n\geqslant 0} \mathcal{P}_n^L$. We introduce a partial order $\leqslant$ on $\mathcal{P}^L$ putting

$$\beta = (\beta_1, \cdots, \beta_c) \leqslant \alpha = (\alpha_1, \cdots, \alpha_d)$$

if β can be obtained from α by partial summation, i.e., there is a surjection $f : \{1, \cdots, d\} \to \{1, \cdots, c\}$ such that $\beta_i = \sum_{j\in f^{-1}(i)} \alpha_j$.

Let $\mathbf{z} = \sum_{x\in X} n_x \cdot x \in Z(X, L)$. Writing the set $\{x \mid n_x \neq 0\}$ as $\{x_1, \cdots, x_d\}$ in some order, we get a partition $\alpha_{\mathbf{z}} = (n_{x_1}, \cdots, n_{x_d})$ of $n = \deg(\mathbf{z})$. For a given partition $\alpha \in \mathcal{P}^L$ we denote $Z(X, L)_\alpha \subset Z(X, L)$ the set of $\mathbf{z}$ such that $\alpha_{\mathbf{z}} = \alpha$. Thus we obtain a disjoint decomposition

$$Z(X,L) = \bigsqcup_{\alpha\in\mathcal{P}^L} Z(X,L)_\alpha \tag{2.8}$$

into locally closed subsets. We denote this decomposition $\mathcal{S}^{(0)}$ and refer to the $Z(X, L)_\alpha$ as the *strata* of $\mathcal{S}^{(0)}$.

PROPOSITION 2.9.

(a) *Each stratum $Z(X, L)_\alpha$, $\alpha = (\alpha_1, \cdots, \alpha_d)$ is a C^∞-manifold of dimension pd.*

(b) *The closure $\overline{Z(X, L)_\alpha}$ is the union of the $Z(X, L)_\beta$ for $\beta \leqslant \alpha$.*

(c) *With respect to the decomposition $\mathcal{S}^{(0)}$, each skeleton $Z_l(X, L)$, has a natural structure of a Whitney stratified space and each $Z_l(X, L) \hookrightarrow Z_{l'}(X, L)$, $r \leqslant r'$ is a closed embedding of stratified spaces.*

Thus $(Z(X, L), \mathcal{S}^{(0)})$ has the structure of an *ind-stratified space*, a filtered inductive limit of stratified spaces and their closed stratified embeddings.

PROOF.

(a) Let us write us write $\alpha \in \mathcal{P}_n^L$ as consisting of d_1 repetitions of n_1, d_2 repetitions of n_2, ..., d_q repetitions of n_q, where $n_1, \cdots, n_q$ are distinct. So $d_1 n_1 + \cdots + d_q n_q = n$. Then it is immediate that $Z(X, L)_{\mathbf{n}}$ is identified with the open subset in $\mathrm{Sym}^{d_1}(X) \times \cdots \times \mathrm{Sym}^{d_q}(X)$ formed by tuples

$$(\{x_1, \cdots x_{d_1}\}, \{x_{d_1+1}, \cdots, x_{d_1+\cdots+d_2}\}, \cdots, \{x_{d_1+\cdots+d_{q-1}+1}, \cdots, x_{d_1+\cdots+d_n}\})$$

such that all the x_i are distinct.

(b) This is clear from the definition of the topology on $Z(X, L)$, see §2.1

(c) We consider the Cartesian power X_K^l with its natural stratification by multiplicities (patterns of coincidence of points from X_K). The map p_l of Proposition 2.6 takes this stratification to the decomposition of $\mathrm{Im}(p_l)$ induced by $\mathcal{S}^{(0)}$. Since p_l is a finite map, the identifications on X_K^l induced by the p_l define the stratified structure on $\mathrm{Im}(p_l)$. Since $Z_l(X, L)$ is the union of the $\mathrm{Im}(p_i)$, $i \leqslant l$, we conclude the same for $Z_l(X, L)$. □

EXAMPLE 2.10. Let $L = \mathbb{T} = \{\overline{0}, \overline{1}\}$, so $Z(X, \mathbb{T})_{\overline{1}} = \mathrm{Ran}(X)$. The set $\mathcal{P}_{\overline{1}}^{\mathbb{T}}$ consists of partitions $\overline{1}^d = (\overline{1}, \cdots, \overline{1})$ (d times) of arbitrary length $d \geqslant 1$. The corresponding stratum $\mathrm{Ran}(X)_{\overline{1}^d}$ is identified with $\mathrm{Sym}^d_{\neq}(X)$, the open part in $\mathrm{Sym}^d(X)$ consisting of collections of d distinct points. Further, the natural set of generators of $\mathbb{T}$ consists of one element $\overline{1}$. This gives the skeleta $\mathrm{Ran}_r(X) = \{I \subset X : |I| \leqslant r\}$. Each such skeleton is a stratified space, and the ind-stratified structure on $\mathrm{Ran}(X)$ is given by

$$X = \mathrm{Ran}_1(X) \subset \mathrm{Ran}_2(X) \subset \cdots \subset \mathrm{Ran}(X), \quad \mathrm{Ran}_r(X) \setminus \mathrm{Ran}_{r-1}(X) \simeq \mathrm{Sym}^r_{\neq}(X).$$

3. Contingency matrices with values in a monoid

3.1. p-dimensional contingency matrices. Let L be a commutative monoid satisfying Assumptions 2.2 and let $p, d_1, \cdots, d_p \geqslant 1$ be integers. By a *p-dimensional matrix* of size $d_1 \times \cdots \times d_p$ with values in L we will mean simply an array

$$M = \|m_{i_1, \cdots, i_p}\|, \quad m_{i_1, \cdots, i_p} \in L, \quad 1 \leqslant i_\nu \leqslant d_\nu$$

of elements of L. Thus for $p = 1$ we have vectors $\mathbf{m} = (m_1, \cdots, m_d)$, for $p = 2$ matrices $M = \|m_{ij}\|$ in the usual sense and so on. The *content* of a p-dimensional matrix M is defined as $|M| = \sum m_{i_1, \cdots, i_p} \in L$.

Let $1 \leqslant \nu \leqslant p$ and $1 \leqslant i \leqslant d_\nu$. The *ith slice* of M in the direction ν is the $(p-1)$-dimensional matrix $\sigma_i^\nu(M)$ formed by elements $m_{i_1, \cdots, i_{\nu-1}, i, i_{\nu+1}, \cdots, i_p}$ with the νth index fixed and equal to p. Thus for $p = 2$ we obtain the rows and columns of a usual matrix. For $p = 1$ we define the ith slice $\sigma_i^1(M)$ to be the ith element of the vector $M = (m_1, \cdots, m_d)$.

We say that M is a *contingency matrix*, if no slice $\sigma_i^\nu(M)$ consists entirely of zeroes. Thus a 1-dimensional contingency matrix is a vector $\mathbf{m} = (m_1, \cdots, m_d)$ with all $m_i \neq 0$ (we will also refer to such vectors as *ordered partitions*). A 2-dimensional contingency matrix is a matrix $M = \|m_{ij}\|$ with all the rows and columns being nonzero. We denote by $\mathrm{CM}_n(d_1, \cdots, d_p; L)$ the set of contingency matrices of size $d_1 \times \cdots \times d_p$ with values in L which have content n. Proposition 2.4 implies that each $\mathrm{CM}_n(d_1, \cdots, d_p; L)$ is a finite set. We put

$$\mathrm{CM}_n^p(L) = \bigsqcup_{d_1, \cdots, d_p \geqslant 1} \mathrm{CM}_n(d_1, \cdots, d_p; L).$$

Here $n \neq 0$. We also introduce formally the *empty p-dimensional contingency matrix* $\varnothing$ of content 0 and put $\mathrm{CM}_0^p(L) = \{\varnothing\}$. We then put

$$\mathrm{CM}^p(L) = \bigsqcup_{n \in L} \mathrm{CM}_n^p(L).$$

Let $M \in \mathrm{CM}_n(d_1, \cdots, d_p; L)$. For $1 \leqslant \nu \leqslant p$ and $0 \leqslant i \leqslant d_{\nu-1}$ we define the contingency matrix

$$\partial_i^\nu M \in \mathrm{CM}_n^p(d_1, \cdots, d_{\nu-1}, d_\nu - 1, d_{\nu+1}, \cdots, d_p; L)$$

by putting

(3.1) $(\partial_i^\nu M)_{i_1, \cdots, i_p}$

$$= \begin{cases} m_{i_1, \cdots, i_p}, & \text{if } i_\nu \leqslant i, \\ m_{i_1, \cdots, i_{nu-1}, i+1, i_{\nu+2}, \cdots, i_p} + m_{i_1, \cdots, i_{nu-1}, i+2, i_{\nu+2}, \cdots, i_p}, & \text{if } i_\nu = i+1, \\ m_{i_1, \cdots, i_{\nu-1}, i_\nu - 1, i_{\nu+1}, \cdots, i_p}, & \text{if } i_\nu \geqslant i+2, \end{cases}$$

i.e., by adding the $(i+1)$st and $(i+2)$nd slices in direction ν. The fact that ∂_i^ν is again a contingency matrix, follows from Assumptions 2.2.

We refer to the operations ∂_i^ν as *contractions*. It is clear that the ∂_i^ν commute for different ν and satisfy the (semi-)simplicial relations for the same ν:

$$\partial_i^\nu \partial_j^\mu = \partial_j^\mu \partial_i^\nu, \ \mu \neq \nu, \quad \partial_i^\nu \partial_j^\nu = \partial_{j-1}^\nu \partial_i^\nu, \ i < j. \tag{3.2}$$

This extends the definition given in [**21**, §2] for the case $p = 1, 2$ and $L = \mathbb{Z}_+$.

To formulate this more concisely, let Δ_+ be the *augmented semi-simplex category* whose objects are all finite ordinals, including $\varnothing$ and morphisms are monotone embeddings. As usual, we denote the ordinal $\{0, 1, \cdots, m\}$, $m \geqslant 0$, by $[m]$ and $\varnothing$ by $[-1]$. Eq. (3.2) can be restated as:

PROPOSITION 3.3. *The maps ∂_i^ν make $\mathrm{CM}_n^p(L)$ into a p-fold augmented semi-simplicial set. More precisely, they define a functor*

$$\underline{\mathrm{CM}}_n^p(L) = \mathrm{CM}_n^p(\bullet + 2, \cdots, \bullet + 2; L) : (\Delta_+)^{\mathrm{op}} \longrightarrow \mathrm{Set}. \quad \square$$

For future convenience introduce the following.

DEFINITION 3.4. Let $M \in \mathrm{CM}_n^p(d_1, \cdots, d_p; L)$. For $\nu = 1, \cdots, p$ the *νth margin* of M is the ordered partition $\mathrm{Mar}^\nu(M) \in \mathrm{CM}_n^1(d_\nu; L)$ with components

$$\mathrm{Mar}^\nu(M)_j = \sum m_{i_1, \cdots, i_{\nu-1}, j, i_{\nu+1}, \cdots, i_p},$$

the summation being over all collections of $i_\lambda \in \{1, \cdots, d_\lambda\}$, $\lambda \neq \nu$.

3.2. Contingency matrices as objects of a category: The Grothendieck construction. Let $\mathcal{C}$ be a category and $F : \mathcal{C} \to \mathrm{Set}$ be a functor. By the *Grothendieck construction* of F we will mean the category $\mathrm{Gro}(F)$ whose objects are pairs (c, a) with $c \in \mathrm{Ob}(\mathcal{C})$ and $a \in F(c)$. A morphism $(c, a) \to (c', a')$ in $\mathrm{Gro}(F)$ consists of a morphim $\varphi : c \to c'$ in $\mathcal{C}$ such that $F(\varphi)(a) = a'$. Composition of morphisms in $\mathrm{Gro}(F)$ is induced by the composition of morphisms in $\mathcal{C}$. This definition is a particular case of the standard more general concept defined for (pseuso-)functors with values in categories, not just sets (discrete categores), see, e.g., [...].

Alternatively, for $c \in \mathrm{Ob}(\mathcal{C})$ let $h^c = \mathrm{Hom}_{\mathcal{C}}(c, -) : \mathcal{C} \to \mathrm{Set}$ be the covariant functor represented by c. Then elements of $F(c)$ are in bijection with natural transformations $h^c \to F$. Thus $\mathrm{Gro}(F)$ consists of all natural transformations of this kind.

We will apply this to $\mathcal{C} = (\Delta_+^p)^{\mathrm{op}}$ and $F = S : (\Delta_+^p)^{\mathrm{op}} \to \mathrm{Set}$ being a p-fold augmented semi-simplicial set. Objects of Δ_+^p being tuples $[\mathbf{d}] = ([d_1], \cdots, [d_p])$, $\mathbf{d} = (d_1, \cdots, d_p)$, $d_i \geqslant -1$, we denote by

$$\Delta_+^p[\mathbf{d}] = \mathrm{Hom}_{\Delta_+^p}(-, [\mathbf{d}]) : (\Delta_+^p)^{\mathrm{op}} \longrightarrow \mathrm{Set}$$

the functor represented by $[\mathbf{d}]$ and call it the *standard multisimplex* of dimension $\mathbf{d}$. Thus $\mathrm{Gro}(S)$ can be viewed as consisting of all morphisms $\Delta_+^p[\mathbf{d}] \to S$, i.e., of "all multisimplices of S".

We further specialize to $S = \underline{\mathrm{CM}}_n^p(L)$ being the p-fold augmented semi-simplicial set of p-dimensional contingency matrices of content n. Thus $\mathrm{Ob}(\mathrm{Gro}(\underline{\mathrm{CM}}_n^p(L))) = \mathrm{CM}_n^p(L)$ is the set of contingency matrices and morphisms are given by contractions and their compositions. That is, $\mathrm{Mor}(\mathrm{Gro}(\underline{\mathrm{CM}}_n^p(L)))$ is generated by the morphisms $M \to \partial_i^\nu M$ for $M \in \mathrm{CM}_n(d_1, \cdots, d_p)$, $\nu = 1, \cdots, p$ and $i = 0, \cdots, d_p - 2$; we will call these *elementary morphisms.*

For $\nu = 1, \cdots, p$ we denote by $\mathrm{Gro}^\nu(\underline{\mathrm{CM}}_n^p(L))$ the subcategory in $\mathrm{Gro}(\underline{\mathrm{CM}}_n^p(L))$ with the same objects and morphisms induced only by the νth copy of Δ_+^{op}, i.e., by elementary morphisms $M \to \partial_i^\nu M$ with this fixed ν. It is clear that any morphism in $\mathrm{Gro}(\underline{\mathrm{CM}}_n^p(L))$ can be uniquely written as the composition of p morphisms, one from each of the $\mathrm{Gro}^1(\underline{\mathrm{CM}}_n^p(L)), \ldots, \mathrm{Gro}^p(\underline{\mathrm{CM}}_n^p(L))$ (in each of the $p!$ possible orders).

EXAMPLE 3.5. Note that $\varnothing$ is an initial object of Δ_+. Therefore (n), the 1-entry matrix of size $1 \times \cdots \times 1$, is a final object in $\mathrm{Gro}(\underline{\mathrm{CM}}_n^p(L))$: each $M \in \mathrm{CM}_n^p(L)$ has a canonical morphism $M \to (n)$ given by the total contraction (summation of all the entries).

EXAMPLE 3.6. Let $L = \mathbb{Z}_+$. In this case the category $\mathrm{Gro}(\underline{\mathrm{CM}}_n^p(\mathbb{Z}_+))$ reduces to a poset, i.e., for any two $M, M' \in \mathrm{CM}_n^p(\mathbb{Z}_+)$ there is at most one morphism $M \to M'$ in $\mathrm{Gro}(\underline{\mathrm{CM}}_n^p(\mathbb{Z}_+))$.

This fact is immediately seen for $p = 1$, when elements of $\mathrm{CM}_n^1(\mathbb{Z}_+)$, i.e., unordered partitions of $n \in \mathbb{Z}_+$, are in bijection with faces of the $(n-1)$-dimensional simplicial cone $\mathbb{R}_+^{n-1}$, and the contraction $\partial_i = \partial_i^1$ correspond to taking faces of the faces of this cone. The case $p > 1$ is reduced to this. Indeed, given a contraction $M \to M'$ of p-dimensional contingency matrices we have, for each $\nu = 1, \ldots, p$, the contraction $\mathrm{Mar}^\nu(M) \to \mathrm{Mar}^\nu(M')$ of unordered partitions obtained by taking the margins in the νth direction (Definition 3.4). By the above, such a contraction corresponds to a unique morphism of Δ_+. The p-tuple of such morphisms is a morphism of Δ_+^p which is the only possible morphism that can correspond to $M \to M'$.

EXAMPLE 3.7. Let $L = \mathbb{T}$. In this case already for $p = 1$ the category $\mathrm{Gro}(\underline{\mathrm{CM}}_{\bar{1}}^1(\mathbb{T}))$ is not reduced to a poset. In fact, this category is identified with Δ_+^{op}, with the ordered partitions $(\bar{1})$, $(\bar{1}, \bar{1})$, $(\bar{1}, \bar{1}, \bar{1})$ etc. corresponding to the ordinals $[-1] = \varnothing$, $[0]$, $[1]$ etc.

3.3. Anodyne morphisms of contingency matrices. An elementary morphism $M \to \partial_i^\nu M = M'$ in $\mathrm{Gro}(\underline{\mathrm{CM}}_n^p(L))$ will be called *anodyne* if, in each sum in the second line of (3.1), at least one summand is equal to 0. In this case the sets of nonzero entries of M' and M are in bijection. More generally, a morphism

$M \to M'$ will be called anodyne, if it can be decomposed into a chain of elementary anodyne morphisms. In this case again, the sets of nonzero entries of M' and M are in bijection.

We denote by Ano resp. Ano^ν the set of anodyne morphisms in $\mathrm{Gro}(\underline{\mathrm{CM}}_n^p(L))$ resp. $\mathrm{Gro}^\nu(\underline{\mathrm{CM}}_n^p(L))$. We further denote $\equiv_{\mathrm{Ano}}$ resp. $\equiv_{\mathrm{Ano}}^\nu$ the equivalence relation of $\mathrm{CM}_n^p(L)$ generated by the binary relation R_{Ano} resp. R_{Ano}^ν, where $M\,R_{\mathrm{Ano}}N$ resp. $M\,R_{\mathrm{Ano}}^\nu N$ means that there is an anodyne morphis, $M \to N$ in $\mathrm{Gro}(\underline{\mathrm{CM}}_n^p(L))$ resp. $\mathrm{Gro}^\nu(\underline{\mathrm{CM}}_n^p(L))$. Thus, contingency matrices from each equivalence class of $\equiv_{\mathrm{Ano}}$ have the same set of nonzero entries.

4. The contingency cell decomposition of the space of 0-cycles in $\mathbb{R}^p$

4.1. The contingency cell decomposition $\mathcal{S}^{(2)}$. We now consider the case $X = \mathbb{R}^p$ being the Euclidean space. Let $\mathbf{z} = \sum_{x\in\mathbb{R}^p} n_x \cdot x$ be a point of $Z(\mathbb{R}^p, L)$. We associate to $\mathbf{z}$ a p-dimensional contingency matrix $M(\mathbf{z}) = \|m_{i_1,\cdots,i_p}(\mathbf{z})\|$ as follows. Each $x \in \mathbb{R}^p$ is a vector $(x_1, \cdots, x_p)$. For $1 \leqslant \nu \leqslant p$ let $I_\nu \subset \mathbb{R}$ be the set of the νth coordinates x_ν of all the $x \in \mathbb{R}^p$ such that $n_x \neq 0$. Let $d_\nu = |I_\nu|$ and let us write $I_\nu = \{t_1^{(\nu)} < \cdots < t_{d_\nu}^{(\nu)}\}$. By definition, $n_x \neq 0$ implies that $x \in I_1 \times \cdots \times I_p$. We define the entries of $M(\mathbf{z})$ to be the multiplicities of the points of $I_1 \times \cdots \times I_p$ in $\mathbf{z}$:

$$m_{i_1,\cdots,i_p} = n_{(t_{i_1}^{(1)},\cdots,t_{i_p}^{(p)})} \in L.$$

By construction, $M(\mathbf{z})$ is a contingency matrix. Figure 1 which we reproduce with notational changes from [**21**], illustrates this construction for $p = 2$.

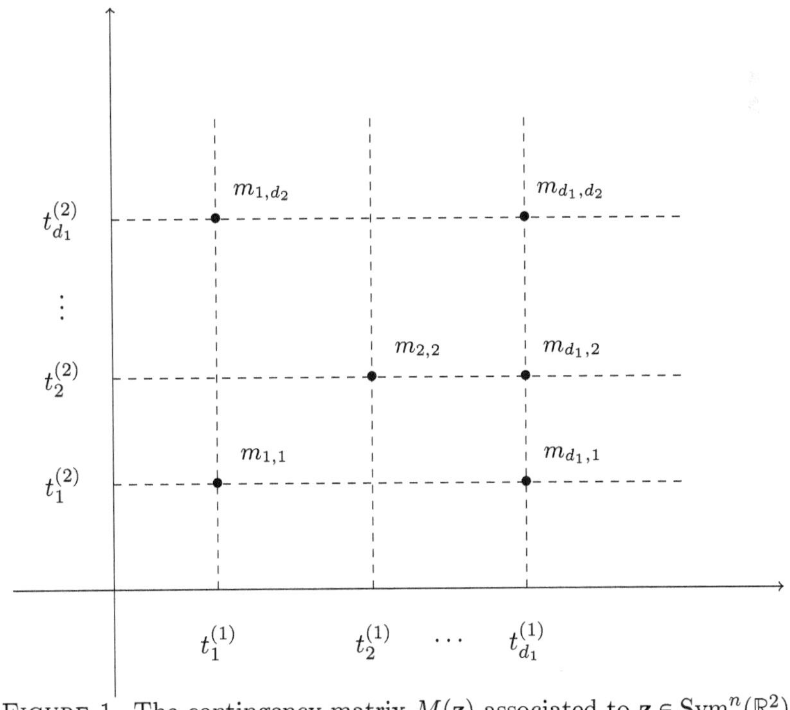

FIGURE 1. The contingency matrix $M(\mathbf{z})$ associated to $\mathbf{z} \in \mathrm{Sym}^n(\mathbb{R}^2)$.

Now let $M \in \mathrm{CM}^p(L)$. Define $U_M = U_M^{\mathbb{R}^p} \subset Z(\mathbb{R}^p, L)$ to be the set of all 0-cycles $\mathbf{a}$ with $M(\mathbf{z}) = M$.

PROPOSITION 4.1.

(a) *U_M is a topological cell, i.e., it is homeomorphic to a Euclidean space. If $M \in \mathrm{CM}_n(d_1, \cdots, d_p; L)$, then $\dim U_M = d_1 + \cdots + d_p$.*

(b) *Let $\mathbf{n}(M)$ be the partition of $|M|$ into the nonzero entries of M. Then U_M lies in the stratum $Z(X,L)_{\mathbf{n}(M)}$ of the stratification $\mathcal{S}^{(0)}$, see Proposition 2.9.*

PROOF. To see (a), note that $\mathbf{z} \in U_M$ is completely determined by the data of subsets $I_\nu \subset \mathbb{R}$, $\nu = 1, \cdots, p$, with $|I_\nu| = d_\nu$. Therefore

$$U_M \simeq \mathbb{R}^{d_1}_< \times \cdots \times \mathbb{R}^{d_p}_<, \text{ where } \mathbb{R}^d_< = \{(t_1, \cdots, t_d) \in \mathbb{R}^d \mid t_1 < \cdots < t_d\}$$

is indeed a cell. Part (b) is clear. □

We will denote the cell decomposition of $Z(\mathbb{R}^p, L)$ into the U_M by $\mathcal{S}^{(2)} = \mathcal{S}^{(2)}_{\mathbb{R}^p}$ and call it the *contingency cell decomposition.* The cells U_M will be called *contingency cells.*

4.2. $\mathcal{S}^{(2)}$ and $\mathcal{S}^{(0)}$. Recall the stratification $\mathcal{S}^{(0)}$ of $Z(\mathbb{R}^p, L)_n$ labelled by unordered partitions; see §2.3 By Proposition 4.1(d), $\mathcal{S}^{(2)}$ refines $\mathcal{S}^{(0)}$, which we write as $\mathcal{S}^{(2)} \prec \mathcal{S}^{(0)}$. Let us describe this refinement more explicitly.

PROPOSITION 4.2. *Assume $p \geqslant 2$. Then, U_M and $U_{M'}$ lie in the same stratum of $\mathcal{S}^{(0)}$ if an only if $M \equiv_{\mathrm{Ano}} N$; see §3.3*

PROOF. Let $\equiv$ be the equivalence relation on $\mathrm{CM}^P_n(L)$ given by $M \equiv N$ iff U_M and U_N lie in the same stratum of $\mathcal{S}^{(0)}$. We need to prove that $M \equiv_{\mathrm{Ano}} N$ if and only if $M \equiv N$.

In one direction, if $M \to N = \partial_i^\nu M$ is an elementary anodyne contraction, then the nonzero entries of M and N are in bijection, so the unordered partitions $\mathbf{n}(M)$ and $\mathbf{n}(N)$ are the same, which means that U_M and U_N lie in the same stratum of $\mathcal{S}^{(0)}$, i.e., $M \equiv N$. Since $\equiv_{\mathrm{Ano}}$ is the equivalence relation generated by the existence of an elementary anodyne contraction, we conclude that $M \equiv_{\mathrm{Ano}} N$ implies $M \equiv N$.

In the other direction, let C be an equivalence class of $\equiv$. Let us prove that C is also an equivalence class of $\equiv_{\mathrm{Ano}}$. Call a matrix $M \in C$ *maximal*, if there is no anodyne contraction $M' \to M$. In this each slice $\sigma_i^\nu M$ contains exactly one nonzero entry. Indeed, otherwise we can split this slice into two parts and still get a contingency matrix M' with an anodyne contraction $M' \to M = \partial_j^\nu M$. This means that M is a "permutation" p-dimensional matrix of size $r \times \cdots \times r$ for some r, with the only nonzero entries being $m_{i,w_1(i),\cdots,w_{p-1}(i)}$, $i = 1, \cdots, r$, where $W - \nu \in S_r$ are some permutations of $\{1, \cdots, r\}$. By the above, each $M \in C$ is $\equiv_{\mathrm{Ano}}$-equivalent to a maximal matrix. It remains to show that any two maximal matrices are $\equiv_{\mathrm{Ano}}$-equivalent. To see this, note that any maximal matrix M can be contracted in an anodyne fashion to a matrix of size $r \times 1 \times \cdots \times 1$ which can be viewed as an ordered partition $(a_1, \cdots, a_r)$. The a_i are therefore all the nonzero entries of M, in some order. It remains to show that any two $r \times 1 \times \cdots \times 1$ contingency matrices different by a reordering, e.g., $(a_1, \cdots, a_r)$ and $(a_{w(1)}, \cdots, a_{w(r)})$, $w \in S_r$, are $\equiv_{\mathrm{Ano}}$-equivalent. But this is a standard manipulation involving matrices of size $r \times 2 \times 1 \times \cdots \times 1$, .e.g.,

$$(a_1, a_2, a_3) \leftarrow \begin{bmatrix} a_1 & 0 & a_3 \\ 0 & a_2 & 0 \end{bmatrix} \rightarrow \begin{bmatrix} a_1 & a_2 \\ a_2 & 0 \end{bmatrix} \leftarrow \begin{bmatrix} 0 & a_1 & a_3 \\ a_2 & 0 & 0 \end{bmatrix} \rightarrow (a_2, a_1, a_3),$$

the arrows representing anodyne contractions. □

EXAMPLE 4.3. Let $L = \mathbb{T} = \{\overline{0}, \overline{1}\}$. Elements of $\mathrm{CM}^p(\mathbb{T})$ will be called (*p*-dimensional) *Boolean contingency matrices.* Such matrices consist of $\overline{0}$'s and $\overline{1}$'s and the simplicial contractions (3.1) are defined using Boolean addition. The content of any Boolean contingency matrix other then $\varnothing$ is $\overline{1}$. Taking into account Example 2.1(e), we obtain a cell decomposition of the Ran space $\mathrm{Ran}(\mathbb{R}^p) = Z(\mathbb{R}^p, \mathbb{T})_{\overline{1}}$ labelled by nonempty *p*-dimensional Boolean contigency matrices.

4.3. $\mathcal{S}^{(2)}$ as a twisted realization. We upgrade the argument in the proof of Proposition 4.1(a) into viewing $Z(\mathbb{R}^p, L)_n$ with its cell decomposition $\mathcal{S}^{(2)}$ as a sort of geometric realization of the *p*-fold augmented semi-simplicial set $\underline{\mathrm{CM}}^p_n(L)$. Consider the closed cone

$$\mathbb{R}^m_{\leqslant} = \{(t_1, \cdots, t_m) \in \mathbb{R}^m \mid t_1 \leqslant \cdots \leqslant t_m\}$$

with its stratification into 2^{m-1} faces on which some of the inequalities become equalities. Note that

$$\mathbb{R}^{\bullet+2}_{\leqslant} : [d] \mapsto \mathbb{R}^{d+2}_{\leqslant}, \quad d = -1, 0, 1, \cdots$$

is an augmented co-semi-simplicial object in the category STop of stratified spaces, i.e., a functor $\Delta_+ \to \mathrm{STop}$.

For each $\mathbf{d} = (d_1, \cdots, d_p) \in \mathbb{Z}^p_{\geqslant -1} = \mathrm{Ob}(\Delta^p_+)$ we denote $\mathbb{R}^{\mathbf{d}+2}_{\leqslant} = \mathbb{R}^{d_1+2}_{\leqslant} \times \cdots \times \mathbb{R}^{d_p+2}_{\leqslant}$. Let $X : (\Delta^p_+)^{\mathrm{op}} \to \mathrm{Set}$ be any *p*-fold augmented semi-simplicial set, whose value on an object $[\mathbf{d}] \in \mathrm{Ob}(\Delta^p_+)$ will be denoted $X_{\mathbf{d}}$. We define the *augmented realization* of X to be the coend [**26**]

$$\|\!|X|\!\| = \int^{\mathbf{d} \in \Delta^p_+} \mathbb{R}^{\mathbf{d}+2}_{\leqslant} \times X_{\mathbf{d}} = \underrightarrow{\lim}^{\mathcal{T}op}_{(\Delta^p_+[\mathbf{d}] \to X) \in \mathrm{Gro}(X)^{\mathrm{op}}} \mathbb{R}^{\mathbf{d}+2}_{\leqslant},$$

the colimit being taken over the (opposite to the) Grothendieck construction. Note that $\|\!|X|\!\|$ comes with a natural stratifiction $\mathcal{S}_{|||}$ glued out of the face stratifications of the $\mathbb{R}^{\mathbf{d}+2}_{\leqslant}$.

PROPOSITION 4.4. *We have a stratified homeomorphism*

$$(Z(\mathbb{R}^p, L)_n, \mathcal{S}^{(2)}) \simeq (\|\!|\underline{\mathrm{CM}}^p_n(L)|\!\|, \mathcal{S}_{|||}).$$

PROOF. As observed in the proof of Proposition 4.1(a), the cell $U_M \subset Z(\mathbb{R}^p, L)_n$, $M \in \mathrm{CM}_n(d_1, \cdots, d_p; L)$, is identified with $\mathbb{R}^{d_1}_{<} \times \cdots \times \mathbb{R}^{d_p}_{<}$, the open stratum of $\mathbb{R}^{d_1}_{\leqslant} \times \cdots \times \mathbb{R}^{d_p}_{\leqslant}$. It remains to notice that under these identifications, the embedding $U_{\partial^\nu_i M} \subset \overline{U}_M$ is identified with the embedding

$$\mathbb{R}^{d_1}_{\times} \cdots \times \mathbb{R}^{d_i-1}_{<} \times \cdots \times \mathbb{R}^{d_p}_{<} \subset \mathbb{R}^{d_1}_{\leqslant} \times \cdots \times \mathbb{R}^{d_p}_{\leqslant} = \overline{\mathbb{R}^{d_1}_{<} \times \cdots \times \mathbb{R}^{d_p}_{<}},$$

and similarly for iterated contractions. □

4.4. The exit path category of the contingency cell decomposition. Let $(Z, \mathcal{S})$ be a (Whitney) stratified space. We have a partial order or Z given by $z \leqslant z'$ if the stratum containing z lies in the closure of the stratum containing z'. An *exit path* for $(Z, \mathcal{S})$ is a continuous path $\gamma : [0,1] \to Z$ which is monotonous, i.e., $\gamma(t) \leqslant \gamma(t')$ for $t \leqslant t'$. We refer to [**8**, **9**, **31**] for the definition of the *exit path category* $\mathrm{Exit}(Z, \mathcal{S})$ (or simply $\mathrm{Exit}(Z)$, if $\mathcal{S}$ is clear from the context). Thus, objects of $\mathrm{Exit}(Z, \mathcal{S})$ are points of Z and morphisms are appropriate isotopy classes of exit paths.

It was a fundamental observation of MacPherson that morphisms of $\mathrm{Exit}(Z, \mathcal{S})$ act on the stalks of $\mathcal{S}$-constructible sheaves (see below).

The part of $\mathrm{Exit}(Z, \mathcal{S})$ formed by exit paths staying in a fixed stratum is just the fundamental groupoid of this stratum. So one can replace $\mathrm{Exit}(Z, \mathcal{S})$ by an equivalent subcategory with the objects corresponding to connected components of the strata.

The concept of exit paths categories is applicable, in particular, to cell decompositions. In this case the fundamental groupoid of each stratum (cell) is trivial, so $\mathrm{Exit}(Z, \mathcal{S})$ describes different ways ("branches") in which a cell of higher dimension can approach a cell of lower dimension. We are especially interested in the contingency cell decomposition $\mathcal{S}^{(2)}$ of the space $Z(\mathbb{R}^p, L)_n$.

PROPOSITION 4.5. *We have an equivalence*

$$\mathrm{Exit}(Z(\mathbb{R}^p, L)_n, \mathcal{S}^{(2)}) \simeq \mathrm{Gro}(\underline{\mathrm{CM}}^p_n(L))^{\mathrm{op}}.$$

PROOF. We establish the following general fact which in virtue of Proposition 4.4, will imply our statement.

PROPOSITION 4.6. *Let* $X : (\Delta^p_+)^{\mathrm{op}} \to \mathrm{Set}$ *be any p-fold augmented semi-simplicial set. Then we have an equivalence*

$$\mathrm{Exit}(|||X|||, \mathcal{S}_{|||}) \simeq \mathrm{Gro}(X)^{\mathrm{op}}.$$

PROOF OF PROPOSITION 4.6. Consider first the case when X is a standard multi-simplex, i.e., a representable functor

$$\Delta^p_+[\mathbf{d}] = \mathrm{Hom}_{\Delta^p_+}(-, [\mathbf{d}]), \quad \mathbf{d} \in \mathbb{Z}^p_{\geqslant -1}.$$

In this case $|||\Delta^p_+[\mathbf{d}]||| = \mathbb{R}^{\mathbf{d}+2}_{\leqslant}$ with its stratification (cell decomposition) by faces. This cell decomposition is quasi-regular, see [**19**, §1D] and so the corresponding category of exit paths is just the poset of cells ordered by inclusion of closures. This poset is canonically identified with $\mathrm{Gro}(\Delta^p_+[\mathbf{d}])^{\mathrm{op}}$. Indeed, the objects of the latter category are, by definition, all the multi-simplices of $\Delta^p_+[\mathbf{d}]$, i.e., faces of $\mathbb{R}^{\mathbf{d}+2}_{\leqslant}$, and the morphism between two such multisimplices (a contraction) is unique if it exists, which happens precisely when we have an inclusion of face closures.

Let now $X : (\Delta^p_+)^{\mathrm{op}} \to \mathrm{Set}$ be arbitrary. Note that

$$\mathrm{Gro}(X)^{\mathrm{op}} = \int^{\mathbf{d} \in \Delta^p_+} X_{\mathbf{d}} \times \mathrm{Gro}(\Delta^p_+[\mathbf{d}])^{\mathrm{op}} = \varinjlim^{\mathrm{Cat}}_{(\Delta^p_+[\mathbf{d}] \to X) \in \mathrm{Gro}(X)^{\mathrm{op}}} \mathrm{Gro}(\Delta^p_+[\mathbf{d}])^{\mathrm{op}},$$

the coend of colimit being taken in the category Cat of small categories and $X_{\mathbf{d}}$ begin considered as a discrete category. This is a general fact exhibiting the Grothendieck construction of any Set-valued functor as the colimit (over itself) of the Grothendieck constructions of representable functors.

Now, considering Exit as a functor STop $\to$ Cat, the above analysis of Exit $|||\Delta_+^p[\mathbf{d}]|||$ gives, by the universal property of $\varinjlim$, a functor $\xi : \mathrm{Gro}(X)^{\mathrm{op}} \to$ Exit $|||X|||$. We claim that ξ is an equivalence. This is a van Kampen-type statement: that the functor Exit commutes with colimits of a certain type. It does not seem to be covered by the standard formulations of the "van Kampen theorem for exit paths" [**8**, **25**, **31**] which treat colimits of spaces arising from open covers. So we give a direct proof.

Indeed, ξ is a bijection on objects (we take one base point in each cell of $|||X|||$, and these cells are in bijection with multi-simplices of X). Let us prove that ξ is bijective on morphisms. Let $\sigma \in X_{\mathbf{d}}$ be a multisimplex of X and $U_\sigma \subset |||X|||$ be the corresponding cell. Then $\overline{U}_\sigma \subset |||X^{\leqslant\sigma}|||$, where $X^{\leqslant\sigma} \subset X$ is the p-fold augmented semi-simplicial subset whose multisimplices have the form $\varphi(\sigma)$, $\varphi \in \mathrm{Hom}_{(\Delta_+^p)^{\mathrm{op}}}([\mathbf{d}], [\mathbf{d}'])$. Thus, for $\tau \in X_{\mathbf{d}'}$ we have $U_\tau \subset \overline{U}_\sigma$ if and only if $\tau \in X_{\mathbf{d}'}^{\leqslant\sigma}$. Now, $\mathrm{Hom}_{\mathrm{Exit}\,|||X|||}(U_\tau, U_\sigma)$ can be calculated inside $\overline{U}_\sigma$. So to prove that

$$\xi_{\tau,\sigma} : \mathrm{Hom}_{\mathrm{Gro}(X)^{\mathrm{op}}}(\tau, \sigma) \longrightarrow \mathrm{Hom}_{\mathrm{Exit}\,|||X|||}(U_\tau, U_\sigma)$$

is a bijection, we can assume $X = X^{\leqslant\sigma}$. Thus $|||X|||$ is obtained from $\mathbb{R}_{\leqslant}^{\mathbf{d}} = |||\Delta_+^p[\mathbf{d}]|||$ by identifying some faces (according to the identifications of the faces of $\Delta_+^p[\mathbf{d}]$ in $X^{\leqslant\sigma}$). Let $\rho : \mathbb{R}_{\leqslant}^{\mathbf{d}} \to |||X|||$ be the resulting map. Since in our setting there are no degeneracies (all morphisms in Δ_+ are monomorphisms), ρ does not contract any faces to lower dimension but can only identify several faces of the same dimension. In particular, ρ is proper with finite fibers.

Let us now construct an inverse $\eta_{\tau,\sigma}$ to $\xi_{\tau,\sigma}$. Let $\gamma : [0,1] \to |||X|||$ be an exit path with $\gamma(0) \in U_\tau$, $\gamma(1) \in U_\sigma$. By definition, this means that there are finitely many "exit times" $0 = t_0 < t_1 < \cdots < t_k = 1$ such that $\gamma|_{(t_i,t_{i+1}]}$ lies in the same cell, say, U_{τ_i}, $\tau_i \in X_{\mathbf{d}^{(i)}}$, and

$$U_\tau \subset \overline{U}_{\tau_1}, U_{\tau_1} \subset \overline{U}_{\tau_2}, \cdots, U_{\tau_{k-1}} \subset \overline{U}_{\tau_k}, \ \tau_k = \sigma.$$

Now, since ρ induces a homeomorphism $\mathbb{R}_{<}^{\mathbf{d}} \to U_\sigma$ and $\gamma(t) \in U_\sigma$ for $t \in (t_{k-1}, 1]$, we consider the path $\rho^{-1}(\gamma) : (t_{k-1}, 1] \to \mathbb{R}_{<}^{\mathbf{d}}$. Because the the above properties of ρ we conclude that $\rho^{-1}(\gamma)$ has a well defined limit $\lim_{t\to t_{k-1}^+} \rho^{-1}(\gamma(t)) \in \mathbb{R}_{\leqslant}^{\mathbf{d}}$, which we denote $(\rho^{-1}\gamma)(t_{k-1})$. It lies in some face of $\mathbb{R}_{\leqslant}^{\mathbf{d}}$, which corresponds to some $\varphi_{k-1} : \Delta_+^p[\mathbf{d}^{(k-1)}] \to \Delta_+^p[\mathbf{d}]$, so we denote this face $\varphi_{k-1}(\mathbb{R}_{\leqslant}^{\mathbf{d}})$. Further, ρ induces a homeomorphism of the interior of this face and the face $U_{\tau_{k-1}}$, so we extend the lift of γ to $\gamma_{k-2} : (t_{k-2}, 1] \to \mathbb{R}_{\leqslant}^{\mathbf{d}}$ such that $\rho(\gamma_{k-2}(t)) = \gamma(t)$. Then, as before, we find $\lim_{t\to t_{k-2}^+} \gamma_{k-2}(t) \in \mathbb{R}_{\leqslant}^{\mathbf{d}}$ and so on. In this way we lift γ to $\gamma_0 : [0,1] \to \mathbb{R}_{\leqslant}^{\mathbf{d}}$ which is an exit path for the standard (face) stratification of $\mathbb{R}_{\leqslant}^{\mathbf{d}}$. By the above, such path gives a morphism $\varphi : [\mathbf{d}'] \to [\mathbf{d}]$ in $(\Delta_+^p)^{\mathrm{op}}$ with $\varphi(\tau) = \sigma)$. So φ can be seen as a morphism from τ to σ in $\mathrm{Gro}(X)^{\mathrm{op}}$. We define $\eta_{\tau,\sigma}(\gamma) = \varphi$.

It is clear that this construction is invariant under admissible isotopies of exit paths, so we get a well defined map

$$\eta_{\tau,\sigma} : \mathrm{Hom}_{\mathrm{Exit}\,|||X|||}(U_\tau, U_\sigma) \longrightarrow \mathrm{Hom}_{\mathrm{Gro}(X)^{\mathrm{op}}}(\tau, \sigma).$$

It is further clear that $\eta_{\tau,sigma}$ is inverse to $\xi_{\tau,\sigma}$.

EXAMPLES 4.7.

(a) Let $L = \mathbb{Z}_+$, so $Z(\mathbb{R}^p, \mathbb{Z}_+)_n = \mathrm{Sym}^n(\mathbb{R}^p)$. In this case the contingency cell decomposition $\mathcal{S}^{(2)}$ is quasi-regular (see [**21**] for the case $p = 2$, the case $p > 2$ is similar). So $Exit(\mathrm{Sym}^n(\mathbb{R}^p), \mathcal{S}^{(2)})$ is identified with the poset of cells. This is in agreement with Example 3.6 which identifies $\mathrm{Gro}(\underline{\mathrm{CM}}^p_n(\mathbb{Z}_+))^{\mathrm{op}}$ as a poset.

(b) Let $L = \mathbb{T} = \{\bar{0}, \bar{1}\}$ and $p = 1$. For $d \geqslant 1$ let $(\bar{1}^d)$ be the ordered partition $(\bar{1}, \cdots, \bar{1})$ (d times). The cell $U_{(\bar{1}^d)} \subset Z(\mathbb{R}, \mathbb{T})_{\bar{1}} = \mathrm{Ran}(\mathbb{R})$ consists of all d-element subsets $\{t_1, \cdots, t_d\} \subset \mathbb{R}$, where we can assume $t_1 < \cdots < t_d\}$. We see that $U_{(\bar{1}^d)}$ can approach $U_{(\bar{1}^{d-1})}$ along $d-1$ "branches" which correspond to t_i merging with t_{i+1}, $i = 1, \cdots, d-1$. These "branches" correspond to face morphisms $[d-3] \to [d-2]$ in $\Delta_+ = \mathrm{Gro}(\underline{\mathrm{CM}}^1_{\bar{1}}(\mathbb{T}))^{\mathrm{op}}$.

5. 0-cycles in $\mathbb{R}^2 = \mathbb{C}$: Further features

5.1. The case $p = 2$, or $X = \mathbb{C}$. From now on we further specialize to $X = \mathbb{R}^2$ which we identify with the complex plane $\mathbb{C}$ via $(x_1, x_2) \mapsto x_1 + ix_2$. Accordingly, all contingency matrices will be assumed 2-dimensional, unless indicated otherwise and we use the notation $\mathrm{CM}_n(L)$ for $\mathrm{CM}^2_n(L)$ etc. We will denote the simplicial contractions $\partial^1_i, \partial^2_i$ by $\partial'_i, \partial''_i$ and write $\mathrm{Gro}'(\underline{\mathrm{CM}}_n(L))$, $\mathrm{Gro}''(\underline{\mathrm{CM}}_n(L))$ for $\mathrm{Gro}^1(\underline{\mathrm{CM}}^2_n(L))$ and $\mathrm{Gro}^2(\underline{\mathrm{CM}}^2_n(L))$ respectively. Similarly, the equivalence relations $\equiv^1_{\mathrm{Ano}}, \equiv^2_{\mathrm{Ano}}$ will be denoted by $\equiv'_{\mathrm{Ano}}, \equiv''_{\mathrm{Ano}}$. This agrees with the notation of [**21, 23**] as well as of [**22**]. We adapt some further considerations of [**22**, §1] to our situation.

Let $\varphi : [d] \to [d']$, $d, d' \geqslant -1$, be a morphism in Δ^{op}_+ (i.e., a composition of the face maps ∂_j). Let $M \in \mathrm{CM}_n(d+2, q; L)$, $N \in \mathrm{CM}_n(d'+2, q; L)$ be contingency matrices such that $(\varphi \times \mathrm{Id}) \in \mathrm{Mor}(\Delta^{\mathrm{op}}_+ \times \Delta^{\mathrm{op}}_+)$ takes M to N. We will denote the morphism $M \to N$ in $\mathrm{Gro}'(\underline{\mathrm{CM}}_n(L))$ represented by $\varphi \times \mathrm{Id}$ by $M \xrightarrow{\varphi'} N$. Similarly, if $M \in \mathrm{CM}_n(p, d_2; L)$ abd $N \in \mathrm{CM}_n(p, d'+2; L)$ we such that $(\mathrm{Id} \times \varphi) \in \mathrm{Mor}(\Delta^{\mathrm{op}}_+ \times \Delta^{\mathrm{op}}_+$ takes M to N, then we denote the resulting morphism $M \to N$ in $\mathrm{Gro}''(\underline{\mathrm{CM}}_N(L))$ by $M \xrightarrow{\varphi''} N$. This is in agreement with the notation $\partial'_j, \partial''_j$ when $\varphi = \partial_j$ is a generating morphism of Δ^{op}_+.

5.2. Mixed forks and their suprema. By a *mixed fork* in $\mathrm{Gro}(\underline{\mathrm{CM}}_n(L))$ we will mean a diagram $P = \{M' \xrightarrow{\varphi'} N' \xleftarrow{\psi''} N\}$ in which the first morphism lies in $\mathrm{Gro}'(\underline{\mathrm{CM}}_n(L))$ and the second one in $\mathrm{Gro}''(\underline{\mathrm{CM}}_n(L))$. We call the *mixed supremum* of P and denote $\mathrm{Sup}(P)$ the set of contingency matrices M fitting into a commutative diagram in $\mathrm{Gro}(\underline{\mathrm{CM}}_n(L))$:

$$
\begin{array}{ccc}
M & \xrightarrow{\varphi'} & N \\
{\scriptstyle \psi''}\downarrow & & \downarrow{\scriptstyle \psi''} \\
M' & \xrightarrow[\varphi']{} & N'.
\end{array}
\tag{5.1}
$$

This means that N is obtained from M by applying the same horizontal simplicial contraction (namely $\varphi' = \varphi \times \mathrm{Id}$ corresponding to $\varphi \in \mathrm{Mor}(\Delta^{\mathrm{op}}_+)$) as N is obtained from M', and similarly in the vertical direction. See §5.1 for the notation φ', φ'' for the horizontal and vertical contractions associated to φ.

Thus, if N' is of size $p \times q$, then M' is of size $p_1 \times q$, $p_1 \geqslant p$, while N is of size $p \times q_1$, $q_1 \geqslant q$ and M must of of size $p_1 \times q_1$. The following is an analog of [**22**, Proposition 1.11].

PROPOSITION 5.2. *Suppose that in a mixed fork* $P = \{M' \xrightarrow{\varphi'} N' \xleftarrow{\psi''} N\}$ *one of the morphisms is anodyne. Then* $\mathrm{Sup}(P)$ *consists of precisely one element, i.e., there is a unique diagram* (5.1). *Further, if* $M' \xrightarrow{\varphi'} N'$ *is anodyne, then* $M \xrightarrow{\varphi'} N$ *is anodyne. If* $N \xrightarrow{\psi''} N'$ *is anodyne, then* $M \xrightarrow{\psi''} M'$ *is anodyne.*

PROOF. We first consider the case when φ and ψ are generating morphisms of Δ_+^{op}. That is, $\varphi = \partial_i$ for some i, so $\varphi' = \partial_i'$ sums the $(i+1)$st and $(i+2)$nd rows of a matrix, while $\psi = \partial_j$ for some j, so $\psi'' = \partial_j''$ sums the $(j+1)$st and $(j+2)$nd columns. Suppose that $M' \xrightarrow{\partial_i'} N'$ is anodyne (the other case is treated similarly). Note that all the entries of any possible $M \in \mathrm{Sup}(P)$ are determined uniquely except for the 2×2 submatrix

$$
\begin{gathered}
\begin{bmatrix} m_{i,j} & m_{i,j+1} \\ m_{i+1,j} & m_{i+1,j+1} \end{bmatrix}, \qquad \begin{aligned} m_{i,j} + m_{i,j+1} &= n_{i,j}, \\ m_{i+1,j+1} + m_{i+1,j+1} &= n_{i+1,j}, \end{aligned} \\
m_{i,j} + m_{i+1,j} = m'_{i,j}, \quad m_{i,j+1} + m_{i+1,j+1} = m'_{i+1,j}, \\
m'_{i,j} + m'_{i+1,j} = n_{i,j} + n_{i,j+1} = n'_{i,j}.
\end{gathered} \tag{5.3}
$$

By the assumption that $M' \xrightarrow{\partial_i'} N'$ is anodyne, one of the $m'_{i,j}, m'_{i+1,j}$ is 0. Suppose $m'_{i,j} = 0$, the other case is treated similarly. Then $m_{i,j} + m_{i,+1j} = m'_{i,j} = 0$ which implies $m_{i,j} = m_{i+1,j} = 0$. The conditions (5.3) imply that $m_{i,j+1} = n_{i,j}$ and $m_{i+1,j+1} = n_{i,j+1}$. This determines the 2×2 matrix uniquely. Further, if we define it by the above rules and so define the matrix M, then M will be a contingency matrix. Indeed, the only possible thing to check is that the ith column of M, where we put two zero entries, is not entirely 0. But the rest of this column is identical with the rest of the ith column of M, and this rest cannot be 0 since $m'_{i,j} = 0$ and M' is a contingency matrix. This shows that there exists exactly one M fitting into a diagram (5.1). Further, the fact that $M \xrightarrow{\partial_i'} N$ is anodyne, follows from $m_{i,j} = m_{i,j+1} = 0$. This proves the proposition in the case when $\varphi = \partial_i, \psi = \partial_j$.

In the general case we write $\varphi = \partial_{i_1} \cdots \partial_{i_p}$, $\psi = \partial_{j_1} \cdots \partial_{j_q}$. If $M' \xrightarrow{\varphi'} N'$ is anodyne (the case when $N \xrightarrow{\psi''} N'$ is anodyne, is treated similarly), then all the contraction induced by the ∂_{i_ν} are anodyne. So using the particular case treated already we recover, in a unique way, a commutative diagram in the form of a $p \times q$ grid of rectangles with horizontal arrows being anodyne ∂_{i_ν} and vertical arrows being the ∂_{j_μ}. Our required diagram is then recovered as the outside border of this grid with arrows in each direction composed. □

5.3. The Fox-Neuwirth-Fuchs cell decomposition $\mathcal{S}^{(1)}$ of $Z(\mathbb{C}, L)$. For an ordered partition $\mathbf{m} = (m_1, \cdots, m_d) \in \mathrm{CM}^1(L)$ we write $l(\mathbf{m}) = d$ and call this number the *length* of α.

Let such an $\mathbf{m}$ be given. Further, for each $i = 1, \cdots, d$ let $\mathbf{n}^{(i)} \in \mathrm{CM}^1_{\alpha_i}(L)$ be an ordered partition of m_i. We write $\vec{\mathbf{n}}$ for the sequence (vector) of ordered partitions $(\mathbf{n}^{(1)}, \cdots, \mathbf{n}^{(d)})$.

Denote by $\mathrm{CM}_{[\mathbf{m}|\vec{\mathbf{n}}]}(L)$ the set of contingency matrices $M = \|m_{ij}\| \in \mathrm{CM}_n(L)$ with the following properties:

(1) M has exactly d rows (and some number q of columns) and for each $i = 1$, ..., d the ith row sum is equal to $\sum_j m_{ij} = m_i$.

(2) The ordered partition of m_i given by omitting the zero entries from among the $m_{i1}, \cdots, m_{iq}$, is equal to $\mathbf{n}^{(i)}$.

Let $U_{[\mathbf{m}|\vec{\mathbf{n}}]} \subset Z(\mathbb{C}, L)$ be the union of the contingency cells U_M for $M \in \mathrm{CM}_{[\mathbf{m}|\vec{\mathbf{n}}]}(L)$.

PROPOSITION 5.4. *Each* $U_{[\mathbf{m}|\vec{\mathbf{n}}]}$ *is a topological cell of dimension* $l(\mathbf{m})+\sum l(\mathbf{n}^{(i)})$.

PROOF. Indeed, $U_{[\mathbf{m}|\vec{\mathbf{n}}]}$ fibers over the contingency cell $U^{\mathbb{R}}_{\mathbf{m}} \subset Z(\mathbb{R}, L)$ with fibers homeomorphic to the product $\prod U^{\mathbb{R}}_{\mathbf{n}^{(i)}}$ of similar cells, i.e., to a cell. Therefore $U_{[\mathbf{m}|\vec{\mathbf{n}}]}$ is a cell itself. □

Extending the terminology of [**21**], we will call the $U_{[\mathbf{m}|\vec{\mathbf{n}}]}$ the *Fox-Neuwirth-Fuchs cells* of $Z(\mathbb{C}, L)$. The stratification (cell decomposition) of $Z(\mathbb{C}, L)$ given by these cells will be denoted $\mathcal{S}^{(1)}$.

PROPOSITION 5.5. *The sets* $\mathrm{CM}_{[\mathbf{m}|\vec{\mathbf{n}}]}(L)$ *are precisely the equivalence classes of* $\equiv'_{\mathrm{Ano}}$.

PROOF. We first prove that $M \equiv'_{\mathrm{Ano}} N$ implies that M and N lie in the same $\mathrm{CM}_{[\mathbf{m}|\vec{\mathbf{n}}]}(L)$. For this it suffices to assume that there is anodyne contraction $N \xrightarrow{\varphi'} M$, $\varphi \in \mathrm{Mor}(\Delta_+^{\mathrm{op}})$. This implies that M is obtaines from N by adding some columns. This does not affect the row sums, so such sums for M and N are equal, giving the same ordered partition $\mathbf{m}$. Further, any anodyne contraction ∂'_i does not change the pattern of the nonzero entries of the rows of the matrix. Thus the ordered partitions $\mathbf{n}^{(i)}$ associated to the rows of M and N are the same. This shows that M, N lie in the same $\mathrm{CM}_{[\mathbf{m}|\vec{\mathbf{n}}]}(L)$.

Let us prove the converse statement: that whenever $M, N \in \mathrm{CM}_{[\mathbf{m}|\vec{\mathbf{n}}]}(L)$, we necessarily have $M \equiv'_{\mathrm{Ano}} N$. For this, we look at the cell decomposition of $U_{[\mathbf{m}|\vec{\mathbf{n}}]}$ into the contingency cells U_M (labelled by the set $\mathrm{CM}_{[\mathbf{m}|\vec{\mathbf{n}}]}(L)$). By Proposition 4.5 any inclusion of the cell closures $U_M \subset \overline{U}_N$ for $M, N \in \mathrm{CM}_{[\mathbf{m}|\vec{\mathbf{n}}]}(L)$ means that there is a morphism from N to M in $\mathrm{Gro}(\underline{\mathrm{CM}}_n(L))$. Because M, N both lie in $\mathrm{CM}_{[\mathbf{n}|\vec{\mathbf{n}}]}(L)$, such a morphism must have the form of an anodyne contraction $N \xrightarrow{\varphi'} M$, $\varphi \in \mathrm{Mor}(\Delta_+^{\mathrm{op}})$. Since $U_{[\mathbf{m}|\vec{\mathbf{n}}]}$ (being itself a cell) is connected, the equivalence relation on cells of its cell decomposition generated by the relation of inclusion of the cell closures, has only one equivalence class. □

REMARK 5.6. One can generalize the Fox-Neuwirth-Fuchs cell decomposition to $Z(\mathbb{R}^p, L)$ for any $p \geqslant 2$, following the approach of Vassiliev [**32**, §I.3.3]. Since in this paper we are interested only in the case $p = 2$, we do not pursue this generalization.

5.4. Imaginary strata of $Z(\mathbb{C}, L)$**.** A continuous map $f : X \to Y$ of topological spaces gives the direct image map $f_* : Z(X, L) \to Z(Y, L)$ given by

$$f_*\Big(\sum_{x\in X} n_x \cdot x\Big) = \sum_{y\in Y} m_y \cdot y, \quad m_y = \sum_{f(x)=y} n_x.$$

In particular, the imaginary part map $\mathrm{Im} : \mathbb{C} \to \mathbb{R}$ gives $\mathrm{Im}_* : Z(\mathbb{C}, L) \to Z(\mathbb{R}, L)$. Now, $Z(\mathbb{R}, L)$ has its own cell decomposition $\mathcal{S}^{(2)} = \mathcal{S}^{(2)}_{\mathbb{R}}$, see §4.1 It consists of cells $U_\alpha^{\mathbb{R}}$ labelled by ordered partitions (1-dimensional contingency matrices) $\mathbf{n} = (n_1, \cdots, n_d)$. We denote

$$Z_{\mathbf{n}}^{\mathrm{Im}} = \mathrm{Im}_*^{-1}(U_{\mathbf{n}}^{\mathbb{R}}) \subset Z(\mathbb{C}, L) \tag{5.7}$$

the preimage of the cell $U_{\mathbf{n}}$ under Im_*. It is a locally closed subset but not generally a manifold. We will call the $Z_{\mathbf{n}}^{\mathrm{Im}}$ the *imaginary strata* of $Z(\mathbb{C}, L)$. If $n = |\mathbf{n}| = \sum n_i$, then $Z_{\mathbf{n}}^{\mathrm{Im}}$ lies in the component $Z(\mathbb{C}, L)_n$. We denote the stratification (not satisfying the Whitney conditions) of $Z(\mathbb{C}, L)$ into imaginary strata by $\mathcal{I}$.

For $\mathbf{n}$ as above let $\mathrm{CM}_{\mathbf{n}}(L) \subset \mathrm{CM}(L)$ be the set of contingency matrices $M = \|m_{ij}\| \in \mathrm{CM}(L)$ with exactly d rows summing to $\sum_j m_{ij} = n_i$.

PROPOSITION 5.8.

(a) *Suppose* $\mathbf{n} = (n_1, \cdots, n_d)$. *The set* $Z_{\mathbf{n}}^{\mathrm{Im}}$ *is the union of contingency cells* U_M *for* $M \in \mathrm{CM}_{\mathbf{n}}(L)$

(b) *Let* $\equiv'$ *be the equivalence relation on* $\mathrm{CM}(L)$ *generated b y the relation of existence of a morphism in* $\mathrm{Gro}'(\underline{\mathrm{CM}}_n(L))$. *The sets* $\mathrm{CM}_{\mathbf{n}}(L)$ *for all* $\mathbf{m} \in \mathrm{CM}^1(L)$ *are precisely the equivalence classes of* $\equiv'$, *i.e., the connected components of* $\mathrm{Gro}'(\underline{\mathrm{CM}}_n(L))$.

(c) *We have the following refinement relations*

$$\mathcal{S}^{(2)} \prec \mathcal{S}^{(1)} \prec \mathcal{I}, \mathcal{S}^{(0)}.$$

PROOF. (a) is clear from the definition. Part (b) is proved similarly to Proposition 5.5. Finally, (c) follows from (a) and from the definition of the cells $U_{[\mathbf{m}|\vec{\mathbf{n}}]}$ comprising $\mathcal{S}^{(1)}$. □

EXAMPLES 5.9.

(a) The smallest imaginary stratum in $Z(\mathbb{C}, L)_n$ is $Z_{(n)}^{\mathrm{Im}} = \mathbb{R} \times Z(\mathbb{R}, L)_n$. It corresponds to the ordered partition (n) consisting of a single element n and consists of 0-cycles $\sum_{x \in \mathbb{C}} n_x \cdot x$, $\sum n_x = n$, with the property that all the x such that $n_x \neq 0$ have the same imaginary part.

(b) Let $L = \mathbb{T} = \{\bar{0}, \bar{1}\}$, so $Z(\mathbb{C}, \mathbb{T})_{\bar{1}} = \mathrm{Ran}(\mathbb{C})$ is the Ran space of $\mathbb{C}$. Similarly to Example 2.10, ordered partitions of $\bar{1}$ have the form $\bar{1}^d = (\bar{1}, \cdots, \bar{1})$ (d times), $d \geqslant 1$. The corresponding imaginary stratum $Z_{\bar{1}^d}^{\mathrm{Im}}$ will be denoted $\mathrm{Ran}(\mathbb{C})_d^{\mathrm{Im}}$. It consists of finite nonempty subsets $I \subset \mathbb{C}$ such that the set of imaginary parts $\mathrm{Im}(I) \subset \mathbb{R}$ has cardinality exactly d. Note that each $\mathrm{Ran}(\mathbb{C})_d^{\mathrm{Im}}$ is infinite-dimensional and has infinite codimension in $\mathrm{Ran}(\mathbb{C})$.

6. Bialgebras graded by a monoid and their PROB

6.1. L-graded bialgebras. Let L be a monoid satisfying Assumptions 2.2. Let also $(\mathcal{V}, \otimes, \mathbf{1}, R)$ be a $\mathbf{k}$-linear braided monoidal category. By an *L-graded bialgebra* we mean a collection $A = (A_n)_{n \in L}$ of objects of $\mathcal{V}$ together with multiplication and comultiplication morphisms

$$\mu_{m,n} : A_m \otimes A_n \longrightarrow A_{m+n}, \ \Delta_{m,n} : A_{m+n} \longrightarrow A_m \otimes A_n, \ m, n \in L,$$

satisfying the following axioms;

(LB0) (Primitivity) $A_0 = \mathbf{1}$ and the morphisms

$$\mu_{0,n} : \mathbf{1} \otimes A_n \to A_n, \ \mu_{n,0}; A_n \otimes \mathbf{1} \to A_n, \ \Delta_{n,0} : A_n \to A_n \otimes \mathbf{1}, \ \Delta_{0,n} : A_n \to \mathbf{1} \otimes A_n$$

are the identities.

(LB1) The $\mu_{m,n}$ are associative.
(LB2) The $\Delta_{m,n}$ are coassociative.
(LB3) (Compatibillity) Suppose $l_1 + l_2 = m_1 + m_2 = n$. Then

$$\Delta_{l_1,l_2}\,\mu_{m_1,m_2} = \sum_O (\mu_{o_{11},o_{21}} \otimes \mu_{o_{12},o_{22}}) \circ (\mathrm{Id} \otimes R_{A_{o_{12}},A_{o_{21}}} \otimes \mathrm{Id}) \circ (\Delta_{o_{11},o_{12}} \otimes \Delta_{o_{21},o_{22}}),$$

the summation being over 2×2 matrices

$$O = \begin{pmatrix} o_{11} & o_{12} \\ o_{21} & o_{22} \end{pmatrix}, o_{ij} \in L,$$

such that

$$o_{11} + o_{21} = l_1,\ o_{12} + o_{22} = l_2, \quad o_{11} + o_{12} = m_1,\ o_{21} + o_{22} = m_2.$$

REMARKS 6.1.

(a) Note that Assumptions 2.2 imply that the summation in (LB3) is over a finite set.

(b) Suppose $\mathcal{V}$ has countable direct sums (which we do not assume in general). Then, instead of understanding A as the collection (A_n), we can form the direct sum $A^{\oplus} = \bigoplus_{n \in L} A_n$ which is a bialgebra in $\mathcal{V}$ graded by L. The data of $A^{\oplus}$ as a bialgebra with grading determines the $\mu_{m,n}$ and $\Delta_{m,n}$ as the graded components of the multiplication and comultiplication in $A^{\oplus}$. In this way we can view L-graded bialgebras as bialgebras graded by L and satisfying the additional condition (LB0).

EXAMPLE 6.2. Suppose that $\mathcal{V}$ is abelian, with $\otimes$ being exact in each argument. Let B be an ungraded bialgebra in $\mathcal{V}$ with unit $e : \mathbf{1} \to B$ and counit $\eta : B \to \mathbf{1}$. Let $B_+ = \mathrm{Ker}(\eta)$. Then B_+ is an ideal in B as an algebra, since η is an algebra homomorphism. Let us define $\overline{\Delta} : B \to B \otimes B$ by $\overline{\Delta} = \Delta - \mathrm{Id} \otimes e - e \otimes \mathrm{Id}$, i.e., $\overline{\Delta}(x) = \Delta(x) - x \otimes 1 - 1 \otimes x$ in the element notation. Then $\overline{\Delta}$ takes B_+ to

$$B_+ \otimes B_+ = \mathrm{Ker}\{\eta \otimes \mathrm{Id}\} \cap \mathrm{Ker}\{\mathrm{Id} \otimes \eta\}.$$

Indeed, using the element notation again, $(\eta \otimes \mathrm{Id})(\Delta(x)) = x$ by the axiom of counit. At the same time, if x is an element of A_+, then

$$(\eta \otimes \mathrm{Id})(x \otimes 1) = \eta(x) = 0, \quad (\eta \otimes \mathrm{Id})(1 \otimes x) = x,$$

so $\Delta(x) - x \otimes 1 - 1 \otimes x$ is annihilated by $\eta \otimes \mathrm{Id}$. Similarly for $\mathrm{Id} \otimes \eta$.

Now, putting $A_{\overline{0}} = \mathbf{1}$, $A_{\overline{1}} = B_+$, we associate to B a $\mathbb{T}$-graded bialgebra A, where $\mathbb{T}$ is the Boolean algebra of truth values, see Example 2.1(e). Its only nontrivial data are:

$$\mu_{\overline{1},\overline{1}} = \mu : B_+ \otimes B_+ \longrightarrow B_+, \quad \Delta_{\overline{1},\overline{1}} = \overline{\Delta} : B_+ \longrightarrow B_+ \otimes B_+.$$

This defines an equivalence of categories

$$\{\text{Ungraded bialgebras in } \mathcal{V} \text{ with unit and counit}\} \xrightarrow{\simeq} \{\mathbb{T}\text{-graded bialgebras in } \mathcal{V}\}.$$

The functor in the opposite direction is given by

$$(\mathbf{1} = A_{\overline{0}}, A_{\overline{1}}) \mapsto B = \mathbf{1} \oplus A_{\overline{1}}.$$

The two functors are quasi-inverse to each other because for any unital and counital bialgebra B the endomorhism $P = e\eta : B \to B$ is a projector with image B_+ and $Q = \mathrm{Id} - P$ is a projector with image $\mathrm{Im}(e) \simeq \mathbf{1}$.

6.2. The PROB of L-graded bialgebras. We denote by $\mathfrak{B}^L$ the k-linear braided monoidal category generated by the *universal braided L-graded bialgebra* $\mathbf{a} = (\mathbf{a}_n)_{n\in L}$. In other words, for any other $\mathbf{k}$-linear braided category $\mathcal{V}$, we have a bijection between braided monoidal functors $F : \mathfrak{B}^L \to \mathcal{V}$ and L-graded bialgebras A in $\mathcal{V}$ given by $A = F(\mathbf{a})$.

Thus, similarly to [**16**] and [**23**], $\mathfrak{B}^L$ is generated, as a braided monoidal category, by the generating objects which are formal symbols $\mathbf{a}_n$, $n \in L$, with $\mathbf{a}_0 = \mathbf{1}$. In other words, objects of $\mathfrak{B}^L$ are formal tensor products $\mathbf{a_n} = \mathbf{a}_{n_1}\otimes\cdots\otimes\mathbf{a}_{n_d}$ corresponding to all ordered partitions $\mathbf{n} = (n_1, \cdots, n_d) \in \mathrm{CM}^1(L)$, see §3.1 Morphisms are generated by the elementary formal morphisms

$$\mu_{m,n} : \mathbf{a}_m \otimes \mathbf{a}_n \to \mathbf{a}_{m+n}, \quad \Delta_{m,n} : \mathbf{a}_{m+n} \to \mathbf{a}_m \otimes \mathbf{a}_n$$

as well as the braidings, which are subject only to the relations following from (LB0)–(LB3) and the axioms of a braided category.

We refer to $\mathfrak{B}^L$ as the *(colored) PROB governing L-graded bialgebras*, see [**23**] for the discussion of terminology. By construction, $\mathfrak{B}^L$ splits, as a $\mathbf{k}$-linear category, into an orthogonal direct sum of blocks $\mathfrak{B}^L_n$, $n \in L$, where $\mathfrak{B}^L_n$ is the full subcategory on objects $\mathbf{a_n} = \mathbf{a}_{n_1} \otimes \cdots \otimes \mathbf{a}_{n_d}$ corresponding to partitions $\mathbf{n} = (n_1, \cdots, n_d)$ from $\mathrm{CM}^1_n(L)$, i.e., such that $\sum n_i = n$. The tensor product gives functors $\otimes$: $\mathfrak{B}^L_m \times \mathfrak{B}^L_n \to \mathfrak{B}^L_{m+n}$.

EXAMPLES 6.3.

(a) Let $L = \mathbb{T}$ be the Boolean algebra of truth values, see Example 2.1(e). Since $\mathbb{T}$ has only one non-zero element $\bar{1}$, the braided category $\mathfrak{B}^{\mathbb{T}}$ is generated by one object $\mathbf{a}_{\bar{1}}$. Thus $\mathfrak{B}^{\mathbb{T}}$ is an uncolored PROB: its objects are $\mathbf{a}_{\bar{1}}^{\otimes d}$, $d \geqslant 0$.

(b) Let also $\mathfrak{B}^{\mathrm{cu}}$ be the uncolored PROB governing (ungraded) counital and unital bialgebras. Thus $\mathfrak{B}^{\mathrm{cu}}$ is generated by one formal object $\mathbf{b}$ and morphisms

$$\mu : \mathbf{b}\otimes\mathbf{b} \to \mathbf{b}, \quad \Delta : \mathbf{b} \to \mathbf{b}\otimes\mathbf{b}, \quad \eta : \mathbf{b} \to \mathbf{1}, \quad e : \mathbf{1} \to \mathbf{b}$$

subject to the axioms of a counital and unital bialgebra.

It is instructive to compare $\mathfrak{B}^{\mathbb{T}}$ and $\mathfrak{B}^{\mathrm{cu}}$. Let $\mathfrak{B}^{\mathbb{T}\oplus}$ be the additive envelope of $\mathfrak{B}^{\mathbb{T}}$, i.e., the braided monoidal category formed by formal direct sums of objects of $\mathfrak{B}^{\mathbb{T}}$.

PROPOSITION 6.4. *We have a braided monoidal functor*

$$\mathfrak{B}^{\mathrm{cu}} \to \mathfrak{B}^{\mathbb{T}\oplus}, \quad \mathbf{b} \mapsto 1 \oplus \mathbf{a}_{\bar{1}},$$

which identifies $\mathfrak{B}^{\mathrm{cu}}$ with the full subcategory in $\mathfrak{B}^{\mathbb{T}\oplus}$ on the objects $(1 \oplus \mathbf{a}_{\bar{1}})^{\otimes d}$.

PROOF. This is an upgrade of the reasoning of Example 6.2. That is, the idempotents $P = e\eta$, $Q = \mathrm{Id} - P$ in $\mathrm{Hom}_{\mathfrak{B}^{\mathrm{cu}}}(\mathbf{b}, \mathbf{b})$ identify $\mathbf{b}$ with $\mathbf{1} \oplus \mathbf{a}_{\bar{1}}$. This gives an explicit identification

$$\mathrm{Hom}_{\mathfrak{B}^{\mathbb{T}}}(\mathbf{a}_{\bar{1}}^{\otimes m}, \mathbf{a}_{\bar{1}}^{\otimes n}) = Q^{\otimes n}\,\mathrm{Hom}_{\mathfrak{B}^{\mathrm{cu}}}(\mathbf{b}^{\otimes m}, \mathbf{b}^{\otimes n}) \subset \mathrm{Hom}_{\mathfrak{B}^{\mathrm{cu}}}(\mathbf{b}^{\otimes m}, \mathbf{b}^{\otimes n}). \tag{6.5}$$ □

7. Perverse sheaves on the spaces of 0-cycles in $\mathbb{C}$

7.1. The constructible derived category. We fix $n \in L$ and denote the component $Z(\mathbb{C}, L)_n$ simply by Z. Note that Z can be infinite dimensional and we consider it with the exhaustion by finite dimensional skeleta, see (2.7)

$$Z_1 \subset Z_2 \subset \cdots \subset Z = \varinjlim{}^{Top} Z_r, \quad Z_r := Z_r(\mathbb{C}, L) \cap Z(\mathbb{C}, L)_n.$$

We denote by $i_r : Z_N \hookrightarrow Z$ the (closed) embeddings.

Let $\mathcal{V}$ be a **k**-linear Grothendieck abelian category (not assumed monoidal). We can speak about $\mathcal{V}$-valued sheaves on Z on any topological space T. The category of such sheaves will be denoted $\mathrm{Sh}(T, \mathcal{V})$.

Note that $i_{r*} : \mathrm{Sh}(Z_r, \mathcal{V}) \to \mathrm{Sh}(Z, \mathcal{V})$ is an embedding of a full subcategory (whose objects are sheaves supported on Z_r). Let $\mathcal{F} \in \mathrm{Sh}(Z, \mathcal{V})$. We say that $\mathcal{F}$ is *finitely supported*, if $\mathcal{F}$ lies in the image of one of the i_{r*}. We denote $\mathrm{Sh}^{<\infty}(Z, \mathcal{V})$ the full category in $\mathrm{Sh}(Z, \mathcal{V})$ formed by finitely supported sheaves. Thus $\mathrm{Sh}^{<\infty}(Z, \mathcal{V}) = \mathrm{Sh}(Z, \mathcal{V})$, if Z is finite dimensional.

For any abelian category $\mathcal{A}$ we denote by $C^b(\mathcal{A})$ the category of bounded complexes over $\mathcal{A}$ and by $D^b(\mathcal{A})$ the bounded derived category of $\mathcal{A}$.

The category $\mathrm{Sh}^{<\infty}(Z, \mathcal{V})$ is abelian and we denote $D^{<\infty}(Z, \mathcal{V}) = D^b\, \mathrm{Sh}^{<\infty}(Z, \mathcal{V})$ its bounded derived category.

Recall the stratifications $\mathcal{S}^{(i)}$ of $Z(\mathbb{C}, L)$, $i \in \{0, 1, 2\}$. We denote $\mathrm{Sh}^{<\infty}_{\mathcal{S}^{(i)}}(Z, \mathcal{V}) \subset \mathrm{Sh}^{<\infty}(Z, \mathcal{V})$ the full subcategory formed by $\mathcal{S}^{(i)}$*-constructible sheaves*, i.e., sheaves whose restriction to any stratum of $\mathcal{S}^{(i)}$ is locally constant. Further, let $D^{<\infty}_{\mathcal{S}^{(i)}}(Z, \mathcal{V}) \subset D^{<\infty}(Z, \mathcal{V})$ and be the full subcategory formed by $\mathcal{S}^{(i)}$*- constructible complexes*, i.e., complexes whose cohomology sheaves are $\mathcal{S}^{(i)}$-constructible. Since $\mathcal{S}^{(2)} \prec \mathcal{S}^{(1)} \prec \mathcal{S}^{(0)}$, we have embeddings of full subcategories

$$\mathrm{Sh}^{<\infty}_{\mathcal{S}^{(0)}}(Z, \mathcal{V}) \subset \mathrm{Sh}^{<\infty}_{\mathcal{S}^{(1)}}(Z, \mathcal{V}) \subset \mathrm{Sh}^{<\infty}_{\mathcal{S}^{(2)}}(Z, \mathcal{V}),$$

$$D^{<\infty}_{\mathcal{S}^{(0)}}(Z, \mathcal{V}) \subset D^{<\infty}_{\mathcal{S}^{(1)}}(Z, \mathcal{V}) \subset D^{<\infty}_{\mathcal{S}^{(2)}}(Z, \mathcal{V}).$$

7.2. Cellular description of constructible categories. Recall the category $\mathrm{Gro}(\underline{\mathrm{CM}}_n(L))$, see §33.2 Let $\mathrm{Fun}^{<\infty}(\mathrm{Gro}(\underline{\mathrm{CM}}_n(L))^{\mathrm{op}}, \mathcal{V})$ be the category of functors $\mathrm{Gro}(\underline{\mathrm{CM}}_n(L))^{\mathrm{op}} \to \mathcal{V}$ which vanish on almost all objects.

PROPOSITION 7.1.

(a) *We have equivalences of categories*

$$\mathrm{Sh}_{\mathcal{S}^{(2)}}(Z, \mathcal{V}) \simeq \mathrm{Fun}(\mathrm{Gro}(\underline{\mathrm{CM}}_n(L))^{\mathrm{op}}, \mathcal{V}),$$

$$\mathrm{Sh}^{<\infty}_{\mathcal{S}^{(2)}}(Z, \mathcal{V}) \simeq \mathrm{Fun}^{<\infty}(\mathrm{Gro}(\underline{\mathrm{CM}}_n(L))^{\mathrm{op}}, \mathcal{V}).$$

(b) *The category* $\mathrm{Sh}^{<\infty}_{\mathcal{S}^{(0)}}(Z, \mathcal{V})$ *is equivalent to the full subcategory in*

$$\mathrm{Fun}^{<\infty}(\mathrm{Gro}(\underline{\mathrm{CM}}_n(L))^{\mathrm{op}}, \mathcal{V})$$

consisting of functors that take any anodyne morphism to an isomorphism.

(c) *The category* $\mathcal{S}^{<\infty}_{\mathcal{S}^{(1)}}(Z, \mathcal{V})$ *is equivalent to the full subcategory in*

$$\mathrm{Fun}^{<\infty}(\mathrm{Gro}(\underline{\mathrm{CM}}_n(L))^{\mathrm{op}}, \mathcal{V})$$

consisting of functors that take any anodyne morphism in the subcategory $\mathrm{Gro}'(\underline{\mathrm{CM}}_n(L))^{\mathrm{op}}$ *to an isomorphism.*

PROOF. (a) The first equivalence follows from Proposition 4.5 and the general result [**8, 9, 31**] describing constructible sheaves on any stratified space in terms of representations of the category of exit paths. The second equivalence is the restriction of the first one. Part (b) follows from (a) and from Proposition 4.1, since the strata of $\mathcal{S}^{(0)}$ are connected. Part (c) follows from (a) and from Proposition 5.5 in a similar way. □

PROPOSITION 7.2.

(a) *The category* $D^{<\infty}_{\mathcal{S}^{(2)}}(Z,\mathcal{V})$ *is equivalent to* $D^b\,\mathrm{Fun}^{<\infty}(\mathrm{Gro}(\underline{\mathrm{CM}}_n(L))^{\mathrm{op}},\mathcal{V})$, *the bounded derived category of the abelian category* $\mathrm{Fun}^{<\infty}(\mathrm{Gro}(\underline{\mathrm{CM}}_n(L))^{\mathrm{op}},\mathcal{V})$.

(b) *The category* $D^{<\infty}_{\mathcal{S}^{(0)}}(Z,\mathcal{V})$ *is equivalent to the full subcategory in*

$$D^b\,\mathrm{Fun}^{<\infty}(\mathrm{Gro}(\underline{\mathrm{CM}}_n(L))^{\mathrm{op}},\mathcal{V})$$

consisting of complexes of functors (or, equivalently, functors $\mathrm{Gro}(\underline{\mathrm{CM}}_n(L))^{\mathrm{op}}\to C^b(\mathcal{V})$*) taking all anodyne morphisms to quasi-isomorphisms.*

(c) *The category* $D^{<\infty}_{\mathcal{S}^{(1)}}(Z,\mathcal{V})$ *is equivalent to the full subcategory in*

$$D^b\,\mathrm{Fun}^{<\infty}(\mathrm{Gro}(\underline{\mathrm{CM}}_n(L))^{\mathrm{op}},\mathcal{V})$$

consisting of functors $\mathrm{Gro}(\underline{\mathrm{CM}}_n(L))^{\mathrm{op}}\to C^b(\mathcal{V})$ *taking anodyne morphisms from* $\mathrm{Gro}'(\underline{\mathrm{CM}}_n(L))^{\mathrm{op}}$ *to quasi-isomorphisms.*

PROOF. It suffices to establish (a), as (b) and (c) are formal consequences. Now, the proof of (a) goes along the standard lines of comparing Ext's. we present it in some detail since we are dealing with sheaves with values in an arbitrary Grothendieck abelian category $\mathcal{V}$.

Any complex of $\mathcal{S}^{(2)}$-constructible sheaves has, obviously, $\mathcal{S}^{(2)}$-constructible cohomology sheaves. This gives an exact functor of triangulated categories

$$\xi : D^b\,\mathrm{Fun}^{<\infty}(\mathrm{Gro}(\underline{\mathrm{CM}}_n(L))^{\mathrm{op}},\mathcal{V})\simeq D^b\,\mathrm{Sh}^{<\infty}(Z,\mathcal{V})\longrightarrow D^{<\infty}_{\mathcal{S}^{(2)}}(Z,\mathcal{V}),$$

the first identification being a consequence of Proposition 7.1(a). Let us prove that ξ is an equivalence. We separate this into two lemmas. □

LEMMA 7.3. ξ *is fully faithful.*

LEMMA 7.4. ξ *is essentially surjective.*

7.3. Proof of Lemma 7.3. The statement means that for any bounded complexes $\mathcal{F}^\bullet,\mathcal{G}^\bullet$ from $\mathrm{Sh}^{<\infty}_{\mathcal{S}^{(2)}}(Z,\mathcal{V})$ the canonical map

$$\xi_{\mathcal{F}^\bullet,\mathcal{G}^\bullet}:\mathrm{Ext}^i_{\mathrm{Sh}^{<\infty}_{\mathcal{S}^{(2)}}(Z,\mathcal{V})}(\mathcal{F}^\bullet,\mathcal{G}^\bullet)\longrightarrow \mathrm{Hom}_{D^{<\infty}_{\mathcal{S}^{(2)}}(Z,\mathcal{V})}(\mathcal{F}^\bullet,\mathcal{G}^\bullet[i])$$

is an isomorphism for all i. Since $\mathcal{F}^\bullet$ and $\mathcal{G}^\bullet$ are supported on finitely many cells, dévissage reduces this question to the case when $\mathcal{F}^\bullet$ and $\mathcal{G}^\bullet$ are sheaves in degree 0 and, moreover, are single-cell sheaves. That is, for $M\in\mathrm{CM}_n(L)$ let $j_M:U_M\to Z$ be the embedding of the corresponding contingency cell. Then we can assume that

$$\mathcal{F}^\bullet=\mathcal{F}=j_{M*}\underline{V}_{U_M},\quad \mathcal{G}^\bullet=\mathcal{G}=j_{N*}\underline{W}_{U_N},\quad M,N\in\mathrm{CM}_n(L),$$

where $V,W\in\mathrm{Ob}(\mathcal{V})$ and $\underline{V}_{U_M}$, $\underline{W}_{U_N}$ are the corresponding constant sheaves on U_M and U_N respectively.

LEMMA 7.5. *Under the equivalence of Proposition* 7.1*(a), the sheaf* $j_{M*}\underline{V}_{U_M}$ *corresponds to the functor*

$$V^{h^M}=\mathrm{Map}(\mathrm{Hom}_{\mathrm{Gro}(\underline{\mathrm{CM}}_n(L))}(M,-),V);\mathrm{Gro}(\underline{\mathrm{CM}}_n(L))^{\mathrm{op}}\longrightarrow\mathcal{V}.$$

To explain the notation, for a set I and an object $V\in\mathcal{V}$ we have the object

$$V^I=\mathrm{Map}(I,V):=\prod_{i\in I}V\quad\in\quad\mathcal{V},$$

the product of I copies of V. So the covariant set-valued functor

$$h^M=\mathrm{Hom}(M,-):\mathrm{Gro}(\underline{\mathrm{CM}}_n(L))\longrightarrow\mathrm{Set}$$

gives a contravariant $\mathcal{V}$-valued functor $V^{h^M} : \mathrm{Gro}(\underline{\mathrm{CM}}_n(L))^{\mathrm{op}} \longrightarrow \mathcal{V}$.

PROOF OF LEMMA 7.5. This is a generality. Denote for short

$$\mathcal{E} = \mathrm{Gro}(\underline{\mathrm{CM}}_n(L))$$

(the "entrance path" category of $\mathcal{S}^{(2)}$). Let $G : \mathcal{E}^{\mathrm{op}} \to \mathcal{V}$ be the functor corresponding to $j_{M*}\underline{V}_{U_M}$ by Proposition 7.1(a). Note that $j_{M*}\underline{V}_{U_M}$ is characterized by the natural identification

$$\mathrm{Hom}(\mathcal{F}, j_{M*}\underline{V}_{U_M}) \simeq \mathrm{Hom}(j_M^*\mathcal{F}, \underline{V}_{U_M})$$

for any $\mathcal{F} \in \mathrm{Sh}_{\mathcal{S}^{(2)}}(Z, \mathcal{V})$. Such $\mathcal{F}$ are in equivalence with functors $F : \mathcal{E}^{\mathrm{op}} \to \mathcal{V}$ and

$$\mathrm{Hom}(j_M^*\mathcal{F}, \underline{V}_{U_M}) = \mathrm{Hom}_{\mathcal{V}}(F(M), V).$$

So G is characterized by the natural (in F) isomorphism

$$\mathrm{Hom}_{\mathrm{Fun}(\mathcal{E}^{\mathrm{op}},\mathcal{V})}(F, G) \simeq \mathrm{Hom}_{\mathcal{V}}(F(M), V).$$

The functor $G = V^{h^M}$ obviously admits such an isomorphism (a standard property of representable functors). □

Continuing with the proof of Lemma 7.5, we note that

(7.6)

$$\mathrm{Ext}^\bullet_{\mathrm{Fun}(\mathrm{Gro}(\underline{\mathrm{CM}}_n(L))^{\mathrm{op}},\mathcal{V})}(V^{h^M}, W^{h^N}) = \mathrm{Map}\big(\mathrm{Hom}_{\mathrm{Gro}(\underline{\mathrm{CM}}_n(L))}(M,N), \mathrm{Ext}^\bullet_{\mathcal{V}}(V,W)\big).$$

Indeed, by passing to injective resolutions (which is possible since $\mathcal{V}$, being Grothendieck, has enough injectives), this statement reduces to the case when W is injective, in which case W^{h^N} is also an injective object of the functor category. So one can replace $\mathrm{Ext}^\bullet$ by Hom on both sides of (7.6), after which the statement becomes a standard property of representable functors.

So to prove that ξ is fully faithful, it suffices to show, again under the assumption that W is injective, the identification

$$\mathrm{Ext}^q_{\mathrm{Sh}(Z,\mathcal{V})}(j_{M*}\underline{V}_{U_M}, j_{N*}\underline{W}_{U_N}) = \begin{cases} \mathrm{Hom}_{\mathcal{V}}(V,W)^{\mathrm{Hom}_{\mathrm{Gro}(\underline{\mathrm{CM}}_n(L))}(M,N)}, & \text{if } q = 0; \\ 0, & \text{if } q > 0. \end{cases}$$

Here, the case $q = 0$ is a rephrasing of the fact that the (abelian) category $\mathrm{Sh}_{\mathcal{S}^{(2)}}(Z, \mathcal{V})$ is equivalent to $\mathrm{Fun}(\mathrm{Gro}(\underline{\mathrm{CM}}_n(L))^{\mathrm{op}}, \mathcal{V})$., i.e., of Proposition 7.1(a). The vanishing of Ext^q for $q > 0$ is a consequence of the following fact which alone uses the specifics of the situation.

LEMMA 7.7. *The sheaf $j_{N*}\underline{W}_{U_N}$ is quasi-isomorphic to the complex $Rj_{N*}\underline{W}_{U_N}$, i.e., the higher direct images $R^q j_{N*}\underline{W}_{U_N}$, $q > 0$, are zero.*

To see the deduction, notice that by adjunction in the derived category,

$$\begin{aligned}\mathrm{Ext}^q_{\mathrm{Sh}(Z,\mathcal{V})}(j_{M*}\underline{V}_{U_M}, Rj_{N*}\underline{W}_{U_N}) &= \mathrm{Ext}^q_{\mathrm{Sh}(U_N,\mathcal{V})}(j_N^* j_{M*}\underline{V}_{U_M}, \underline{W}_{U_N}) \\ &= \mathrm{Ext}^q_{\mathcal{V}}(V', W) = 0,\end{aligned}$$

where V' is the stalk at U_N of the constant sheaf $j_N^* j_{M*}\underline{V}_{U_M}$. This last Ext^q vanishes since W is injective.

PROOF OF LEMMA 7.7. Choose a finite set I of generators of the monoid L, so we have a surjective morphism of monoids $p : \mathbb{Z}_+^I \to L$. For any $\tilde{n} \in \mathbb{Z}_+^I$ we then have the induced map of cycle spaces $\pi_{\tilde{N}} : Z(\mathbb{C}, \mathbb{Z}_+^I)_{\tilde{n}} \to Z(\mathbb{C}, L)_{p(\tilde{n})}$. Note that $\pi_{\tilde{n}}$ is a proper morphism with finite fibers. The space $Z(\mathbb{C}, \mathbb{Z}_+^I)_{\tilde{n}}$ has its own contingency cell decomposition into cells $U_{\tilde{N}}$, $\tilde{N} \in \mathrm{CM}_{\tilde{n}}(\mathbb{Z}_+^I)$. As in Example 4.7(a), this decomposition is quasi-regular, i.e., $\underline{U}_{\tilde{N}}$ is homeomorphic to a part of a closed Euclidean ball. In particular, $R^q j_{\tilde{N}*} \underline{W}_{U_{\tilde{N}}} = 0$ for $q > 0$. Now, given $N \in \mathrm{CM}_n(L)$, we can find $\tilde{n} \in \mathbb{Z}_+^I$ with $p(\tilde{n}) = n$ and $\tilde{N} \in \mathrm{CM}_{\tilde{n}}(\mathbb{Z}_+^I)$ such that $\pi_{\tilde{n}}$ maps $U_{\tilde{N}}$ to U_N in a homeomorphic way and $\overline{U}_{\tilde{N}}$ to $\overline{U}_N$ in a surjective, finite-to one (by the above) way. Thus

$$Rj_{N*}\underline{W}_{U_N} = R\pi_{\tilde{n}*}Rj_{\tilde{N}*}\underline{W}_{U_{\tilde{N}}} = \pi_{\tilde{n}*}j_{\tilde{N}*}\underline{W}_{U_{\tilde{N}}}$$

is situated in degree 0. □

This finishes the proof of Lemma 7.3.

7.4. Proof of Lemma 7.4. By construction, the essential image of ξ contains all single sheaves from $\mathrm{Sh}_{\mathcal{S}^{(2)}}(Z, \mathcal{V})$ as well as theirn shifts. Now, every $\mathcal{F}^\bullet \in D^{<\infty}_{\mathcal{S}^{(2)}}(Z, \mathcal{V})$ admits cohomological truncations $\tau_{\leqslant r}\mathcal{F}^\bullet$, $r \in \mathbb{Z}$, fitting into exact triangles

$$\underline{H}^r(\mathcal{F}^\bullet)[-r-1] \xrightarrow{\varepsilon_r} \tau_{\leqslant r-1}\mathcal{F}^\bullet \to \tau_{\leqslant r}\mathcal{F}^\bullet \to \underline{H}^r(\mathcal{F}^\bullet)[-r],$$

with $\tau_{\leqslant r}\mathcal{F}$ being 0 for $r \ll 0$ and $\mathcal{F}^\bullet$ for $r \gg 0$. Using Lemma 7.3 we represent, inductively, the object $\tau_{\leqslant r-1}\mathcal{F}^\bullet$ and the morphism ε_r as the images under ξ of an object and a morphism (denote this morphis m e_r) in the source of ξ, after which $\tau_{\leqslant r}\mathcal{F}^\bullet$ is exhibited as the image under ξ of the cone of e_r. This finishes the proof of Lemma 7.4 and of Proposition 7.2.

7.5. More general abelian categories. Criterion of $\mathcal{S}^{(0)}$-constructibility. Propositions 7.1 and 7.2 allow us to eliminate, in the remainder of the paper, the assumption that $\mathcal{V}$ is Grothendieck (needed, a priori, to handle descent in infinite coverings). Indeed, we can (and will) treat all the constructible sheaves or complexes on Z in a purely combinatorial way, as certain functors from $\mathrm{Gro}(\underline{\mathrm{CM}}_n(L))^{\mathrm{op}}$ to $\mathcal{V}$ or to the category of complexes over $\mathcal{V}$. For this, $\mathcal{V}$ can be any abelian category.

Firther, the functors Rf_*, $f^!$ etc. associated to various embeddings f of cellular subspaces in $(Z, \mathcal{S}^{(2)})$ can be defined in purely cellular terms, i.e., directly at the level of functors from $\mathrm{Gro}(\underline{\mathrm{CM}}_n(L))^{\mathrm{op}}$ or its subcategories, so we will use them, often without further explanation.

Let $\tau : \mathbb{C} \to \mathbb{C}$ be the permutation of coordinates: $\tau(x + iy) = y + ix$. The stratifications $\mathcal{S}^{(0)}$ and $\mathcal{S}^{(2)}$ are preserved by τ but $\mathcal{S}^{(1)}$ is taken into a different stratification $\tau\mathcal{S}^{(1)}$. Similarly to [**21**, Corollary 5.9], we have the following.

PROPOSITION 7.8. *Let $\mathcal{F}$ be an $\mathcal{S}^{(2)}$-constructible complex on Z. The following are equivalent:*

(i) *$\mathcal{F}$ is $\mathcal{S}^{(0)}$-constructible.*
(ii) *$\mathcal{F}$ is both $\mathcal{S}^{(1)}$-constructible and $\tau\mathcal{S}^{(1)}$-constructible.*

PROOF. This is a direct consequence of Proposition 5.5 and the fact that the class of anodyne morphism in $\mathrm{Gro}(\underline{\mathrm{CM}}_n(L))^{\mathrm{op}}$ is generated by the subclasses of horizontal and vertical anodyne morphisms. □

7.6. Verdier duality. Let $\mathcal{V}^{\mathrm{op}}$ be the category opposite to $\mathcal{V}$. It is again abelian. Since $(Z_r, \mathcal{S}^{(i)})$, $i = 0,1,2$, is a finite-dimesional stratified space, it possesses Verdier duality which is an anti-equivalence between the bounded derived categories of constructible complexes of sheaves with values in $\mathcal{V}$ and $\mathcal{V}^{\mathrm{op}}$. Because $D^{<\infty}_{\mathcal{S}^{(i)}}(Z,\mathcal{V})$ consists of bounded complexes of sheaves supported on finite-dimensional skeleta, these dualities unite into equivalences

$$\mathbb{D} : D^{<\infty}_{\mathcal{S}^{(i)}}(Z,\mathcal{V}) \longrightarrow D^{<\infty}_{\mathcal{S}^{(i)}}(Z,\mathcal{V}^{\mathrm{op}})^{\mathrm{op}}.$$

Let us describe $\mathbb{D}$ at the level of functors on the exit path category.

Let $F : \mathrm{Gro}(\underline{\mathrm{CM}}_n(L))^{\mathrm{op}} \to C^b(\mathcal{V})$ be a functor equal to 0 on almost all objects. Let us define a functor $F^! : \mathrm{Gro}(\underline{\mathrm{CM}}_n(L)) \to C^b(\mathcal{V})$. First, introduce the following notation. For $M \in \mathrm{CM}_n(L)$ let $d_M = \dim_{\mathbb{R}} U_M$, i.e., $d_M = p+q$, if M is of size $p \times q$. Let aslo $\mathrm{OR}_M = H^{d_m}_c(U_M, \mathbf{k})$ be the 1-dimensional orientation space of U_M. Further, let us write $\varphi : M' \to_d M$ to signify that φ is a morphism $M' \to M$ in $\mathrm{Gro}(\underline{\mathrm{CM}}_n(L))$ and $d_{M'} = d_M + d$. Now put

$$F^!(M) = \mathrm{Tot}\Big\{ \mathrm{OR}_M \otimes F(M) \xrightarrow{\delta} \bigoplus_{\varphi_1 : M_1 \to_1 M} \mathrm{OR}_{M_1} \otimes F(M_1) \xrightarrow{\delta} \bigoplus_{\varphi_2 : M_1 \to_2 M} \mathrm{OR}_{M_2} \otimes F(M_2) \xrightarrow{\delta} \cdots \Big\}[-d],$$

where:

- The matrix element
$$\delta_{\varphi_p,\varphi_{p+1}} : \mathrm{OR}_{M_p} \otimes F(M_p) \longrightarrow \mathrm{OR}_{M_{p+1}} F(M_{p+1})$$
of δ is equal to
$$\sum_{\varphi : M_{p+1} \to M_p,\ \varphi_p \varphi = \varphi_{p+1}} \varepsilon_\varphi \otimes F(\varphi),$$
where $\varepsilon_\varphi : \mathrm{OR}_{M_p} \to \mathrm{OR}_{M_{p+1}}$ is the identification (coorientation) induced by the embedding of the $(d_M + p)$-dimensional cell U_{M_p} to the closure of the $(d_M + p + 1)$-dimensional cell $U_{M_{p+1}}$.

PROPOSITION 7.9. *The differential δ thus defined, satisfies $\delta^2 = 0$.*

PROOF. If $N \in \mathrm{CM}_n(L)$ has size $r \times s$, then we have an identification $U_N \simeq \mathbb{R}^r_< \times \mathbb{R}^s_< \subset \mathbb{R}^{r+s}$, see the proof of Proposition 4.1(a). For each such N let us trivialize OR_{U_N}, i.e., identify it with $\mathbf{k}$ using the above identification of U_N and the standard orientation of $\mathbb{R}^{r+s}$. After this, we write the summand of $F^!(M)$ corresponding to $\varphi_p : M_p \to_p M$ as $F(M_p)$.

Let now $\varphi_p : M_p \to_p M$, $\varphi_{p+1} : M_{p+1} \to_{p+1} M$ be given. Suppose M_p is of size $r \times s$. The existence of at least one $\varphi : M_{p+1} \to M_p$ means that one of the two cases hold:

(1) M_{p+1} is of size $(r+1) \times s$ and $\varphi = \partial'_i$ for some $i = 0, \cdots, r$.
(2) M_{p+1} is of size $r \times (s+1)$ and $\varphi = \partial''_j$ for some $j = 0, \cdots, s$.

For such φ the identification $\varepsilon(\varphi)$ of already trivialized orientation spaces is given by the standard simplicial alternating sign pattern:

$$\varepsilon(\partial'_i) = (-1)^i, \quad \varepsilon(\partial''_j) = (-1)^{r+j}.$$

So we can write $\delta_{\varphi_p,\varphi_{p+1}} : F(M_p) \to F(M_{p+1})$ as

$$\delta_{\varphi_p,\varphi_{p+1}} = \sum_{\substack{i:\, \partial'_i M_{p+1}=M_p \\ \varphi_p\partial'_i=\varphi_{p+1}}} (-1)^i \delta'_i + \sum_{\substack{j:\, \partial''_j M_{p+1}=M_p \\ \varphi_p\partial''_j=\varphi_{p+1}}} (-1)^{r+j}\delta''_j,$$

where $\delta'_i = F(\partial'_i)$ and $\delta''_j = F(\partial''_j)$ are the values of F on elementary morphisms in $\mathrm{Gro}(\underline{\mathrm{CM}}_n(L))$. By the above, only one of the two sums can be nonzero.

Now the property $\delta^2 = 0$ follows, in the usual way, from the standard simplicial identities $\partial'_{i_1}\partial'_{i_2} = \partial'_{i_2-1}\partial_{i_1}$, $i_1 < i_2$, from the similar identities for the ∂''_j and from the commutativity of ∂'_i and ∂''_j. □

Let $\psi : M \to_q N$ be a morphism in $\mathrm{Gro}(\underline{\mathrm{CM}}_n(L))$. Any $\varphi : M_p \to_p M$ gives $\psi_{p+q} = \psi\varphi_p : M_p \to_{p+q} N$. Because of the grading convention of the complex $F^!(M)$, the correspondence $\varphi_p \mapsto \psi_{p+q}$ defines a morphism of cmplexes $\psi_* : F^!(M) \to F^!(N)$. This defines a functor $F^! : \mathrm{Gro}(\underline{\mathrm{CM}}_n(L)) \to C^b(\mathcal{V})$.

PROPOSITION 7.10. *Suppose $\mathcal{F} \in D^{<\infty}_{\mathcal{S}^{(2)}}(Z, \mathcal{V})$ is represented by*

$$F : \mathrm{Gro}(\underline{\mathrm{CM}}_n(L))^{\mathrm{op}} \to C^b(\mathcal{V}).$$

Then the Verdier dual complex $\mathbb{D}(\mathcal{F}) \in D^{<\infty}_{\mathcal{S}^{(2)}}(Z, \mathcal{V}^{\mathrm{op}})$ is represented by $F^!$ considered as a functor $\mathrm{Gro}(\underline{\mathrm{CM}}_n(L))^{\mathrm{op}} \to C^b(\mathcal{V}^{\mathrm{op}})$.

PROOF. For a point $x \in Z$ let $j_x : \{x\} \to Z$ be the embedding. By general properties of Verdier duality, the stalk of $\mathbb{D}(\mathcal{F})$ at x is $j_x^!\mathcal{F}^\bullet$, the "!-stalk" of $\mathcal{F}$. If $M \in \mathrm{CM}_n(L)$, then the $j_x^!\mathcal{F}$ for all $x \in U_M$ can be identified with

$$R\Gamma_c(U_M, j_M^!\mathcal{F}) = R\Gamma_c(U_M, \mathbf{k}) \otimes R\Gamma(U_M, j_M^!\mathcal{F}) = \mathrm{OR}_{U_M}[-d_M] \otimes R\Gamma(U_M, j_M^!\mathcal{F}).$$

Now, $j_M^!\mathcal{F} = \underline{R\Gamma}_{U_M}(\mathcal{F})$ is the complex calculating the (hyper)cohomology with support in U_M. If $\mathcal{F}$ is a single sheaf, corresponding to $F : \mathrm{Gro}(\underline{\mathrm{CM}}_n(L))^{\mathrm{op}} \to \mathcal{V}$, then the space of *sections* with support in U_M is found as

$$\Gamma_{U_M}(Z, \mathcal{F}) = \mathrm{Ker}\Big\{F(M) \longrightarrow \bigoplus_{\varphi: M_1 \to M} F(M_1)\Big\},$$

where the direct sum is over all non-unit morphisms $M_1 \to M$ in $\mathrm{Gro}(\underline{\mathrm{CM}}_n(L))$ (not necessarily $M_1 \to_1 M$). We take the derived functor of this by using the standard simplicial resolution of any functors in terms of injective functors V^{h^M} discussed in §7.3 This gives an explicit complex

$$R\Gamma_{U_M}(Z, \mathcal{F}) = \Big\{F(M) \to \bigoplus_{M^{(1)} \to M} F(M^{(1)}) \to \bigoplus_{M^{(2)} \to M^{(1)} \to M} F(M^{(2)}) \to \cdots\Big\},$$

where the direct sum in the qth term is over all chains of non-unit morphisms $M^{(q)} \to \cdots \to M$. Denote this complex by $R^\bullet$. It has a decreasing filtration V where $V^pR^\bullet$ consists of summands corresponding to chains $M^{(q)} \to \cdots \to M$ with $d_{M^{(q)}} \geqslant d_M + p$. The quotient $V^pR^\bullet/V^{p+1}R^\bullet$ is the direct sum, over all $\varphi_p : M_p \to_p M$, of tensor products $O^\bullet_{\varphi_p} \otimes F(M_p)$, where $O^\bullet_{\varphi_p}$ is a complex of vector spaces defined as follows. Let $\mathrm{Dec}_\bullet(\varphi_p)$ be the augmented semi-simplicial set with $\mathrm{Dec}_n(\varphi_p)$ being the set of decomposition sof φ_p into the composition of $n+1$ non-unit morphisms $M_p = M^{(n+1)} \to M^{(n)} \to \cdots \to M^{(0)} = M$ (so $\mathrm{Dec}_{-1}(\varphi_p)$ consists of one element). Then $O^\bullet_{\varphi_p} = \mathbf{k}[\mathrm{Dec}_\bullet(\varphi_p)]^*[-2]$ is the cochain complex of $\mathrm{Dec}_\bullet(\varphi_p)$ with coefficients in $\mathbf{k}$, shifted so that its lowest term (dual to $\mathbf{k}[\mathrm{Dec}_{-1}(\varphi_p)] = \mathbf{k}$) appears in degree $+1$.

We note that $O^\bullet_{\varphi_p}$ is quasi-isomorphic to $\mathrm{OR}_{M_p/M}[-d_{M_p}+d_M]$, the relative orientation space $\mathrm{OR}_{M_p/M}=\mathrm{OR}_{M_p}\otimes\mathrm{OR}_M^{-1}$ in degree $D_M-d_{M_p}$. Indeed, suppose M_p is of size $a\times b$. Then morphisms in $\mathrm{Gro}(\underline{\mathrm{CM}}_n(L))$ with source M_p are in bijection with faces of the polyhedral cone $\mathbb{R}^a_\leqslant\times\mathbb{R}^b_\leqslant$, see §4.3 Thus $\varphi_p: M_p\to M$ corresponds to some face Γ, and $\mathrm{Dec}_\bullet(\varphi_p)$ is the augmented nerve of the poset of faces containing Γ, i.e., the augmented simplicial set whose n-simplices, $n\geqslant -1$, are chains of faces $\Gamma\subsetneq\Gamma_0\subsetneq\cdots\subsetneq\Gamma_n\subsetneq\mathbb{R}^a_\leqslant\times\mathbb{R}^b_\leqslant$ (the empty chain for $n=-1$). Now, removing the augmentation, we obtain a semi-simplicial set whose realization is a sphere of dimension $\mathrm{codim}(\Gamma)-2$. The top cohomology of this realization is identified with the relative orientation space $\mathrm{OR}_{(\mathbb{R}^a_\leqslant\times\mathbb{R}^b_\leqslant)/\Gamma}=\mathrm{OR}_{M_p/M}$ while the 0th cohomology is killed by the augmentation. Thus

$$\mathrm{gr}^\bullet_V R^\bullet \simeq \bigoplus_{p\geqslant 0}\bigoplus_{\varphi_p:M_p\to_p M} O^\bullet_{\varphi_p}\otimes F(M)$$

$$\overset{\text{qis.}}{\simeq}\bigoplus_{p\geqslant 0}\bigoplus_{\varphi_p:M_p\to_p M}\mathrm{OR}_{M_p/M}[-d_{M_p}+d_M]\otimes F(M_p)$$

which, as a graded object (without differential) is identified with the complex $F^!(M)$ but without the factor $\mathrm{OR}_M[-d_M]$. Further, it is straightforward to check that the induced differential on the associated graded (i.e., the "differential δ_1 in the E_1-term of the spectral sequence of the filtered complex") is given by the same formula $\delta_{\varphi_p,\varphi_{p+1}}=\sum\varepsilon_\varphi\otimes F(\varphi)$ as in the definition of $F^!(M)$. So restoring the factor $\mathrm{OR}_M[-d_M]$, we get

$$R\Gamma_c(U_M,j^!_M\mathcal{F})\simeq\mathrm{OR}_M[-d_M]\otimes H^\bullet(gr^\bullet_V R^\bullet,\delta_1)=F^!(M).$$

We omit further details. □

REMARK 7.11. Proposition 7.10 is an extension of the well known cellular interpretation of $\mathbb{D}$ for cellular spaces for which the exit path category is a poset, see, e.g., [**8**, Chapter 12]. Our argument is likewise general though we restricted the attention to our concrete situation.

7.7. The perverse t-structure. For any stratum S of the stratification $\mathcal{S}^{(0)}$ of Z we denote by $j_S: S\hookrightarrow Z$ the embedding. We denote by $\dim(S)$ the complex dimension of S. Consider the following standard perversity conditions on $\mathcal{F}\in D_{\mathcal{S}^{(0)}}(Z,\mathcal{V})$:

(P^-) For any stratum S of $\mathcal{S}^{(0)}$ we have $\underline{H}^q(j_S^*\mathcal{F})=0$ for $q>-\dim(S)$.

(P^+) For any stratum S of $\mathcal{S}^{(0)}$ we have $\underline{H}^q(j_S^!\mathcal{F})=0$ for $q<-\dim(S)$.

PROPOSITION 7.12.

(a) *Let $D^{<\infty}_{\mathcal{S}^{(0)}}(Z,V))^-, D^{<\infty}_{\mathcal{S}^{(0)}}(Z,V))^+\subset D^!_{\mathcal{S}^{(0)}}(Z,V))$ be the full subcategories consisting of objects satisfying the conditions (P^-) and (P^+) respectively. The pair $(D^{<\infty}_{\mathcal{S}^{(0)}}(Z,V))^-, D^{<\infty}_{\mathcal{S}^{(0)}}(Z,V))^+)$ is a t-structure on the triangulated category $D^{<\infty}_{\mathcal{S}^{(0)}}(Z,V))$.*

(b) *Verdier duality $\mathbb{D}$ sends $\mathcal{D}^{<\infty}_{\mathcal{S}^{(0)}}(Z,\mathcal{V})^-$ to $\mathcal{D}^{<\infty}_{\mathcal{S}^{(0)}}(Z,\mathcal{V})^+$ and vice versa.*

PROOF. (a) This follows, in a formal and standard way, from the fact that the conditions (P^-) and (P^+) define a t-structure on each $D_{\mathcal{S}^{(0)}}(Z_r,\mathcal{V})$ and that these t-structures are compatible under the direct images with respect to the embeddings $Z_r\hookrightarrow Z_{r'}$, $r\leqslant r'$. Part (b) is deduced from the statements about the Z_r in a similar way. □

We denote by $\mathrm{Perv}^{<\infty}(Z, \mathcal{V})$ the heart of the above t-structure on $D^{<\infty}_{\mathcal{S}^{(0)}}(Z, \mathcal{V})$, i.e., the full subcategory of objects satisfying both (P^-) and (P^+). It is an abelian category whose objects will be called *perverse sheaves on Z with finite support.* By Proposition 7.12(b), Verdier duality identifies $\mathrm{Perv}^{<\infty}(Z, \mathcal{V})$ with $\mathrm{Perv}^{<\infty}(Z, \mathcal{V}^{\mathrm{op}})^{\mathrm{op}}$.

REMARK 7.13. The conditions ($\mathrm{P}^{\pm}$) make sense for arbitrary (not necessarily bounded or finitely supported) $\mathcal{S}^{(0)}$-constructible complexes on Z. We do not know whether they define a t-structure on the triangulated category formed by such complexes. More specifically, it is not clear whether any such complex $\mathcal{F}$ includes into a triangle

$$\mathcal{F}^{\leqslant 0} \longrightarrow \mathcal{F} \longrightarrow \mathcal{F}^{\geqslant 1}$$

with $\mathcal{F}^{\leqslant 0}$ satisfying (P^-) and $\mathcal{F}^{\geqslant 1}[-1]$ satisfying (P^+). The standard argument for this (in the case of finite-dimensional stratified spaces) starts from the highest (open) stratum and goes downward, each step making correction at lower and lower strata, see, e.g., [**15**, Th. 8.1.27]. In particular, we can retain the name "perverse sheaves" for complexes satisfying (P^-) and (P^+) but it is not clear whether the full subcategory formed by such objects is abelian.

The question about the "correct" category of perverse sheaves not necessarily supported on a finite skeleton can be studied for more general ind-stratified spaces and we plan to address it is a separate paper.

7.8. The main result: Perverse sheaves with finite support on $Z(\mathbb{C}, L)_n$ and the block $\mathfrak{B}^L_n$. Recall also the block $\mathfrak{B}^L_n$ of the PROB $\mathfrak{B}^L$ of L-graded bialgebras, see §6.2 A functor $F : \mathfrak{B}^L_n \to \mathcal{V}$ will be said to have *finite support,* if F vanishes on almost all objects of $\mathfrak{B}^L_n$. We denote $\mathrm{Fun}^{<\infty}(\mathfrak{B}^L_n, \mathcal{V}) \subset \mathrm{Fun}(\mathfrak{B}^L_n, \mathcal{V})$ the full subcategory formed by functors with finite support. The following is our first main result whose proof will occupy the rest of the paper.

THEOREM 7.14. *We have an equivalence of categories*

$$\mathrm{Perv}^{<\infty}(Z, \mathcal{V}) \simeq \mathrm{Fun}^{<\infty}(\mathfrak{B}^L_n, \mathcal{V}).$$

REMARK 7.15. Continuing Remark 7.13, it is natural to expect the existence of a "correct" category $\mathrm{Perv}(Z(\mathbb{C}, L)_n, \mathcal{V})$ (with no assumption on finite support) which is identified with the category $\mathrm{Fun}(\mathfrak{B}^L_n, \mathcal{V})$ of all functors $\mathfrak{B}^L_n \to \mathcal{V}$. As we shall see, the arguments proving Theorem 7.14 will provide a certain complex $\mathcal{F}^\bullet_F$ of sheaves for an arbitrary functor $F : \mathfrak{B}^L_n \to \mathcal{V}$ but we do not know a good sense in which it is perverse. In particular, if $\mathcal{V}$ is braided monoidal and B is an L-graded bialgebra in $\mathcal{V}$, then we have a braided functor $F : \mathfrak{B}^L \to \mathcal{V}$ which gives, for each n, a functor $F_n : \mathfrak{B}^L_n \to \mathcal{V}$. These functors typically do not have finite support. The complexes $\mathcal{F}^\bullet_{F_n}$, taken together, carry a natural factorization structure (see [**20**] for the case $L = \mathbb{Z}_+$ when all the $Z(\mathbb{C}, L)_n$ are finite dimensional and so the concept of a perverse sheaf is unambiguous). Thus the role of the correct category $\mathrm{Perv}(Z(\mathbb{C}, L)_n, \mathcal{V})$ is that it should contain the components of the "factorizable perverse sheaves" associated to L-graded bialgebras.

8. Janus sheaves, graded bialgebras, and perverse sheaves

8.1. Janus sheaves. Here we adapt [**22**, Definition 2.1] to our situation. We use the notation and conventions of §5.1 Fix $n \in L$, so $\mathrm{CM}_n(L)$ is the set of contingency matrices of content n and $\underline{\mathrm{CM}}_n(L)$ is the corresponding double augmented semi-simplicial set. We recall the notations $\mathrm{Gro}'(\underline{\mathrm{CM}}_n(L))$ and $\mathrm{Gro}''(\underline{\mathrm{CM}}_n(L))$ for the horizontal and vertical Grothemdieck constructions of $\underline{\mathrm{CM}}_n(L)$. These are categories with the same set of objects, namely $\mathrm{CM}_n(L)$.

In addition to the standard notation $f : M \to N$ for a morphism in $\mathrm{Gro}'(\underline{\mathrm{CM}}_n(L))$ we will use the notation expained in §55.1 That is, if $N = (\varphi \times \mathrm{Id})(M)$ where $\varphi : [m] \to [n]$ is a morphism in Δ_+^{op} and M has $m+2$ columns, then we denote the morphism $f : M \to N$ associated to $\varphi \times \mathrm{Id}$ by $f = \varphi'$. Similarly, if $N = (\mathrm{Id} \times \varphi)M$ where M has $m+2$ rows, then we denote the morphism $f : M \to N$ in $\mathrm{Gro}''(\underline{\mathrm{CM}}_n(L))$ associated to $\mathrm{Id} \times \varphi$ by $f = \varphi''$.

Let $\mathcal{V}$ be a $\mathbf{k}$-linear category (not assumed abelian or monoidal).

Definition 8.1. A *Janus presheaf* of type (L, n) with values in $\mathcal{V}$ is a set E of data consisting of an object $E(M) \in \mathcal{V}$ for each $M \in \mathrm{CM}_n(L)$ and morphisms

- $\delta'_f = \delta'_{f,E} : E(M) \to E(N)$ for any morphism $f : N \to M$ in $\mathrm{Gro}'(\underline{\mathrm{CM}}_n(L))$.
- $\delta''_g = \delta''_{g,E} : E(N) \to E(M)$ for any morphism $g : N \to M$ in $\mathrm{Gro}''(\underline{\mathrm{CM}}_n(L))$,

satisfying the conditions:

(JS1) The δ'_f form a functor $\mathrm{Gro}'(\underline{\mathrm{CM}}_n(L))^{\mathrm{op}} \longrightarrow \mathcal{V}$. Similarly, the δ''_g form a functor $\mathrm{Gro}''(\underline{\mathrm{CM}}_n(L)) \longrightarrow \mathcal{V}$.

(JS2) The δ'- and δ''-maps commute with each other. That is, suppose $P = \{M' \xrightarrow{\varphi'} N' \xleftarrow{\psi''} N\}$ be a mixed fork in $\mathrm{Gro}(\underline{\mathrm{CM}}_n(L))$, see §55.2 Then

$$\delta'_{\varphi' : M' \to N} \delta''_{\psi'' : N \to N'} = \sum_{M \in \mathrm{Sup}(P)} \delta''_{\psi'' : M \to M'} \delta'_{\varphi' : M \to N}.$$

where the sum os over M fitting into a diagram (5.1).

A Janus presheaf is called a *Janus sheaf*, if it satisfies the following:

(JS3) If $f : N \to M$ is an anodyne morphism in $\mathrm{Gro}'(\underline{\mathrm{CM}}_n(L))$, then δ'_f is an isomorphism. If $g : N \to M$ is is an anodyne morphism in $\mathrm{Gro}''(\underline{\mathrm{CM}}_n(L))$, then δ''_g is an isomorphism.

We denote by $\mathrm{JPS}(\mathcal{V}) = \mathrm{JPS}_{L,n}(\mathcal{V})$ the category of Janus presheaves of type (L, n) with values in $\mathcal{V}$ and by $\mathrm{JS}(\mathcal{V}) = \mathrm{JS}_{L,n}(\mathcal{V})$ the full subcategory formed by Janus sheaves. We say that a Janus sheaf E has *finite support*, if $E(M) = 0$ for almost all $M \in \mathrm{CM}_n(L)$. Denote $\mathrm{JS}^{<\infty}(\mathcal{V}) = \mathrm{JS}^{<\infty}_{L,n}(\mathcal{V})$ the full subcategory in $\mathrm{JS}_{L,n}(\mathcal{V})$ formed by Janus sheaves with finite support.

Example 8.2. The concept of a Janus sheaf is analogous to that of a mixed Bruhat sheaf [**22**, Def. 2.1] which is defined for a pair (W, S) consisting of a finite reflection group W (acting on a real vector space $\mathfrak{h}_{\mathbb{R}}$) and a set S of simple reflections generating W. In fact, the two concepts have a nonempty intersection, appearing when $L = \mathbb{Z}_+$, so $n \in \mathbb{Z}_+$. A Janus sheaf of type $(\mathbb{Z}_+, n)$ is the same as a mixed Bruhat sheaf of type (S_n, S) where S_n is the symmetric group acting on $\mathbb{R}^n$ by permutation and S is the set of simple transpositions $(i, i+1)$.

8.2. Janus sheaves and graded bialgebras. Let $\mathcal{V}$ be a braided monoidal category. let I be a finite set and $(V_i)_{i \in I}$ be a family of objects of $\mathcal{V}$ labelled by I. Unless $\mathcal{V}$ is symmetric, the notation $\bigotimes_{i \in I} V_i$ is ambiguous: although the tensor products of the V_i in different orders are isomorphic via the braidings, the isomorphisms are not canonical since there are many braids lifting a given permutation. It has been pointed out by Deligne that a choice of an embedding $\varphi : I \hookrightarrow \mathbb{C}$ (a "2-dimensional ordering") fixes the issues, defining an object which can be denoted

$\bigotimes_{i\in I}^{\varphi} V_i$ uniquely up to a unique isomorphism. The braiding isomorphisms in the approach arise from the monodromy of the local system on the space of embeddings $\varphi : I \hookrightarrow \mathbb{C}$ formed by the $\bigotimes_{i\in I}^{\varphi} V_i$. A detailed treatment of this construction can be found in [**20**, §3.2A] and [**23**, §2B]. When the embedding $I \hookrightarrow \mathbb{C}$ is clear from the context or explained verbally, we will use the simple notation $\bigotimes_{i\in I} V_i$.

Let now $A = (A_n)_{n\in L}$ be a graded bialgebra in $\mathcal{V}$. For any contingency matrix $M = \|m_{ij}\|$ of size $p \times q$ we define

$$A_M = \bigotimes_{(i,j)\in\{1,\cdots,p\}\times\{1,\cdots,q\}} A_{m_{ij}},$$

where the set $\{1,\cdots,p\} \times \{1,\cdots,q\}$ is embedded into $\mathbb{C}$ by $(i,j) \mapsto i + j\sqrt{-1}$.

Let $i \in \{0,\cdots,p-2\}$, so $\partial_i' M$ is the contingency matrix of size $(p-1) \times q$ obtained from M by adding the $(i+1)$st ant $(i+2)$nd columns:

$$(\partial_i' M)_{rs} = \begin{cases} m_{rs}, & \text{if } r \leqslant i' \\ m_{i_1,s} + m_{i+2,s}, & \text{if } r = i+1, \\ m_{r-1,s} < & \text{if } r \geqslant i+2. \end{cases}$$

We define the. *ith comultiplication map*

$$\delta_{i,M}' : A_{\partial_i' M} = \bigotimes_{(r,s)} A_{(\partial_i' M)_{rs}} \longrightarrow \bigotimes_{(r,s)} A_{m_{rs}} = A_M$$

as the tensor product of the identities of the $A_{m_{rs}}$ for $r \leqslant i$ or $r \geqslant i+2$ and of the comultiplication maps

$$\Delta_{m_{i+1,s},m_{i+2,s}} : A_{m_{i+1,s}+m_{i+2,s}} \longrightarrow A_{m_{i+1,s}} \otimes A_{m_{i+2,s}}.$$

See [**23**, §2F] for a more detailed treatment in the case $L = \mathbb{Z}_+$. The details in the general case are identical.

Similarly, let $j \in \{0,\cdots,q-2\}$ so $\partial_j'' M$ is obtained from M by adding the $(j+1)$st and $(j+2)$nd columns:

$$(\partial_j'' M)_{rs} = \begin{cases} m_{rs}, & \text{if } s \leqslant j, \\ m_{r,j+1} + m_{r,j+2}, & \text{if } s = j+1, \\ m_{r,s-1}, & \text{if } s \geqslant j+2. \end{cases}$$

We define the *jth multiplication map*

$$\delta_{j,M}'' : A_M = \bigotimes_{(r,s)} A_{m_{rs}} \longrightarrow \bigotimes_{(r,s)} A_{(\partial_j'' M)_{rs}} = A_{\partial_j'' M}$$

as the tensor product of the identities of the $A_{m_{rs}}$ for $s \leqslant j$ or $s \geqslant j+2$ and of the multiplication maps

$$\mu_{m_{s,j+1},m_{s,j+2}} : A_{m_{s,j+1}} \otimes A_{m_{s,j+2}} \longrightarrow A_{m_{s,j+1}+m_{s,j+2}}.$$

Again, the details are identical to those for $L = \mathbb{Z}_+$ explained in [**23**, §2F].

PROPOSITION 8.3. *Let $n \in L$. The correspondence taking:*

- *each $M \in \mathrm{CM}_n(L)$ to A_M;*
- *each generating morphism $M \to \partial_i' M$ in $\mathrm{Gro}'(\underline{\mathrm{CM}}_n(L))$ to $\delta_{i,M}'$;*
- *each generating morphism $M \to \partial_j'' M$ in $\mathrm{Gro}''(\underline{\mathrm{CM}}_n(L))$ to $\delta_{j,M}''$,*

extends to a Janus sheaf $E_{A,n}$ on type (L, n) with values in $\mathcal{V}$.

PROOF. Similar to the proof given in [**23**] to Proposition 2.7 and Corollary 3.2 for the case $L = \mathbb{Z}_+$. To summarize the main points: the functoriality in $\mathrm{Gro}'(\underline{\mathrm{CM}}_n(L))$ follows from coassociativity, while the functoriality in $\mathrm{Gro}''(\underline{\mathrm{CM}}_n(L))$ follows from associativity of A. This shows that we get a datum satisfying (JS1). The axiom (JS2) follows by induction from the axiom (LB3) (compatibility of Δ and μ) for the graded bialgebra A, see §6 6.1 In fact, the sum over 2×2 matrices in (LB3) is nothing but an instance of (JS2) corresponding to the mixed fork

$$(l_1, l_2) \xrightarrow{\partial_0'} (n) \xleftarrow{\partial_0''} \begin{pmatrix} m_1 \\ m_2 \end{pmatrix}$$

in $\mathrm{Gro}(\underline{\mathrm{CM}}_n(L))$. Finally, (JS3) is obvious since $A_0 = \mathbf{1}$ and so the tensor product for A_M and A_N for M, N connected by an anodyne morphism, contain the same non-unit factors. □

8.3. Janus sheaves and perverse sheaves. The role of Janus sheaves for us stems from the following statement which is the second main result of this paper and will be used to prove Theorem 7.14.

THEOREM 8.4. *We have an equivalence of categories*

$$\mathrm{Perv}^{<\infty}(Z(\mathbb{C}, L)_n, \mathcal{V}) \simeq \mathrm{JS}^{<\infty}_{L,n}(\mathcal{V}).$$

The proof will occupy the next section.

9. Janus sheaves and Cousin complexes

In this section we prove Theorem 8.4. by identifying Janus sheaves of type (L, n) with some explicit complexes of sheaves on $Z(\mathbb{C}, L)_n$ which turn out to represent perverse sheaves in a one-to-one way.

9.1. Sheaves on imaginary strata. Recall that $Z = Z(\mathbb{C}, L)_n$ is subdivided into the imaginary strata $Z^{\mathrm{Im}}_{\mathbf{n}}$ labelled by ordered partitions $\mathbf{n} = (\mathbf{n}_1, \cdots, \mathbf{n}_d) \in \mathrm{CM}^1_n(L)$ of n. For such $\mathbf{n}$ let $\varepsilon_{\mathbf{n}} : Z^{\mathrm{Im}}_{\mathbf{n}} \hookrightarrow Z$ be the embedding. Recall that $\mathrm{CM}_{\mathbf{n}}(L) \subset \mathrm{CM}_n(L)$ denotes the set of contingency matrices with exactly d rows and the sum of the ith row equal to n_i. The contingency cells U_M, $M \in \mathrm{CM}_{\mathbf{n}}(L)$ form a cell decomposition of $Z^{\mathrm{Im}}_{\mathbf{n}}$ which we still denote $\mathcal{S}^{(2)}$ or by $\mathcal{S}^{(2)}_{\mathbf{n}}$. Combining them together, we form a disjoint union

$$Z^{\{\mathrm{Im}\}} = \bigsqcup_{\mathbf{n} \in \mathrm{CM}^1_n(L)} Z^{\mathrm{Im}}_{\mathbf{n}} \xrightarrow{\varepsilon} Z$$

with the map ε being a set-theoretic bijection but not a homeomorphism. We denote $\mathcal{S}^{(2)} = \mathcal{S}^{(2)}_{\{\mathrm{Im}\}}$ the cell decomposition of $Z^{\{\mathrm{Im}\}}$ formed by the $\mathcal{S}^{(2)}_{\mathbf{n}}$.

Let $\mathcal{V}$ be a Grothendieck abelian category. Let us introduce the following notation, analogous to that of §77.1

$\mathrm{Sh}_{\mathcal{S}^{(2)}}(Z^{\{\mathrm{Im}\}}, \mathcal{V}) \subset \mathrm{Sh}(Z^{\{\mathrm{Im}\}}, \mathcal{V})$: the full subcategory of $\mathcal{S}^{(2)}$-cellular sheaves.

$\mathrm{Sh}^{<\infty}(Z^{\{\mathrm{Im}\}}, \mathcal{V}) \subset \mathrm{Sh}(Z^{\{\mathrm{Im}\}}, \mathcal{V})$: the full subcategory of sheaves supported on a finite skeleton.

$$\mathrm{Sh}^{<\infty}_{\mathcal{S}^{(2)}}(Z^{\{\mathrm{Im}\}}, \mathcal{V}) := \mathrm{Sh}^{<\infty}(Z^{\{\mathrm{Im}\}}, \mathcal{V}) \cap \mathrm{Sh}_{\mathcal{S}^{(2)}}(Z^{\{\mathrm{Im}\}}, \mathcal{V}).$$

$D^{<\infty}_{\mathcal{S}^{(2)}}(Z^{\{\mathrm{Im}\}}, \mathcal{V})$: the full subcategory in $D^b\,\mathrm{Sh}^{<\infty}(Z^{\{\mathrm{Im}\}}, \mathcal{V})$ formed by complexes whose cohomology sheaves lie in $\mathrm{Sh}_{\mathcal{S}^{(2)}}(Z^{\{\mathrm{Im}\}}, \mathcal{V})$.

Note that an object of $\mathrm{Sh}(Z^{\{\mathrm{Im}\}}, \mathcal{V})$ is simply a collection $(\mathcal{F}_{\mathbf{n}})_{\mathbf{n}\in \mathrm{CM}^1_n(L)}$ of sheaves $\mathcal{F}_{\mathbf{n}}$ on the $Z^{\mathrm{Im}}_{\mathbf{n}}$.

PROPOSITION 9.1.

(a) *The exit path category of* $(Z^{\{\mathrm{Im}\}}, \mathcal{S}^{(2)})$ *is identified with* $\mathrm{Gro}'(\underline{\mathrm{CM}}_n(L))^{\mathrm{op}}$, *the opposite of the horizontal Grothendieck construction of* $\underline{\mathrm{CM}}_n(L)$. *In particular,* $\mathrm{Sh}_{\mathcal{S}^{(2)}}(Z^{\{\mathrm{Im}\}}, \mathcal{V})$ *is identified with* $\mathrm{Fun}(\mathrm{Gro}'(\underline{\mathrm{CM}}_n(L))^{\mathrm{op}}, \mathcal{V})$.

(b) *The category* $D^{<\infty}_{\mathcal{S}^{(2)}}(Z^{\{\mathrm{Im}\}}, \mathcal{V})$ *is identified with*

$$D^b\,\mathrm{Fun}^{<\infty}(\mathrm{Gro}'(\underline{\mathrm{CM}}_n(L))^{\mathrm{op}}, \mathcal{V}),$$

where $\mathrm{Fun}^{<\infty}$ *means the category of functors equal to* 0 *on almost all objects.*

PROOF.

(a) This is a consequence of Proposition 4.5 and the following fact: if $M, N \in \mathrm{CM}_{\mathbf{n}}(L)$ are such that $U_N \subset \overline{U}_M$, then any morphism $M \to N$ in $\mathrm{Gro}(\underline{\mathrm{CM}}_n(L))$ actually lies in $\mathrm{Gro}'(\underline{\mathrm{CM}}_n(L))$.

(b) This is a consequence of Proposition 7.2. □

As before, we use this proposition to eliminate assumption that $\mathcal{V}$ is Grothendieck, so in the sequel $\mathcal{V}$ is an arbitrary abelian category.

PROPOSITION 9.2. *Let* $\mathcal{F} \in \mathrm{Sh}_{\mathcal{S}^{(2)}}(Z^{\{\mathrm{Im}\}}, \mathcal{V})$ *be any* $\mathcal{S}^{(2)}$*-cellular sheaf on* $Z^{\{\mathrm{Im}\}}$. *Then* $R^q\varepsilon_*\mathcal{F} = 0$ *for* $q > 0$.

Equivalently, for any $\mathcal{S}^{(2)}$-cellular sheaf $\mathcal{F}_{\mathbf{n}}$ on $Z^{\mathrm{Im}}_{\mathbf{n}}$ we have $R^q\varepsilon_{n*}\mathcal{F}_{\mathbf{n}} = 0$ for $q > 0$. Here, as before, $\varepsilon_{\mathbf{n}} : Z^{\mathrm{Im}}_{\mathbf{n}} \to Z$ denotes the embedding.

PROOF. Let $e : \mathrm{Gro}'(\underline{\mathrm{CM}}_n(L))^{\mathrm{op}} \to \mathrm{Gro}(\underline{\mathrm{CM}}_n(L))^{\mathrm{op}}$ be the embedding of the horizontal Grothendieck construction to the full one (a bijection on objects). In the identifications of Propositions 4.5, 7.2 and 9.1, the pullback functor

$$\varepsilon^* : \mathrm{Sh}_{\mathcal{S}^{(2)}}(Z, \mathcal{V}) \longrightarrow \mathrm{Sh}_{\mathcal{S}^{(2)}}(Z^{\{\mathrm{Im}\}}, \mathcal{V})$$

corresponds to the functor given by the pullback (composition with e):

$$e^* : \mathrm{Fun}(\mathrm{Gro}(\underline{\mathrm{CM}}_n(L))^{\mathrm{op}}, \mathcal{V}) \longrightarrow \mathrm{Fun}(\mathrm{Gro}'(\underline{\mathrm{CM}}_n(L))^{\mathrm{op}}, \mathcal{V}).$$

Therefore $R^q\varepsilon_*$ is given by the qth derived right Kan extension functor

$$R^q e_* : \mathrm{Fun}(\mathrm{Gro}'(\underline{\mathrm{CM}}_n(L))^{\mathrm{op}}, \mathcal{V}) \longrightarrow \mathrm{Fun}(\mathrm{Gro}(\underline{\mathrm{CM}}_n(L))^{\mathrm{op}}, \mathcal{V}).$$

Explicitly, the value of $R^q e_* F$ on $M \in \mathrm{CM}_n(L)$ is given by the qth derived functor of the projective limit functor

$$(R^q e_* F)(M) = R^q \varprojlim_{\{M \to e(N)\} \in M/e} F(M), \tag{9.3}$$

where M/e is the comma category whose objects are pairs formed by $N \in \mathrm{CM}_n(L)$ and a morphism $M \to e(N)$ in $\mathrm{Gro}(\underline{\mathrm{CM}}_n(L))^{\mathrm{op}}$ or, in plainer terms, morphisms φ from $e(N) = N$ to M in $\mathrm{Gro}(\underline{\mathrm{CM}}_n(L))$. In the latter interpretation, morphisms from $(N_1, \varphi_1 : N_1 \to M)$ to $(N_2, \varphi_2 : N_2 \to M)$ are morphisms $\psi : N_2 \to N_1$ in $\mathrm{Gro}'(\underline{\mathrm{CM}}_n(L))$ such that $\varphi_1\psi = \varphi_2$.

Let us use the notation $N \overset{\prime}{\to} M$ resp. $N \overset{\prime\prime}{\to} M$ to indicate a morphism $N \to M$ in $\mathrm{Gro}'(\underline{\mathrm{CM}}_n(L))$ resp. $\mathrm{Gro}''(\underline{\mathrm{CM}}_n(L))$. Any object of M/e figuring in (9.3) can be uniquely represented as a composition $N \overset{\prime}{\to} P \overset{\prime\prime}{\to} M$. This means that the non-derived ($q = 0$) instance of (9.3) can be represented as a double projective limit

$$\varprojlim_{\{P \overset{''}{\to} M\}} \varprojlim_{\{N \overset{'}{\to} P\}} F(N). \tag{9.4}$$

To prove that the derived functor in (9.3) is zero for $q > 0$, it suffices to show that each limit in (9.4) has higher derived functors equal to 0. Now, the inner limit over $N \overset{'}{\to} P$ is over a conical diagram, since F is a functor on $\mathrm{Gro}'(\underline{\mathrm{CM}}_n(L))$. Therefore $\varprojlim_{\{N \overset{'}{\to} P\}} F(N) = F(P)$, while higher derived functors of this $\varprojlim$ vanish. So it remains to show that $R^q \varprojlim_{\{P \overset{''}{\to} M\}} F(P) = 0$ for $q > 0$. But the comma category formed by arrows $P \overset{''}{\to} M$ is discrete. Indeed, morphisms between such arrows are given by morphisms $P_1 \to P_2$ in $\mathrm{Gro}'(\underline{\mathrm{CM}}_n(L))$ making a triangle commutative. So the $R^q \varprojlim$ over this discrete category vanish for $q > 0$. □

Since the limit over a discrete category is the direct product, we arrive at the following complement to Proposition 9.2.

PROPOSITION 9.5. *Let $\mathcal{F} \in \mathrm{Sh}_{\mathcal{S}^{(2)}}(Z^{\{\mathrm{Im}\}}, \mathcal{V})$ and $F : \mathrm{Gro}'(\underline{\mathrm{CM}}_n(L))^{\mathrm{op}} \to \mathcal{V}$ be the corresponding functor so that $F(M)$ is the stalk of $\mathcal{F}$ at U_M. Then $\varepsilon_* \mathcal{F}$ corresponds to the functor $e_* F : \mathrm{Gro}(\underline{\mathrm{CM}}_n(L))^{\mathrm{op}} \to \mathcal{V}$ given on objects by*

$$(e_* F)(M) = \prod_{N \overset{''}{\to} M} F(N).$$

The values of $e_* F$ on morphisms from $\mathrm{Gro}'(\underline{\mathrm{CM}}_n(L))$ are induced by those of F, while the value on a morphism $g : M_1 \to M_2$ from $\mathrm{Gro}''(\underline{\mathrm{CM}}_n(L))$ is the morphism

$$(e_* F)(g) : \prod_{\varphi_2 : N_2 \overset{''}{\to} M_2} F(N_2) \longrightarrow \prod_{\varphi_1 : N_1 \overset{''}{\to} M_1} F(N_1)$$

with matrix elements

$$(e_* F)(g)_{\varphi_1, \varphi_2} = \begin{cases} \mathrm{Id}_{F(N_2)}, & \text{if } N_1 = N_2 \text{ and } \varphi_2 = g\varphi_1, \\ 0, & \text{otherwise.} \end{cases}$$

□

9.2. Cousin-type complexes.

DEFINITION 9.6. A *Cousin-type complex* on Z (with values in $\mathcal{V}$) is a datum consisting of $\mathcal{S}^{(2)}$-cellular sheaves $\mathcal{F}_{\mathbf{n}}$ on $Z^{\mathrm{Im}}_{\mathbf{n}}$, $\mathbf{n} \in \mathrm{CM}^1_n(L)$ and a complex of sheaves of the form

$$\mathcal{F}^\bullet = \left\{ \cdots \overset{d}{\to} \bigoplus_{l(\mathbf{n}) = r} \varepsilon_{\mathbf{n}*} \mathcal{F}_{\mathbf{n}} \overset{d}{\to} \cdots \overset{d}{\to} \bigoplus_{\mathbf{n} = (n_1, n_2)} \varepsilon_{(n_1, n_2)*} \mathcal{F}_{(n_1, n_2)} \overset{d}{\to} \varepsilon_{(n)*} \mathcal{F}_{(n)} \right\}$$

graded so that the sum over $l(\mathbf{n}) = r$ is in degree $-r$ (so the complex terminates in degree -1). Here $\varepsilon_{\mathbf{n}*}$ is understood as the non-derived direct image functor, i.e., as $R^0 \varepsilon_{\mathbf{n}*}$.

PROPOSITION 9.7. *The category formed by Cousin-type complexes (with morphisms being morphisms of complexes of sheaves) is equivalent to the category of Janus presheaves of type (L, n).*

PROOF. Let $E = (E(M), \delta'_f, \delta''_g)$ be a Janus presheaf of type (L, n). For $\mathbf{n} \in \mathrm{CM}_n(L)$ we form the cellular sheaf $\mathcal{F}_{\mathbf{n}}$ on $Z^{\mathrm{Im}}_{\mathbf{n}}$ via Proposition 9.1(a). That is, the stalk of $\mathcal{F}_{\mathbf{n}}$ at the cell $U_M, M \in \mathrm{CM}_{\mathbf{n}}(L)$, is $E(M)$ and the generalization maps are given by the δ'_f.

Next, let $l(\mathbf{n}) = r$ and $N \in \mathrm{CM}_{\mathbf{n}}(L)$. For any $i = 0, \cdots, d-2$ we have the ordered partition $\partial_i \mathbf{n}$ obtained by adding the $(i+1)$st and $(i+2)$nd entries of $\mathbf{n}$ and the contingency matrix $\partial''_i N \in \mathrm{CM}_{\partial_i \mathbf{n}}(L)$ obtained by adding he $(i+1)$st and $(i+2)$nd rows of N. Thus $N \to \partial''_i N$, an elementary morphism in $\mathrm{Gro}''(\underline{\mathrm{CM}}_n(L))$. Let

$$\delta''_{i,N} = \delta''_{N \to \partial''_n N} : E(N) \longrightarrow E(\partial''_i N).$$

Let $\underline{\mathrm{CM}}_{\mathbf{n}}(L) \subset \underline{\mathrm{CM}}_n(L)$ be the (single, not double) augmented semi-simplicial set formed by $N \in \mathrm{CM}_{\mathbf{n}}(L)$ and the face morphisms δ'_i and their compositions. Let $\mathrm{Gro}'(\underline{\mathrm{CM}}_{\mathbf{n}}(L)) \subset \mathrm{Gro}'(\underline{\mathrm{CM}}_n(L))$ be the corresponding Grothendieck construction. Taken for all $\mathbf{n} \in \mathrm{CM}^1_n(L)$, the $\mathrm{Gro}'(\underline{\mathrm{CM}}_{\mathbf{n}}(L))$ form the connected components of $\mathrm{Gro}'(\underline{\mathrm{CM}}_n(L))$.

Let $e_{\mathbf{n}} : \mathrm{Gro}'(\underline{\mathrm{CM}}_{\mathbf{n}}(L)) \to \mathrm{Gro}(\underline{\mathrm{CM}}_n(L))$ be the embedding functor. The functor $E_{\mathbf{n}} : \mathrm{Gro}'(\underline{\mathrm{CM}}_{\mathbf{n}}(L))^{\mathrm{op}} \to \mathcal{V}$ representing the cellular sheaf $\mathcal{F}_{\mathbf{n}}$ is the restriction of (E, δ') to $\mathrm{Gro}'(\underline{\mathrm{CM}}_{\mathbf{n}}(L)$. So $\varepsilon_{\mathbf{n}*}\mathcal{F}_{\mathbf{n}}$ is represented by the right Kan extension $e_{\mathbf{n}*}E_{\mathbf{n}} : \mathrm{Gro}(\underline{\mathrm{CM}}_n(L))^{\mathrm{op}} \to \mathcal{V}$ whose values on objects and morphisms are given by Proposition 9.5. Let us define a morphism of functors

$$D_{i,\mathbf{n}} : e_{\mathbf{n}*}E_{\mathbf{n}} \longrightarrow (e_{\partial_i \mathbf{n}})_* E_{\partial_i \mathbf{n}}$$

by defining the functor

$$D^*_{i,\mathbf{n}} : e^*_{\partial_i \mathbf{n}} e_{\mathbf{n}*} E_{\mathbf{n}} \longrightarrow E_{\partial_i \mathbf{n}}$$

corresponding to it by adjunction. The value of $D^*_{i,\mathbf{n}}$ on an object $M \in \mathrm{CM}_{\partial_i \mathbf{n}}(L)$ is the morphism

$$(D^*_{i,\mathbf{n}})_M : \prod_{\substack{h : N \overset{''}{\to} M \\ N \in \mathrm{CM}_{\mathbf{n}}(L)}} E(N) \longrightarrow E(M),$$

whose component $(D^*_{i,\mathbf{n}})_{M,h} : E(N) \to E(M)$ is 0 unless $M = \partial''_i N$ and $g = \partial''_i$, in which case it is $\delta''_{i,N}$.

Let us show that $D^*_{i,\mathbf{n}}$ is indeed a morphism of functors, i.e., that it commutes with the action of each morphism $f : M_1 \to M_2$ from $\mathrm{Gro}'(\underline{\mathrm{CM}}_{\partial_i \mathbf{n}}(L))$. In other words, we prove that the diagram

$$\begin{array}{ccc}
\prod_{\substack{h_1 : N_1 \overset{''}{\to} M_1 \\ N_1 \in \mathrm{CM}_{\mathbf{n}}(L)}} E(N_1) & \xrightarrow{(D^*_{i,\mathbf{n}})_{M_1}} & E(M_1) \\
\Big\uparrow {\scriptstyle (e^*_{\partial_i \mathbf{n}} e_{\mathbf{n}*} E_{\mathbf{n}})(f)} & & \Big\uparrow {\scriptstyle \delta'_f} \\
\prod_{\substack{h_2 : N_2 \overset{''}{\to} M_2 \\ N_2 \in \mathrm{CM}_{\mathbf{n}}(L)}} E(N_2) & \xrightarrow[(D^*_{i,\mathbf{n}})_{M_2}]{} & E(M_2)
\end{array}$$

is commutative. Let us check it on a factor $E(N_2)$ in the lower left corner, corresponding to any given h_2. The composition $\delta'_f (D^*_{i,\mathbf{n}})_{M_2}$ takes $E(N_2)$ to $E(\partial''_i N_2)$ by $\delta''_{i,N}$, if $M_2 = \partial''_i N_2$ (and to 0 otherwise). After this, application of δ'_f gives $\delta'_f \delta''_{i,N}$.

This gives a mixed fork $P = \{N_2 \overset{\partial_i''}{\to} M_2 \overset{f}{\leftarrow} M_1\}$ in the sense of §55.2 Now, the other path in the diagram corresponds to the sum over $\mathrm{Sup}(P)$ as in the condition (JS2). So this condition implies that $D_{i,\mathbf{n}}^*$ and thus $D_{i,\mathbf{n}}$ is indeed a morphism of functors.

Let $d_{i,\mathbf{n}} : \varepsilon_{\mathbf{n}*}\mathcal{F}_{\mathbf{n}} \to (\varepsilon_{\partial_i\mathbf{n}})_*\mathcal{F}_{\partial_i\mathbf{n}}$ be the morphism of sheaves corresponding to $D_{i,\mathbf{n}}$ and let

$$d = \sum_{l(\mathbf{n})=r} \sum_{i=0}^{r-2} (-1)^i d_{i,\mathbf{n}} : \bigoplus_{l(\mathbf{n})=r} \varepsilon_{\mathbf{n}*}\mathcal{F}_{\mathbf{n}} \longrightarrow \bigoplus_{l(\mathbf{n})=r-1} \varepsilon_{\mathbf{n}*}\mathcal{F}_{\mathbf{n}}.$$

We then have $d^2 = 0$ because E is a functor $\mathrm{Gro}''(\underline{\mathrm{CM}}_n(L)) \to \mathcal{V}$. So we have a Cousin-type complex associated to E. Denote it $\mathcal{F}^\bullet(E)$.

Conversely, let $\mathcal{F}^\bullet$ be a Cousin-type complex. We recover the Janus sheaf E from $\mathcal{F}^\bullet$ by reversing the above considerations. That is, we define $E(M)$ as the stalk of $\mathcal{F}_{\mathbf{n}}$ at U_M, where $\mathbf{n}$ is such that $M \in \mathrm{CM}_{\mathbf{n}}(L)$. The data of the δ'_f, i.e., the contravariant functoriality of E on $\mathrm{Gro}'(\underline{\mathrm{CM}}_n(L))$, are given by the sheaf structure of the $\mathcal{F}_{\mathbf{n}}$. The differential d gives the covariant functoriality on $\mathrm{Gro}''(\underline{\mathrm{CM}}_n(L))$ for elementary morphisms ∂_i''. The condition $d^2 = 0$ translates into simplicial identities for the morphisms $\delta''_{i,N}$ associated by E to the ∂_o'', i.e., they define E as a functor on all morphisms from $\mathrm{Gro}''(\underline{\mathrm{CM}}_n(L))$. The condition that d is a morphism of sheaves, translates into (JS2). Proposition 9.7 is proved.

9.3. The Cousin-type complex as the horizontal Verdier dual. The Cousin-type complex $\mathcal{F}^\bullet = \mathcal{F}^\bullet(E)$ associated to a Janus presheaf E, is a complex of $\mathcal{S}^{(2)}$-cellular sheaves. As such, it is represented by a functor which we denote $\mathbb{D}''E : \mathrm{Gro}(\underline{\mathrm{CM}}_n(L))^{\mathrm{op}} \to C^b(\mathcal{V})$. In particular, $\mathbb{D}''E$ is contravariant on both subcategories $\mathrm{Gro}'(\underline{\mathrm{CM}}_n(L))$ and $\mathrm{Gro}''(\underline{\mathrm{CM}}_n(L))$, while E itself is contravariant on Gro' and covariant on Gro'' (so it is a "sheaf in the horizontal direction" and a "cosheaf in the vertical direction"). The following result describes $\mathbb{D}''E$ as a kind of relative Verdier dual to E in the vertical direction so that the cosheaf–type behavior in that direction is transformed into the sheaf-type behavior.

Given $M \in \mathrm{CM}_n(L)$ of size $p \times q$, the cell U_M is identified with $\mathbb{R}^p_< \times \mathbb{R}^q_<$, see §44.1 So the orientation space OR_M splits naturally into tensor product $\mathrm{OR}_M = \mathrm{OR}'_M \otimes \mathrm{OR}''_M$, where OR'_M is the orientation space of $\mathbb{R}^p_<$ and OR''_M is that of $\mathbb{R}^q_<$. We also write $d'_M = p$, $d''_M = q$, so $d_M = \dim_{\mathbb{R}} U_M = d'_M + d''_M$. Further, we will write $f : N \overset{''}{\to}_r M$ to signify that f is a morphism in $\mathrm{Gro}''(\underline{\mathrm{CM}}(L))$ and $d_N = d_M + r$ (or, equivalently, $d''_N = d''_M + r$). We also write $\mathrm{OR}''_{N/M} = \mathrm{OR}''_M \otimes (\mathrm{OR}''_M)^{-1}$.

PROPOSITION 9.8. *We have*

$$(\mathbb{D}''E)(M) = \Bigg\{ \cdots \overset{d}{\to} \bigoplus_{g_2 : M_2 \overset{''}{\to}_2 M} E(M_2) \otimes \mathrm{OR}''_{M_2} \overset{d}{\to} \bigoplus_{g_1 : M_1 \overset{''}{\to}_1 M} E(M_1) \otimes \mathrm{OR}''_{M_1} \overset{d}{\to} E(M) \otimes \mathrm{OR}''_M \Bigg\},$$

where:

- *The rightmost term $E(M) \otimes \mathrm{OR}''_M$ is in degree $(-d''_M)$.*

- *The matrix element*
$$d_{g_r,g_{r-1}} : E(M_r)\otimes \mathrm{OR}''_{M_r} \longrightarrow E(M_{r-1})\otimes \mathrm{OR}''_{M_{r-1}}$$
is equal to
$$\sum_{\substack{\varphi: M_r \overset{''}{\to} M_{r-1}\\ g_{r-1}\varphi = g_r}} \varepsilon_\varphi \otimes \delta''_\varphi,$$
where, in its turn:
  - δ''_φ *is the value of* E *on* φ*;*
  - $\varepsilon_\varphi : \mathrm{OR}''_{M_r} \to \mathrm{OR}''_{M_{r-1}}$ *is the identification (coorientation) induced by the embedding of* $U_{M_{r-1}}$ *as a codimension* 1 *cell in the closure of* U_{M_r}.

PROOF. Direct comparison with the construction of $\mathcal{F}^\bullet(E)$. We need to note that any φ in the sum for $d_{g_r,g_{r-1}}$ has the form ∂''_i for some i. Furtherm the space OR''_{M_r} can be trivialized using the identification $U_{M_r} \simeq \mathbb{R}^{d'_{M_r}}_< \times \mathbb{R}^{d''_{M_r}}_<$ and the canonical orientation of $\mathbb{R}^{d''_{M_r}}$. After this, the identification ε_φ is transformed into the standard simplicial sign $(-1)^i$, thus matching with the differential constructed in Proposition 9.7. □

9.4. Cousin-type complexes and (absolute) Verdier duality. Let $\mathcal{V}$ be an abelian category and $\mathcal{V}^{\mathrm{op}}$ its opposite category. For convenience we will denote by V^* the object of V^{op} corresponding to $V \in \mathrm{Ob}(\mathcal{V})$ and by $f^* : W^* \to V^*$ the morphism in $\mathcal{V}^{\mathrm{op}}$ corresponding to a morphism $f : V \to W$ in $\mathcal{V}$.

Let $E = (E(M), \delta'_{f,E}, \delta''_{g,E})$ be a Janus presheaf of type (L, n) with values in $\mathcal{V}$. We define a Janus presheaf $E^\tau = (E^\tau(M), \delta'_{f,E^\tau}, \delta''_{g,E^\tau})$ with values in $\mathcal{V}^{\mathrm{op}}$ as follows:

- $E^\tau(M) = E(M^t)^*$, where $M^t = \|m_{ji}\|$ is the transpose of $M = \{m_{ij}\|$. Note that a morphism $f : M \to N$ in $\mathrm{Gro}'(\underline{\mathrm{CM}}_n(L))$ resp. $\mathrm{Gro}''(\underline{\mathrm{CM}}_n(L))$ gives a morphism $f^t : M^\tau \to N^\tau$ in $\mathrm{Gro}''(\underline{\mathrm{CM}}_n(L))$ resp. $\mathrm{Gro}'(\underline{\mathrm{CM}}_n(L))$.
- For $f : M \to N$ a morphism in $\mathrm{Gro}'(\underline{\mathrm{CM}}_n(L))$ we put $\delta'_{f,E^\tau} = (\delta''_{f^t,E})^*$.
- Similarly, for $g : M \to N$ a morphism in $\mathrm{Gro}''(\underline{\mathrm{CM}}_n(L))$ we put $\delta''_{g,E^\tau} = (\delta'_{g^t,E})^*$.

It is clear that E^τ is again a Janus presheaf. Further, if E is a Janus sheaf, then so is E^τ.

As before, we denote by τ the involution on $Z = Z(\mathbb{C}, L)_n$ induced by permutation of coordinates $(x + iy) \mapsto (y + ix)$ on $\mathbb{C}$. Clearly τ preserves the cell dcomposition $\mathcal{S}^{(2)}$, with $\tau(U_M) = U_{M^t}$. Recall that we have Verdier duality which is an anti-equivalence
$$\mathbb{D} : D^{<\infty}_{\mathcal{S}^{(2)}}(Z, \mathcal{V}) \longrightarrow D^{<\infty}_{\mathcal{S}^{(2)}}(Z, \mathcal{V}^{\mathrm{op}}).$$

PROPOSITION 9.9. *Let* E *be a Janus presheaf of type* (L, n) *with values in* $\mathcal{V}$ *and with finite support. Let* $\mathcal{F}^\bullet(E)$ *be the corresponding Cousin-type complex. Then we have an isomorphism in the derived category:*
$$\mathbb{D}(\mathcal{F}^\bullet(E)) \simeq \tau^*\mathcal{F}^\bullet(E^\tau).$$

PROOF. Let us denote by the same letter $\mathbb{D}$ the effect of Verdier duality on functors from the category of exit paths, i.e., the horizontal arrow in the following diagram:

$$\begin{array}{ccc} D^b\operatorname{Fun}^{<\infty}(\operatorname{Gro}(\underline{\mathrm{CM}}_n(L))^{\mathrm{op}}, \mathcal{V}) & \xrightarrow{\mathbb{D}} & D^b\operatorname{Fun}^{<\infty}(\operatorname{Gro}(\underline{\mathrm{CM}}_n(L)), \mathcal{V}) \\ \big\downarrow \mathbb{D}'' & \nearrow_{\mathbb{D}'} & \\ D^b\operatorname{JPS}^{<\infty}(\mathcal{V}) & & \end{array}$$

The description of $\mathbb{D}$ given in Proposition 7.10 implies that it can be decomposed as the composition $\mathbb{D} = \mathbb{D}' \circ \mathbb{D}''$ of the horizontal and vertical Verdier dualities. Here $\mathbb{D}''$ changes contravriant functoriality on $\operatorname{Gro}''(\underline{\mathrm{CM}}_n(L))$ to covariant one, thus producing a complex of Janus sheaves while $\mathbb{D}'$ changes contravariant functoriality on $\operatorname{Gro}'(\underline{\mathrm{CM}}_n(L))$ to covariant one, thus producing a complex of cosheaves with values in $\mathcal{V}$, i.e., of sheaves with values in $\mathcal{V}^{\mathrm{op}}$. More precisely, $\mathbb{D}'$ is the analog of the functor described on objects in Proposition 9.8 but with $\operatorname{Gro}'(\underline{\mathrm{CM}}_n(L))$ playing the role of $\operatorname{Gro}''(\underline{\mathrm{CM}}_n(L))$, while $\mathbb{D}''$ involves Gro''.

This means that the complex of $\mathcal{V}^{\mathrm{op}}$-valued sheaves associated to $\mathbb{D}'E$, $E \in \operatorname{JPS}^{<\infty}(\mathcal{V})$ is $\tau^*\mathcal{F}^\bullet(E^\tau)$, as the rows of the real and imaginary strata are interchanged. Our statement reduces to the claim that vertical Verdier duality is involutive, i.e., the composition

$$D^b\operatorname{JPS}^{<\infty}(\mathcal{V}) \xrightarrow{\mathbb{D}''} D^b\operatorname{Fun}^{<\infty}(\operatorname{Gro}(\underline{\mathrm{CM}}_n(L))^{\mathrm{op}}, \mathcal{V}) \xrightarrow{\mathbb{D}''} D^b\operatorname{JPS}^{<\infty}(\mathcal{V})$$

is isomorphic to the identity. This is proved by the same formal argument as the involutivity of the ordinary (absolute) Verdier duality at the cellular level. More precisely, let $E \in D^b\operatorname{JPS}^{<\infty}(\mathcal{V})$. Then, $\mathbb{D}''E$ being contravariant in $\operatorname{Gro}''(\underline{\mathrm{CM}}_n(L))$, the value $(\mathbb{D}''\mathbb{D}''E)(M)$ is the total complex of

$$(\mathbb{D}''E)(M) \otimes \mathrm{OR}''_M \\ \to \bigoplus_{N_1 \overset{''}{\to}_1 M} (\mathbb{D}''E)(N_1) \otimes \mathrm{OR}''_{N_1} \to \bigoplus_{N_2 \overset{''}{\to}_2 M} (\mathbb{D}''E)(N_2) \otimes \mathrm{OR}''_{N_2} \to \cdots$$

the horizontal grading starting at d''_M. For uniformity, the leftmost term corresponds to the unique arrow (identity) $M = N_0 \overset{''}{\to}_0 M$. Substituting here the expression for each $(\mathbb{D}''E)(N_r)$ from Proposition 9.8, we represent $(\mathbb{D}''\mathbb{D}''E)(M)$ as the total complex of the double complex

$$\begin{array}{ccccc} E(M) & & & & \\ \uparrow & & & & \\ \bigoplus_{M_1 \to_1 N_0 = M} E(M_1) \otimes \mathrm{OR}''_{M_1/N_0} & \longrightarrow & \bigoplus_{M_1 = N_1 \to_1 M} E(M_1) & & \\ \uparrow & & \uparrow & & \\ \bigoplus_{M_2 \to_2 N_0 = M} E(M_2) \otimes \mathrm{OR}''_{M_2/N_0} & \to & \bigoplus_{M_2 \to_1 M_1 \to_1 N_0 = M} E(M_2) \otimes \mathrm{OR}''_{M_2/N_1} & \to & \bigoplus_{M_2 = N_2 \to_2 M} E(M_2) \\ \uparrow & & \uparrow & & \uparrow \\ \vdots & & \vdots & & \vdots \end{array}$$

where $E(M)$ is in degree $(0,0)$ and terms that are 0, are not depicted. All arrows indexing direct sums, are in $\mathrm{Gro}''(\underline{\mathrm{CM}}_n(L))$, so we use the notation $\to_p$ instead of $\overset{''}{\to}_p$. The vertical differential involves the covariant functoriality of E on $\mathrm{Gro}''(\underline{\mathrm{CM}}_n(L))$ while the horizontal differential is purely combinatorial. More precisely, the $(-p)$th term is a direct sum $\bigoplus_{f:M_p \overset{''}{\to}_p M} C_f^\bullet \otimes_{\mathbf{k}} E(M_p)$, where f runs over all morphisms $M_p \to M$ in $\mathrm{Gro}''(\underline{\mathrm{CM}}_n(L))$ of relative dimension p, and $C_f^\bullet$ is the complex of $\mathbf{k}$-vector spaces with

$$C_f^r = \bigoplus_{\substack{M_p \overset{h}{\to}_{p-r} N_r \overset{g}{\to}_r M \\ gh=f}} \mathrm{OR}''_{M_p/N_r},$$

the sum being over all factorizations of f into the compostion gh where g has relative dimension r and h has relative dimension $p-r$.

We now prove that each row of our double complex except the top one (consisting of $E(M)$ alone) is exact. For this it suffices to prove that each $C_f^\bullet$ is exact, if $p>0$. To see this, note that f, being a morphism in the Grothendieck construction, corresponds to a morphism $\varphi : [d''_M + p - 2] \to [d''_M - 2]$ in Δ_+^{op} (here d''_M is the number of columns of M). In particular, $M = \partial''_\varphi(M_p)$ with respect to the double augmented semi-simplicial structure on $\underline{\mathrm{CM}}_n(L)$. This means that factorizations of f are in bijection with factorizations of φ in Δ_+^{op}. Let $m = d''_M + p - 2$. Morphisms in Δ_+^{op} with source M are in bijection with faces $\Gamma \subset \Delta^m$ of the standard m-simplex. Factorizations of φ correspond to intermediate faces $\Gamma \subset \Gamma' \subset \Delta^m$, where γ is the face corresponding to φ. So we find that $C_f^\bullet$ is the augmented cellular cochain complex of the link of Γ in Δ^m. Since this link is itself a simplex, $C_f^\bullet$ is exact and so $(\mathbb{D}''\mathbb{D}''E)(M)$ is quasi-isomorphic to $E(M)$. This finishes the proof of Proposition 9.9.

9.5. Fron a Janus sheaf to a perverse sheaf. Let E be a Janus sheaf with finite support. The associated Cousin-type complex $\mathcal{F}^\bullet(E) \in D^{<\infty}_{\mathcal{S}^{(2)}}(Z, \mathcal{V})$ will be called the *Cousin complex* of E.

PROPOSITION 9.10. $\mathcal{F}^\bullet(E) \in \mathrm{Perv}^{<\infty}(Z, E)$.

PROOF. This statement consists of three parts:

(1) $\mathcal{F}^\bullet(E)$ is an $\mathcal{S}^{(0)}$-constructible complex, i.e., the $\underline{H}^r\mathcal{F}^\bullet(E)$ are $\mathcal{S}^{(0)}$-constructible sheaves.
(2) $\mathcal{F}^\bullet(E)$ satisfies (P^-), see §7.7
(3) $\mathcal{F}^\bullet(E)$ satisfies (P^+).

To prove (1) it suffices, by Proposition 7.8, that $\mathcal{F}^\bullet(E)$ is both $\mathcal{S}^{(1)}$-constructible and $\tau\mathcal{S}^{(1)}$-constructible. We note that each term $\mathcal{F}^{-r}(E) = \bigoplus_{l(\mathbf{n})=r} \varepsilon_{\mathbf{n}*}\mathcal{F}_{\mathbf{n}}$ is $\mathcal{S}^{(1)}$-constructible (and so the cohomology sheaves of $\mathcal{F}^\bullet(E)$ are $\mathcal{S}^{(1)}$-constructible as well). This is because each sheaf $\mathcal{F}_{\mathbf{n}} \in \mathrm{Sh}_{\mathcal{S}^{(2)}}(Z_{\mathbf{n}}^{\mathrm{Im}})$ is $\mathcal{S}^{(1)}$-constructible. This last statement is a consequence of the axiom (JS3) for E which says that anodyne morphisms in $\mathrm{Gro}''(\underline{\mathrm{CM}}_n(L))$ (corresponding to inclusions of $\mathcal{S}^{(2)}$-cells within one $\mathcal{S}^{(1)}$-cell) are taken to isomorphisms.

To prove that $\mathcal{F}^\bullet(E)$ is $\tau\mathcal{S}^{(1)}$-constructible, , it suffices to show that $\mathbb{D}\mathcal{F}^\bullet(E)$, its Verdier dual, is $\tau\mathcal{S}^{(1)}$-constructible. But by Proposition 9.9, $\mathbb{D}\mathcal{F}^\bullet(E) \simeq \tau^*\mathcal{F}^\bullet(E^\tau)$. Since E^τ is a Janus sheaf, $\mathcal{F}^\bullet(E^\tau)$ is $\mathcal{S}^{(1)}$-constructible by the above, and so $\tau\mathcal{F}^\bullet(E^\tau)$ is $\tau\mathcal{S}^{(1)}$-constructible. This proves (1).

To prove (2), we need to show that for each $r \geqslant 1$ the sheaf $\underline{H}^{-r}\mathcal{F}^\bullet(E)$ (already known to be $\mathcal{S}^{(0)}$-constructible) haas support on the union of $\mathcal{S}^{(0)}$-strata of complex dimension $\leqslant r$. The strata Z_α of $\mathcal{S}^{(0)}$ are labelled by unordered partitions α of n. If $\mathbf{n} \in \mathrm{CM}^1_n(L)$ is an ordered partition of n, we denote $\alpha(\mathbf{n})$ the corresponding unordered partition. Now, if $Z_\alpha \subset \operatorname{supp} \underline{H}^{-r}\mathcal{F}^\bullet(E)$ then $Z_\alpha \subset \operatorname{supp} \mathcal{F}^{-r}(E) = \bigoplus_{l(\mathbf{n})=r} Z_{\mathbf{n}}^{\mathrm{Im}}$. By irreducibility of Z_α this implies that $Z_\alpha \subset Z_{\mathbf{n}}^{\mathrm{Im}}$ for some $\mathbf{n} \in \mathrm{CM}^1_n(L)$ with $l(\mathbf{n}) = r$. But such an ainclusion is possible only if α is a coarsening of $\alpha(\mathbf{n})$, so α consists of $\leqslant r$ parts and therefore $\dim_{\mathbb{C}} Z_\alpha \leqslant r$. This proves (2).

Finally, the condition (P^+) is Verdier dual to (P^-). So to show that $\mathcal{F}^\bullet(E)$ satisfies (P^+), it is enough to show that $\mathbb{D}\mathcal{F}^\bullet(E)$ satisfies (P^-). But $\mathbb{D}\mathcal{F}^\bullet(E) \simeq \tau^*\mathcal{F}^\bullet(E^\tau)$, and $\mathcal{F}^\bullet(E^\tau)$ satisfies (P^-) by the above. Since $\tau : Z \to Z$ preserves the strata of $\mathcal{S}^{(0)}$, we see that $\tau^*\mathcal{F}^\bullet(E^\tau)$ satisfies (P^-) as well. This shows (3) and finishes the proof of Proposition 9.10. □

9.6. From a perverse sheaf to a Janus sheaf. Proposition 9.10 defined a functor

$$c : \mathrm{JS}^{<\infty}_{L,n}(Z, \mathcal{V}) \longrightarrow \mathrm{Perv}^{<\infty}(Z, \mathcal{V}), \quad E \mapsto \mathcal{F}^\bullet(E). \tag{9.11}$$

In order to prove Theorem 8.4, we construct the inverse functor.

PROPOSITION 9.12. *Let $\mathcal{G}^\bullet \in \mathrm{Perv}^{<\infty}(Z, \mathcal{V})$ and $\mathbf{n} \in \mathrm{CM}^1_n(L)$ be an ordered partition of length r. Then the complex $\varepsilon^!_{\mathbf{n}}\mathcal{G}^\bullet$ on $Z^{\mathrm{Im}}_{\mathbf{n}}$ is quasi-isomorphic to a single sheaf $\mathcal{F}_{\mathbf{n}}$ in degree $(-r)$.*

PROOF. As in [**22**, §4.1], we deduce the statement from a result of [**19**] about hyperplane arrangements which we recall for convenience of the reader.

Let $\mathcal{H}$ be an arrangement of real hyperplanes (containing 0) in a Euclidean space $\mathbb{R}^D$. It subdivides $\mathbb{R}^D$ into a disjoint union of *faces* which are locally closed conic subsets. For a face C we denote $\rho_C : \mathbb{R}^D + C\sqrt{-1} \hookrightarrow \mathbb{C}^D$ the emdedding of the corresponding tube domain. The complexification of the hyperlplanes from $\mathcal{H}$ define a complex stratification of $\mathbb{C}^D$ into generic points of complex flats. Recall that a *complex flat* is the intersection of the complexifications of some hyperplanes from $\mathcal{H}$. We denote $\mathrm{Perv}_{\mathcal{H}}(\mathbb{C}^D)$ the category of perverse sheaves on $\mathbb{C}^D$, constructible with respect to this stratification.

PROPOSITION 9.13. *Let C be a face of $\mathcal{H}$ of real dimension r. Then for any $\mathcal{G}^\bullet \in \mathrm{Perv}_{\mathcal{H}}(\mathbb{C}^D)$ we have $\underline{H}^{\neq -r}\rho^!_C\mathcal{G}^\bullet = 0$.*

PROOF. This is a consequence of [**19**, Prop. 3.10] (reduction to the case $r = 0$) and [**19**, Prop. 4.9(a)] (case $r = 0$). For another transparent proof of the case $r = 0$ see [**20**, Prop. 1.3.3]. These references deal with the case $\mathcal{V} = \mathrm{Vect}_{\mathbf{k}}$. The case of an arbitrary abelian category $\mathcal{V}$ presents no difference. □

To reduce Proposition 9.12 from 9.13, we choose a set $(\xi_k)_{k\in K}$ of monoid generators of L and for each $l \geqslant 0$ consider the map $p_l : (\mathbb{C} \times K)^l \to Z(\mathbb{C}, L)$ described in (2.5). The product $(\mathbb{C} \times K)^l = \mathbb{C}^l \times K^l$ is the disjoint union of K^l copies of $\mathbb{C}^l$. Denote the copy corresponding to $(k_1, \cdots, k_l) \in K^l$ by $\mathbb{C}^l_{(k_1,\cdots,k_l)}$. The restriction of p_l to $\mathbb{C}^l_{(k_1,\cdots,k_l)}$ is the map

$$p_{(k_1,\cdots,k_l)} : \mathbb{C}^l_{(k_1,\cdots,k_l)} \longrightarrow Z(\mathbb{C}, L), \quad (z_1, \cdots, z_l) \mapsto \sum \xi_{k_i} \cdot z_i.$$

Its image lies in $Z(\mathbb{C}, L)_n$, where $n = \sum \xi_{k_i}$. The stratification of $(\mathbb{C} \times K)^l$ by multiplicities, when restricted to $\mathbb{C}^l_{(k_1,\cdots,k_l)}$, gives the stratification given by the arrangement $\mathcal{H}'$ of hyperplanes $H_{ij} = \{z_i = z_j\}$ for all pairs $(i < j)$ such that $k_i = k_j$. Let $\mathcal{H} \supset \mathcal{H}'$ be the diagonal arrangement formed by all H_{ij}, $i < j$.

Let us now choose l so that $\mathrm{Im}(p_l) \supset \mathrm{Supp}(\mathcal{G})$. Since $\mathcal{G}$ is a pervese sheaf on $Z(\mathbb{C}, L)_n$, it follows tnat there is $(k_1, \cdots, k_l)$ such that $\mathrm{Supp}(\mathcal{G}) \subset \mathrm{Im}(p_{(k_1,\cdots,k_l)})$. Thus

$$p^*\mathcal{G} \in \mathrm{Perv}_{\mathcal{H}'}(\mathbb{C}^l_{(k_1,\cdots,k_l)}) \subset \mathrm{Perv}_{\mathcal{H}}(\mathbb{C}^l_{(k_1,\cdots,k_l)}).$$

It further follows that

$$\mathrm{Supp}(\varepsilon^!_{\mathbf{n}}\mathcal{G}) \subset Z^{\mathrm{Im}}_{\mathbf{n}} \cap \mathrm{Im}(p_{(k_1,\cdots,k_l)}).$$

let F be the fiber product fitting into the Cartesian diagram

$$\begin{array}{ccc}
F & \xrightarrow{\ \pi\ } & Z^{\mathrm{Im}}_{\mathbf{n}} \\
{\scriptstyle\rho}\downarrow & & \downarrow{\scriptstyle\varepsilon_{\mathbf{n}}} \\
\mathbb{C}^l_{(k_1,\cdots,k_l)} & \xrightarrow[p_{(k_1,\cdots,k_l)}]{} & Z(\mathbb{C}, L)_n.
\end{array}$$

To show that the sheaf $\underline{H}^j(\varepsilon^!_{\mathbf{n}})\mathcal{G})$ vanishes when $j \neq -r$, it suffices to prove that the sheaf $\pi^*\underline{H}^j(\varepsilon^!_{\mathbf{n}}\mathcal{G}) = \underline{H}^j(\pi^*\varepsilon^!_{\mathbf{n}}\mathcal{G})$ vanishes for $j \neq -r$. But $p_{(k_1,\cdots,k_l)}$ is finite and so, by Cartesianity, is π. Therefore $\pi^* = \pi^!$ and $p^*_{(k_1,\cdots,k_l)} = p^!_{(k_1,\cdots,k_l)}$. Furthermore, $\mathcal{G}' := p^*_{(k_1,\cdots,k_l)}\mathcal{G}$ lies in $\mathrm{Perv}_{\mathcal{H}}(\mathbb{C}^l_{(k_1,\cdots,k_l)})$, so it is enough to show that

$$\pi^*\varepsilon^!_{\mathbf{n}}\mathcal{G} = \pi^!\varepsilon^!_{\mathbf{n}}\mathcal{G} = \rho^! p^!_{(k_1,\cdots,k_l)}\mathcal{G} = \rho^! p^*_{(k_1,\cdots,k_l)}\mathcal{G} = \rho^!\mathcal{G}'$$

has $\underline{H}^{\neq -r} = 0$. But this folllows from Proposition 9.13 once we notice that F is a disjoint union of tube domains $\mathbb{R}^l + iC$ for some faces C of $\mathcal{H}$ of dimension r. Indeed, let $(z_1, \cdots, z_l) \in \mathbb{C}^l_{(k_1,\cdots,k_l)}$ with $z_\nu = x_\nu + y_\nu\sqrt{-1}$, $x_\nu, y_\nu \in \mathbb{R}$. The condition that $p_{(k_1,\cdots,k_l)}(z_1, \cdots, z_l) = \sum \xi_{k_i} \cdot z_i$ lies in $Z^{\mathrm{Im}}_{\mathbf{n}}$ is a condition on $(y_1, \cdots, y_l) \in \mathbb{R}^l$. More precisely, there should be exactly r distinct values among the $y_1, \cdots, y_l$, so $\{1, \cdots, l\}$ splits into a disjoint union of r nonempty subsets $I_1, \cdots, I_r$ with $y_i = y_j$ for any $i, j \in I_\nu$. Denoting this common value y_{I_ν} and numbering the I_ν by $y_{I_1} < \cdots < y_{I_r}$, the final condition for $p_{(k_1,\cdots,k_l)}(z_1, \cdots, z_l) \in Z^{\mathrm{Im}}_{\mathbf{n}}$ is

$$\sum_{i \in I_1} \xi_{k_i} = n_1, \cdots, \sum_{i \in I_r} \xi_{k_i} = n_r. \tag{9.14}$$

Now, faces of $\mathcal{H}$ of dimension r are precisely labelled by ordered sequences $\mathbf{I} = (I_1, \cdots, I_r)$ of nonempty subsets of $\{1, \cdots, l\}$, forming a disjoint decomposition. The face $C_{\mathbf{I}}$ associated to I consists of $(y_1, \cdots, y_l)$ with the y_i for i in each I_ν being equal to some number y_{I_ν} such that $y_{I_1} < \cdots < y_{I_r}$. Thus F is the union of the tube domains $\mathbb{R}^l + C_{\mathbf{I}}\sqrt{-1}$ for $\mathbf{I}$ satisfying (9.14). Proposition 9.12 is proved.

We now construct a canonical Cousin resolution of any $\mathcal{G} \in \mathrm{Perv}^{<\infty}(Z, \mathcal{V})$, $Z = Z(\mathbb{C}, L)_n$ by the method of [**19**]. That is, we have a filtration of Z by closed subsets

$$F_1 \subset F_2 \subset \cdots \subset Z, \quad F_r = \bigsqcup_{l(\mathbf{n}) \leqslant r} Z^{\mathrm{Im}}_{\mathbf{n}},$$

with $\mathbf{n}$ running over ordered partitions of n. This filtration may be infinite, but $\mathcal{G}$ being supported on a finite skeleton of $\mathcal{S}^{(2)}$, is supported on some finite F_r. Denote

$\varepsilon_r : F_r \hookrightarrow Z$ the embedding. Then we have a Postnikov system for $\mathcal{G}$: the sequence

$$0 \xrightarrow{\eta_0} \varepsilon_{1*}\varepsilon_1^! \mathcal{G} \xrightarrow{\eta_1} \varepsilon_{2*}\varepsilon_2^! \mathcal{G} \xrightarrow{\eta_2} \cdots \tag{9.15}$$

stabilizes to $\mathcal{G}$ (so $\eta_r = \mathrm{Id}_{\mathcal{G}}$ for $r \gg 0$) and

$$\mathrm{Cone}(\eta_r) \simeq \bigoplus_{l(\mathbf{n})=r+1} \varepsilon_{b\mathbf{n}*}\varepsilon_{\mathbf{n}}^! \mathcal{G} \simeq \bigoplus_{l(\mathbf{n})=r+1} \varepsilon_{\mathbf{n}*}\mathcal{F}_{\mathbf{n}}[-r], \tag{9.16}$$

where $\mathcal{F}_{\mathbf{n}} = \underline{H}^{-r-1}(\varepsilon_{\mathbf{n}}^! \mathcal{G})$, the last identification in (9.16) following from Proposition 9.12. In fact, if $\mathcal{G}$ is represented as a complex of injective sheaves, then (9.15) is a filtration on $\mathcal{G}$, eventually stabilizing to $\mathcal{G}$ and with quotients given by (9.16). This gives a complex of sheaves

$$\begin{aligned}\mathcal{F}^\bullet(\mathcal{G}) &:= \{\cdots \to \mathrm{Cone}(\eta_2)[-2] \xrightarrow{d} \mathrm{Cone}(\eta_1)[-1] \xrightarrow{d} \mathrm{Cone}(\eta_0)\} \simeq \\ &\simeq \{\cdots \xrightarrow{d} \bigoplus_{l(\mathbf{n})=3} \varepsilon_{\mathbf{n}*}\mathcal{F}_{\mathbf{n}} \xrightarrow{d} \bigoplus_{l(\mathbf{n})=2} \varepsilon_{\mathbf{n}*}\mathcal{F}_{\mathbf{n}} \xrightarrow{d} \varepsilon_{(n)*}\mathcal{F}_{(n)}\}\end{aligned}$$

whose differential d is "the differential d_1 in the spectral sequence of the filtered complex (9.15)" and which is quasi-isomorphic to $\mathcal{G}$. In other words, we constructed a canonical resolution $\mathcal{F}^\bullet(\mathcal{G})$ of $\mathcal{G}$ which is a Cousin-type complex.

Denote by $\mathbb{E}(\mathcal{G})$ the Janus sheaf associatedto the Cousin-type complex $\mathcal{F}^\bullet(\mathcal{G})$ by Proposition 9.7.

PROPOSITION 9.17. *Let E be a Janus sheaf and $\mathcal{F}^\bullet(E)$ be the perverse sheaf associated to E by Proposition* 9.10. *Then $\mathbb{E}(\mathcal{F}^\bullet(E)) \simeq E$.*

PROOF. We show the identification on the level of objects of $\underline{\mathrm{CM}}_n(L)$. The comparison of values on morphisms is similarly straightforward.

The value $\mathbb{E}(\mathcal{F}(E))(M)$, $M \in \mathrm{CM}_n(L)$ is the stalk at U_M of the sheaf $\varepsilon_{\mathbf{n}}^! \mathcal{F}^\bullet(E)$. Now, the terms of the complex $\mathcal{F}^\bullet(E)$ have the form $\varepsilon_{\mathbf{m}*}\mathcal{F}_{\mathbf{m}}$, where $\mathcal{F}_{\mathbf{m}}$ is the sheaf on $Z_{\mathbf{m}}^{\mathrm{Im}}$ associated to the δ'_f data of E. So our statement will follow from the following identification:

$$\varepsilon_{\mathbf{n}}^! \varepsilon_{\mathbf{m}*}\mathcal{F}_{\mathbf{m}} = \begin{cases} \mathcal{F}_{\mathbf{m}}, & \text{if } \mathbf{n} = \mathbf{m}; \\ 0, & \text{otherwise.} \end{cases}$$

To see this last identification, consider the Cartesian diagram of closed embeddings

$$\begin{array}{ccc} I & \xrightarrow{\zeta_{\mathbf{m}}} & Z_{\mathbf{m}}^{\mathrm{Im}} \\ {\scriptstyle \zeta_{\mathbf{n}}}\downarrow & & \downarrow{\scriptstyle \varepsilon_{\mathbf{m}}} \\ Z_{\mathbf{n}}^{\mathrm{Im}} & \xrightarrow[\varepsilon_{\mathbf{n}}]{} & Z, \end{array} \qquad I := Z_{\mathbf{n}}^{\mathrm{Im}} \cap Z_{\mathbf{m}}^{\mathrm{Im}}.$$

By base change, $\varepsilon_{\mathbf{n}}^! \varepsilon_{\mathbf{m}*}\mathcal{F}_m \simeq \zeta_{\mathbf{n}*}\zeta_{\mathbf{m}}^! \mathcal{F}_{\mathbf{m}}$. If $\mathbf{m} \neq \mathbf{n}$, then $I = \varnothing$ and we get 0. If $\mathbf{m} = \mathbf{n}$, then $I = Z_{\mathbf{n}}^{\mathrm{Im}} = Z_{\mathbf{m}}^{\mathrm{Im}}$ and we get $\mathcal{F}_{\mathbf{m}}$. □

PROPOSITION 9.18. *If $\mathcal{G} \in \mathrm{Perv}^{<\infty}(Z, \mathcal{V})$, then the Janus presheaf $\mathbb{E}(\mathcal{G})$ is a Janus sheaf.*

PROOF. The stratification induced by $\mathcal{S}^{(0}$ on the imaginary strata is the restriction of $\mathcal{S}^{(1)}$, so each $\mathcal{F}_{\mathbf{n}}$ is $\mathcal{S}^{(1)}$-constructible. This means that anodyne morphisms from $\mathrm{Gro}'(\underline{\mathrm{CM}}_n(L))$ are taken by $\mathbb{E}(\mathcal{G})$ into isomorphisms. This is one hanf of the conditions (JS3). As for the other half – that anodyne morphisms in $\mathrm{Gro}''(\underline{\mathrm{CM}}_n(L))$ are taken by $\mathbb{E}(\mathcal{G})$ into isomorphisms, it follows from the first half applied to the perverse sheaf $\tau^*\mathbb{D}(\mathcal{G})$. Indeed, Propositions 9.17 and 9.9 imply that $\mathbb{E}(\tau^*\mathbb{D}(\mathcal{G})) \simeq \mathbb{E}(\mathcal{G})^\tau$. So $\mathbb{E}(\tau^*\mathbb{D}(\mathcal{G}))$ taking anodyne morphisms in $\mathrm{Gro}'(\underline{\mathrm{CM}}_n(L))$ into isomorphisms implies that so does $\mathbb{E}(\mathcal{G})^\tau$, i.e., that $\mathbb{E}(\mathcal{G})$ takes anodyne morphisms in $\mathrm{Gro}''(\underline{\mathrm{CM}}_n(L))$ into isomorphisms. □

The above proposition means that we have a functor

$$e : \mathrm{Perv}^{<\infty}(Z, \mathcal{V}) \longrightarrow \mathrm{JS}^{<\infty}_{L,n}(Z, \mathcal{V}), \quad \mathcal{G} \mapsto \mathbb{E}(\mathcal{G}).$$

We claim that e is quasi-inverse to the functor c from (9.11). Indeed, Proposition 9.17 means that $ec \simeq \mathrm{Id}$. The isomorphism $ce \simeq \mathrm{Id}$ follows directly from the construction of $\mathbb{E}(\mathcal{G})$, as $e(\mathcal{G})$ is the Janus presheaf (in fact sheaf) associated to the Cousin resolution $\mathcal{F}^\bullet(\mathcal{G})$, so $c(e(\mathcal{G})) = \mathcal{F}^\bullet(\mathcal{G})$ is quasi-isomorphic to $\mathcal{G}$.

This finishes the proof of Theorem 8.4.

10. The universal Janus sheaf

In this section we prove Theorem 7.14. The argument, reducing it to Theorem 8.4, proceeds by the same route as used in [**23**] for the case $L = \mathbb{Z}_+$. So we indicate the main steps but will be brief on details which are similar to [**23**].

10.1. The k-linear category of contingency matrices. Let $n \in L$. We introduce a $\mathbf{k}$-linear category $\mathfrak{CM}_n(L)$ whose objects are formal symbols $[M]$, $M \in \mathrm{CM}_n(L)$. Morphisms in $\mathfrak{CM}_n(L)$ are generated by the generating morphisms

- $\delta'_f : [M] \to [N]$ for any morphism $f : N \to M$ in $\mathrm{Gro}'(\underline{\mathrm{CM}}_n(L))$,
- $\delta''_g : [N] \to [M]$ for any morphism $g : N \to M$ in $\mathrm{Gro}''(\underline{\mathrm{CM}}_n(L))$,

subject to the following relations:

($\mathfrak{CM}$1) If f is an identity morphism, then $\delta'_f = \mathrm{Id}$. Further, $\delta'_{f_1 f_2} = \delta'_{f_2}\delta'_{f_1}$ for any two composable morphisms f_1, f_2 in $\mathrm{Gro}'(\underline{\mathrm{CM}}_n(L))$. Similarly, if g is an identity morphism, then $\delta''_g = \mathrm{Id}$. Further, $\delta''_{g_1 g_2} = \delta''_{g_1}\delta''_{g_2}$ for any two composable morphisms g_1, g_2 in $\mathrm{Gro}''(\underline{\mathrm{CM}}_n(L))$.

($\mathfrak{CM}$2) For any mixed fork $P = \{M' \xrightarrow{\varphi'} N' \xleftarrow{\psi''} N\}$ in $\mathrm{Gro}(\underline{\mathrm{CM}}_n(L))$, we have the relation

$$\delta'_{\varphi' : M' \to N}\delta''_{\psi'' : N \to N'} = \sum_{M \in \mathrm{Sup}(P)} \delta''_{\psi'' : M \to M'}\delta'_{\varphi' : M \to N}.$$

where the sum os over M fitting into a diagram (5.1).

($\mathfrak{CM}$3) If $f : N \to M$ is an anodyne morphism in $\mathrm{Gro}'(\underline{\mathrm{CM}}_n(L))$, then δ'_f is an isomorphism. If $g : N \to M$ is is an anodyne morphism in $\mathrm{Gro}''(\underline{\mathrm{CM}}_n(L))$, then δ''_g is an isomorphism.

The meaning of ($\mathfrak{CM}$3) is that $\mathfrak{CM}_n(L)$ is defined as the localization $\widetilde{\mathfrak{CM}}_n(L)[\mathrm{Ano}^{-1}]$, where $\widetilde{\mathfrak{CM}}_n(L)$ is the k-linear category with relations ($\mathfrak{CM}1-2$) and Ano is the set of morphisms in $\widetilde{\mathfrak{CM}}_n(L)$ consisting of δ'_f and δ''_g for f or g anodyne.

10.2. The universal Janus sheaf. The relations in $\mathfrak{CM}_n(L)$ mimick the axioms (JS1-3) of a Janus sheaf. This gives the following fact.

PROPOSITION 10.1.

(a) *The correspondence*

$$E^{\mathrm{un}} : M \mapsto [M], \ \mapsto \delta'_f, \ g \mapsto \delta''_g$$

is a Janus sheaf of type (L, n) *with values in* $\mathfrak{CM}_n(L)$.

(b) *Let* $\mathcal{V}$ *be any* $\mathbf{k}$*-linear category and* E *be a Janus sheaf of type* (L, n) *with values in* $\mathcal{V}$. *Then there is a unique functor* $F_E : \mathfrak{CM}_n(L) \to \mathcal{V}$ *such that*

$$F(E^{\mathrm{un}}) = E.$$

□

10.3. The monoidal structure. Let $\mathfrak{CM}(L) = \bigoplus_{n\in L} \mathfrak{CM}_n(L)$ be the orthogonal direct sum of the $\mathfrak{CM}_n(L)$. Explicitly

$$\mathrm{Ob}(\mathfrak{CM}(L)) = \bigsqcup_{n\in L} \mathrm{Ob}(\mathfrak{CM}_n(L)) = \{[M],\, M \in \mathrm{CM}(L)\}.$$

If $M \in \mathrm{CM}_n(L)$ and $N \in \mathrm{CM}_n(L)$, then

$$\mathrm{Hom}_{\mathfrak{CM}(L)}([M],[N]) = \begin{cases} 0, & \text{if } m \neq n, \\ \mathrm{Hom}_{\mathfrak{CM}_n(L)}([M],[N]), & \text{if } m = n. \end{cases}$$

We introduce a monoidal structure $\otimes$ on $\mathfrak{CM}(L)$ to be given on objects by

$$[M] \otimes [N] = [M \oplus N], \quad \text{where} \quad M \oplus N = \begin{pmatrix} M & 0 \\ 0 & N \end{pmatrix}.$$

Further, let $f_1 : M_1 \to N_1$ and $f_2 : M_2 \to N_2$ be two morphisms in $\mathrm{Gro}'(\underline{\mathrm{CM}}_{n_1}(L))$ and $\mathrm{Gro}'(\underline{\mathrm{CM}}_{n_2}(L))$ respectively. Then we have a natural morphism $f_1 \oplus f_2 : M_1 \oplus M_2 \to N_1 \oplus N_2$ in $\mathrm{Gro}'(\underline{\mathrm{CM}}_{n_1+n_2}(L))$. More precisely, f_1 expresses the fact that N_1 is obtained from M_1 by summing some sets $J_1, \cdots, J_p$ of adjacent columns, while f_2 expresses the fact that N_2 is obtained from M_2 by summing some sets K_1, $\cdots$, K_q of djacent columns. Now, the set of columns of $M_1 \oplus M_2$ is identified with the disjoint union of the sets of columns of M_1 and M_2. With respect to this identification, after summing the sets of columns $J_1, \cdots, J_p, K_1, \cdots, K_q$ in $M_1 \oplus M_2$, we get $N_1 \oplus N_2$ and this gives $f_1 \oplus f_2$.

Similarly, if $g_1 : M_1 \to N_1$ and $g_2 : M_2 \to N_2$ are two morphisms in $\mathrm{Gro}''(\underline{\mathrm{CM}}_{n_1}(L))$ and $\mathrm{Gro}''(\underline{\mathrm{CM}}_{n_2}(L))$ respectively, we have a natural morphism $g_1 \oplus g_2 : M_1 \oplus M_2 \to N_1 \oplus N_2$ in $\mathrm{Gro}''(\underline{\mathrm{CM}}_{n_1+n_2}(L))$.

We now define the monoidal structure $\otimes$ on generating morphisms as follows:

- If f_1, f_2 are two morphisms in $\mathrm{Gro}'(\underline{\mathrm{CM}}(L))$, then we put $\delta'_{f_1} \otimes \delta'_{f_2} = \delta'_{f_1 \oplus f_2}$.
- If g_1, g_2 are two morphisms in $\mathrm{Gro}''(\underline{\mathrm{CM}}(L))$, then we put $\delta''_{g_1} \otimes \delta''_{g_2} = \delta''_{g_1 \oplus g_2}$.
- Let $f : M_1 \to N_1$ be a morphism in $\mathrm{Gro}'(\underline{\mathrm{CM}}(L))$ and $g : M_2 \to N_2$ be a morphism in $\mathrm{Gro}''(\underline{\mathrm{CM}}(L))$. We have a commutative diagram in $\mathrm{Gro}(\underline{\mathrm{CM}}(L))$:

$$\begin{array}{ccc} M_1 \oplus M_2 & \xrightarrow[\ \prime\ ]{f\oplus \mathrm{Id}_{M_2}} & N_1 \oplus M_2 \\ {\scriptstyle \mathrm{Id}_{M_1}\oplus g}\Big\downarrow{\scriptstyle \prime\prime} & & {\scriptstyle \prime\prime}\Big\downarrow{\scriptstyle \mathrm{Id}_{N_1}\oplus g} \\ M_1 \oplus N_2 & \xrightarrow[f\oplus \mathrm{Id}_{N_2}]{\prime} & N_1 \oplus N_2 \end{array}$$

which is the unique diagram (5.1) containing the mixed form obtained by omitting $M_1 \oplus M_2$. Therefore by ($\mathfrak{CM}2$)

$$\delta''_{\mathrm{Id}_{M_1} \oplus g} \delta'_{f \oplus \mathrm{Id}_{M_2}} = \delta'_{f \oplus \mathrm{Id}_{M_1}} \delta''_{\mathrm{Id}_{N_1} \oplus g}$$

and we define $\delta'_f \otimes \delta''_g$ to be equal to this common value.

- The definition of $\delta''_g \otimes \delta'_f$ is similar.

PROPOSITION 10.2. *The above rules for the tensor products of generating morphisms extend to a monoidal structure on the category $\mathfrak{CM}(L)$ with the unit object $[\varnothing]$, where $\varnothing$ is the empty (size 0×0) contingency matrix.*

PROOF. Similar to [**23**, Prop.3.5]. □

10.4. The braiding. We now define a braiding on the monoidal category $(\mathfrak{CM}(L), \otimes)$. The construction is an adaptation of that of [**23**, §3B-C].

Let $M = \|m_{ij}\|_{i=1\cdots,r}^{j=1,\cdots,s} \in \mathrm{CM}(L)$. We denote its rows and columns by

$$M_{i,\bullet} = (m_{i1}, \cdots, m_{is}), \quad M_{\bullet j} = (m_{1j}, \cdots, m_{rj})^t.$$

Denote by $\sigma''_{i,i+1}M$ the matrix obtained from M by interchanging the i and $(i+1)$st rows, i.e., $M_{i,\bullet}$ and $M_{i+1,\bullet}$. Similarly, let $\sigma'_{j,j+1}M$ be the matrix obtained from M by interchanging the jth and $(j+1)$st columns, i.e., $M_{\bullet,j}$ and $M_{\bullet,j+1}$.

Suppose that $M_{i,\bullet}$ and $M_{i+1,\bullet}$ are *disjoint*, i.e., for each j at least one among $\{m_{ij}, m_{i+1,j}\}$ is equal to 0. then we have anodyne morphisms in $\mathrm{Gro}''(\underline{\mathrm{CM}}(L))$:

$$M \longrightarrow \partial''_{i-1}(M) = \partial''_{i-1}(\sigma''_{i,i+1}M) \longleftarrow \sigma''_{i,i+1}(M) \tag{10.3}$$

and we define the *row exhange isomorphism* $\varepsilon''_{i,i+1} : [M] \to [\sigma''_{i,i+1}M]$ in $\mathfrak{CM}(L)$ as the composition of the δ''-morphism and the inverse of another such corresponding to (10.3).

Similarly, if $M_{\bullet,j}$ and $M_{\bullet,j+1}$ are disjoint, i.e., for each i at least one among $\{m_{ij}, m_{i,j+1}\}$ is equal to 0, we have anodyne morphisms in $\mathrm{Gro}''(\underline{\mathrm{CM}}(L))$

$$M \longrightarrow \partial'_{j-1}(M) = \partial'_{j-1}(\sigma'_{j,j+1}M) \longleftarrow \sigma'_{j,j+1}M \tag{10.4}$$

and we define the *column exchange isomorphism* $\varepsilon'_{j,j+1} : [M] \to [\sigma'_{j,j+1}M]$ in $\mathfrak{CM}(L)$ by composing the δ'-morphism and the inverse of another such corresponding to (10.4).

Let now $M, N \in \mathrm{CM}(L)$ be two contingency matrices of sizes $p \times q$ and $r \times s$ respectively. We define the braiding isomorphism

$$R_{[M],[N]} : [M] \otimes [N] = [M \oplus N] \longrightarrow [N \oplus M] = [N] \otimes [M]$$

as the composition

$$[M \oplus N] = \left[\begin{pmatrix} M & 0 \\ 0 & N \end{pmatrix}\right] \xrightarrow{R''_{M,N}} \left[\begin{pmatrix} 0 & N \\ M & 0 \end{pmatrix}\right] \xrightarrow{R'_{M,N}} \left[\begin{pmatrix} N & 0 \\ 0 & M \end{pmatrix}\right] = [N \oplus M],$$

where:

(B'') $R''_{M,N}$ is the composition of pr row exchange isomorphisms moving r rows of $(0N)$ past the p rows of $(M0)$.

(B') $R'_{M,N}$ is the composition of qs column exchange isomorphisms moving s columns of $\begin{pmatrix} N \\ 0 \end{pmatrix}$ past q columns of $\begin{pmatrix} 0 \\ M \end{pmatrix}$.

Note that there are several ways to move the two groups of rows in (B'') or of columns in (B') past each other but the result is independent of the choice, similar to [**23**, Proposition 3.4].

PROPOSITION 10.5. *The isomorphisms $R_{[M],[N]}$ define a braiding on the monoidal category $\mathfrak{CM}(L)$.*

PROOF. Similar to [**23**, Proposition 3.6]. □

10.5. A graded bialgebra in $\mathfrak{CM}(L)$. We now define a graded bialgebra $\mathfrak{a} = (\mathfrak{a}_n)_{n\geqslant 0}$ in the braided category $\mathfrak{CM}(L)$ with components $\mathfrak{a}_n = [n]$ (the object corresponding to the 1×1 contingency matrix (n)) for $n > 0$ and $\mathfrak{a}_0 = [\varnothing]$. The multiplication and comultiplication are given by

$$\mu_{m,n} : [m] \otimes [n] = \left[\begin{pmatrix} m & 0 \\ 0 & n \end{pmatrix}\right] \xrightarrow{(\delta')^{-1}} \left[\begin{pmatrix} m \\ n \end{pmatrix}\right] \xrightarrow{\delta''} [m+n],$$

$$\Delta_{m,n} : [m+n] \xrightarrow{\delta'} [(m,\, n)] \xrightarrow{(\delta'')^{-1}} \left[\begin{pmatrix} m & 0 \\ 0 & n \end{pmatrix}\right] = [m] \otimes [n].$$

PROPOSITION 10.6. *The morphisms $\mu_{m,n}$, $\Delta_{m,n}$ make $\mathfrak{a}$ into a graded bialgebra in $\mathfrak{CM}$.*

PROOF. Similar to [**23**, Prop.4.1]. □

10.6. The category $\mathfrak{CM}(L)$ and the PROB $\mathfrak{B}^L$. Because $\mathfrak{CM}(L)$ is the target of the universal Janus sheaf of type (L, n), to deduce Theorem 7.14 from Theorem 8.4, it suffices to show the following.

THEOREM 10.7. *We have an equivalence of braided monoidal categories ξ : $\mathfrak{CM}(L) \to \mathfrak{B}^L$ which for any $n \in L$ induces an eqiuvalence of categories ξ_n : $\mathfrak{CM}_n(L) \to \mathfrak{B}^L_n$.*

PROOF. The graded bialgebra $\mathfrak{a}$ in $\mathfrak{CM}(L)$ gives, by the universal property of $\mathfrak{B}^L$, a braided monoidal functor $F_{\mathfrak{a}} : \mathfrak{B}^L \to \mathfrak{CM}(L)$. On the other hand, consider the universal graded bialgebra $\mathbf{a} = (\mathbf{a}_n)_{n\in L}$ in $\mathfrak{B}^L$. By Proposition 8.3 $\mathbf{a}$ gives, for each $n \in L$, a Janus sheaf $E_{\mathbf{a},n}$ of type (L, n) with values in $\mathfrak{B}^L$. By Proposition 10.1, $E_{\mathbf{a},n}$ is unduced from the universal Janus sheaf E^{un} by a unique (up to a unique isomorphism) functor $\xi_{\mathbf{a},n} : \mathfrak{CM}_n(L) \to \mathfrak{B}^L$. It is straightforward to check that the $\xi_{\mathbf{a},n}$ for various $n \in L$ unite into a monoidal functor $\xi_{\mathbf{a}} : \mathfrak{CM}(L) \to \mathfrak{B}^L$ which is quasi-inverse to $F_{\mathfrak{a}}$. □

This finishes the proof of Theorem 7.14.

Acknowledgment

We are grateful to D. Gaitsgory for correspondence about the issues related to defining perverse sheaves on the Ran space.

References

[1] John Frank Adams, *Infinite loop spaces*, Annals of Mathematics Studies, No. 90, Princeton University Press, Princeton, NJ; University of Tokyo Press, Tokyo, 1978. MR505692

[2] Bojko Bakalov and Alexander Kirillov Jr., *Lectures on tensor categories and modular functors*, University Lecture Series, vol. 21, American Mathematical Society, Providence, RI, 2001, DOI 10.1090/ulect/021. MR1797619

[3] A. A. Beĭlinson, *How to glue perverse sheaves*, K-theory, arithmetic and geometry (Moscow, 1984), Lecture Notes in Math., vol. 1289, Springer, Berlin, 1987, pp. 42–51, DOI 10.1007/BFb0078366. MR923134

[4] Alexander Beilinson and Vladimir Drinfeld, *Chiral algebras*, American Mathematical Society Colloquium Publications, vol. 51, American Mathematical Society, Providence, RI, 2004, DOI 10.1090/coll/051. MR2058353

[5] Roman Bezrukavnikov, Michael Finkelberg, and Vadim Schechtman, *Factorizable sheaves and quantum groups*, Lecture Notes in Mathematics, vol. 1691, Springer-Verlag, Berlin, 1998, DOI 10.1007/BFb0092343. MR1641131

[6] Neil Chriss and Victor Ginzburg, *Representation theory and complex geometry*, Modern Birkhäuser Classics, Birkhäuser Boston, Ltd., Boston, MA, 2010. Reprint of the 1997 edition, DOI 10.1007/978-0-8176-4938-8. MR2838836

[7] Kevin Costello and Owen Gwilliam, *Factorization algebras in quantum field theory. Vol. 1*, New Mathematical Monographs, vol. 31, Cambridge University Press, Cambridge, 2017, DOI 10.1017/9781316678626. MR3586504

[8] Justin Michael Curry, *Sheaves, cosheaves and applications*, ProQuest LLC, Ann Arbor, MI, 2014. Thesis (Ph.D.)–University of Pennsylvania. MR3259939

[9] Justin Curry and Amit Patel, *Classification of constructible cosheaves*, Theory Appl. Categ. **35** (2020), Paper No. 27, 1012–1047. MR4117065

[10] Michael Finkelberg and Ivan Mirković, *Semi-infinite flags. I. Case of global curve* $\mathbf{P}^1$, Differential topology, infinite-dimensional Lie algebras, and applications, Amer. Math. Soc. Transl. Ser. 2, vol. 194, Amer. Math. Soc., Providence, RI, 1999, pp. 81–112, DOI 10.1090/trans2/194/05. MR1729360

[11] D. Gaitsgory, *Notes on factorizable sheaves*, `http://www.math.harvard.edu/~gaitsgde/GL/FS.pdf` (2008)

[12] Dennis Gaitsgory and Jacob Lurie, *Weil's conjecture for function fields. Vol. 1*, Annals of Mathematics Studies, vol. 199, Princeton University Press, Princeton, NJ, 2019. MR3887650

[13] A. Galligo, M. Granger, and Ph. Maisonobe, *$\mathcal{D}$-modules et faisceaux pervers dont le support singulier est un croisement normal* (French), Ann. Inst. Fourier (Grenoble) **35** (1985), no. 1, 1–48, DOI 10.5802/aif.996. MR781776

[14] Michel Granger and Philippe Maisonobe, *Faisceaux pervers relativement à un point de rebroussement* (French, with English summary), C. R. Acad. Sci. Paris Sér. I Math. **299** (1984), no. 12, 567–570. MR770448

[15] Ryoshi Hotta, Kiyoshi Takeuchi, and Toshiyuki Tanisaki, *D-modules, perverse sheaves, and representation theory*, Progress in Mathematics, vol. 236, Birkhäuser Boston, Inc., Boston, MA, 2008. Translated from the 1995 Japanese edition by Takeuchi, DOI 10.1007/978-0-8176-4523-6. MR2357361

[16] Kazuo Habiro, *Bottom tangles and universal invariants*, Algebr. Geom. Topol. **6** (2006), 1113–1214, DOI 10.2140/agt.2006.6.1113. MR2253443

[17] André Joyal and Ross Street, *Braided tensor categories*, Adv. Math. **102** (1993), no. 1, 20–78, DOI 10.1006/aima.1993.1055. MR1250465

[18] A. Kalugin, *A note on a quantization via the Ran space*, `arXiv:1911.05424`.

[19] Mikhail Kapranov and Vadim Schechtman, *Perverse sheaves over real hyperplane arrangements*, Ann. of Math. (2) **183** (2016), no. 2, 619–679, DOI 10.4007/annals.2016.183.2.4. MR3450484

[20] Mikhail Kapranov and Vadim Schechtman, *Shuffle algebras and perverse sheaves*, Pure Appl. Math. Q. **16** (2020), no. 3, 573–657, DOI 10.4310/PAMQ.2020.v16.n3.a9. MR4176533

[21] Mikhail Kapranov and Vadim Schechtman, *Contingency tables with variable margins (with an appendix by Pavel Etingof)*, SIGMA Symmetry Integrability Geom. Methods Appl. **16** (2020), Paper No. 062, 22, DOI 10.3842/SIGMA.2020.062. MR4119960

[22] Mikhail Kapranov and Vadim Schechtman, *Parabolic induction and perverse sheaves on* $W\backslash\mathfrak{h}$, Adv. Math. **398** (2022), Paper No. 108200, 52, DOI 10.1016/j.aim.2022.108200. MR4367797

[23] Mikhail Kapranov and Vadim Schechtman, *PROBs and perverse sheaves I: symmetric products*, Selecta Math. (N.S.) **31** (2025), no. 2, Paper No. 23, 32, DOI 10.1007/s00029-025-01016-x. MR4861254

[24] Masaki Kashiwara and Pierre Schapira, *Sheaves on manifolds*, Grundlehren der mathematischen Wissenschaften [Fundamental Principles of Mathematical Sciences], vol. 292, Springer-Verlag, Berlin, 1990. With a chapter in French by Christian Houzel, DOI 10.1007/978-3-662-02661-8. MR1074006
[25] J. Lurie, *Higher Algebra*, Available on the author's home page.
[26] Saunders MacLane, *Categories for the working mathematician*, Graduate Texts in Mathematics, Vol. 5, Springer-Verlag, New York-Berlin, 1971. MR354798
[27] Martin Markl, *Operads and PROPs*, Handbook of algebra. Vol. 5, Handb. Algebr., vol. 5, Elsevier/North-Holland, Amsterdam, 2008, pp. 87–140, DOI 10.1016/S1570-7954(07)05002-4. MR2523450
[28] I. Mirković and K. Vilonen, *Geometric Langlands duality and representations of algebraic groups over commutative rings*, Ann. of Math. (2) **166** (2007), no. 1, 95–143, DOI 10.4007/annals.2007.166.95. MR2342692
[29] Teimuraz Pirashvili, *On the PROP corresponding to bialgebras* (English, with French summary), Cah. Topol. Géom. Différ. Catég. **43** (2002), no. 3, 221–239. MR1928233
[30] Paolo Salvatore, *Configuration spaces with summable labels*, Cohomological methods in homotopy theory (Bellaterra, 1998), Progr. Math., vol. 196, Birkhäuser, Basel, 2001, pp. 375–395. MR1851264
[31] David Treumann, *Exit paths and constructible stacks*, Compos. Math. **145** (2009), no. 6, 1504–1532, DOI 10.1112/S0010437X09004229. MR2575092
[32] V. A. Vassiliev, *Complements of discriminants of smooth maps: topology and applications*, Translations of Mathematical Monographs, vol. 98, American Mathematical Society, Providence, RI, 1992. Translated from the Russian by B. Goldfarb, DOI 10.1090/mmono/098. MR1168473

Kavli IPMU (WPI), UTIAS, University of Tokyo, 5-1-5 Kashiwanoha, Kashiwa, Chiba, 277-8583 Japan
Email address: mikhail.kapranov@protonmail.com

Institut de Mathématiques de Toulouse, Université Paul Sabatier, 118 route de Narbonne, 31062 Toulouse, France; and Kavli IPMU (WPI), UTIAS, University of Tokyo, 5-1-5 Kashiwanoha, Kashiwa, Chiba, 277-8583 Japan
Email address: schechtman@math.ups-tlse.fr

Contemporary Mathematics
Volume **841**, 2026
https://doi.org/10.1090/conm/841/16838

Gauss-Manin connection in noncommutative geometry

Boris Tsygan

To Ezra on his sixtieth birthday

Abstract. This paper is a review of results on the Gauss-Manin connection in noncommutative geometry. The Gauss-Manin connection in periodic cyclic homology was introduced by Ezra Getzler in 1991, then generalized to a superconnection by the author in a joint work with Dolgushev and Tamarkin. The key to these constructions is the Cartan calculus in noncommutative geometry. The original results of this article, namely the comparisons between the noncommutative Cartan calculus and its classical version, are contained in Section 5. The rest of the paper is mostly a review of results from [**13**], [**9**], [**27**], [**33**], although the approach is somewhat new.

1. Introduction

In [**25**], Yu. I. Manin, then nineteen years old, considered the following question. Let $\{X_s | s \in S\}$ be a family of algebraic curves over $\mathbb{C}$. Fix an element γ in homology of X_s where s is a point of S. Let ω_s be an algebraic differential on X_s that depends on s algebraically. Locally in s, we can identify homologies of different fibers. Therefore, locally in s, the period $s \mapsto \int_\gamma \omega_s$ over a cycle γ is a function in s. It turns out that this function satisfies the Picard-Fuchs differential equation. Why is that so?

Manin's answer: the period actually depends on the class of ω_s modulo exact differentials. But the space of such classes is finite dimensional. So, when you start to differentiate the period, you will eventually get a linear relation, very much the same way as a matrix has to satisfy a polynomial equation.

But the space of classes of differentials depends on s. How do you differentiate a function with values in such a space? Manin provided an answer as follows. Let $\mathfrak{X}$ be the total space of the family, with X_s being fibers of the morphism $\mathfrak{X} \to S$. We work Zariski locally in $\mathfrak{X}$, actually passing to fields K and R of rational functions on S and $\mathfrak{X}$ respectively. A derivation of K can always be extended to a derivation of R. Manin gives a way to differentiate using such an extension, and then shows that the result does not depend on an extension.

Soon after, Grothendieck and then Katz and Oda re-interpreted and generalized Manin's construction in terms of the following. The assignment $s \mapsto H^\bullet_{\mathrm{DR}}(X_s)$ is actually a vector bundle on the base, and it admits a flat connection (the Gauss-Manin connection). This is explained by the Cartan calculus on differential forms.

2020 *Mathematics Subject Classification.* Primary 14A22, 16E40.

©2026 American Mathematical Society

Namely, a vector field ξ defines the operator $L_\xi : \Omega^\bullet_X \to \Omega^\bullet_X$ as well as the operator $\iota_\xi : \Omega^\bullet_X \to \Omega^{\bullet-1}_X$ satisfying relations

$$\iota_\xi^2 = 0;\ d\iota_\xi + \iota_\xi d = L_\xi \tag{1.1}$$

Now replace X by an associative (possibly noncommutative) algebra A. De Rham cohomology gets replaced by periodic cyclic homology $\mathrm{HC}^{\mathrm{per}}_\bullet(A)$. In [**13**], Ezra Getzler constructed a flat connection in the bundle $s \mapsto \mathrm{HC}^{\mathrm{per}}_\bullet(A_s)$ for a family of algebras over a smooth base S. This is based on the following generalization of the Cartan calculus. Let $C_\bullet(A)$ and $C^\bullet(A)$ be the standard chain and cochain complexes computing the Hochschild homology $\mathrm{HH}_\bullet(A)$ and the Hochschild cohomology $\mathrm{HH}^\bullet(A)$ [**3**], [**23**]. For $\varphi \in C^m(A)$ there are two operators

$$L_\varphi : \mathrm{C}_\bullet(A) \to \mathrm{C}_{\bullet-m}(A);\ \iota_\varphi : \mathrm{C}_\bullet(A) \to \mathrm{C}_{\bullet-m+1}(A) \tag{1.2}$$

Those operators induce pairings at the level of Hochschild (co)homology that satisfy the standard Cartan relations [**10**].

It is natural to ask what replaces the Cartan relations at the level of Hochschild complexes rather than at the level of (co)homology. Let us try to predict the answer. For any algebra A and any algebra endomorphism $f : A \to A$ we define a complex

$$\mathrm{TR}_A(f) = C_\bullet(A, {}_f A) \tag{1.3}$$

(Here ${}_f A$ is A viewed as an A-bimodule, with the action $a_0 \cdot a \cdot a_1 = f(a_0)aa_1$ and the right hand side of (1.3) is the Hochschild complex with coefficients in this bimodule. Roughly, this is a noncommutative analog of differential forms on the space of fixed points of an endomorphisms of a variety). By functoriality, there are two endomorphisms of $\mathrm{TR}_A(f)$: id_* and f_*. We predict that they should be chain homotopic. In fact one can deduce that from the fact that they coincide at the level of

$$\mathrm{tr}_A(f) = \mathrm{HH}_0(A, {}_f A) = {}_f A/[A, {}_f A] = A/\langle f(a_0)a_1 - a_1 a_0\rangle \tag{1.4}$$

(here $\langle\rangle$ stands for linear span). So there should be an operator

$$B_f : \mathrm{TR}_A(f) \to \mathrm{TR}_A(f)[1] \tag{1.5}$$

satisfying

$$[B_f, b_f] = f_* - \mathrm{id}_* \tag{1.6}$$

(here b_f is the differential on the complex $\mathrm{TR}_A(f)$). Now, B_f^2 is a natural transformation $\mathrm{TR}_A(f) \to \mathrm{TR}_A(f)[2]$; we expect it to be chain homotopic to zero. Proceeding by induction, we expect to have natural operators $B_f^{(n)} : \mathrm{TR}_A(f) \to \mathrm{TR}_A(f)[1+2n], n \geq 0$, such that $B_f = B_f^{(0)}$ and

$$(b_f + u\mathbb{B}_f)^2 = f_* - \mathrm{id}_* \tag{1.7}$$

where u is a formal variable of cohomological degree 2 and

$$\mathbb{B}_f = \sum_{n=0}^{\infty} u^n B_f^{(n)}$$

This would have the following consequences. First, put $f = \mathrm{id}$. Put $\mathrm{TR}(A) = \mathrm{TR}_A(\mathrm{id})$, $b = b_{\mathrm{id}}$ and $\mathbb{B} = \mathbb{B}_{\mathrm{id}}$. Then (1.7) tells us that $\mathbb{B}$ is a differential on $\mathrm{TR}(A)[[u]]$. And indeed, such a complex does exist; it is the negative cyclic complex

$CC^-_\bullet(A)$. Its localization $TR(A)((u))$ is the periodic cyclic complex computing $HC^{per}_\bullet(A)$ as mentioned above.

Second, let D be a derivation of A (this is a noncommutative analog of a vector field on a variety). Applying the above formally to $f = \exp(D)$, we get natural operators

$$I_{D^n} : TR(A)[[u]] \to TR(A)[[u]][-1] \tag{1.8}$$

such that for

$$I(D) = \sum_{n=1}^{\infty} \frac{1}{n!} I_{D^n}$$

we have

$$(b + u\mathbb{B} + I(D))^2 = \exp(L_D) - 1 \tag{1.9}$$

Here $L_D = D_*$ is the action of D on $TR(A)[[u]]$. This is equivalent to

$$[b + u\mathbb{B}, I_{D^n}] + \sum_{k=1}^{n-1} \binom{n}{k} I_{D^k} I_{D^{n-k}} = L_D^n \tag{1.10}$$

Note that the above equation has no denominators and in fact holds over $\mathbb{Z}$. Note also that in reality, on the standard Hochschild complex we already have $B^2 = 0$ so we can put $\mathbb{B} = B$.

Let us compare this to the classical Cartan calculus. There, we already have $\iota_\xi^2 = 0$. So we can put

$$J(\xi) = \iota_\xi; \quad b = 0; \quad \mathbb{B} = d \tag{1.11}$$

and rewrite the Cartan relations as

$$(b + u\mathbb{B} + J(\xi))^2 = uL_\xi \tag{1.12}$$

The discrepancy between the classical Cartan relations (1.12) and the noncommutative Cartan relations (1.9) can be removed (Lemma 4.5). Therefore, when A is commutative, we have two generalized Cartan calculi: the classical one on forms, and the new one on Hochschild chains. But (in characteristic zero) one can compare the two via the HKR map. We show that this comparison agrees with the Cartan calculus structure (Theorem 5.1).

Now let $\mathcal{A}$ be a sheaf of $\mathcal{O}_S$-algebras on an algebraic variety S. Assume that ∇ is a connection on $\mathcal{A}$. *In the case when the connection ∇ preserves the algebra structure*, this allows to construct a Getzler-Gauss-Manin superconnection as follows:

$$\nabla^{GM} = b + u\mathbb{B} + L_\nabla - \sum_{n=1}^{\infty} \frac{u^{-n}}{n!} J(R^n)$$

where $R = \nabla^2$. Note that the formula makes sense in characteristic zero when R is nilpotent (i.e. when the base is smooth).

But the above condition on ∇ is too restrictive, even locally in S. In the classical case, the interpretation of the Gauss-Manin connection due to Grothendieck and Katz-Oda starts with a connection defined locally in the total space of the family $\mathfrak{X}$ (cf. Section 2). But in the noncommutative case there is no such thing. Instead, following [**13**], we look at the discrepancy together with the curvature:

$$R = \nabla m + \nabla^2$$

where m is the product on $\mathcal{A}$. This is a form with values in the Hochschilld cochain complex of $\mathcal{A}$. Hochschild cochains are noncommutative analogs of multivector fields, therefore we have to generalize to the noncommutative case the *extended Cartan calculus* between forms and multivectors. Here we do this using the approach to Cartan calculus originated by Khalkhali in [**21**] and [**22**]. Namely, the DG Lie algebra of Hochschild cochains is quasi-isomorphic to the algebra of coderivations of the cobar construction, and we can apply the dual version of the Cartan calculus as above to the DG coalgebra $\mathrm{Bar}(A)$ instead of A. Due to a theorem from [**31**] and [**11**], the dual version of the periodic cyclic complex of $\mathrm{Bar}(A)$ is quasi-isomorphic to the periodic cyclic complex of A. We call it *the big periodic cyclic complex.* The above allows us to construct the Gauss-Manin superconnection on the big complex and prove its main properties (Theorem 8.5).

REMARK 1.1. The result holds not only in characteristic zero but also over the p-adics, provided the differential of the product is p-adically small. Moreover, the deviation of the multiplication from associativity only needs to be p-adically small, which allows to define a noncommutative crystalline complex of an algebra over $\mathbb{F}_p$. This is done in [**27**] and [**33**]; we plan to extend this paper's approach to the p-adic case elsewhere.

It would be interesting to investigate convergence of our formulas in other topologies. Note that those formulas involve exponentials which suggest some convergence. However, those are exponentials of the form $\exp(\frac{1}{u}S)$ where u is a formal parameter. This means that one can hope for convergence in some refined versions of the periodic cyclic complex in which some convergence property on chains $\sum_{j=-m}^{\infty} u^j c_j$ is imposed (cf. e.g. [**7**], [**26**]).

REMARK 1.2. As shown in [**16**], in characteristic zero the periodic cyclic complex $(C_\bullet(A)((u)), b+uB)$ can be replaced by the double complex $(\Omega^\bullet_A((u)), d+u\iota_\Delta)$. Here $\Omega^\bullet$ stands for noncommutative forms, d is the De Rham differential, and ι_Δ the Ginzburg-Schedler differential. For any derivation ξ of A one defines L_ξ and ι_ξ in a straightforward way, subject to strict Cartan relations (1.1). Passing from A to its cofibrant resolution, we get another version of a big periodic cyclic complex on which the Gauss-Manin superconnection can be constructed. We do not know how this can be adapted to the p-adic situation, although it might be interesting to investigate.

This paper greatly benefited from my multiple conversations with Ezra Getzler and from presentations and discussions in the Gauss-Manin seminar at Northwestern, as well as from my collaboration and discussions with Ryszard Nest. Finally, I am grateful to Anton Alekseev for fruitful discussions, in particular for turning my attention to [**1**], [**2**], and [**15**] that contain generalizations of Cartan calculus that are similar to the ones considered here.

2. The Gauss-Manin connection in algebraic geometry

We recall the classical facts about the Gauss-Manin connection [**25**], [**17**], [**20**]. Our exposition follows [**24**].

2.1. Calc 1. Let $p: X \to S$ be a smooth morphism of smooth algebraic varieties over $\mathbb{C}$. The structure sheaf $\mathcal{O}_{X/S}$ is automatically a sheaf of $p^{-1}\mathcal{O}_S$-algebras; $\Omega^1_{X/S}$ denotes the module of Kähler differentials over $\mathcal{O}_S$; $\mathcal{T}_{X/S}$ is the Lie algebra of $\mathcal{O}_S$-derivations of $\mathcal{O}_{X/S}$. Let

$$\Omega^\bullet_{X/S} = \wedge^\bullet_{\mathcal{O}_{X/S}} \Omega^1_{X/S}$$

be the differential graded algebra of relative differential forms on X/S. The Lie algebra $\mathcal{T}_{X/S}$ acts on $\Omega^\bullet_{X/S}$ by Lie derivatives; the Abelian Lie algebra $\mathcal{T}_{X/S}[1]$ acts on the same sheaf by contractions. We denote these actions by

$$\xi \mapsto \mathrm{Lie}_\xi$$

and

$$\xi \mapsto \iota_\xi$$

for a vector field ξ. These operators satisfy the classical Cartan identities

$$[\mathrm{Lie}_\xi, \mathrm{Lie}_\eta] = \mathrm{Lie}_{[\xi,\eta]},\ [\mathrm{Lie}_\xi, \iota_\eta] = \iota_{[\xi,\eta]},\ [\iota_\xi, \iota_\eta] = 0\ \forall \xi, \eta \in \mathcal{T}_{X/S}. \tag{2.1}$$

The De Rham differential, $d_{X/S}$, satisfies

$$[d_{X/S}, d_{X/S}] = 0,\ [d_{X/S}, \iota_\xi] = \mathrm{Lie}_\xi,\ [d_{X/S}, \mathrm{Lie}_\xi] = 0. \tag{2.2}$$

To conclude:

LEMMA 2.1. *The formula* $\xi + \eta\epsilon \mapsto \mathrm{Lie}_\xi + \iota_\eta$ *defines an action of the differential graded Lie algebra* $(\mathcal{T}_{X/S}[\epsilon], \frac{\partial}{\partial\epsilon})$ *on the differential graded algebra* $(\Omega^\bullet_{X/S}, d_{X/S})$. *Here* ϵ *is a formal parameter of degree* -1 *and square zero.*

2.2. Given the morphism $p : X \to S$ as in Section 2.1, define a connection to be a linear map

$$\nabla : \mathcal{O}_{X/S} \to \Omega^1_S \otimes_{\mathcal{O}_S} \mathcal{O}_{X/S} \tag{2.3}$$

such that

$$\nabla(fg) = f\nabla(g) + g\nabla(f), \nabla(h) = d_S h\ \forall f, g \in \mathcal{O}_{X/S}, h \in \mathcal{O}_S, \tag{2.4}$$

where $d_S h$ is the differential of h. Geometrically, ∇ is a rule that lifts a vector field on S to that on X/S. Such connections exist at least locally on X.

The curvature of ∇ is an element

$$R = \frac{1}{2}[\nabla, \nabla] \tag{2.5}$$

of $\Omega^2_S \otimes_{\mathcal{O}_S} \mathcal{T}_{X/S}$.

We will now extend the scalars from $\mathcal{O}_S$ to the entire de Rham complex $\Omega^\bullet_S$, thus replacing $\Omega^\bullet_{X/S}$ with $\Omega^\bullet_S \otimes_{\mathcal{O}_S} \Omega^\bullet_{X/S}$. The connection ∇ extends to a derivation Lie_∇ of degree one of $\Omega^\bullet_S \otimes_{\mathcal{O}_S} \Omega^\bullet_{X/S}$ satisfying

$$\mathrm{Lie}_\nabla\, \alpha = d_S\alpha,\ \alpha \in \Omega^\bullet_S \tag{2.6}$$

For $\xi \in \Omega^\bullet_S \otimes_{\mathcal{O}_{X/S}} \mathcal{T}_{X/S}$, Lie_ξ and ι_ξ define derivations of $\Omega^\bullet_S \otimes_{\mathcal{O}_S} \Omega^\bullet_{X/S}$; we have

$$\mathrm{Lie}^2_\nabla = \mathrm{Lie}_R \tag{2.7}$$

Lemma 2.1 is still valid if we replace the Lie algebra $\mathcal{T}_{X/S}$ by the differential graded Lie algebra $\Omega^\bullet_S \otimes_{\mathcal{O}_S} \mathcal{T}_{X/S}$.

The Cartan identities (2.1) and (2.2) allow us to flatten Lie_∇. Namely, define the derivation

$$D_\nabla = \mathrm{Lie}_\nabla + d_{X/S} - \iota_R \in \mathrm{End}_{\mathbb{C}}(\Omega^\bullet_S \otimes_{\mathcal{O}_S} \Omega^\bullet_{X/S})$$

LEMMA 2.2. *The pair* $(\Omega_S^\bullet \otimes_{\mathcal{O}_S} \Omega_{X/S}^\bullet, D_\nabla)$ *is a complex, i.e.,*

$$D_\nabla^2 = 0.$$

PROOF. The result follows from the following computation:

$$\begin{aligned}
&2D_\nabla^2 = [D_\nabla, D_\nabla] = \\
&[\mathrm{Lie}_\nabla, \mathrm{Lie}_\nabla] + [d_{X/S}, d_{X/S}] + [\iota_R, \iota_R] \\
&+ 2[\mathrm{Lie}_\nabla, d_{X/S}] - 2[\mathrm{Lie}_\nabla, \iota_R] - 2[d_{X/S}, \iota_R] = \\
&2\,\mathrm{Lie}_R - 2\iota_{[\nabla,R]} - 2\,\mathrm{Lie}_R = 0,
\end{aligned}$$

which consists in a repeated application of (2.1), (2.2), including

$$[\mathrm{Lie}_\nabla, \iota_R] = \iota_{[\nabla,R]} = 0$$

thanks to the Bianchi identity $[\nabla, R] = 0$. □

Various choices of ∇ give rise to canonically isomorphic complexes:

LEMMA 2.3. *The map*

$$\exp(\iota_{\nabla_2 - \nabla_1}) : (\Omega_S^\bullet \otimes_{\mathcal{O}_S} \Omega_{X/S}^\bullet, D_{\nabla_1}) \to (\Omega_S^\bullet \otimes_{\mathcal{O}_S} \Omega_{X/S}^\bullet, D_{\nabla_2})$$

is an isomorphism of complexes.

PROOF. It is clear that the operator $\iota_{\nabla_2-\nabla_1} \in \mathrm{End}_{\mathbb{C}}(\Omega_S^\bullet \otimes_{\mathcal{O}_S} \Omega_{X/S}^\bullet)$ is nilpotent and is an even derivation of the algebra $\Omega_S^\bullet \otimes_{\mathcal{O}_S} \Omega_{X/S}^\bullet$. Therefore it defines an automorphism of $\Omega_S^\bullet \otimes_{\mathcal{O}_S} \Omega_{X/S}^\bullet$. What we need to check then is that

$$\exp(ad_{\iota_{\nabla_2-\nabla_1}})(D_{\nabla_1}) = D_{\nabla_2},$$

where $ad_{\iota_{\nabla_2-\nabla_1}}$ means the operation of taking the bracket, $[\iota_{\nabla_2-\nabla_1}, .]$. Indeed, let A be an element of $\Omega_S^1 \otimes_{\mathcal{O}_S} \mathcal{T}_{X/S}$. Then

$$\exp(ad_{\iota_A})(\mathrm{Lie}_\nabla) = \mathrm{Lie}_\nabla - \iota_{\nabla A}; \tag{2.8}$$

$$\exp(ad_{\iota_A})(d_{X/S}) = d_{X/S} + \mathrm{Lie}_A + \frac{1}{2}[\iota_A, \mathrm{Lie}_A] \;= d_{X/S} + \mathrm{Lie}_A - \frac{1}{2}\iota_{[A,A]}; \tag{2.9}$$

Also $[\iota_A, \iota_R] = 0$, so

$$\exp(ad_{\iota_A})(d_{X/S} + \mathrm{Lie}_\nabla + \iota_R) = d_{X/S} + \mathrm{Lie}_\nabla - (\iota_R + \iota_{\nabla A} + \frac{1}{2}\iota_{[A,A]}) \tag{2.10}$$

But $R + \nabla A + \frac{1}{2}[A, A]$ is the curvature of the connection $\nabla + A$. □

2.3. Comparison with Katz-Oda. The complex $(\Omega_S^\bullet \otimes_{\mathcal{O}_S} \Omega_{X/S}^\bullet, D_\nabla)$ is filtered, $F^p(\Omega_S^\bullet \otimes_{\mathcal{O}_S} \Omega_{X/S}^\bullet) = (\Omega_S^{\geq p} \otimes_{\mathcal{O}_S} \Omega_{X/S}^\bullet)$. It is easy to see that, locally in X, a connection ∇ identifies $(\Omega_S^\bullet \otimes_{\mathcal{O}_S} \Omega_{X/S}^\bullet, D_\nabla)$ with the absolute De Rham complex $(\Omega_X^\bullet, d)$. The filtration gets identified with the filtration from [**20**]. We see that our construction reproduces the Katz-Oda construction of the Gauss-Manin connection.

2.4. Globalization. We will now show that Lemmas 2.2 and 2.3 imply the existence of a globally defined complex; such a complex is not unique, but different such complexes are canonically isomorphic.

The variety S being smooth, it can be covered by open subsets S_α étale over $\mathbb{C}^n$, hence each having a coordinate system $\{u_i, \partial_{u_i}\}$, $\partial_{u_i}(u_j) = \delta_{ij}$. The morphism $p : X \to S$ being smooth, the preimage $p^{-1}(S_\alpha)$ has an open cover $\{U_{\alpha\beta}\}$ so that each $U_{\alpha\beta}$ has a coordinate system $\{\{u_i, x_j, \partial_{u_i}, \partial_{x_j}\}$. This means that over each $U_{\alpha\beta}$, the restriction $\Omega^\bullet_S \otimes_{\mathcal{O}_S} \Omega^\bullet_{X/S}|_{U_{\alpha\beta}}$ regarded as a sheaf over S_α carries a connection, $\nabla_{\alpha\beta}$.

To streamline the notation: we have a sheaf $\Omega \stackrel{\text{def}}{=} \Omega^\bullet_S \otimes_{\mathcal{O}_S} \Omega^\bullet_{X/S}$ on X and an open cover $\{U_i\}$ of X; on each restriction $\Omega_i \stackrel{\text{def}}{=} \Omega|_{U_i}$ we have chosen a connection ∇_i and hence a differential $D_{\nabla_i} : \Omega_i \to \Omega_i$, Lemma 2.2. According to Lemma 2.3, over the intersection $U_i \cap U_j$ there is an isomorphism

$$G_{ji} \stackrel{\text{def}}{=} \exp\left(\iota_{\nabla_j - \nabla_i}\right) : (\Omega_i, D_{\nabla_i}) \to (\Omega_j, D_{\nabla_j}).$$

It is clear that on triple intersections we have the consistency condition:

$$G_{ki} = G_{kj} \circ G_{ji}.$$

Therefore, having fixed a cover $\{U_i\}$ and a bunch of connections $\{\nabla_i\}$, which we will denote simply by ∇, we obtain a new globally defined complex, Ω_∇, by tearing the original $\Omega^\bullet_S \otimes_{\mathcal{O}_S} \Omega^\bullet_{X/S}$ apart and regluing the pieces $\{(\Omega_i, \nabla_i)\}$ using $\{G_{ij}\}$ as transition functions. This proves the existence.

Two distinct such complexes, Ω_∇ and $\Omega_{\nabla'}$ (attached to distinct collections ∇ and ∇') are canonically isomorphic. To see this, assume, as we can, that both are defined using the same open cover $\{U_i\}$. Then the collection of isomorphisms

$$\exp\left(\iota_{\nabla'_i} - \iota_{\nabla_i}\right) : \Omega_\nabla|_{U_i} \to \Omega_{\nabla'}|_{U_i}$$

defines an isomorphism of complexes $\Omega_\nabla \to \Omega_{\nabla'}$.

3. Noncommutative calculus

We closely follow [**27**] and [**28**].

3.1. Cyclic complexes. For an associative unital algebra A over a commutative unital ring k, define

$$C_\bullet(A) = A \otimes (A/k)^{\otimes \bullet} \tag{3.1}$$

$$b : C_\bullet(A) \to C_{\bullet-1}(A);\ B : C_\bullet(A) \to C_{\bullet+1}(A); \tag{3.2}$$

(3.3)

$$b(a_0 \otimes \ldots \otimes a_n) = \sum_{i=0}^{n-1} (-1)^i a_0 \otimes \ldots \otimes a_i a_{i+1} \otimes a_n + (-1)^n a_n a_0 \otimes \ldots \otimes a_{n-1};$$

$$B(a_0 \otimes \ldots \otimes a_n) = \sum_i (-1)^{i(n-i+1)} 1 \otimes a_i \otimes \ldots \otimes a_n \otimes a_0 \otimes \ldots a_{i-1}. \tag{3.4}$$

One has

$$b^2 = 0;\ bB + Bb = 0;\ B^2 = 0. \tag{3.5}$$

The homology of the differential b is called the Hochschild homology of A and is denoted by $\mathrm{HH}_\bullet(A)$.

DEFINITION 3.1. The negative cyclic homology $HC_\bullet^-(A)$ (resp. periodic cyclic homology $HC_\bullet^{per}(A)$) of an associative algebra A is the homology of the complex

$$\mathrm{CC}_\bullet^-(A) = (C_\bullet(A)[[u]], b + uB); \tag{3.6}$$

resp.

$$\mathrm{CC}_\bullet^{\mathrm{per}}(A) = (C_\bullet(A)((u)), b + uB), \tag{3.7}$$

where u is a formal variable of degree -2.

3.1.1. *Motivation: the case of a commutative algebra.* Suppose that A is a commutative algebra and k contains the field of rational numbers. Define the Hochschild-Kostant-Rosenberg map

$$\mathrm{HKR} : C_\bullet(A) \to \Omega^\bullet_{A/k};\ a_0 \otimes \ldots \otimes a_n \mapsto \frac{1}{n!} a_0 da_1 \ldots da_n,\ n \geq 0 \tag{3.8}$$

It is easy to see that HKR intertwines b with 0 and B with d. Moreover it extends to

$$\mathrm{HKR} : CC_\bullet^-(A) \to (\Omega^\bullet_{A/k}[[u]], ud);\quad CC_\bullet^{\mathrm{per}}(A) \to (\Omega^\bullet_{A/k}((u)), ud) \tag{3.9}$$

THEOREM 3.2 ([**19**]). *Suppose that A is a commutative regular algebra. Then* HKR *induces an isomorphism* $\mathrm{HH}_\bullet(A) \longrightarrow \Omega^\bullet_{A/k}$ *and quasi-isomorphisms*

$$CC_\bullet^-(A) \longrightarrow (\Omega^\bullet_{A/k}[[u]], ud)$$
$$CC_\bullet^{\mathrm{per}}(A) \longrightarrow (\Omega^\bullet_{A/k}((u)), ud)$$
$$(\wedge^\bullet T_{A/k}, 0) \longrightarrow (C^\bullet(A), \delta).$$

REMARK 3.3. In the case of $A = C^\infty(M)$, the same result holds after replacing algebraic tensor products with projective tensor products.

4. Operations on cyclic complexes

4.1. The Cartan calculus of derivations.

DEFINITION 4.1. For a derivation D of A and for $m \geq 1$, set

$$L_D(a_0 \otimes \ldots \otimes a_n) = \sum_{j=0}^{n} a_0 \otimes \ldots \otimes D(a_j) \otimes \ldots \otimes a_n$$

$$\iota_{D^m}(a_0 \otimes \ldots \otimes a_n) = a_0 D^m(a_1) \otimes a_2 \otimes \ldots \otimes a_n$$

$$S_{D^m}(a_0 \otimes \ldots \otimes a_n) = \sum_{j=1}^{n} (-1)^{nj} a_j \otimes \ldots \otimes a_n \otimes a_0 \otimes L_D^m(a_1 \otimes \ldots \otimes a_{j-1})$$

$$I_D = \iota_D + uS_D.$$

We will often denote L_D simply by D.

LEMMA 4.2. *The following Cartan relations are satisfied:*

$$[b + uB, L_D] = 0;\ [L_D, L_E] = L_{[D,E]};\ [L_D, I_E] = I_{[D,E]};\ [b + uB, I_D] = uL_D \tag{4.1}$$

The missing Cartan relation

$$[I_D, I_E] = 0 \tag{4.2}$$

is true only at the level of homology. More precisely,

PROPOSITION 4.3. *Put*

$$I_{D^n} = \iota_{D^n} + uS_{D^n},\ n \geq 1. \tag{4.3}$$

For all $n > 0$.

$$[b + uB, I_{D^n}] + \sum_{k=1}^{n-1} \binom{n}{k} I_{D^k} I_{D^{n-k}} = uD^n \tag{4.4}$$

REMARK 4.4. For comparison, define in the commutative case the operations

$$J_{D^n} = \iota_D,\ n = 1;\ J_{D^n} = 0,\ n > 1 \tag{4.5}$$

then

$$[ud, J_D] = uD;\ [ud, J_{D^n}] + \sum_{k=1}^{n-1} \binom{n}{k} J_{D^k} J_{D^{n-k}} = 0,\ n > 1; \tag{4.6}$$

and, as a consequence, setting $\mathcal{J}(D) = \sum_{n=1}^{\infty} \frac{1}{n!} J_{D^n}$, we get

$$(ud + \mathcal{J}(D))^2 = uD \tag{4.7}$$

On the other hand, the relations from Proposition 4.3 are equivalent to

$$(b + uB + \mathcal{I}(D))^2 = u(e^D - 1) \tag{4.8}$$

where again

$$\mathcal{I}(D) = \sum_{n=1}^{\infty} \frac{1}{n!} I_{D^n} \tag{4.9}$$

This seems to be another instance of the appearance of the inverse Todd series $\frac{e^D - 1}{D}$, as it often happens in the comparison between commutative and non-commutative contexts.

One can pass from $\mathcal{I}(D)$ to $\mathcal{J}(D)$ as follows.

LEMMA 4.5. *Define the Stirling numbers as the coefficients of the power series*

$$\sum_{k,l \geq 0} c_{k,l} x^k y^l = \sum_{n=1}^{\infty} \frac{1}{n!} y(y - x) \ldots (y - (n-1)x). \tag{4.10}$$

Let I_{D^n}, $n \geq 1$, *satisfy the relations from the proposition* 4.3 *and set*

$$\mathcal{J}(D) = \sum_{k,l \geq 0} c_{k,l} L_D^k I_{D^l}.$$

Then

$$(b + uB + \mathcal{J}(D))^2 = uL_D$$

EXAMPLE 4.6. Let $\mathcal{A}$ be a differential graded algebra whose differential we denote by $d_{\mathcal{A}}$. Let α be a derivation of degree one of $\mathcal{A}$ and set

$$R = (d_{\mathcal{A}} + \alpha)^2 = [d_{\mathcal{A}}, \alpha] + \frac{1}{2}[\alpha, \alpha].$$

Consider first the commutative case and the de Rham complex $\Omega^{\bullet}_{\mathcal{A}/k}$ with the differential $d_{\mathcal{A}} + d$. Then

$$D^u_{\alpha} = d_{\mathcal{A}} + ud - \frac{1}{u}\iota_R \tag{4.11}$$

is a differential on $\Omega^{\bullet}_{\mathcal{A}/k}((u))$.

Now, more generally, assume that, in addition to the Lie algebra $\mathrm{Der}(\mathcal{A})$, a collection of J_{R^n} acts on a complex with differential $b+uB$, subject to (4.6). Then, formally, set

$$\mathbf{D}^u_\alpha = d_{\mathcal{A}} + ud - \Psi(R) \text{ where } \Psi(R) = \sum_{n=1}^{\infty} \frac{u^{-n}}{n!} J_{R^n}. \tag{4.12}$$

One checks that $(\mathbf{D}^u_\alpha)^2 = 0$.

Let us now assume that instead of a collection of operators J_{R^n}, a collection of operators I_{R^n} acts subject equations (4.3). For example, the complex could be $\mathrm{CC}^{\mathrm{per}}_\bullet(\mathcal{A})$. Looking for $\Phi(R)$ such that

$$(d_{\mathcal{A}} + b + uB - \Phi(R))^2 = 0, \tag{4.13}$$

we find

$$\Phi(R) = \sum_{k,l} c_{k,l} R^l I_{R^k} \tag{4.14}$$

where

$$\sum_{k,l} c_{k,l} y^l x^k = (1 + \frac{y}{u})^{\frac{x}{y}} - 1 = \sum_{n=1}^{\infty} \frac{1}{n! u^n} x(x-y)\dots(x-(n-1)y) \tag{4.15}$$

We see that $\exp(\frac{x}{u})$ gets replaced by $(1+\frac{y}{u})^{\frac{x}{y}}$.

Both operators require a convergence condition. For example, one might assume that k contains the rationals, the image of R is inside an ideal of $\mathcal{A}$. Or, one assumes that the image of R is contained in $p\mathcal{A}$ where $p > 2$ is a prime. In both cases, $b + uB + d_{\mathcal{A}} - \Phi(R)$ is a well-defined differential on the periodic cyclic complex completed with respect to the filtration induced by powers of the ideal (in the second case, this means p-adic completion); cf. [**27**], [**33**].

REMARK 4.7. As shown in [**1**], given a manifold with an action of a Lie group and a closed odd equivariant differential form, the space of equivariant forms acquires a generalized Cartan calculus structure as in (4.6). There are more examples of such generalized calculus: one in [**2**] (again as in (4.6)), another in [**15**] (as in (4.4)). The author is grateful to Anton Alekseev for pointing this out.

5. Compatibility with the Hochschild-Kostant-Rosenberg map

As we have seen above, when A is a commutative algebra and D is its derivation, one can define formal operators

$$\mathcal{J}(D) = \sum_{n=1}^{\infty} \frac{1}{n!} J_{D^n}$$

in two ways. One, on $\Omega^\bullet_{A/k}[[u]]$, is defined in Remark 4.4. The other, on $C_\bullet(A)[[u]]$, is constructed by Lemma 4.5 from (4.3). Both satisfy

$$(b + uB + \mathcal{J}(D))^2 = uD$$

where $b + uB$ on forms stands for ud.

THEOREM 5.1. *Suppose that A is commutative. Let k be of characteristic zero. Let D be a derivation of A. There exist natural (homogeneous of degree m in D) morphisms of degree zero*

$$\mathbf{HKR}_{D^m} : C_\bullet(A) \to \Omega^\bullet_{A/k}$$

such that $\mathbf{HKR}_0 = \mathrm{HKR}$ is the quasi-isomorphism given in Theorem 3.2 and

$$(ud + \mathcal{J}(D))\mathbf{HKR}(D) = \mathbf{HKR}(D)(b + uB + \mathcal{J}(D))$$

where

$$\mathbf{HKR}(D) = \sum_{n=0}^{\infty} \mathbf{HKR}_{D^n}$$

The proof occupies the rest of Section 5.

5.1. Natural operators $\mathcal{J}(D)$ on forms. Let us look at all possible natural systems of operators J_{D^n} on forms that satisfy (4.6). Any such system is of the form

$$\mathcal{J}(D) = A(D)\iota_D + B(D)ud \tag{5.1}$$

The condition (4.6) becomes

$$A(B+1) = 1$$

We claim that any two such $\mathcal{J}_1(D)$ and $\mathcal{J}_2(D)$ with $A(0) = B(0)$ are equivalent, meaning that there exists an invertible formal operator $F(D)$ such that

$$(ud + \mathcal{J}_2(D))F(D) = F(D)(ud + \mathcal{J}_1(D)) \tag{5.2}$$

The only natural candidate for such an operator is

$$F(D) = P(D) + Q(D)d\iota_D \tag{5.3}$$

where P and Q are some power series to be determined. The condition (5.2) becomes

$$-(P + DQ)A_1 + A_2P = 0; \quad B_2(P + DQ) = PB_1 - QD \tag{5.4}$$

Since $B_i = A_i^{-1} - 1$, those two equations are the same. We can start with any invertible $P(D)$ and find $Q(D)$ uniquely out of the first equation. We will be interested in a special choice of $\mathcal{J}(D)$.

LEMMA 5.2. *Define*

$$I_D = \iota_D + \frac{D}{2}ud; \; I_{D^n} = \frac{D^n}{n+1}ud, \; n \geq 2. \tag{5.5}$$

and

$$\mathcal{J}_Q(D) = \frac{\log(1+D)}{D}\iota_D + (\frac{D}{\log(1+D)} - 1)ud \tag{5.6}$$

In other words:

$$\frac{1}{n!}J_{Q,D^n} = \frac{(-1)^{n-1}}{n}D^{n-1}\iota_D + c_nD^n ud \tag{5.7}$$

where the c_n are defined by the generating function

$$1 + \sum_{n=1}^{\infty} c_n z^n = \frac{z}{\log(1+z)} \tag{5.8}$$

Then the I_{D^n} satisfy (4.4) *(with ud instead of $b+uB$); the J_{D^n} are the result of applying the construction of Lemma* 4.5. *In particular, the J_{D^n} satisfy* (4.6).

(In other words: (5.5) provides the simplest family of I_{D^n} on forms; applying the construction of Lemma 4.5 to it, we get a family of J_{D^n}, but this is not the simplest one possible).

PROOF.

$$[ud, I_{D^n}] + \sum_{k=1}^{n-1} \binom{n}{k} I_{D^k} I_{D^{n-k}} = n([\iota_D, \frac{u}{n} dD^{n-1}] = uD^n;$$

$$\sum_{k+l=n} c_{k,l} L_D^k I_{D^l} = c_{n-1,1}\iota_D D^{n-1} + \sum_{k+l=n} c_{k,l} D^k \frac{D^l}{l+1} =$$

$$\frac{(-1)^{n-1}}{n} D^{n-1}\iota_D + \frac{u}{n!}\frac{1}{D}\int_0^D y(y-D)\dots(y-(n-1)D)dy$$

Note also

$$\sum_{n=1}^{\infty} \frac{1}{xn!}\int_0^x y(y-x)\dots(y-(n-1)x)dy = \frac{1}{x}\int_0^x (1+x)^{\frac{y}{x}} dy - 1 = \frac{x}{\log(1+x)} - 1$$

This proves our statement. □

5.2. The operator HKR(D). We will construct a formal operator

$$\text{HKR}(D) = \text{HKR} + \sum_{n=1}^{\infty} \text{HKR}_{D^n} \tag{5.9}$$

such that HKR_{D^n} is a homogeneous map of degree n from $\text{Der}(A)$ to operators $C_\bullet(A, A) \to \Omega^\bullet_{A/k}$ satisfying

$$[b + uB + \mathcal{I}(D), \text{HKR}(D)] = 0 \tag{5.10}$$

Here $\mathcal{I}(D)$ on Hochschild chains is defined in Proposition 4.3 and on forms in (5.5). Having done that, we will be able to construct $\text{HKR}_Q(D)$ as in Theorem 5.1 but with $\mathcal{J}_Q(D)$ instead of $\mathcal{J}(D)$:

$$\text{HKR}_{Q,D^n} = \sum_{k+l=n} c_{k,l} D^k \text{HKR}_{D^l}$$

Since $\mathcal{J}(D)$ and $\mathcal{J}_Q(D)$ are equivalent, this will prove the theorem.

The obstructions to existence of HKR(D) lie in the cohomology of an easy-to-compute complex. We will show that those obstructions all vanish.

5.2.1. *The complex of operations.* Let $n \geq 0$. Consider a collection $m_i \geq 0$ and $\epsilon_i = 0$ or 1, $i = 0, \dots, n$, such that for any $i > 0$, m_i and ϵ_i are not both equal to zero. Define the morphism $C_\bullet(A) \to \Omega^\bullet_{A/k}$

$$a_0 \otimes \dots \otimes a_n \mapsto d^{\epsilon_0} D^{m_0}(a_0) \dots d^{\epsilon_n} D^{m_n}(a_n) \tag{5.11}$$

Denote this operation by $\eta^{\epsilon_0} t^{m_0} \otimes \dots \otimes \eta^{\epsilon_n} t^{m_n}$.

LEMMA 5.3. *The linear span of all operations* (5.11) *with the differential $\circ b$ is isomorphic to the Hochschild cochain complex of the coalgebra $k[\eta, t]$ where $|\eta| = -1$, $|t| = 0$, $\eta^2 = 0$ and the comultiplication is the algebra morphism such that $t \mapsto t \otimes 1 + 1 \otimes t$, $\eta \mapsto \eta \otimes 1 + 1 \otimes \eta$.*

PROOF. Straightforward. □

We denote our coalgebra by C and the complex of operations by $C^\bullet(C)$. Inside this complex there is a subcomplex $\mathrm{Cobar}(C)$ spanned by operations for which $m_0 = \epsilon_0 = 0$.

The cohomology of $C^\bullet(C)$ is straightforward. First, $C^\bullet(C) \xrightarrow{\sim} C^\bullet(k[t]) \otimes C^\bullet(k[\eta])$ by means of the coshuffle product: $t^{m_0}\eta^{\epsilon_0} \otimes \ldots \otimes t^{m_n}\eta^{\epsilon_n}$ maps to zero if for some $i > 0$ both m_i and ϵ_i are nonzero; otherwise, it maps to

$$(t^{m_0} \otimes t^{m_{i_1}} \otimes \ldots \otimes t^{m_{i_k}}) \otimes (\eta^{\epsilon_0} \otimes \eta^{\epsilon_{j_1}} \otimes \ldots \otimes \eta^{\epsilon_{j_{n-k}}})$$

where $m_{j_p} > 0$ for all p, $\epsilon_{j_q} = 1$ for all q, $i_1 < \ldots < i_k$, $j_1 < \ldots < j_{n-k}$, and $\{1, \ldots, n\} = \{i_1, \ldots, i_k, j_1 \ldots, j_{n-k}\}$. Furthermore: $C^\bullet(k[t])$ projects quasi-isomorphically to its cohomology which is spanned over k by t^m and $t^m \otimes t$, $m \geq 0$. The projection sends a monomial t^{m_0} or $t^{m_0} \otimes t$ to itself, and any other monomial $t^{m_0} \otimes \ldots \otimes t^{m_n}$ to zero. Note also that the differential on $C^\bullet(k[\eta])$ is zero.

We have constructed an explicit quasi-isomorphism

$$C^\bullet(C) \to (\oplus_{m \geq 0}(kt^m \oplus kt^m \otimes t)) \otimes (\oplus_{n \geq 0}(k1 \otimes \eta^{\otimes n} \oplus k\eta \otimes \eta^{\otimes n})) \tag{5.12}$$

which restricts to

$$\mathrm{Cobar}(C) \to (\oplus_{m \geq 0}(k1 \oplus k1 \otimes t)) \otimes (\oplus_{n \geq 0}(k1 \otimes \eta^{\otimes n}) \tag{5.13}$$

5.2.2. *The existence for* $\mathrm{HKR}(D)$. First observe that $[\mathrm{HKR}, \iota_D]$ maps to zero by (5.12). Therefore there is h_D in $\mathrm{Cobar}(C)$ such that $[\mathrm{HKR}, \iota_D] + h_D \circ b = 0$. We want to construct HKR_{D^m} such that

$$[\mathrm{HKR}, \iota_{D^m}] + \sum_{k=1}^{m-1} \binom{m}{k} [\mathrm{HKR}_{D^k}, \iota_{D^{m-k}}] = -\mathrm{HKR}_{D^m} \circ b \tag{5.14}$$

for all m. Assume we have constructed all D^k for $k < n$. Then the left hand side of (5.14) is an operation in $\mathrm{Cobar}(C)$ that vanishes under (5.13) (indeed, it has weight > 1 in t). This shows that HKR_{D^n} exists.

We claim that

$$\mathrm{HKR}(D) = \mathrm{HKR} + \sum_{n \geq 1} \frac{1}{n!} \mathrm{HKR}_{D^n} \tag{5.15}$$

is the one we are looking for (i.e. that no higher terms in u are needed; note also that, if needed, they would not be available because there are no natural operations $C_\bullet \to \Omega^\bullet$ of degree bigger that one). To proceed, we do need some explicit knowledge of HKR_{D^m}.

Let h_{D^m} be the principal term of HKR_{D^m} in the following sense: it is the part involving only the monomials (5.11) where only one m_j is nonzero.We claim that

$$h_{D^m}(a_0 \otimes \ldots \otimes a_n) = \sum_{k=1}^{n} \frac{(n-k+m)!}{(n+m)!(n-k)!} a_0 da_1 \ldots dD^m(a_k) \ldots da_n \tag{5.16}$$

or equivalently
(5.17)

$$h_{D^m}(a_0 \otimes \ldots \otimes a_n) = \mathrm{HKR}(\sum_{k=1}^{n} \frac{n!(n-k+m)!}{(n+m)!(n-k)!} a_0 \otimes a_1 \ldots \otimes D^m(a_k) \otimes \ldots \otimes a_n)$$

To see that, write

$$h_{D^m}(a_0 \otimes \ldots \otimes a_n) = \sum_{k=1}^{n} c_j^m(n) a_0 da_1 \ldots dD^m(a_k) \ldots da_n \tag{5.18}$$

Now apply (5.14) to $a_0 \otimes \ldots \otimes a_{n+1}$. Compute the coefficients of

$$a_0 D^m(a_1) da_2 \ldots da_{n+1}.$$

For $m = 1$, we get

$$c_1^1(n) + \frac{1}{(n+1)!} - \frac{1}{n!} = 0 \tag{5.19}$$

For $m > 1$, we get

$$c_1^m(n) + m c_1^{m-1}(n+1) - \frac{1}{n!} = 0 \tag{5.20}$$

Computing the coefficients of $a_0 da_1 \ldots D^m(a_j) da_{j+1} \ldots da_{n+1}$ for $j > 1$, we get

$$c_j^1(n) - c_{j-1}^1(n) + \frac{1}{(n+1)!} = 0 \tag{5.21}$$

for $m = 1$ and

$$c_j^m(n) - c_{j-1}^m(n) + m c_1^{m-1}(n+1) = 0 \tag{5.22}$$

for $m > 1$. In other words, we have recursive relations (5.20) and (5.21) where $c_j^0(n) = \frac{1}{n!}$ (this may look at first a bit strange as compared to (5.18)). These relations have the unique solution

$$c_j^m(n) = \frac{(n+m-k)!}{(n+m)!(n-k)!} \tag{5.23}$$

Now we have to show that

$$[\mathrm{HKR}, S_{D^m}] + \sum_{k=1}^{m-1} \binom{m}{k} [\mathrm{HKR}_{D^k}, S_{D^{m-k}}] + [\mathrm{HKR}_{D^m}, B] = 0 \tag{5.24}$$

Let us show that the composition of the left hand side with b is zero. Denote

$$\iota(D) = \sum_{m=1}^{\infty} \iota_{D^m}; \; S(D) = \sum_{m=1}^{\infty} S_{D^m}$$

We have

$$[b + \iota(D), \mathrm{HKR}(D)] = 0 \tag{5.25}$$

and

$$[b + \iota(D), B + S(D)] = e^D - 1$$

Therefore

$$[b + \iota(D), [B + S(D), \mathrm{HKR}(D)]] = 0 \tag{5.26}$$

Let us write

$$[B + S(D), \mathrm{HKR}(D)] = \sum_{n=1}^{\infty} A_{D^n} \tag{5.27}$$

By induction in m, we see that $[b, A_{D^m}] = 0$ for all m.

Next, let us show that A_{D^m} map to zero under (5.12). The only terms that do not automatically map to zero are $dD^m(a_0) da_1 \ldots da_n$, corresponding to the operation $\eta t^m \otimes \eta^{\otimes m}$. There are two places in (5.14) in which this may appear. One is $-S_{D^m}\mathrm{HKR}$ in the first summand; the other is $h_{D^m}B$ in the last summand. Recall that

$$S_{D^m} = \frac{dD^m}{m+1}$$

on forms. Therefore the total coefficient of $\eta t^m \otimes \eta^{\otimes m}$ in the left hand side of (5.14) is

$$-\frac{1}{(m+1)n!} + \sum_{j=1}^{n+1} c_j^m(n+1)$$

But

$$\sum_{j=1}^{n} c_j^m(n) = \sum_{j=1}^{n} \frac{(n+m-j)!}{(n+m)!(n-j)!} = \frac{m!}{(n+m)!}\sum_{j=1}^{n}\binom{n+m-j}{m} =$$

$$\frac{m!}{(n+m)!}\sum_{k=0}^{n-1}\binom{m+k}{m} = \frac{m!}{(n+m)!}\binom{n+m}{m+1} = \frac{1}{(m+1)(n-1)!}$$

Therefore every A_{D^m} vanishes under (5.12). This means that it is in the image of the differential. But it corresponds to a linear combination of terms $\eta t^{m_0}\otimes \ldots \otimes \eta t^{m_n}$ in $C^\bullet(C)$ which cannot be in the image of the differential. Indeed, the differential preserves the weight in η and increases n by one. Thus, all A_{D^m} are equal to zero. This concludes the proof of Theorem **??**.

6. Extended Cartan calculus, I

6.1. Algebras $\mathcal{A}_0$ and $\mathcal{A}_1$.

DEFINITION 6.1. For any differential graded Lie algebra $\mathfrak{g}$, let $U^+(\mathfrak{g})$ be the kernel of the augmentation $U(\mathfrak{g}) \to k$. Let $\mathrm{Cobar}(U^+(\mathfrak{g}))$ be the free associative algebra generated by $U^+(\mathfrak{g})[-1]$ (the degree shift by one). We denote the free generator corresponding to $x \in U^+(\mathfrak{g})$ by (x). Define

$$\partial_{\mathrm{Cobar}}(x) = \sum (-1)^{|x^{(1)}|}(x^{(1)})(x^{(2)}) \tag{6.1}$$

where the comultiplication is defined by

$$\Delta x = \sum x^{(1)} \otimes x^{(2)}$$

In addition, the differential $d_\mathfrak{g}$ induces a differential on $\mathrm{Cobar}(U^+(\mathfrak{g}))$. Now define the dg algebra

$$\mathcal{A}_1(\mathfrak{g}) = U(\mathfrak{g}) \ltimes_1 \mathrm{Cobar}(U^+(\mathfrak{g})) \tag{6.2}$$

as follows. It is an algebra over $k[u]$ generated by the DG subalgebra $(U(\mathfrak{g}), d_\mathfrak{g})$ and the subalgebra $\mathrm{Cobar}(U^+(\mathfrak{g}))$. The only additional relations are

$$[X, (x)] = (-1)^{|X|}(\mathrm{ad}_X(x)),\ X \in \mathfrak{g}, x \in U^+(\mathfrak{g}).$$

The differential acts as follows:

$$x \mapsto d_\mathfrak{g}x, x \in U(\mathfrak{g});\ (x) \mapsto (-d_\mathfrak{g}x) + \partial_{\mathrm{Cobar}}(x) + ux,\ x \in U^+(\mathfrak{g}). \tag{6.3}$$

Define also

$$\mathcal{A}_0(\mathfrak{g}) = U(\mathfrak{g}) \ltimes_0 \mathrm{Cobar}(\mathrm{Sym}^+(\mathfrak{g})) \tag{6.4}$$

in the same way as above, the differential being

$$x \mapsto d_{\mathfrak{g}}x, x \in U(\mathfrak{g});\ (x) \mapsto (-d_{\mathfrak{g}}x) + \partial_{\mathrm{Cobar}}(x) + uB_0(x) \tag{6.5}$$

where

$$B_0(x) = x,\ x \in \mathrm{Sym}^1(\mathfrak{g}); B_0(x) = 0, x \in \mathrm{Sym}^{>1}(\mathfrak{g})$$

Lemma 4.2 can be strengthened as follows.

PROPOSITION 6.2. *There is a natural action of $\mathcal{A}_1(\mathrm{Der}(A))$ on $\mathrm{CC}^-_\bullet(A)$ where $\mathfrak{g} = \mathrm{Der}(A)$ such that for any derivation D, D acts by L_D and (D^n) acts by I_{D^n}* [**32**], [**9**].

Let us define this action expicitly. Recall for $D \in \mathrm{Der}(A)$

$$L_D(a_1 \otimes \ldots \otimes a_n) = \sum_{j=1}^{n} a_1 \otimes \ldots \otimes D(a_j) \otimes \ldots \otimes a_n = \lambda_D(a_1 \otimes \ldots \otimes a_n) \tag{6.6}$$

(to reconcile this with our notation later on, we use two different letters for the same operator). Let $X_1, \ldots, X_m \in \mathrm{Der}(A)$. For $X = X_1 \ldots X_m \in U^+(\mathrm{Der}(A))$, we put

$$\lambda_X = \lambda_{X_1} \ldots \lambda_{X_m} \tag{6.7}$$

DEFINITION 6.3. For $X \in U^+(\mathrm{Der}(A)$

$$I_X = \iota_{\overline{X}} + uS_X \tag{6.8}$$

where

$$\overline{X_1 \ldots X_m} = X_1 \circ \ldots \circ X_m; \tag{6.9}$$

$$\iota_{\overline{X}}(a_0 \otimes \ldots \otimes a_n) = a_0 \overline{X}(a_1) \otimes a_2 \otimes \ldots \otimes a_n \tag{6.10}$$

$$S_X(a_0 \otimes \ldots \otimes a_n) = \sum_{j=0}^{n} (-1)^{nj} a_j \otimes \ldots \otimes a_n \otimes a_0 \otimes \lambda_X(a_1 \otimes \ldots \otimes a_{j-1}) \tag{6.11}$$

LEMMA 6.4. *The above formulas define an action of $\mathcal{A}_1(\mathrm{Der}(A))$ on $\mathrm{CC}^-_\bullet(A)$.*

PROOF. This can be shown by direct verification [**32**]; it follows also from considerations as in the introduction (namely, the argument establishing (1.10)). □

7. Extended Cartan calculus, II

7.1. Noncommutative calculus of multivector fields and forms. Return to the case of a commutative algebra A. Recall that $\wedge^\bullet T_{A/k}[1]$ carries a graded Lie algebra structure. The action by operators ι_D, $D \in Der(A)$ extends to an action by contraction of multivectors (multivector fields) so that $\Omega^{-\bullet}$ is a graded module over the graded algebra $\wedge^\bullet T_{A/k}$. Set, for $\alpha \in \Lambda^m T_{A/k}$,

$$L_\alpha = [d, \iota_\alpha] : \Omega^\bullet_{A/k} \to \Omega^{\bullet - m + 1}_{A/k} \tag{7.1}$$

THEOREM 7.1. *The operation of Lie-derivative L makes $\Omega^{\bullet - m+1}_{A/k}$ into a module over the Lie algebra $\wedge^\bullet T_{A/k}[1]$. The following identities hold.*

(1) $[L_\alpha, L_\beta] = L_{[\alpha,\beta]}$;
(2) $[L_\alpha, \iota_\beta] = (-1)^{|\alpha|-1}\iota_{[\alpha,\beta]}$;
(3) $[\iota_\alpha, \iota_\beta] = 0$;
(4) $[d, \iota_\alpha] = L_\alpha$.

Moreover,

$$\iota_{\alpha\beta} = \iota_\alpha \iota_\beta; \; L_{\alpha\beta} = L_\alpha \iota_\beta + (-1)^{|\alpha|} \iota_\alpha L_\beta \tag{7.2}$$

7.2. Hochschild cochain complex. The role of noncommutative multivector fields is played by Hochschild cochains.

DEFINITION 7.2. Let A be a unital associative algebra over k. Set

$$C^\bullet(A) = \mathrm{Hom}_k(A^{\otimes \bullet}, A) \tag{7.3}$$

with the differential $\delta : C^\bullet(A) \to C^{\bullet+1}(A)$ given by

$$\delta\phi(a_1, \ldots, a_{n+1}) = a_1\phi(a_2, \ldots, a_{n+1})$$

$$+ \sum_{k=1}^{n} (-1)^k \phi(a_1, \ldots, a_k a_{k+1} \ldots a_{n+1}) + \phi(a_1, \ldots, a_n)a_{n+1}$$

The complex $(C^\bullet(A), \delta)$ computes the groups $Ext^\bullet_{A\otimes A^{op}}(A, A)$, usually called the Hochschild cohomology of A and denoted by $\mathrm{HH}^\bullet(A)$ or $\mathrm{HH}^\bullet(A, A)$.

Note that the space of cocycles in $C^1(A)$ is $\mathrm{Der}(A)$. There is a differential graded Lie algebra structure on $C^\bullet(A)[1]$ that extends the commutator of derivations (the Gerstenhaber bracket). For later use, let us be more explicit. Let $\phi \in C^k(A)$ and $\psi \in C^l(A)$. Set

$$\phi \circ \psi = \phi \circ (\psi \otimes id \ldots \otimes id) + (-1)^{|\psi|} \phi \circ (id \otimes \psi \otimes id \ldots \otimes id) +$$
$$\ldots + (-1)^{|\psi||\phi|} \phi \circ (id \otimes \ldots \otimes \psi)$$

and

$$[\phi, \psi] = \phi \circ \psi - (-1)^{|\psi||\phi|} \psi \circ \phi.$$

Here $|\cdot|$ refers to the Lie algebra degree, i. e., for $\phi \in C^k$, $|\phi| = k - 1$. $[\cdot,\cdot]$. The bracket is called the Gerstenhaber bracket and the following holds.

THEOREM 7.3. *$(C^\bullet(A)[1], [\cdot,\cdot], \delta)$ is a differential graded Lie algebra.*

7.3. Hochschild cochains and the bar construction. Recall the approach to Cartan calculus originated in **[21]** and **[22]**. For a graded k-module A, let $\mathrm{Bar}(A) = T(A[1])$ be the cofree coassociative coalgebra. As usual, we denote $a_1[1] \otimes \ldots \otimes a_n[1]$ by $(a_1| \ldots |a_n)$. An A_∞ structure on A is by definition a coderivation of degree $+1$ and of square zero of $\mathrm{Bar}(A)$. Any coderivation is uniquely determined by its composition with the projection onto $A[1]$. In other words, an A_∞ structure on A is a collection of k-linear maps $m_n : A^{\otimes n} \to A[2-n]$ for $n \geq 1$. Any DG algebra is an A_∞ algebra if we put $m_1(a_1) = da_1$, $m_2(a_1, a_2) = (-1)^{|a_1|} a_1 a_2$ for $n = 2$, and $m_n = 0$ for $n > 2$.

LEMMA 7.4. *The DG Lie algebra* $\mathrm{Coder}(\mathrm{Bar}(A))$ *is isomorphic to the DG Lie algebra $(C^\bullet(A)[1], [\cdot,\cdot], \delta)$ from* 7.3.

7.4. Hochschild and cyclic complexes of coalgebras. For a counital DG coalgebra B, let

$$C^\bullet_{\mathrm{II}}(B) = \bigoplus_{n=0}^{\infty} B \otimes \bar{B}[-1]^{\otimes n} \tag{7.4}$$

Here $\bar{B}$ is the kernel of the counit $B \to k$. The above complex carries two differentials: one coming from the differential on B, the other defined in the dual way to the Hochschild differential B. Cf. [8], [18], [33]. We view $C^\bullet_{\rm II}(B)$ as a complex whose differential is the sum of these two differentials.

REMARK 7.5. If we followed definition of the Hochschild complex $C_\bullet(A)$ of a DG algebra A and applied the dual construction to a coalgebra B, we would get the direct product totalization, not the direct sum. This is why we use the subscript II and use the term complex of the second kind (compare to [30]). Our definition is *not* invariant under quasi-isomorphisms of DG coalgebras (which is good because we are going to apply it to an acyclic coalgebra $\mathrm{Bar}(A)$).

Similarly, we define the differential B by a formula dual to the one on the chain complex. Then we define cyclic complexes of a DG coalgebra by

$$\mathrm{CC}^\bullet_{\rm II}(B) = (C^\bullet_{\rm II}(B)[[v^{-1}]], d_B + b + v^{-1}B) \tag{7.5}$$

$$\mathrm{CC}^\bullet_{\rm per,II}(B) = (C^\bullet_{\rm II}(B)((v^{-1})), d_B + b + v^{-1}B) \tag{7.6}$$

$$\mathrm{CC}^\bullet_{-,\rm II}(B) = (C^\bullet_{\rm II}(B)((v^{-1}))/v^{-1}C^\bullet_{\rm II}(B)[[v^{-1}]], d_B + b + v^{-1}B) \tag{7.7}$$

Finally, for any DG coalgebra B (not necessarily counital), we denote by B^+ the coalgebra with counit adjoined, and but

$$\mathrm{C}^\bullet_{?,\rm II}(B) = \ker(\mathrm{C}^\bullet_{?,\rm II}(B^+) \to \mathrm{C}^\bullet_{?,\rm II}(k)) \tag{7.8}$$

$$\mathrm{CC}^\bullet_{?,\rm II}(B) = \ker(\mathrm{CC}^\bullet_{?,\rm II}(B^+) \to \mathrm{CC}^\bullet_{?,\rm II}(k)) \tag{7.9}$$

where ? stands for $-$, per, or nothing.

THEOREM 7.6 (cf. [31], [11]). *There is a natural chain of quasi-isomorphisms*

$$\mathrm{C}_{-\bullet}(A) \xrightarrow{\sim} \mathrm{C}^\bullet_{\rm II}(\mathrm{Bar}(A))$$

$$\mathrm{CC}^-_{-\bullet}(A) \xrightarrow{\sim} \mathrm{CC}^\bullet_{\rm II}(\mathrm{Bar}(A))$$

$$\mathrm{CC}_{-\bullet}(A) \xrightarrow{\sim} \mathrm{CC}^\bullet_{-,\rm II}(\mathrm{Bar}(A))$$

The proof will be given in Subsections 7.5 and 7.6 below.

7.5. Short chain complexes. For an algebra A, let $C^{\rm sh}(A)$ be the truncated Hochschild complex

$$0 \to C_1(A)/bC_2(A) \xrightarrow{b} C_0(A) \to 0 \tag{7.10}$$

Equivalently,

$$C_1^{\rm sh}(A) \xrightarrow{\sim} \Omega^1_A/[A, \Omega^1_A] \tag{7.11}$$

where Ω^1_A is the bimodule of noncommutative one-forms: it is generated by symbols $da, a \in A/k\cdot 1$, linear in a and subject to relations $d(ab) = da\cdot b + a\cdot db$. The differential B induces $B : A \to C_1^{\rm sh}(A)$ which under (7.11) becomes the De Rham differential $d : a \mapsto da$. Factoring out all components $u^pC_n(A)$, $n \geq 2$, and $u^pbC_1(A)$, we get the short cyclic complexes

$$\mathrm{CC}^{\rm sh}_\bullet(A),\ \mathrm{CC}^{-,\rm sh}_\bullet(A),\ \mathrm{CC}^{\rm per,sh}_\bullet(A) \tag{7.12}$$

Dually, for a coalgebra B define

$$C^0_{\rm II,sh}(B) = B;\ C^1_{\rm II,sh}(B) = \ker(C^1_{\rm II}(B) \xrightarrow{b} C^2_{\rm II}(B));$$

$C^n_{\rm II,sh}(B) = 0$ for $n \geq 2$. Replacing $\mathrm{C}^\bullet_{\rm II}$ by $\mathrm{C}^\bullet_{\rm II,sh}$ in (7.5), (7.6), (7.7), we get complexes

$$\mathrm{CC}^\bullet_{\rm II,sh}(B),\ \mathrm{CC}^\bullet_{-,\rm II,sh}(B),\ \mathrm{CC}^\bullet_{\rm per,II,sh}(B). \tag{7.13}$$

LEMMA 7.7. *If A is free as an algebra then the projections*

$$\mathrm{C}_\bullet(A) \to \mathrm{C}_\bullet^{\mathrm{sh}}(A)$$

and

$$\mathrm{CC}_\bullet^?(A) \to \mathrm{CC}_\bullet^{?,\mathrm{sh}}(A)$$

are quasi-isomorphisms. Dually, if B is cofree as a coalgebra then the embeddings

$$\mathrm{C}_{\mathrm{II}}^\bullet(B) \to \mathrm{C}_{\mathrm{II,sh}}^\bullet(B)$$
$$\mathrm{CC}_{?,\mathrm{II}}^\bullet(B) \to \mathrm{CC}_{?,\mathrm{II,sh}}^\bullet(B)$$

are quasi-isomorphisms. Here, as above, ? may stand for $-$, per, *or nothing.*

Now apply the lemma to $B = \mathrm{Bar}(A)$ and observe that:
$\mathrm{C}_{\mathrm{II,sh}}^\bullet$ is isomorphic to the cone of

$$1-\tau : (A^{\otimes \bullet+1}, b') \to (A^{\otimes \bullet+1}, b);$$

$\mathrm{CC}_{\mathrm{II,sh}}^\bullet$ is isomorphic to the standard $(b, b', 1-\tau, N)$ complex for computing negative cyclic homology; $\mathrm{CC}_{\mathrm{perII,sh}}^\bullet$ is isomorphic to the standard $(b, b', 1-\tau, N)$ complex for computing periodic cyclic homology. This proves Theorem 7.6.

7.6. The big Hochschild and cyclic complexes.

DEFINITION 7.8. Put

$$\mathrm{C}_\bullet^{\mathrm{big}}(A) = \mathrm{C}_{\mathrm{II}}^{-\bullet}(\mathrm{Bar}(A))$$
$$\mathrm{CC}_\bullet^{\mathrm{big}}(A) = \mathrm{CC}_{-,\mathrm{II}}^{-\bullet}(\mathrm{Bar}(A))$$
$$\mathrm{CC}_\bullet^{\mathrm{NEG}}(A) = \mathrm{CC}_{\mathrm{II}}^{-\bullet}(\mathrm{Bar}(A))$$
$$\mathrm{CC}_\bullet^{\mathrm{PER}}(A) = \mathrm{CC}_{\mathrm{per,II}}^{-\bullet}(\mathrm{Bar}(A))$$

Due to Theorem 7.6, these complexes are natural deformation retracts of their standard counterparts. When A is an ordinary algebra then

$$\mathrm{C}_\bullet^{\mathrm{big}}(A) = 0,\ \bullet < 0;$$

$$\mathrm{C}_0^{\mathrm{big}}(A) \xrightarrow{\sim} \bigoplus_{k=0}^{\infty} A^{\otimes(k+1)};$$

$$\mathrm{C}_n^{\mathrm{big}}(A) \xrightarrow{\sim} \bigoplus_{k=0}^{\infty} (A^{\otimes(k+1)})^{m_k}$$

for some integers m_k (dependent on n). Similarly to the standard case, one has

$$\mathrm{CC}_\bullet^{\mathrm{NEG}}(A) = \mathrm{C}_\bullet^{\mathrm{big}}(A)[[u]];\ \mathrm{CC}_\bullet^{\mathrm{PER}}(A) = \mathrm{C}_\bullet^{\mathrm{big}}(A)((u));$$
$$\mathrm{CC}_\bullet^{\mathrm{big}}(A) = \mathrm{C}_\bullet^{\mathrm{big}}(A)((u))/u\mathrm{C}_\bullet^{\mathrm{big}}(A)[[u]]$$

The differential is of the form $\mathbf{b} + u\mathbf{B}$.

THEOREM 7.9. *There is a natural action of the DG algebra $\mathcal{A}_1(\mathfrak{g}_A)$ on the big negative cyclic complex $\mathrm{CC}_\bullet^{\mathrm{NEG}}(A)$.*

PROOF. The action is defined by operators dual to the ones in Definition 6.3.
□

7.6.1. *Extended noncommutative Cartan calculus for the short cochain complex.* In the rest of this section we will briefly outline another way to extend the Cartan calculus. Namely, we will observe that it extends to the short cochain complex $\mathfrak{g}_A^{\mathrm{sh}}$. After that, one can replace A by its cofibrant DG resolution R; the inclusion $\mathfrak{g}_R^{\mathrm{sh}} \to \mathfrak{g}_R$ is a quasi-isomorphism; it is also known that $\mathfrak{g}_R$ and $\mathfrak{g}_A$ are equivalent as DG Lie algebras. This yields an action of a DG Lie algebra equivalent to $\mathfrak{g}_A$ on another version of a big negative cyclic complex, namely $\mathrm{CC}_\bullet^-(R)$. We are not using this anywhere else in the article.

DEFINITION 7.10. Let A be a unital algebra. Define a differential graded Lie subalgebra of $\mathfrak{g}_A$ (consisting of zero-cochains and one-cocycles)

$$\mathfrak{g}_A^{\mathrm{sh}} = A[1] \rtimes Der(A)$$

Here $A[1]$ is Abelian. The differential is zero on $\mathrm{Der}(A)$ and sends $a \in A[1]$ to $\mathrm{ad}(a)$.

PROPOSITION 7.11. *The action from Proposition* 6.2 *naturally extends to an action of the algebra* $U(\mathfrak{g}_A^{\mathrm{sh}}) \ltimes_1 \mathrm{Cobar}(U^+(\mathfrak{g}_A^{\mathrm{sh}}))$ *on* $\mathrm{CC}_\bullet^-(A)$.

PROOF. For $x \in A$, we define

$$\lambda_x(a_1 \otimes \ldots \otimes a_n) = \sum_{j=0}^{n} \pm a_1 \otimes \ldots \otimes a_j \otimes x \otimes a_{j+1} \otimes \ldots \otimes a_n; \tag{7.14}$$

$$L_x(a_0 \otimes \ldots \otimes a_n) = \pm a_0 \otimes \lambda_x(a_1 \otimes \ldots \otimes a_n); \tag{7.15}$$

$$\iota_x(a_0 \otimes \ldots \otimes a_n) = a_0 x \otimes a_1 \otimes \ldots \otimes a_n \tag{7.16}$$

For $X_1, \ldots, X_m \in \mathfrak{g}_A^{\mathrm{sh}}$ and $X = X_1 \ldots X_m \in U(\mathfrak{g}_A^{\mathrm{sh}})$, put

$$L_X = L_{X_1} \ldots L_{X_m} \tag{7.17}$$

Extend the composition operation from $\mathrm{Der}(A)$ to $\mathfrak{g}_A^{\mathrm{sh}}$ as follows. For $D \in \mathrm{Der}(A)$ and $x \in A$, $D \circ x = D(x)$ and $x \circ D = 0$; for $x, y \in A$, $x \circ y = 0$. For $X_1, \ldots, X_m \in \mathfrak{g}_A^{\mathrm{sh}}$ and $X = X_1 \ldots X_m \in U^+(\mathfrak{g}_A^{\mathrm{sh}})$, put

$$\overline{X} = \overline{X_1 \ldots X_m} = ((\ldots(X_1 \circ X_2) \circ \ldots) \circ X_m) \tag{7.18}$$

For $X_1, \ldots, X_m \in \mathfrak{g}_A^{\mathrm{sh}}$ and $X = X_1 \ldots X_m \in U^+(\mathfrak{g}_A^{\mathrm{sh}})$ set

$$I_X = \iota_{\bar{X}} + uS(X) \tag{7.19}$$

where

$$S_X(a_0 \otimes \ldots \otimes a_n) = \sum_{j=0}^{n} (-1)^j 1 \otimes a_j \otimes \ldots \otimes a_n \otimes a_0 \otimes \lambda_X(a_1 \otimes \ldots \otimes a_{j-1}) \tag{7.20}$$

Note that $\overline{X}$ is either a linear mar $A \to A$ (if all X_i are derivations) or an element of A. In the former case $\iota_{\overline{X}}$ is defined by (6.10), in the latter case by (7.16).

Let $X \in U(\mathfrak{g}_A^{\mathrm{sh}})$ act by L_X (7.17) and let (X), $X \in U^+(\mathfrak{g}_A^{\mathrm{sh}})$, act by I_X (7.19). We claim that $U(\mathfrak{g}_A^{\mathrm{sh}}) \ltimes_1 \mathrm{Cobar}(U^+(\mathfrak{g}_A^{\mathrm{sh}}))$ defines an action of $U(\mathfrak{g}_A^{\mathrm{sh}}) \ltimes_1 \mathrm{Cobar}(U^+(\mathfrak{g}_A^{\mathrm{sh}}))$. The proof will be sketched below. $\square$

8. The Gauss-Manin superconnection

8.1. The universal formulas. Recall the algebras

$$\mathcal{A}_0(\mathfrak{g}) = U(\mathfrak{g}) \ltimes_0 \mathrm{Cobar}(\mathrm{Sym}(\mathfrak{g})^+) \tag{8.1}$$

$$\mathcal{A}_1(\mathfrak{g}) = U(\mathfrak{g}) \ltimes_1 \mathrm{Cobar}(U(\mathfrak{g})^+) \tag{8.2}$$

defined in 6.1. Those are algebras over $k[u]$; as we have seen, $\mathcal{A}_1(\mathfrak{g}_A)$ acts on the big negative cyclic complex of A. This action obviously extends to the (u)-adic completion of $\mathcal{A}_1(\mathfrak{g}_A)$. However, for applications we will rather need the (u^{-1})-adic completion of the localized algebra $\mathcal{A}_0[u^{-1}]$.

LEMMA 8.1. *Let k contain the rationals. There is a natural isomorphism of DG algebras*

$$\mathcal{A}_0(\mathfrak{g}) \xrightarrow{\sim} \mathcal{A}_1(\mathfrak{g})$$

PROOF. For D_1, ..., D_k in $\mathfrak{g}$ define $(D_1 \ldots D_k)$ to be the coefficient at $t_1 \ldots t_k$ in

$$\frac{1}{k!} J(t_1 D_1 + \ldots + t_k D_k) \tag{8.3}$$

where J is as in Lemma 4.5. Here t_1, ..., t_k are formal variables that are central and such that $|t_i| = -|D_i|$. We view $(D_1 \ldots D_k)$ as an element of $\mathcal{A}_0\mathfrak{g}$. We get an $\mathfrak{g}$-equivariant map $\mathrm{Sym}(\mathfrak{g})^+ \to \mathcal{A}_1(\mathfrak{g})$ that extends to a DG algebra isomorphism $\mathcal{A}_0(\mathfrak{g}) \xrightarrow{\sim} \mathcal{A}_1(\mathfrak{g})$. □

Let $\mathfrak{a}_n$ be the free graded Lie algebra with generators λ_0, ..., λ_n of degree one.

LEMMA 8.2. *Let k be of characteristic zero. Then $k[u, u^{-1}] \to \mathcal{A}_0(\mathfrak{a}_n)[u^{-1}]$ is a quasi-isomorphism.*

PROOF. The cohomology of the differential $\partial_{\mathrm{Cobar}}$ is

$$\mathrm{Sym}(\mathfrak{a}_n[-1]) \otimes U(\mathfrak{a}_n)[u, u^{-1}] \tag{8.4}$$

The differential induced by B_0 is the standard Chevalley-Eilenberg differential computing the homology $H_\bullet(\mathfrak{a}_n, U(\mathfrak{a}_n))$ which is isomorphic to k. The corresponding spectral sequence proves the statement. □

DEFINITION 8.3. Let $\widehat{\mathcal{A}}_0(\mathfrak{a}_n)$ be the completion of $\mathcal{A}_0(\mathfrak{a}_n))[u^{-1}]$ with respect to the increasing filtration induced by the grading of $\mathfrak{a}_n$. Define

$$D_{\lambda_j} = \lambda_j - J(\frac{1}{u} R_j) = \lambda_j - (\exp(\frac{R_j}{u}) - 1) \in \widehat{\mathcal{A}}_0(\mathfrak{a}_n)^1$$

where $R_j = \lambda_j^2$.

LEMMA 8.4. *Let k be of characteristic zero.*

(1) *For any j one has*

$$(\partial_{\mathrm{Cobar}} + uB_0) D_{\lambda_0} + D_{\lambda_0}^2 = 0$$

in $\widehat{\mathcal{A}}_0(\mathfrak{a}_n)$.

(2) *There exist elements* $T(\lambda_0, \ldots, \lambda_m)$ *of degree* $1-m$ *in* $\widehat{A}_0(\mathfrak{a}_m)$ *for all* $m > 1$ *such that*

$$(\partial_{\text{Cobar}} + uB_0)T(\lambda_0, \ldots, \lambda_n) + D_{\lambda_0}T(\lambda_0, \ldots, \lambda_n) +$$
$$\sum_{j=1}^{n-1}(-1)^{j-1}T(\lambda_0, \ldots, \lambda_j)T(\lambda_j, \ldots, \lambda_n) + (-1)^n T(\lambda_0, \ldots, \lambda_n)D_{\lambda_n}$$
$$+\sum_{j=1}^{n-1}(-1)^j T(\lambda_0, \ldots, \widehat{\lambda}_j, \ldots, \lambda_n) = 0$$

PROOF. Follows from Lemma 8.2. Indeed, let us look for

$$T(\lambda_0, \lambda_1) = 1 + \sum_{k=1}^{\infty} T_k(\lambda_0, \lambda_1) \tag{8.5}$$

where T_k is homogenous of degree k. By induction in k, we find T_k as the solution for $(\partial_{\text{Cobar}} + uB_0)T_k = U_k$ where U_k is some given cocycle. This gives us $T(\lambda_0, \lambda_1)$. Similarly, assume we have found all $T(\lambda_0, \ldots, \lambda_m)$ for $m < n$.

$$T(\lambda_0, \ldots, \lambda_n) = \sum_{k=1}^{\infty} T_k(\lambda_0, \ldots, \lambda_n) \tag{8.6}$$

Then we can find each homogenous component T_k using induction in k. □

8.2. The main statement. Let S be an algebraic variety over a unital commutative ring k. Let $\mathcal{A}$ be a sheaf of $\mathcal{O}_S$-algebras. We assume that locally in S there is a connection

$$\nabla : \mathcal{A} \to \Omega^1_S \otimes_{\mathcal{O}_S} \mathcal{A} \tag{8.7}$$

on the family $\mathcal{A}$. We do not require the connection to preserve the algebra structure.

Let $\mathrm{CC}^{\mathrm{PER}}_\bullet(\mathcal{A}/\mathcal{O}_S)$ be the sheaf of big periodic cyclic complexes where $\mathcal{O}_S$ is regarded as the ring of scalars. Then ∇ extends to an operator of degree one on $\Omega^\bullet_S \otimes_{\mathcal{O}_S} \mathrm{CC}^{\mathrm{PER}}_\bullet(\mathcal{A}/\mathcal{O}_S)$.

THEOREM 8.5. *Let k be of characteristic zero. Let S be an algebraic variety over k and let $\mathcal{A}$ be a sheaf of $\mathcal{O}_S$-algebras.*

(1) *For any connection ∇ on $\mathcal{A}$ there exists canonically defined element of degree 1*

$$A(\nabla) \in \Omega^\bullet_S \otimes_{\mathcal{O}_S} \mathrm{End}^\bullet_{\mathcal{O}_S}(\mathrm{CC}^{\mathrm{PER}}_\bullet(\mathcal{A}/\mathcal{O}_S))$$

such that

$$(\nabla^{\mathrm{GM}})^2 = 0$$

where $\nabla^{\mathrm{GM}} = \mathbf{b} + u\mathbf{B} + \nabla + A(\nabla)$.

(2) *For any connections $\nabla_0, \ldots, \nabla_n$ on $\mathcal{A}$, there is a canonically defined element of degree* $2-n$

$$T(\nabla_0, \ldots, \nabla_n) \in \Omega^\bullet_S \otimes_{\mathcal{O}_S} \mathrm{End}^\bullet_{\mathcal{O}_S}(\mathrm{CC}^{\mathrm{PER}}_\bullet(\mathcal{A}/\mathcal{O}_S))$$

such that

$$(\mathbf{b}+u\mathbf{B})T(\nabla_0,\dots,\nabla_n)+\nabla_0^{\mathrm{GM}}T(\nabla_0,\dots,\nabla_n)+$$

$$\sum_{j=1}^{n-1}(-1)^jT(\nabla_0,\dots,\nabla_j)T(\nabla_j,\dots,\nabla_n)+(-1)^nT(\nabla_0,\dots,\nabla_n)\nabla_n^{\mathrm{GM}}=0$$

PROOF. Put

$$R_\nabla=\nabla m+\nabla^2; \tag{8.8}$$

then

$$(\mathbf{b}+u\mathbf{B}+\nabla)^2=L_{R_\nabla} \tag{8.9}$$

where m is the Hochschild cochain defining the product on $\mathcal{A}$.

Now put

$$\nabla^{\mathrm{GM}}=\mathbf{b}+u\mathbf{B}+\nabla-J(\frac{1}{u}R_\nabla) \tag{8.10}$$

and

$$T(\nabla_0,\dots,\nabla_n)=T(m+\nabla_0,\dots,m+\nabla_n) \tag{8.11}$$

as in Lemma 8.4. The same lemma implies the statement. □

8.3. Explicit formulas and integrality. Lemma 8.4 can be improved to provide explicit formulas for $T(\lambda_0,\dots,\lambda_n)$. Those formulas have an integrality property; namely, if λ_0, ..., λ_n are divisible by a prime $p>3$ then $T(\lambda_0,\dots,\lambda_n)$ are defined in the p-adic completion of $\mathcal{A}_0(\mathfrak{a}_n)$ (as are D_{λ_i}). This is not surprising. Indeed, in the proof of the lemma we used the fact that in characteristic zero, both the cobar construction and the chain complex $C_\bullet(\mathfrak{a}_n,U(\mathfrak{a}_n))$ are (almost) acyclic. In fact this is true over the integers for certain divided power versions of these complexes. We were not able to deduce the p-adic analog of Lemma 8.4 just from that observation. Instead, it is possible to follow this logic more closely and to write explicit formulas. These formulas are themselves of interest. They resemble certain WKB-type expansions in a Fock space, figuring fast oscillating exponentials where the Planck constant $\hbar$ is replaced by the formal parameter u of degree two. We will give details in the forthcoming article.

References

[1] Anton Alekseev and Pavol Ševera, *Equivariant cohomology and current algebras*, Confluentes Math. **4** (2012), no. 2, 1250001, 40, DOI 10.1142/S1793744212500016. MR2982768

[2] F. Bonechi, A. S. Cattaneo, and M. Zabzine, *Towards equivariant Yang-Mills theory*, J. Geom. Phys. **189** (2023), Paper No. 104836, 17, DOI 10.1016/j.geomphys.2023.104836. MR4580437

[3] Henri Cartan and Samuel Eilenberg, *Homological algebra*, Princeton University Press, Princeton, NJ, 1956. MR77480

[4] Hiroshi Takai and Toshikazu Natsume, *A. Connes' noncommutative differential geometry* (Japanese), Sūgaku **35** (1983), no. 2, 97–112. MR732432

[5] A. Connes, *Introduction to noncommutative differential geometry*, Workshop Bonn 1984 (Bonn, 1984), Lecture Notes in Math., vol. 1111, Springer, Berlin, 1985, pp. 3–16, DOI 10.1007/BFb0084582. MR797413

[6] Alain Connes, *Noncommutative differential geometry*, Inst. Hautes Études Sci. Publ. Math. **62** (1985), 41–144. MR823176

[7] A. Connes, *Entire cyclic cohomology of Banach algebras and characters of θ-summable Fredholm modules*, K-Theory **1** (1988), no. 6, 519–548, DOI 10.1007/BF00533785. MR953915

[8] Yukio Doi, *Homological coalgebra*, J. Math. Soc. Japan **33** (1981), no. 1, 31–50, DOI 10.2969/jmsj/03310031. MR597479

[9] V. A. Dolgushev, D. E. Tamarkin, and B. L. Tsygan, *Noncommutative calculus and the Gauss-Manin connection*, Higher structures in geometry and physics, Progr. Math., vol. 287, Birkhäuser/Springer, New York, 2011, pp. 139–158, DOI 10.1007/978-0-8176-4735-3_7. MR2762543

[10] I. M. Gel′fand, Yu. L. Daletskiĭ, and B. L. Tsygan, *On a variant of noncommutative differential geometry* (Russian), Dokl. Akad. Nauk SSSR **308** (1989), no. 6, 1293–1297; English transl., Soviet Math. Dokl. **40** (1990), no. 2, 422–426. MR1039918

[11] B. L. Feĭgin and B. L. Tsygan, *Cyclic homology of algebras with quadratic relations, universal enveloping algebras and group algebras*, K-theory, arithmetic and geometry (Moscow, 1984), Lecture Notes in Math., vol. 1289, Springer, Berlin, 1987, pp. 210–239, DOI 10.1007/BFb0078369. MR923137

[12] Murray Gerstenhaber and Alexander A. Voronov, *Homotopy G-algebras and moduli space operad*, Internat. Math. Res. Notices **3** (1995), 141–153, DOI 10.1155/S1073792895000110. MR1321701

[13] Ezra Getzler, *Cartan homotopy formulas and the Gauss-Manin connection in cyclic homology*, Quantum deformations of algebras and their representations (Ramat-Gan, 1991/1992; Rehovot, 1991/1992), Israel Math. Conf. Proc., vol. 7, Bar-Ilan Univ., Ramat Gan, 1993, pp. 65–78. MR1261901

[14] E. Getzler and J. D. S. Jones, *Operads, homotopy algebra, and iterated integrals for double loop spaces*, `arXiv:hep-th/9305013`, 1–70 (1993).

[15] A. Gerasimov, *Localization in GWZW and Verlinde formula*, `arXiv:hep-th/9305090`.

[16] Victor Ginzburg and Travis Schedler, *Free products, cyclic homology, and the Gauss-Manin connection*, Adv. Math. **231** (2012), no. 3-4, 2352–2389, DOI 10.1016/j.aim.2012.06.016. MR2964640

[17] A. Grothendieck, *Crystals and the de Rham cohomology of schemes*, Dix exposés sur la cohomologie des schémas, Adv. Stud. Pure Math., vol. 3, North-Holland, Amsterdam, 1968, pp. 306–358. Notes by I. Coates and O. Jussila. MR269663

[18] Kathryn Hess, Paul-Eugène Parent, and Jonathan Scott, *CoHochschild homology of chain coalgebras*, J. Pure Appl. Algebra **213** (2009), no. 4, 536–556, DOI 10.1016/j.jpaa.2008.08.001. MR2483836

[19] G. Hochschild, Bertram Kostant, and Alex Rosenberg, *Differential forms on regular affine algebras*, Trans. Amer. Math. Soc. **102** (1962), 383–408, DOI 10.2307/1993614. MR142598

[20] Nicholas M. Katz and Tadao Oda, *On the differentiation of de Rham cohomology classes with respect to parameters*, J. Math. Kyoto Univ. **8** (1968), 199–213, DOI 10.1215/kjm/1250524135. MR237510

[21] Masoud Khalkhali, *An approach to operations on cyclic homology*, J. Pure Appl. Algebra **107** (1996), no. 1, 47–59, DOI 10.1016/0022-4049(95)00026-7. MR1377655

[22] Masoud Khalkhali, *On Cartan homotopy formulas in cyclic homology*, Manuscripta Math. **94** (1997), no. 1, 111–132, DOI 10.1007/BF02677842. MR1468938

[23] Jean-Louis Loday, *Cyclic homology*, Grundlehren der mathematischen Wissenschaften [Fundamental Principles of Mathematical Sciences], vol. 301, Springer-Verlag, Berlin, 1992. Appendix E by María O. Ronco, DOI 10.1007/978-3-662-21739-9. MR1217970

[24] F. Malikov, V. Schechtman, and B. Tsygan, *Chiral Cartan Calculus*, `arXiv:2312.01834`.

[25] Ju. I. Manin, *Algebraic curves over fields with differentiation* (Russian), Izv. Akad. Nauk SSSR Ser. Mat. **22** (1958), 737–756. MR103889

[26] Ralf Meyer, *Local and analytic cyclic homology*, EMS Tracts in Mathematics, vol. 3, European Mathematical Society (EMS), Zürich, 2007, DOI 10.4171/039. MR2337277

[27] R. Nest and B. Tsygan, *Cyclic Homology*, book in progress, available at: `https://sites.math.northwestern.edu/~tsygan/Part1.pdf`

[28] Ryszard Nest and Boris Tsygan, *Noncommutative calculus*, Encyclopedia of mathematical physics. Vol. 3. Integrable systems & algebra, Academic Press, Amsterdam, [2025] ©2025, pp. 704–715. MR4975791

[29] George S. Rinehart, *Differential forms on general commutative algebras*, Trans. Amer. Math. Soc. **108** (1963), 195–222, DOI 10.2307/1993603. MR154906

[30] Alexander Polishchuk and Leonid Positselski, *Hochschild (co)homology of the second kind I*, Trans. Amer. Math. Soc. **364** (2012), no. 10, 5311–5368, DOI 10.1090/S0002-9947-2012-05667-4. MR2931331

[31] Daniel Quillen, *Algebra cochains and cyclic cohomology*, Inst. Hautes Études Sci. Publ. Math. **68** (1988), 139–174 (1989). MR1001452

[32] Boris Tsygan, *On the Gauss-Manin connection in cyclic homology*, Methods Funct. Anal. Topology **13** (2007), no. 1, 83–94. MR2308582

[33] B. Tsygan, *Noncommutative crystalline cohomology*, Proceedings of Symposia in Pure and Applied Mathematics vol. **105** (2023), 491–522, available at `https://sites.math.northwestern.edu/~tsygan/Oberwolfachtext.pdf`

[34] Boris Tsygan, *On noncommutative differential forms*, Higher structures in topology, geometry, and physics, Contemp. Math., vol. 802, Amer. Math. Soc., [Providence], RI, [2024] ©2024, pp. 1–22, DOI 10.1090/conm/802/16070. MR4773873

Department of Mathematics, Northwestern University, 2033 Sheridan Road, Evanston, Illinois 60208

Email address: `b-tsygan@northwestern.edu`

Contemporary Mathematics
Volume **841**, 2026
https://doi.org/10.1090/conm/841/16839

On simplicial principal bundles in descent categories

Jesse Wolfson

For Ezra Getzler, on the occasion of his sixtieth birthday

ABSTRACT. Simplicial objects sC in descent categories C, as introduced by Behrend and Getzler, provide a context in which to study higher stacks. In this note, we extend the construction of the canonical cocycle of a smooth principal G-bundle to the context of principal G-bundles in $\mathsf{sC}_{/X}$. As an application, we show how this specializes to $\mathsf{C} = \mathbf{Sets}$ to give a streamlined construction of k-invariants of reduced Kan complexes. We adapt this to give a similarly streamlined construction of minimal Kan complexes, with the goal of clarifying the role of the axiom of choice; Postnikov towers of minimal Kan complexes provide examples of towers of simplicial principal bundles of the type we consider.

1. Introduction

Let C be a subcanonical descent category in the sense of [**BG17**], or a "category with covers" in the sense of [**Wol16**]. Let sC denote the category of simplicial objects in C. Following [**Hen08, Pri13, Wol16, BG17, RZ20**], we can view sC as a setting in which to study "higher geometry", i.e. higher Lie groupoids or geometric higher stacks in the geometry modeled by C. Geometric groups G (Lie groups, algebraic groups, group schemes, etc.) and principal G-bundles form a basic tool in classical geometries, and (discrete) simplicial groups and their bundles have served a similar role in classical homotopy theory since their introduction in [**BGM59**]. The goal of this short note is to flesh out a "convex hull" of these approaches. Concretely, given a principal bundle

$$P \longrightarrow X$$

for a Lie group G over a smooth manifold X, the *nerve* NP of the bundle carries a canonical trivializing cocycle

$$P \times_X P \cong P \times G.$$

Via the implicit function theorem, it is enough to require that a principal bundle $P \longrightarrow X$ in smooth manifolds be a surjective submersion in order to conclude that it is locally trivializeable in the open ball topology. Our goal is to extend this package to sC. Actually, we fix $X \in \mathsf{sC}$ and work in the category $\mathsf{sC}_{/X}$ of maps $Y \longrightarrow X$.

2020 *Mathematics Subject Classification.* Primary 55U10.

The author was supported in part by NSF Grant DMS-1944862.

©2026 by the author

We refer to objects in this category as "X-manifolds", group objects $G \longrightarrow X$ as "X-groups", etc. In this setting, we take *covering left fibrations*, i.e. left fibrations

$$f\colon Z \longrightarrow Y$$

with $f_0\colon Z_0 \longrightarrow Y_0$ a cover, as the analogue of surjective submersions.[1] We now consider *Lie X-groups*, i.e. X-groups with $G \longrightarrow X$ a level-wise cover.

EXAMPLE 1.1. Let G be a Lie group acting by automorphisms on an abelian Lie group A. Let G act on itself by left translations and let EG be the nerve of the associated action groupoid. Let $K(A,n)$ denote the simplicial abelian group whose normalized chain complex is A in degree n and 0 elsewhere. Let $K(A,n)//G := (K(A,n) \times EG)/G$ and consider the canonical map $K(A,n)//G \longrightarrow EG/G = NG$. Then $K(A,n)//G \longrightarrow NG$ is a Lie NG-group in the sense above.

DEFINITION 1.2. Let $Y \longrightarrow X$ be an X-manifold, and $G \longrightarrow X$ a Lie X-group. A *principal G-bundle* is a covering left fibration $P \longrightarrow Y$ in $\mathsf{sC}_{/X}$ equipped with a G-action $\rho\colon G \times_X P \longrightarrow P$ over Y which gives an isomorphism of coequalizer diagrams

$$\begin{array}{ccccc}
G \times_X P & \overset{\rho}{\underset{\pi_P}{\rightrightarrows}} & P & \longrightarrow & Y \\
{\scriptstyle \rho\times\pi_P}\big\downarrow{\scriptstyle \cong} & & \big\| & & \big\| \\
P \times_Y P & \overset{\pi_1}{\underset{\pi_2}{\rightrightarrows}} & P & \longrightarrow & Y
\end{array}$$

To define the nerve, we view both Lie X-groups $G \longrightarrow X$ and the groupoid objects $Z \times_Y Z \rightrightarrows Y$, for $Z \longrightarrow Y$ a covering left fibration, as "simplicial Lie groupoids"–i.e. groupoid objects in sC for which the source and target maps $d_1^h, d_0^h\colon \mathcal{G}_1 \longrightarrow \mathcal{G}_0$ are covers–and we denote these simplicial Lie groupoids by G/X and Z/Y. Note that we do not require $\mathcal{G}_0$ to be a constant simplicial diagram, as is typical in the literature on simplicially enriched categories/groupoids. Following [**BGM59**], we introduce functorial nerve $\overline{W}$ and and universal bundle W constructions for simplicial Lie groupoids $\mathcal{G}$

$$W\mathcal{G} \longrightarrow \overline{W}\mathcal{G}.$$

For $p\colon Z \longrightarrow Y$, the nerve comes with a natural augmentation

$$\overline{W}(p)\colon \overline{W}(Z/Y) \longrightarrow Y$$

We can now state our main results.

PROPOSITION 1.3. *Let $f\colon Z \longrightarrow Y$ be a covering left fibration. The map*

$$\overline{W}(f)\colon \overline{W}(Z/Y) \longrightarrow Y$$

is a hypercover.

THEOREM 1.4. *Let $G \longrightarrow X$ be a Lie X-group. Let $P \longrightarrow Y$ be a principal G-bundle in $\mathsf{sC}_{/X}$. Then the nerve $\overline{W}(P/Y)$ carries a canonical trivializing cocycle for P, i.e. there exists:*

(1) *a natural map $\overline{W}(P/Y) \longrightarrow \overline{W}(G/X)$, and*

[1] The ordinal category Δ comes equipped with an involution $t\colon \Delta \longrightarrow \Delta$ given by "reversing the orientation of simplices". Precomposing with this involution takes left fibrations into right fibrations and vice versa. For ease of exposition, we focus here on left fibrations. We leave it as an exercise for the motivated reader to carry out the translation of this paper into right fibrations.

(2) *a natural isomorphism*

$$\overline{W}(P/Y) \times_Y P \xrightarrow{\cong} \overline{W}(P/Y) \times_{\overline{W}(G/X)} W(G/X)$$

over $\overline{W}(P/Y)$

where "natural" means, natural in $P \longrightarrow Y$ *and* $G \longrightarrow X$ *in the usual sense.*

REMARK 1.5. At the cost of introducing more formalism, we could equivalently restate Theorem 1.4 as the existence of a functor from principal G-bundles to a cocycle category in the sense of Jardine [**Jar09**].

In a final section, we specialize to the classical setting of Kan complexes, i.e. we take $\mathsf{C} = \mathsf{Sets}$ with surjections as covers, and consider the Moore-Duskin construction of the Postnikov tower of a reduced Kan complex X as in [**Hen08**]. In this context, Proposition 1.3 and Theorem 1.4 give a streamlined construction of k-invariants, which we believe should be adaptable to geometric contexts. In a similar spirit, we give a streamlined construction of minimal models of reduced Kan complexes (Theorem 5.1) aimed at making more transparent the dependency of this construction on the axiom of choice. Reduced minimal Kan complexes provide natural examples of objects built from towers of principal bundles of the type we consider in Theorem 1.4.

2. Preliminaries

Except where otherwise noted, we follow the conventions of [**Wol16**]. Recall from [**BG17**] that a *subcanonical descent category* is a category C with a subcategory of "covers" such that

(1) C is closed under finite limits,
(2) pullbacks of covers are covers,
(3) if f and gf are covers, then so is g,
(4) covers are effective epimorphisms, i.e. if $f\colon X \longrightarrow Y$ is a cover, then $X \times_Y X \rightrightarrows X \xrightarrow{f} Y$ is a coequalizer in C.

Examples of subcanonical descent categories include:

- Sets with surjections as covers,
- $\mathsf{An}_{\mathbb{C}}$, the category of complex analytic varieties, with surjective submersions as covers,
- $\mathsf{Schemes}_S$, the category of schemes over a base S, with Zariski open covers (or surjective étale, fppf, or fpqc morphisms) as covers.

Subcanonical descent categories were introduced as a framework in which to study "smooth" maps (the covers) between possibly singular objects. From this perspective, the "smooth" objects are the objects $X \in \mathsf{C}$ such that the map $X \longrightarrow *$ (from X to the terminal object $*$) is a cover.

If we wish to restrict our attention purely to "smooth" objects, we arrive at the framework of [**Wol16**], which we informally refer to here as a "category with covers." Concretely, this consists of a category C with a sub-category of "covers" subject to the following axioms:

(1) C has a terminal object $*$, and the map $X \longrightarrow *$ is a cover for all $X \in \mathsf{C}$,
(2) pullbacks of covers along arbitrary maps exist and are covers,
(3) if f and gf are covers, then so is g,
(4) covers are effective epimorphisms.

Examples include:

- For any subcanonical descent category C, the full subcategory of objects $X \in \mathsf{C}$ such that $X \longrightarrow *$ is a cover, along with its subcategory of covers,
- Mlfds, the category of smooth Banach manifolds, with surjective submersions as covers,
- Smooth_k, the category of smooth schemes over a field k, with surjective smooth morphisms as covers.

As principal bundles and torsors are important tools for studying both smooth and singular objects, our goal in the present note is to treat both contexts. As a matter of exposition, we will take as a standing assumption that C is a category with covers–the geometric constructions one can make are more restricted in this setting–and we will comment as necessary on any adjustments for the case of a subcanonical descent category.

As in [**Wol16**], the axioms for covers guarantee that they define a Grothendieck topology, and our assumption that covers are effective epimorphisms guarantees that the Yoneda embedding takes values in sheaves. We will freely identify an object in C with its functor of points, i.e. with the sheaf it represents, and we will talk about sheaves and (generalized) elements without further comment. We write sC for the category of simplicial objects in C, i.e. contravariant functors

$$X\colon \Delta^{op} \longrightarrow \mathsf{C}$$

from the category of finite non-empty ordinals Δ into C and natural transformations between them, i.e. simplicial maps. We will identify objects in C as constant simplicial objects without further comment (i.e. we treat C as a full subcategory of sC). For $X, Y \in \mathsf{sC}$, we write $\hom_{\mathsf{sC}}(X, Y)$ to denote the hom-sheaf, i.e. the sheaf on C whose value at $U \in \mathsf{C}$ is the set $\hom_{\mathsf{sC}}(U \times X, Y)$ of simplicial maps $U \times X \longrightarrow Y$.[2] We view bisimplicial objects as "horizontal" simplicial diagrams in sC (where the internal simplicial direction in sC is "vertical"). We write $[n]$ for the ordered set $\{0 < \ldots < n\}$, Δ^n for the standard n-simplex, $\Lambda^n_i \subset \Delta^n$ for its "i^{th} horn" and $\partial\Delta^n \subset \Delta^n$ for its boundary. For $X \in \mathsf{sC}$, we write[3]

$$\begin{aligned} \Lambda^n_i X &:= \hom_{\mathsf{sC}}(\Lambda^n_i, X) \\ M_n X &:= \hom_{\mathsf{sC}}(\partial\Delta^n, X). \end{aligned}$$

For a map $f\colon Z \longrightarrow Y$, we write $\Lambda^n_i(f)$ for the sheaf of commuting squares[4]

$$\begin{array}{ccc} \Lambda^n_i & \longrightarrow & Z \\ \downarrow & & \downarrow \\ \Delta^n & \longrightarrow & Y \end{array}$$

and $\lambda^n_i(f)\colon Z_n \longrightarrow \Lambda^n_i(f)$ for the map induced by the inclusion $\Lambda^n_i \subset \Delta^n$.

[2]Note that we will not consider simplicial hom-sheaves, i.e. $\hom_{\mathsf{sC}}(X, Y)$ is a sheaf on C, not a simplicial sheaf.

[3]Note that the right hand side involves a mild abuse of notation if coproducts of the terminal object $*$ do not define a fully faithful embedding $\mathsf{Sets} \longrightarrow \mathsf{C}$. In this case, we view simplicial sets as simplicial objects in sheaves on C and use $\hom_{\mathsf{sC}}$ to denote $\hom_{\mathsf{sSh(C)}}$.

[4]N.b. equivalently, this can be described as the pullback sheaf $\Lambda^n_i Z \times_{\Lambda^n_i Y} Y_n$.

DEFINITION 2.1. A map $f\colon Z \longrightarrow Y$ in sC is a *left fibration* if for all n and $0 \leq i < n$, the map

$$\lambda_i^n(f)\colon Z_n \longrightarrow \Lambda_i^n(f)$$

is a cover. We say it is a *covering left fibration* if, in addition, the map $f_0\colon Z_0 \longrightarrow Y_0$ is a cover.

REMARK 2.2. Recall that a Kan fibration is a left fibration $f\colon Z \longrightarrow Y$ such that the map

$$\lambda_n^n(f)\colon Z_n \longrightarrow \Lambda_n^n(f)$$

is a cover for all n. It is a *covering Kan fibration* if $f_0\colon Z_0 \longrightarrow Y_0$ is a cover.

Following [**Wol16, BG17**], we develop some useful properties of left fibrations in C.

DEFINITION 2.3. Let $m \geq 0$. A *left m-expansion* $S \hookrightarrow T$ be a map of simplicial sets for which there exists a filtration

$$S = F_{-1}T \subset F_0T \subset \cdots \subset F_rT = T$$

such that there exists a sequence $n_\ell \geq m$, $0 \leq \ell \leq r$, and a sequence $0 \leq i_\ell < n_\ell$ and pushout diagrams

$$\begin{array}{ccc} \Lambda_{i_\ell}^{n_\ell} & \longrightarrow & F_{\ell-1}T \\ \downarrow & & \downarrow \\ \Delta^{n_\ell} & \longrightarrow & F_\ell T \end{array}$$

By a *left expansion* we mean a left 0-expansion.

REMARK 2.4. Note that [**BG17**] imposes the additional requirement that the simplices in the expansion should monotonically increasing in dimension. While this does not change the theory, it leads, at times, to more combinatorially involved proofs (e.g. for Lemma 2.5 below). We follow Whitehead [**Whi39**] and omit this requirement here.

Immediately from Definition 2.3, we have the following.

LEMMA 2.5. *Let $R \hookrightarrow S$ be a left m_1-extension and let $S \hookrightarrow T$ be a left m_2-extension. Let $m = \min\{m_1, m_2\}$. Then $R \hookrightarrow T$ is a left m-extension.*

We now state a mild generalization of [**BG17**, Lemma 3.8].

LEMMA 2.6. *If S is a union of $m \leq n$ faces of the n-simplex Δ^n, and $\partial_n \Delta^n \subset S$, then $S \hookrightarrow \Delta^n$ is a left expansion.*

PROOF. The proof is by induction on n. For $n = 1$ the statement is clear. For the induction step, enumerate the faces of Δ^n not in S as $\{i_0, \ldots, i_{n-m}\}$ where $0 \leq i_{n-m} < \cdots < i_0 < n$. Let

$$F_\ell \Delta^n := S \cup \bigcup_{j \leq \ell} \partial_{i_j} \Delta^n.$$

We claim that $F_{\ell-1}\Delta^n \cap \partial_{i_\ell}\Delta^n \hookrightarrow \partial_{i_\ell}\Delta^n$ is a left expansion for all $\ell < n - m$. To see this, observe that, by the simplicial identities, for any $0 \leq j < k \leq n$,

$$\partial_{k-1}\partial_j \Delta^n = \partial_j \Delta^n \cap \partial_k \Delta^n = \partial_j \partial_k \Delta^n.$$

By our assumption that $i_\ell < n$ for all ℓ, $F_{\ell-1}\Delta^n \cap \partial_{i_\ell}\Delta^n$ is a union of at most $n-1$ faces of the $(n-1)$-simplex $\partial_{i_\ell}\Delta^n$ which contains the $(n-1)^{st}$ face. By the inductive hypothesis, it's a left expansion. We conclude that the inclusion

$$S \hookrightarrow F_{n-m-1}\Delta^n = \Lambda^n_{\iota_{n-m}}$$

is a left expansion. Defining $F_{n-m}\Delta^n = \Delta^n$, we see that $F_{n-m-1}\Delta^n \hookrightarrow F_{n-m}\Delta^n$ is the canonical left expansion

$$\Lambda^n_{\iota_{n-m}} \hookrightarrow \Delta^n.$$

We now conclude that the map $S \hookrightarrow \Delta^n$ factors as a composition of left expansions, and thus by Lemma 2.5, $S \hookrightarrow \Delta^n$ is a left expansion. $\square$

LEMMA 2.7. *Let $S \hookrightarrow T$ be a left expansion. Let $f\colon Z \longrightarrow Y$ be a left fibration in* sC *such that the limit*

$$\hom(S, Z) \times_{\hom(S,Y)} \hom(T, Y)$$

exists in C. *If $S \neq \emptyset$, then the limit* $\hom(T, Z)$ *exists and the map*

$$\hom(T, Z) \longrightarrow \hom(S, Z) \times_{\hom(S,Y)} \hom(T, Y)$$

is a cover. If $S = \emptyset$, then the same conclusion holds provided that f is a covering left fibration.

PROOF. *Mutatis mutandis*, the proof is identical to the proof of [**Wol16**, Lemma 2.16].[5] $\square$

LEMMA 2.8. *Let $f\colon Z \longrightarrow Y$ be a covering left fibration. Then for all n, the map $f_n\colon Z_n \longrightarrow Y_n$ is a cover.*

PROOF. We induct on n. For $n = 0$, this follows by assumption. For the inductive step, for f a covering left fibration, the map $f_n\colon Z_n \longrightarrow Y_n$ factors as

$$Z_n \longrightarrow Z_{n-1} \times^{f,d_n}_{Y_{n-1}} Y_n \longrightarrow Y_n;$$

The second map is a pullback of the map $f_{n-1}\colon Z_{n-1} \longrightarrow Y_{n-1}$, so the pullback $Z_{n-1} \times_{Y_{n-1}} Y_n$ exists in C and the map second map is a cover by the inductive hypothesis. We can now apply Lemma 2.7 to conclude that the first map is a cover. $\square$

For a map $f\colon Z \longrightarrow Y$, we write $M_n(f)$ for the nth *relative matching object*, i.e. the sheaf of commuting squares[6]

$$\begin{array}{ccc} \partial\Delta^n & \longrightarrow & Z \\ \downarrow & & \downarrow \\ \Delta^n & \longrightarrow & Y \end{array}$$

and we write $\mu_n(f)\colon Z_n \longrightarrow M_n(f)$ for the map induced by the inclusion $\partial\Delta^n \subset \Delta^n$.

DEFINITION 2.9. A map $f\colon Z \longrightarrow Y$ in sC is a *hypercover* if for all n, the map

$$\mu_n(f)\colon Z_n \longrightarrow M_n(f)$$

is a cover.

---

[5]Note that in *loc. cit.*, the edge case $S = \emptyset$ was mistakenly omitted from the proof. Treating this edge case requires the hypothesis that f_0 is a cover in order to start the induction.

[6]N.b. equivalently, this can be described as the pullback sheaf $M_n Z \times_{M_n Y} Y_n$.

Hypercovers play the role of trivial fibrations in the internal homotopy theory of sC (cf. [**BG17**, **RZ20**]). In particular, they are covering Kan fibrations [**Wol16**, Theorem 2.17].

We refer the reader to [**Hen08**, **Wol16**, **BG17**, **RZ20**] for additional discussions of Kan fibrations, hypercovers, skeleta/coskeleta, representability issues, etc.

3. Nerves of simplicial Lie groupoids

For a simplicial group G, Barratt, Gugenheim and Moore [**BGM59**] introduced a functorial nerve and universal bundle

$$WG \longrightarrow \overline{W}G$$

with good homotopical properties in both discrete and geometric settings (cf. [**Wol16**, Section 6]).

Concretely:

$$\begin{aligned} WG_n &= G_n \times \cdots \times G_0 \\ \overline{W}G_n &= G_{n-1} \times \cdots \times G_0 \end{aligned}$$

and the face and degeneracy maps are given by multiplying an entry to the face of an adjacent one and inserting degeneracies.[7] From a different perspective, we have natural isomorphisms of functors

$$\begin{aligned} \overline{W} &\cong \mathrm{Tot} \circ N \\ W &\cong \mathrm{Tot} \circ \mathrm{Dec}_1 \circ N \end{aligned}$$

where N denotes the "level-wise" nerve, Tot denotes the Artin-Mazur totalization or "co-diagonal" (see [**AM66**, **CR05**]), and Dec_1 denotes the "level-wise" initial décalage [**Ill72**] corresponding to the endo-functor on the ordinal category Δ

$$[n] \mapsto [0] \sqcup [n].$$

Our goal here is to generalize these to *simplicial Lie groupoids.*

DEFINITION 3.1. A *simplicial Lie groupoid* in C is a groupoid object $\mathcal{G}$ in sC for which both the source and target maps from the "morphisms" $\mathcal{G}_1$ to the "objects" $\mathcal{G}_0$

$$d_1^h, d_0^h \colon \mathcal{G}_1 \longrightarrow \mathcal{G}_0$$

are level-wise covers.

Note that this is more general than the usual "simplicially enriched groupoids", as we do not require $\mathcal{G}_0$ to be simplicially constant.

Our two primary cases of interest are:

(1) Lie X-groups $G \longrightarrow X$, where the objects are just X, and the morphisms are G.[8]
(2) groupoids coming from level-wise covers $Z \longrightarrow Y$, where the objects are just Z, and the morphisms are $Z \times_Y Z \longrightarrow Z$.

---

[7] Cf. [**May67**, Chapter IV.21], [**GJ09**, Chapter V.4] or [**Wol16**, Section 6]. Note that our conventions here follow [**May67**, **GJ09**], while the conventions in [**Wol16**] are equivalent, but distinct. To pass from one to the other, pre-compose Tot by the transpose on bisimplicial objects (i.e. "reflect across the diagonal"), and then interchange Dec_1 with Dec^1 in the definition of W.

[8] In a descent category, i.e. if C is closed under finite limits, then we can drop the assumption that $G \longrightarrow X$ is a cover in the definition of the nerve below, as this is just there to ensure representability.

We recall our convention that all objects in sC are viewed as "vertical" simplicial diagrams.

DEFINITION 3.2. Let $\mathcal{G}$ be a simplicial Lie groupoid. Define its *nerve* $\overline{W}\mathcal{G}$ by

$$\overline{W}\mathcal{G} := \operatorname{Tot} N\mathcal{G},$$

where we apply N level-wise to the simplicial Lie groupoid $\mathcal{G}$.

Similarly, define the *universal bundle* $W\mathcal{G}$ by

$$W\mathcal{G} := \operatorname{Tot} \circ \operatorname{Dec}_1 \circ N\mathcal{G},$$

with the projection

$$W\mathcal{G} \longrightarrow \overline{W}\mathcal{G}$$

given by applying Tot to the component of the natural transformation $\operatorname{Dec}_1 \longrightarrow Id$.

The definition above has convenient functorial properties, which we will use later on. However, for computations, a more explicit presentation is often useful.

LEMMA 3.3. *Let $\mathcal{G}$ be a simplicial Lie groupoid. There exists a natural isomorphism*

$$\overline{W}\mathcal{G}_n \cong \mathcal{G}_{0,n} \times^{d_0^v, d_1^h}_{\mathcal{G}_{0,n-1}} \mathcal{G}_{1,n-1} \times^{d_0^h d_0^v, d_1^h}_{\mathcal{G}_{0,n-2}} \cdots \times^{d_0^h d_0^v, d_1^h}_{\mathcal{G}_{0,0}} \mathcal{G}_{1,0},$$

and under this natural isomorphism, the degeneracy and face maps are given by

$$\begin{aligned}
s_0(g_{0,n}, g_{1,n-1}, \dots, g_{1,0}) &= (s_0^v g_{0,n}, s_0^h g_{0,n}, g_{1,n-1}, \dots, g_{1,0}),\\
s_i(g_{0,n}, g_{1,n-1}, \dots, g_{1,0}) &= (s_i^v g_{0,n},\\
&\qquad s_{i-1}^v g_{1,n-1}, \dots, s_0^v g_{1,n-i}, s_0^h d_0^h g_{1,n-i}, g_{1,n-i-1}, \dots, g_{1,0}),\\
d_0(g_{0,n}, g_{1,n-1}, \dots, g_{1,0}) &= (d_0^h g_{1,n-1}, g_{1,n-2}, \dots, g_{1,0}),\\
d_n(g_{0,n}, g_{1,n-1}, \dots, g_{1,0}) &= (d_n^v g_{0,n}, d_{n-1}^v g_{1,n-1}, \dots, d_1^v g_{1,1}),
\end{aligned}$$

and, for $0 < i < n$,

$$d_i(g_{0,n}, g_{1,n-1}, \dots, g_{1,0}) = (d_i^v g_{0,n}, \dots, d_1^v g_{1,n-i+1}, d_0^v g_{1,n-i} \cdot g_{1,n-i-1}, \dots, g_{1,0}).$$

LEMMA 3.4. *Let $\mathcal{G}$ be a simplicial Lie groupoid. There exists a natural isomorphism*

$$W\mathcal{G}_n \cong \mathcal{G}_{1,n} \times^{d_0^h d_0^v, d_1^h}_{\mathcal{G}_{0,n-1}} \mathcal{G}_{1,n-1} \times^{d_0^h d_0^v, d_1^h}_{\mathcal{G}_{0,n-2}} \cdots \times^{d_0^h d_0^v, d_1^h}_{\mathcal{G}_{0,0}} \mathcal{G}_{1,0},$$

and under this natural isomorphism, the degeneracy and face maps are given by

$$\begin{aligned}
s_i(g_{1,n}, g_{1,n-1}, \dots, g_{1,0}) &= (s_i^v g_{1,n}, \dots, s_0^v g_{1,n-i}, s_0^h d_0^h g_{1,n-i}, g_{1,n-i-1}, \dots, g_{1,0})\\
d_i(g_{1,n}, g_{1,n-1}, \dots, g_{1,0}) &= (d_i^v g_{1,n}, \dots, d_1^v g_{1,n-i+1}, d_0^v g_{1,n-i} \cdot g_{1,n-i-1}, \dots, g_{1,0})\\
d_n(g_{1,n}, g_{1,n-1}, \dots, g_{1,0}) &= (d_n^v g_{1,n}, d_{n-1}^v g_{1,n-1}, \dots, d_1^v g_{1,1}).
\end{aligned}$$

Under this isomorphism, the map $W\mathcal{G} \longrightarrow \overline{W}\mathcal{G}$ is given on n-simplices by

$$(g_{1,n}, \dots, g_{1,0}) \mapsto (d_0^h g_{1,n}, g_{1,n-1}, \dots, g_{1,0}).$$

REMARK 3.5. We see from the lemma that the representability of $W\mathcal{G}$ and $\overline{W}\mathcal{G}$ follows directly from our assumption that source and target maps $d_0^h, d_1^h \colon \mathcal{G}_1 \longrightarrow \mathcal{G}_0$ are level-wise covers.[9]

[9] In a descent category, the same definition will also work for arbitrary groupoid objects in sC.

EXAMPLE 3.6. Let $f\colon G \longrightarrow X$ be a Lie X-group, viewed as a simplicial Lie groupoid. Let $\sigma\colon X \longrightarrow G$ denote the identity section. We denote its nerve by $\overline{W}(G/X)$. Then

$$\overline{W}(G/X)_n \cong X_n \times^{d_0,f}_{X_{n-1}} G_{n-1} \times^{fd_0,f}_{X_{n-2}} \cdots \times^{fd_0,f}_{X_0} G_0$$

and the face and degeneracy maps are given by

$$\begin{aligned} d_0(x, g_{n-1}, \ldots, g_0) &= (f(g_{n-1}), g_{n-2}, \ldots, g_0) \\ d_i(x, g_{n-1}, \ldots, g_0) &= (d_i x, d_{i-1} g_{n-1}, \ldots, d_1 g_{n-i+1}, d_0 g_{n-i} \cdot g_{n-i-1}, g_{n-i-2}, \ldots, g_0) \\ d_n(x, g_{n-1}, \ldots, g_0) &= (d_n x, d_{n-1} g_{n-1}, \ldots, d_1 g_1) \\ s_0(x, g_{n-1}, \ldots, g_0) &= (s_0 x, \sigma(x), g_{n-1}, \ldots, g_0) \\ s_i(x, g_{n-1}, \ldots, g_0) &= (s_i x, s_{i-1} g_{n-1}, \ldots, s_0 g_{n-i}, \sigma f(g_{n-i}), g_{n-i-1}, \ldots, g_0). \end{aligned}$$

Taking $X = *$, we recover the usual formulas for $\overline{W}$ of a simplicial group.

EXAMPLE 3.7. Let $f\colon Z \longrightarrow Y$ be a covering left fibration. Then $Z \times_Y Z \rightrightarrows Z$ gives a simplicial Lie groupoid by Lemma 2.8, and we denote its nerve by $\overline{W}(Z/Y)$. Then

$$\overline{W}(Z/Y)_n \cong Z_n \times^{fd_0,f}_{Y_{n-1}} Z_{n-1} \times^{fd_0,f}_{Y_{n-2}} \cdots \times^{fd_0,f}_{Y_0} Z_0$$

and face and degeneracy maps given by

$$\begin{aligned} d_0(z_n, \ldots, z_0) &= (z_{n-1}, \ldots, z_0) \\ d_i(z_n, \ldots, z_0) &= (d_i z_n, \ldots, d_1 z_{n-i+1}, z_{n-i-1}, \ldots, z_0) \\ s_i(z_n, \ldots, z_0) &= (s_i z_n, \ldots, s_0 z_{n-i}, z_{n-i}, \ldots, z_0). \end{aligned}$$

Note that in this case, we have a natural isomorphism

$$\overline{W}(Z/Y) \cong \mathrm{Tot}(\mathrm{Cosk}_{0,*}(Z) \times_{\mathrm{Cosk}_{0,*} Y} Y)$$

where $\mathrm{Cosk}_{0,*}(-)$ is the "horizontal 0-coskeleton" (i.e., which applies the 0-coskeleton to all the horizontal simplicial objects in a bisimplicial object).

4. Proofs

PROOF OF PROPOSITION 1.3. We claim that, for all n, there exists a commuting square

$$\begin{array}{ccc} \bar{W}(Z/Y)_n & \xrightarrow{\mu_n(\overline{W}(f))} & M_n\bar{W}(f) \\ \downarrow{\cong} & & \downarrow{\cong} \\ Z_n \times^{fd_0,f}_{Y_{n-1}} \bar{W}(Z/Y)_{n-1} & \xrightarrow{\Lambda^n_0(f)\times 1} & \Lambda^n_0(f) \times^{f\partial d_0, f\partial}_{M_{n-1}Y} \bar{W}(Z/Y)_{n-1}. \end{array}$$

The proposition immediately follows from our assumption that f is a covering left fibration (n.b. for $n = 0$, the bottom horizontal arrow just becomes $Z_0 \longrightarrow Y_0$.)

The existence of the above square reduces to an exercise in combinatorics. To see this, note that we have a natural isomorphism

$$\bar{W}(Z/Y) \cong \mathrm{Tot}(\mathrm{Cosk}_{0,*}(Z) \times_{\mathrm{Cosk}_{0,*} Y} Y).$$

The functor Tot is right adjoint to Illusie's "total decalage" Dec [**Ill72, CR05**], while the functor $\mathrm{Cosk}_{0,*}$ is right adjoint to the "horizontal 0-skeleton". Similarly, the functor which takes a simplicial object and views it as a horizontally constant bisimplicial object is right adjoint to the colimit preserving functor given on bisimplices by

$$s \otimes 1\colon \Delta^p \otimes \Delta^q \longrightarrow \Delta^0 \otimes \Delta^q$$

where s is the total degeneracy map.

These natural isomorphisms combine to give a natural isomorphism between $\bar{W}(Z/Y)_n$ and the sheaf of commuting squares[10]

$$\begin{array}{ccc} \mathrm{Sk}_{0,*} \operatorname{Dec} \Delta^n & \longrightarrow & Z \\ \downarrow & & \downarrow \\ (s \otimes 1) \operatorname{Dec} \Delta^n & \longrightarrow & Y \end{array}$$

Similarly, the definition of the matching object gives a natural isomorphism between $M_n \bar{W}(f)$ and the sheaf of commuting squares

$$\begin{array}{ccc} \mathrm{Sk}_{0,*} \operatorname{Dec} \partial\Delta^n & \longrightarrow & Z \\ \downarrow & & \downarrow \\ (s \otimes 1) \operatorname{Dec} \Delta^n & \longrightarrow & Y \end{array}$$

and the map $\mu_n(\bar{W}(f))$ corresponds to pre-composition by the commuting square

$$\begin{array}{ccc} \mathrm{Sk}_{0,*} \operatorname{Dec} \partial\Delta^n & \longrightarrow & \mathrm{Sk}_{0,*} \operatorname{Dec} \Delta^n \\ \downarrow & & \downarrow \\ (s \otimes 1) \operatorname{Dec} \Delta^n & = & (s \otimes 1) \operatorname{Dec} \Delta^n \end{array}$$

An exercise in bisimplicial combinatorics gives an isomorphism of this commuting square with

$$\begin{array}{ccc} \Delta^0 \otimes \Lambda^n_0 \coprod \mathrm{Sk}_{0,*} \operatorname{Dec} \Delta^{n-1} & \longrightarrow & \Delta^0 \otimes \Delta^n \coprod \mathrm{Sk}_{0,*} \operatorname{Dec} \Delta^{n-1} \\ \downarrow & & \downarrow \\ (s \otimes 1) \operatorname{Dec} \Delta^n & = & (s \otimes 1) \operatorname{Dec} \Delta^n \end{array}$$

The original claim, and thus the proposition, immediately follows. □

Proof of Theorem 1.4. Let $f\colon G \longrightarrow X$ be a Lie X-group, $Y \longrightarrow X$ an X-manifold and $P \longrightarrow Y$ a principal G-bundle. Let $\pi\colon P \longrightarrow Y \longrightarrow X$ be the composite. By definition, we have

$$\overline{W}(P/Y) := \operatorname{Tot} N(P/Y).$$

By definition, for all n, $P_n \longrightarrow Y_n$ is a principal G_n-bundle in $\mathsf{C}_{/X}$, with the action compatible with the face and degeneracy maps. The usual canonical cocycle on $N(P_n/Y_n)$ gives a natural isomorphism

$$P_n \times_{X_n} \overbrace{G_n \times_{X_n} \cdots \times_{X_n} G_n}^{m} \xrightarrow{\cong} \overbrace{P_n \times_{Y_n} \cdots \times_{Y_n} P_n}^{m+1} = N(P_n/Y_n)_m =: N(P/Y)_{m,n}$$

[10]N.b. this sheaf is equivalently the pullback

$$\hom(\mathrm{Sk}_{0,*} \operatorname{Dec} \Delta^n, Z) \times_{\hom(\mathrm{Sk}_{0,*} \operatorname{Dec} \Delta^n, Y)} \hom(s \otimes 1 \operatorname{Dec} \Delta^n, Y).$$

given on elements by

$$(p, g_1, \ldots, g_m) \mapsto (p, g_1^{-1}p, \ldots, (g_1 \cdots g_m)^{-1}p).$$

Under these isomorphisms, the horizontal face and degeneracy maps of $N(P/Y)$ become

$$\begin{aligned}
d_0^h(p, g_1, \ldots, g_m) &= (g_1^{-1}p, g_2, \ldots, g_m)\\
d_i^h(p, g_1, \ldots, g_m) &= (p, g_1, \ldots, g_i \cdot g_{i+1}, \ldots, g_n)\\
s_0^h(p, g_1, \ldots, g_m) &= (p, \sigma f(g_1), g_1, \ldots, g_n)\\
s_i^h(p, g_1, \ldots, g_m) &= (p, g_1, \ldots, g_i, \sigma f(g_i), g_{i+1}, \ldots, g_n).
\end{aligned}$$

The naturality of these isomorphisms implies that vertical face and degeneracy maps are computed coordinate-wise. Applying Tot, a straightforward exercise in the formulas gives a natural isomorphism

$$\begin{aligned}
\overline{W}(P/Y)_n &\xrightarrow{\cong} P_n \times_{X_{n-1}}^{\pi d_0, f} G_{n-1} \times_{X_{n-2}}^{f d_0, f} \cdots \times_{X_0}^{f d_0, f} G_0,\\
(p_n, \ldots, p_0) &\mapsto (p_n, g_{n-1}, \ldots, g_0)
\end{aligned}$$

where, for each i, $g_i \in G_i$ is the unique element such that $g_i p_i = d_0 p_{i+1}$ (i.e. the canonical cocycle $(d_0 p_{i+1}, p_i)$ to (g_i, p_i)). Under this isomorphism, the face and degeneracy maps are given by

$$\begin{aligned}
d_0(p, g_{n-1}, \ldots, g_0) &= (g_{n-1}^{-1} \cdot d_0 p, g_{n-2}, \ldots, g_0)\\
d_i(p, g_{n-1}, \ldots, g_0) &= (d_i p, d_{i-1} g_{n-1}, \ldots, d_1 g_{n-i+1}, d_0 g_{n-i} \cdot g_{n-i-1}, g_{n-i-2}, \ldots, g_0)\\
s_0(p, g_{n-1}, \ldots, g_0) &= (s_0 p, \sigma\pi(p), g_{n-1}, \ldots, g_0)\\
s_i(p, g_{n-1}, \ldots, g_0) &= (s_i p, s_{i-1} g_{n-1}, \ldots, s_0 g_{n-i}, \sigma f(g_{n-i}), g_{n-i-1}, \ldots, g_0).
\end{aligned}$$

We conclude that the assignment

$$(p, g_{n-1}, \ldots, g_0) \mapsto (\pi(p), g_{n-1}, \ldots, g_0)$$

gives a natural map

$$\overline{W}(P/Y) \longrightarrow \overline{W}(G/X)$$

as desired.

We now consider $P \times_Y \overline{W}(P/Y)$. Applying the isomorphism above, we obtain a natural isomorphism

$$(P \times_Y \overline{W}(P/Y))_n \cong P_n \times_{Y_n} P_n \times_{X_{n-1}}^{\pi d_0, f} G_{n-1} \times_{X_{n-2}}^{f d_0, f} \cdots \times_{X_0}^{f d_0, f} G_0.$$

Post-composing this with the inverse of the canonical cocycle for the G_n-principal bundle $P_n \longrightarrow Y_n$

$$(g_n, p) \mapsto (g_n p, p),$$

we obtain a natural isomorphism

$$\begin{aligned}
(P \times_Y \overline{W}(P/Y))_n &\xrightarrow{\cong} P_n \times_{X_n} G_n \times_{X_{n-1}} \cdots \times_{X_0} G_0\\
(p, g_n^{-1}p, g_{n-1}, \ldots, g_0) &\mapsto (p, g_n, g_{n-1}, \ldots, g_0)
\end{aligned}$$

which by inspection gives a simplicial isomorphism over $\overline{W}(P/Y)$, as desired. □

5. Applications

We close by showing how Theorem 1.4 gives streamlined constructions of k-invariants and minimal models in the classical context of Kan simplicial sets. Let $\mathsf{C} = \mathsf{Sets}$ with surjections as covers. For each $n \geq 0$, Moore [**Moo58**] and Duskin [**Dus79**] introduced "truncation" functors

$$\tau_{<n}, \tau_{\leq n} \colon \mathsf{sSets} \longrightarrow \mathsf{sSets}$$

such that for, X a Kan complex, we have a natural tower of Kan complexes

$$X \longrightarrow \cdots \longrightarrow \tau_{\leq n}X \longrightarrow \tau_{<n}X \xrightarrow{\sim} \tau_{\leq n-1}X \longrightarrow \cdots \longrightarrow \tau_{<1}X \xrightarrow{\sim} \tau_{\leq 0}X = \pi_0 X,$$

with

$$\tau_{<n+1}X \xrightarrow{\sim} \tau_{\leq n}X$$

a trivial fibration, and

$$\tau_{\leq n}X \longrightarrow \tau_{<n}X$$

a minimal Kan fibration for all n, and with

$$X = \varprojlim_n \tau_{\leq n}X = \varprojlim_n \tau_{<n}X \simeq \operatorname{holim}_n \tau_{\leq n}X \simeq \operatorname{holim}_n \tau_{<n}X.$$

See [**May67**, Section II.8] and [**GJ09**, Chapter IV.3] for a discussion of $\tau_{<n}$ in simplicial sets. See [**Hen08, Wol16**] for a discussion of both functors in geometric contexts.

For X reduced,[11] by [**BGM59**, Proposition V.2.2],

$$\tau_{\leq n}X \longrightarrow \tau_{<n}X$$

is a minimal Kan fiber bundle, and in fact a twisted Cartesian product of the Eilenberg-MacLane space $K(\pi_n X, n) := \overline{W}^n(\pi_n X)$ with twisting function given by the action of $\pi_1 X$ (cf. [**BGM59**] or [**GJ09**, Chapter VI.5]). Checking the definitions, we see this is in fact a principal bundle for the $N\pi_1 X$-group

$$K(\pi_n X, n)//\pi_1 X := (W\pi_1 X \times K(\pi_n X, n))/\pi_1 X.$$

Applying Theorem 1.4, we obtain a commuting diagram in the category $\mathsf{sSets}_{/N\pi_1 X}$ for each $n \geq 1$

$$\begin{array}{ccc}
\tau_{\leq n}X \times_{\tau_{<n}X} \overline{W}(\tau_{\leq n}X/\tau_{<n}X) & \longrightarrow & W(K(\pi_n X, n)//\pi_1 X) \\
\downarrow & & \downarrow \\
\overline{W}(\tau_{\leq n}X/\tau_{<n}X) & \longrightarrow & \overline{W}(K(\pi_n X, n)//\pi_1 X) \\
\sim\downarrow & & \\
\tau_{<n}X & &
\end{array}$$

where the bottom vertical map is a trivial fibration, by Proposition 1.3. Using the canonical isomorphism

$$\overline{W}(K(\pi_n X, n)//\pi_1 X) \cong K(\pi_n X, n+1)//\pi_1 X$$

[11] We assume reduced for simplicity. The discussion goes through *mutatis mutandis* for X connected with a chosen basepoint.

the above diagram is naturally isomorphic to

$$\begin{array}{ccc} \tau_{\leq n}X \times_{\tau_{<n}X} \overline{W}(\tau_{\leq n}X/\tau_{<n}X) & \longrightarrow & WK(\pi_n X, n)//\pi_1 X \ , \\ \downarrow & & \downarrow \\ \overline{W}(\tau_{\leq n}X/\tau_{<n}X) & \longrightarrow & K(\pi_n X, n+1)//\pi_1 X \\ \sim\downarrow & & \\ \tau_{<n}X & & \end{array}$$

i.e. this is precisely an n-dimensional k-invariant for the Kan complex X in the sense of [**GJ09**, Proposition VI.5.1]. Comparing this to the construction in [**GJ09**, Chapter VI.5], we see that the above is essentially the construction of *loc. cit.*, but in a streamlined fashion which makes clear the categorical requirements. In particular, one only needs the representability of the functors $\tau_{\leq n}$ and $\tau_{<n}$, e.g. as discussed in [**Hen08, Wol16**], and the analogue of the results of [**BGM59**, Sections III, IV] to obtain a functorial theory of k-invariants for reduced Kan objects in geometric contexts sC.

To put this in context, we now show the following.

THEOREM 5.1. *Let Kan_*^{split} denote the category whose objects are tuples $(X, \{\sigma_n\}_{n\geq 0})$ where X is a reduced Kan complex, and for all $n \geq 0$, $\sigma_n \colon \tau_{\leq n}X \longrightarrow \tau_{<n+1}X$ is a section of the trivial fibration $\tau_{<n+1}X \xrightarrow{\sim} \tau_{\leq n}X$.*[12] *Then there exists a functor*

$$\mu \colon \mathsf{Kan}_*^{split} \longrightarrow \mathsf{MinKan}_*$$

taking values in reduced minimal *Kan complexes and a natural weak equivalence*

$$\alpha \colon \mu(X, \{\sigma_n\}) \xrightarrow{\sim} X$$

which is an inclusion of simplicial sets.

PROOF. Let $(X, \{\sigma_n\}) \in \mathsf{Kan}_*^{split}$. Let

$$\mu_0(X, \{\sigma_n\}) := \pi_0 X = *$$

and let $\alpha_0 \colon \mu_0(X, \{\sigma_n\}) \xrightarrow{1} \tau_{\leq 0}X$ be the identity map. Now suppose that we have constructed a functor

$$\mu_n \colon \mathsf{Kan}_*^{split} \longrightarrow \mathsf{Min(n - Group)}$$

taking values in the category of minimal n-groups, i.e. the full sub-category MinKan_* consisting of those X such that $X = \tau_{\leq n}X$. Suppose also that we have constructed a natural transformation as above

$$\alpha_n \colon \mu_n(X, \{\sigma_n\}) \longrightarrow \tau_{\leq n}X,$$

[12]Note that such sections always exist by the axiom of choice and the definition of a trivial fibration of simplicial sets.

such that $\tau_{\leq n-k}\alpha_n = \alpha_{n-k}$ for all $k \leq n$. Then we define $\mu_{n+1}(X, \{\sigma_n\})$ to be the pullback

$$\begin{array}{ccc} \mu_{n+1}(X, \{\sigma_n\}) & \longrightarrow & \tau_{\leq n+1}X \\ \downarrow & & \downarrow \\ \mu_n(X, \{(\sigma_n)\}) \xrightarrow[\sim]{\alpha_n} & \tau_{\leq n}X & \xrightarrow[\sim]{\sigma_n} \tau_{<n+1}X \end{array}$$

and we define $\alpha_{n+1}\colon \mu_{n+1}(X, \{\sigma_n\}) \longrightarrow \tau_{\leq n+1}X$ to be the top horizontal map. Because $\mu_n(X, \{(\sigma_n)\})$ is a minimal reduced Kan complex and the right vertical map is a minimal Kan fibration with reduced and n-connected fiber, we conclude that $\mu_{n+1}(X, \{\sigma_n\})$ is a minimal reduced Kan complex and the left vertical map is the n-truncation. Because the right vertical map is a Kan fibration and the bottom horizontal map is a weak equivalence and an injection, we conclude that α_{n+1} is a weak equivalence and an injection. By the universal property of pullbacks, the naturality of α_n and the definition of morphisms in Kan_*^{split}, we conclude that α_{n+1} is natural transformation as claimed. This completes the induction step. Defining

$$\mu(X, \{\sigma_n\}) = \varprojlim_n \mu_n(, \{(\sigma_n)\})$$

and

$$\alpha = \varprojlim_n \alpha_n$$

and noting, from the above square, that $\tau_{\leq n}\mu_{n+1} = \mu_n$, we conclude that μ takes values in reduced minimal Kan complexes, that $\tau_{\leq n}\mu = \mu_n$. Noting that $X = \varprojlim_n \tau_{\leq n}X$, we conclude that

$$\alpha\colon \mu(X, \{\sigma_n\}) \longrightarrow X$$

is a natural weak equivalence and an injection as desired. □

REMARK 5.2. The above presents a streamlined version of the classical construction of minimal models (cf. [**GJ09**, Chapter I.10]) which we believe clarifies the role of the axiom of choice. While it does not strictly require the data of the k-invariants constructed above, we include it here as an example of a tower of principal G-bundles for Z-groups $G \longrightarrow Z$, to which Theorem 1.4 applies.

Acknowledgments

The author thanks Ezra Getzler for helpful comments on a draft, and, beyond this, for introducing him to this circle of ideas, and for innumerable discussions, encouragement and advice over the years. The author thanks the anonymous referee for helpful comments and corrections. This paper arose as an offshoot of an ongoing collaboration with Chris Rogers, and the author thanks him for many helpful conversations and advice.

References

[AM66] M. Artin and B. Mazur, *On the van Kampen theorem*, Topology **5** (1966), 179–189, DOI 10.1016/0040-9383(66)90018-8. MR192495 ↑281

[BGM59] M. G. Barratt, V. K. A. M. Gugenheim, and J. C. Moore, *On semisimplicial fibre-bundles*, Amer. J. Math. **81** (1959), 639–657, DOI 10.2307/2372920. MR111028 ↑275, 276, 281, 286, 287

[BG17] Kai Behrend and Ezra Getzler, *Geometric higher groupoids and categories*, Geometry, analysis and probability, Progr. Math., vol. 310, Birkhäuser/Springer, Cham, 2017, pp. 1–45, DOI 10.1007/978-3-319-49638-2_1. MR3821920 ↑275, 277, 279, 281

[CR05] A. M. Cegarra and Josué Remedios, *The relationship between the diagonal and the bar constructions on a bisimplicial set*, Topology Appl. **153** (2005), no. 1, 21–51, DOI 10.1016/j.topol.2004.12.003. MR2172033 ↑281, 283

[Dus79] J. Duskin, *Higher-dimensional torsors and the cohomology of topoi: the abelian theory*, Applications of sheaves (Proc. Res. Sympos. Appl. Sheaf Theory to Logic, Algebra and Anal., Univ. Durham, Durham, 1977), Lecture Notes in Math., vol. 753, Springer, Berlin, 1979, pp. 255–279. MR555549 ↑286

[GJ09] Paul G. Goerss and John F. Jardine, *Simplicial homotopy theory*, Modern Birkhäuser Classics, Birkhäuser Verlag, Basel, 2009. Reprint of the 1999 edition [MR1711612], DOI 10.1007/978-3-0346-0189-4. MR2840650 ↑281, 286, 287, 288

[Hen08] André Henriques, *Integrating L_∞-algebras*, Compos. Math. **144** (2008), no. 4, 1017–1045, DOI 10.1112/S0010437X07003405. MR2441255 ↑275, 277, 281, 286, 287

[Ill72] Luc Illusie, *Complexe cotangent et déformations. II* (French), Lecture Notes in Mathematics, Vol. 283, Springer-Verlag, Berlin-New York, 1972. MR491681 ↑281, 283

[Jar09] J. F. Jardine, *Cocycle categories*, Algebraic topology, Abel Symp., vol. 4, Springer, Berlin, 2009, pp. 185–218, DOI 10.1007/978-3-642-01200-6_8. MR2597739 ↑277

[May67] J. Peter May, *Simplicial objects in algebraic topology*, Van Nostrand Mathematical Studies, No. 11, D. Van Nostrand Co., Inc., Princeton, N.J.-Toronto, Ont.-London, 1967. MR222892 ↑281, 286

[Moo58] John C. Moore, *Semi-simplicial complexes and Postnikov systems*, Symposium internacional de topología algebraica International symposium on algebraic topology, Universidad Nacional Autónoma de México and UNESCO, México, 1958, pp. 232–247. MR111027 ↑286

[Pri13] J. P. Pridham, *Presenting higher stacks as simplicial schemes*, Adv. Math. **238** (2013), 184–245, DOI 10.1016/j.aim.2013.01.009. MR3033634 ↑275

[RZ20] Christopher L. Rogers and Chenchang Zhu, *On the homotopy theory for Lie ∞-groupoids, with an application to integrating L_∞-algebras*, Algebr. Geom. Topol. **20** (2020), no. 3, 1127–1219, DOI 10.2140/agt.2020.20.1127. MR4105550 ↑275, 281

[Whi39] J. H. C. Whitehead, *Simplicial Spaces, Nuclei and m-Groups*, Proc. London Math. Soc. (2) **45** (1939), no. 4, 243–327, DOI 10.1112/plms/s2-45.1.243. MR1576810 ↑279

[Wol16] Jesse Wolfson, *Descent for n-bundles*, Adv. Math. **288** (2016), 527–575, DOI 10.1016/j.aim.2015.10.024. MR3436392 ↑275, 277, 278, 279, 280, 281, 286, 287

Department of Mathematics, University of California Irvine, Irvine, California 92697

Email address: `wolfson@uci.edu`

SELECTED PUBLISHED TITLES IN THIS SERIES

For a complete list of titles in this series, visit the
AMS Bookstore at **www.ams.org/bookstore/conmseries/**.

FSC
www.fsc.org
MIX
Papier | Fördert gute Waldnutzung
FSC® C083411

Kolibri 360 GmbH
Ferdinand-Jühlke-Straße 7
99095 Erfurt, Deutschland
produktsicherheit@kolibri360.de